A VOLUME IN THE
COMPREHENSIVE DICTIONARY
OF PHYSICS

DICTIONARY OF
MATERIAL SCIENCE AND HIGH ENERGY PHYSICS

COMPREHENSIVE DICTIONARY OF PHYSICS

Dipak Basu
Editor-in-Chief

FORTHCOMING AND PUBLISHED VOLUMES

Dictionary of Pure and Applied Physics
Dipak Basu

Dictionary of Material Science and High Energy Physics
Dipak Basu

Dictionary of Geophysics, Astrophysics, and Astronomy
Richard A. Matzner

A VOLUME IN THE
COMPREHENSIVE DICTIONARY
OF PHYSICS

DICTIONARY OF
MATERIAL SCIENCE
AND
HIGH ENERGY PHYSICS

Edited by
Dipak Basu

CRC Press
Boca Raton   London   New York   Washington, D.C.

Library of Congress Cataloging-in-Publication Data

Dictionary of material science and high energy physics / edited by Dipak Basu.
    p. cm.
    ISBN 0-8493-2889-6 (alk. paper)
    1. Particles (Nuclear physics)—Dictionaries. 2. Quantum theory—Dictionaries. 3. Materials—Dictionaries. I. Basu, Dipak. II. Series.

QC772. D57 2001
539′.03—dc21
                                        00-051950

This book contains information obtained from authentic and highly regarded sources. Reprinted material is quoted with permission, and sources are indicated. A wide variety of references are listed. Reasonable efforts have been made to publish reliable data and information, but the author and the publisher cannot assume responsibility for the validity of all materials or for the consequences of their use.

Neither this book nor any part may be reproduced or transmitted in any form or by any means, electronic or mechanical, including photocopying, microfilming, and recording, or by any information storage or retrieval system, without prior permission in writing from the publisher.

All rights reserved. Authorization to photocopy items for internal or personal use, or the personal or internal use of specific clients, may be granted by CRC Press LLC, provided that $.50 per page photocopied is paid directly to Copyright clearance Center, 222 Rosewood Drive, Danvers, MA 01923 USA. The fee code for users of the Transactional Reporting Service is ISBN 0-8493-2889-6/01/$0.00+$.50. The fee is subject to change without notice. For organizations that have been granted a photocopy license by the CCC, a separate system of payment has been arranged.

The consent of CRC Press LLC does not extend to copying for general distribution, for promotion, for creating new works, or for resale. Specific permission must be obtained in writing from CRC Press LLC for such copying.

Direct all inquiries to CRC Press LLC, 2000 N.W. Corporate Blvd., Boca Raton, Florida 33431, or visit our Web site at www.crcpress.com

Trademark Notice: Product or corporate names may be trademarks or registered trademarks, and are used only for identification and explanation, without intent to infringe.

© 2001 by CRC Press LLC

No claim to original U.S. Government works
International Standard Book Number 0-8493-2889-6
Library of Congress Card Number 00-051950
Printed in the United States of America 1 2 3 4 5 6 7 8 9 0
Printed on acid-free paper

Preface

The *Dictionary of Material Science and High Energy Physics* (DMSHEP) is one of the three major volumes being published by CRC Press, the other two being *Dictionary of Pure and Applied Physics* and *Dictionary of Geophysics, Astrophysics, and Astronomy*. Each of these three dictionaries is entirely self-contained.

The aim of the DMSHEP is to provide students, researchers, academics, and professionals in general with definitions in a very clear and concise form. A maximum amount of information is available in this volume that is still of reasonable size. The presentation is such that readers will not have any difficulty finding any term they are looking for. Each definition is given in detail and is as informative as possible, supported by suitable equations, formulae, and diagrams whenever necessary.

The fields covered in the DMSHEP are condensed matter, fluid dynamics, material science, nuclear physics, quantum mechanics, quantum optics, plasma physics, and thermodynamics. Terms have been chosen from textbooks, professional books, scientific and technical journals, etc. The authors are scientists at research institutes and university professors from around the world.

Like most other branches of science, the field of physics has grown rapidly over the last decade. As such, many of the terms used in older books have become rather obsolete. On the other hand, new terms have appeared in scientific and technical literature. Care has been taken to ensure that old terms are not included in the DMSHEP, and new terminologies are not missed. Some of the terms are related to other fields, e.g., engineering (mostly electrical and mechanical), mathematics, chemistry, and biology.

Readership includes physicists and engineers in most fields, teachers and students in physics and engineering at university, college, and high school levels, technical writers, and, in general, professional people.

The uniqueness of the DMSHEP lies in the fact that it is an extremely useful source of information in the form of meanings of scientific terms presented in a very clear language and written by authoritative persons in the fields. It would be of great aid to students in understanding textbooks, help academics and researchers fully appreciate research papers in professional scientific journals, provide authors in the field with assistance in clarifying their writings, and, in general, benefit enhancement of literacy in physics by presenting scientists and engineers with meaningful and workable definitions.

Dipak Basu

CONTRIBUTORS

Ibrahim H. Adawi
University of Missouri-Rolla
Rolla, Missouri

Kazuhiro Akimoto
Teikyo University
Utsunomiya, Japan

Cetin Aktik
University of Sherbrooke
Sherbrooke, Quebec, Canada

Mooread Alexanian
University of North Carolina
Wilmington, North Carolina

Roger Andrews
University of West Indies
St. Augustine, Trinidad

Supriyo Bandyopadhyay
University of Nebraska
Lincoln, Nebraska

Rama Bansil
Boston University
Boston, Massachusetts

Dipak Basu
Carleton University
Ottawa, Canada

Glenn Bateman
Lehigh University
Bethlehem, Pennsylvania

Subir K. Bose
University of Central Florida
Orlando, Florida

Daniel R. Claes
University of Nebraska
Lincoln, Nebraska

Don Correll
Lawrence Livermore National Laboratory
Livermore, California

Paul Christopher Dastoor
University of Newcastle
Callaghan, NSW, Australia

Anupam Garg
Northwestern University
Evanston, Illinois

Willi Graupner
Virginia Tech
Blacksburg, Virginia

Muhammad R. Hajj
Virginia Tech
Blacksburg, Virginia

Parameswar Hari
California State University
Fresno, California

Robert F. Heeter
Lawrence Livermore National Laboratory
Livermore, California

Ed V. Hungerford
University of Houston
Houston, Texas

Nenad Ilic
University of Manitoba
Winnipeg, Canada

Takeo Izuyama
Toho University
Miyama, Japan

Jamey Jacob
University of Kentucky
Lexington, Kentucky

Yingmei Liu
University of Pittsburgh
Pittsburgh, Pennsylvania

Vassili Papavassiliou
New Mexico State University
Las Cruces, New Mexico

Perry Rice
Miami University
Oxford, Ohio

Francesca Sammarruca
University of Idaho
Moscow, Idaho

Douglas Singleton
California State University-Fresno
Fresno, California

Reeta Vyas
University of Arkansas
Fayetteville, Arkansas

Thomas Walther
Texas A&M University
College Station, Texas

Peter Winkler
University of Nevada
Reno, Nevada

Bernard Zygelman
University of Nevada
Las Vegas, Nevada

Editorial Advisor

Stan Gibilisco

Abelian group Property of a group of elements associated with a binary operation. In an *Abelian group*, the group elements commute under the binary operation. If a and b are any two group elements and if the $(+)$ sign denotes the binary operation, then, for an *Abelian group*, $a + b = b + a$.

absolute plasma instabilities A class of *plasma instabilities* with amplitudes growing with time at a fixed point in the plasma medium. *Compare with* convective instabilities.

absolute temperature (T) Scale of temperature defined by the relationship $1/T = (\partial S/\partial U)_{V,N}$; S denotes entropy, U the internal energy, and V the volume of an isolated system of N particles. The *absolute temperature* scale is same as the Kelvin scale of temperature if $S = k_B \ln \Omega$, where Ω is the number of microstates of the system and k_B is the Boltzmann constant.

absolute viscosity Measure of a fluid's resistance to motion whose constant is given by the relation between the shear stress, τ, and velocity gradient, du/dy, of a flow such that

$$\tau \propto \frac{du}{dy}.$$

The constant of proportionality is the *absolute viscosity*. For Newtonian fluids, the relation is linear and takes the form

$$\tau = \mu \frac{du}{dy}$$

where μ, also known as dynamic viscosity, is a strong function of the temperature of the fluid. For gases, μ increases with increasing temperature; for liquids, μ decreases with increasing temperature. For non-Newtonian fluids, the relation is not linear and apparent viscosity is used.

absolute zero (0K) The lowest temperature on the Kelvin or absolute scale.

absorption A process in which a gas is consumed by a liquid or solid, or in which a liquid is taken in by a solid. In *absorption*, the substance absorbed goes into the bulk of the material. The *absorption* of gases in solids is sometimes called sorption.

absorption band (F) If alkali halides are heated in the alkali vapor and cooled to room temperature, there will be a Farbe center defect. F-center is a halide vacancy with its bound electron. The excitation from ground state to the first excited state in F-center leads to an observable *absorption band*, which is called F-*absorption band*. Because there is an uncoupled electron in F-center, it has paramagnetism.

absorption band (V) If alkali halides are heated in the halide vapor and cooled to room temperature, there will be a V-center defect in it. V-center is an alkali vacancy with its bound hole. The excitation from ground state to the first excited state in V-center causes a V-*absorption band*, which lies in the edge of ultra-vision light.

absorption coefficient A measure of the probability that an atom will undergo a state-transition in the presence of electromagnetic radiation. In modern atomic theory, an atom can make a transition to a quantum state of higher energy by absorbing quanta of photons. The energy defect of the transition is matched by the energy posited in the photons.

absorption of photons The loss of light as it passes through material, due to its conversion to other energy forms (typically heat). Light incident on an atom can induce an upward transition of the atom's state from an energy ε_0 to an energy $\varepsilon_n = \varepsilon_0 + \hbar\omega = \varepsilon_0 + \hbar c k$, where $\omega = (\varepsilon_n - \varepsilon_0)/\hbar$ is the angular frequency of the light, and $k = 2\pi/\lambda$ its propagation number. This is interpreted as the absorption of an individual photon of energy $\hbar\omega = \varepsilon_n - \varepsilon_0$ by the positive frequency component $e^{-i\omega t}$ of a perturbation in the Hamiltonian of the atomic electron. The absorption cross section depends on the direction and polarization of the radiation, and is given by

$$\sigma_{abs}(\omega) = \frac{4\pi^2 e^2}{\omega c} \sum_n \left| \left\langle n \left| \vec{j}_{-\vec{k}} \cdot \vec{\lambda} \right| 0 \right\rangle \right|^2$$

$$\delta(\varepsilon_n \varepsilon_0 - \eta\omega)$$

for a polarization vector $\vec{\lambda}$, wave vector $\vec{k} = (2\pi/\lambda)\vec{p}$ and probability current density $\vec{j}$ ($\vec{r}$, t), and ε_0, ε_n are the energy of the initial $|0>$ and final $|n>$ atomic states.

absorption of plasma wave energy The loss of plasma wave energy to the plasma particle medium. For instance, an electromagnetic wave propagating through a plasma medium will increase the motion of electrons due to electromagnetic forces. As the electrons make collisions with other particles, net energy will be absorbed from the wave.

acceptor A material such as silicon that has a resistivity halfway between an insulator and a conductor (on a logarithmic scale). In a pure semiconductor, the concentrations of negative charge carriers (electrons) and positive carriers (holes) are the same. The conductivity of a semiconductor can be considerably altered by adding small amounts of impurities. The process of adding impurity to control the conductivity is called doping. Addition of phosphorus increases the number of electrons available for conduction, and the material is called n-type semiconductor (i.e., the charge carriers are negative). The impurity, or dopant, is called a donor impurity in this case. Addition of boron results in the removal of electrons. The impurity in this case is called the *acceptor* because the atoms added to the material accept electrons, leaving behind positive holes.

acceptor levels The levels corresponding to acceptors are called *acceptor levels*. They are in the gap and very close to the top of the valence band.

accidental degeneracy Describes a property of a many-particle quantum system. In a quantum system of identical particles, the Hamiltonian is invariant under the interchange of coordinates of a particle pair. Eigenstates of such a system are degenerate, and this property is called exchange symmetry. If a degeneracy exists that is not due to exchange symmetry, it is called *accidental degeneracy*.

acoustic modes The relation between frequency w and wave vector $\mathbf{k}$ is called the dispersion relation. In the phonon dispersion relation, there are optical and acoustical branches. Acoustical branches describe the relative motion among primitive cells in crystal. If there are p atoms in each primitive cell, the number of *acoustical modes* is equal to the degree of freedom of each atom. For example, in three-dimensional space, the number of *acoustical modes* is three.

acoustics The study of infinitesimal pressure waves that travel at the speed of sound. *Acoustics* is characterized by the analysis of linear gas dynamic equations where wave motion is small enough not to create finite amplitude waves. The fluid velocity is assumed to be zero.

acoustic wave *See* sound wave.

action A property of classical and quantum dynamical systems. In Hamilton's formulation of classical dynamics, the quantity $S = \int_{t_1}^{t_2} dt L(q(t), \dot{q}(t))$, where $L(q(t), \dot{q}(t))$ is the Lagrangian, and $q(t), \dot{q}(t)$ is the dynamical variable and its time derivative, respectively, is called the *action* of the motion. In quantum physics, Planck's constant h has the dimensions of an action integral. If the *action* for a classical system assumes a value that is comparable to the value of Planck's constant, the system exhibits quantum behavior. Feynman's formulation of quantum mechanics involves a sum of a function of the *action* over all histories.

activity (λ) The absolute *activity* is defined as $\lambda = \exp(\mu/k_B T)$, where μ is the chemical potential at temperature T, and k_B is the Boltzmann constant.

added mass Refers to the effect of increased drag force on a linearly accelerating body. For a sphere (the simplest case to analyze), the drag force in an ideal (frictionless) flow due to acceleration is

$$D = \frac{2}{3}\pi r^3 \rho \frac{dU}{dt}$$

which is equivalent to increasing the volume of the sphere by exactly 1/2. Thus, the increased drag force may be neglected if the *added mass* is included in the sphere to give a total mass of $(\rho + \frac{1}{2}\rho)V$, where ρ is the fluid density and V is the volume of the sphere. Also referred to as virtual mass.

addition of angular momentum Two angular momenta, J_1 and J_2 (orbital angular momentum and spin, or two distinguishable subsystems with different angular momentum quantum numbers j_1 and j_2), can combine to yield any quantized state with a total angular momentum quantum number in the range $|j_1 - j_2| \leq j \leq (j_1 + j_2)$ but with the J_z projections simply adding as $m = m_1 + m_2$. The addition rules follow from the nature of the angular momentum operator relations.

addition theorem The identity, $P_l[cos(\hat{\mathbf{r}}_1 \cdot \hat{\mathbf{r}}_2)] = \frac{4\pi}{2l+1} \sum_{m=-l}^{m=l} Y_{lm}^* (\theta_1 \phi_1) Y_{lm} (\theta_2 \phi_2)$, where $\theta_1 \phi_1$ and $\theta_2 \phi_2$ are the polar and azimuthal angles of particle 1 and 2, respectively, and P_l is a Legendre polynomial. *See* associated Legendre polynomial.

adiabatic bulk modulus (β_S) The *adiabatic bulk modulus* is a measure of the resistance to volume change without deformation or change in shape in a thermodynamic system in a process with no heat exchange, i.e., at constant entropy. It is the inverse of the adiabatic compressibility:

$$\beta_S = -V \left(\frac{\partial P}{\partial V} \right)_S.$$

adiabatic compressibility (κ_s) The fractional decrease in volume with increase in pressure without exchange of heat, i.e., when the entropy remains constant during the compression:

$$\kappa_s = -\frac{1}{V} \left(\frac{\partial P}{\partial V} \right)_S.$$

adiabatic invariant Characteristic parameter that does not change as a physical system slowly evolves; the most commonly used *adiabatic invariant* in plasma physics is the magnetic moment of a charged particle that is spiraling around a magnetic field line.

adiabatic plasma compression Compression of a gas and/or plasma that is not accompanied by gain or loss of heat from outside the plasma confinement system. For example, plasma in an increasing magnetic field that results in plasma compression slow enough that the magnetic moment, or other adiabatic invariants of the plasma particles, may be taken as constant.

adiabatic process A process in which no heat enters or leaves a system.

adiabatic theorem Describes the behavior of the wave function for a system undergoing adiabatic evolution. Consider a quantum system whose time evolution is governed by a Hamiltonian $H(R(t))$, where $R(t)$ is a non-quantum mechanical parameter and t is the time parameter. In the limit of slow evolution, so that the time derivative of $H(t)$ can be neglected, M. Born and V. Fock showed that $|\Psi(t)> = \exp(-\frac{i}{\hbar} \int^t E(t) dt) |\Psi(0)>$, where $E(t)$ is the instantaneous energy eigenvalue for state $|\Psi(t)>$, is a solution to the time dependent Schrödinger equation. This is a statement of the quantum *adiabatic theorem* that was generalized in 1984 by M.V. Berry. *See* Berry's phase.

adjoint equation A corresponding relationship that results from replacing operators by their Hermitian conjugate, ordinary numbers by their complex conjugate, conjugating bras into kets (and kets into bras), and reversing within each individual term the order of these symbols.

adjoint operator Property associated with a pair of operators. For operator A that has the property $A|\psi> = |\psi'>$, where $|\psi>$, $|\psi'>$ are vectors in Hilbert space, the operator $A^\dagger$ is called the *adjoint operator* of A. It has the following property $<\psi|A^\dagger = <\psi'|$, where $<\psi|$ is the dual to vector $|\psi>$. If A is a square matrix, then $A^\dagger$ is the matrix obtained by taking the transpose and complex conjugate of A, i.e., $A^\dagger = (A^T)^*$. *See also* bra vector.

adjoint spinor   To construct Lorentz-invariant terms for the Lagrangian of solutions to the Dirac equation, the inner product of Dirac spinors is expressed in terms of the 4-column spinors, ψ, and the adjoint, $\bar\psi = \psi^\dagger + \gamma^0$ (distinguished from its Hermitian conjugate $\psi^\dagger$). γ^0 is one of the four 4×4 Dirac matrices. Under this rule, the product $\bar\psi\psi$ yields a simple scalar.

adsorption   A process in which a layer of atoms or molecules of one substance forms on the surface of a solid or liquid. The adsorbed layer may be formed by chemical bonds or weaker Van der Waals forces.

adsorption isotherm   A curve that gives the concentration of adsorbed particles as a function of pressure or concentration of the adsorbant at constant temperature.

advection   The movement of fluid from point to point in a flow field by pressure or other forces (as opposed to convection).

adverse pressure gradient   In a boundary layer, a pressure gradient that is positive ($dp/dx > 0$) rather than negative due to an external decelerating flow ($du/dx < 0$). This condition may lead to flow separation.

Aeolian harp   Wire in a flow that produces sound due to the natural vortex shedding that occurs behind a cylinder. Since the wire is free to oscillate, the wire can resonate at its natural frequency with an amplitude that allows the vortex shedding frequency to match that of the wire. The *Aeolian harp* was originally investigated by Lord Rayleigh. *See* Kármán vortex street.

aerodynamics   The study of the motion of air and the forces acting on bodies moving through air as caused by motion, specifically lift and drag. Typically, gravity forces are neglected and viscosity is considered to be small such that viscous effects are confined to thin boundary layers. *Aerodynamics* is characterized by measurement and calculation of various dimensionless coefficients of forces and moments that remain invariant for a given geometry and flight condition, allowing the use of wind tunnels to study geometrically similar models at different scales. The primary flight conditions of import are the Reynolds and Mach numbers.

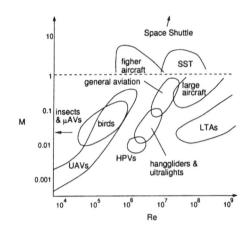

Range of interest in aerodynamics. (Adapted from Lissaman, P.B.S., Low Reynolds number airfoils, Ann. Rev. Fluid Mech., 15, 223, 1983.)

afterglow, or plasma afterglow   Recombination radiation emitted from a cooling plasma when the source of ionization, heating, etc. is removed or turned off.

Aharonov–Bohm effect   Quantum mechanical, topological effect elucidated by David Bohm and Y. Aharonov. Also called the Aharonov–Bohm/Eherenberg–Siday effect. The effect predicts observable consequences that arise when a charged particle interacts with an inaccessible magnetic flux tube. *See also* Berry's phase.

airfoil   Any device used to generate lift in a controlled manner in air; specifically refers to wings on aircraft and blades in pumps and turbines. *Airfoil* geometry and flight regime (as given by Reynolds and Mach numbers) are the primary factors in the creation of lift and drag (*see* hydrofoil).

alcator plasma machine   Name given to a set of tokamaks designed and built at MIT; these

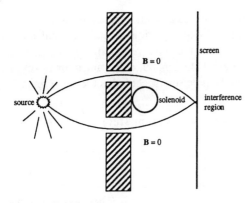

Aharonov–Bohm effect.

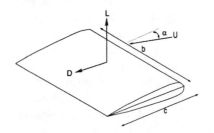

Airfoil geometry.

fusion plasma machines with toroidal magnetic confinement are distinguished by higher magnetic fields with relatively smaller diameters than other toroidal geometries.

Alfvén velocity Phase velocity of the Alfvén wave; equal to the speed of light divided by the square root of 1 plus the ratio of the plasma frequency to the cyclotron frequency. *See also* Alfvén waves.

Alfvén waves Electromagnetic waves that are propagated along lines of magnetic force in a plasma. *Alfvén waves,* named after plasma physicist and Nobel Prize winner Hannes Alfvén, have frequencies significantly less than the ion cyclotron frequency, and are characterized by the fact that the magnetic field lines oscillate with the plasma.

alloy A mixture of two or more metals or of a metal (for example, bronze or brass) and small amounts of a non-metal (for example, steel).

alpha particle A positively charged particle emitted from the nucleus of some unstable isotopes. The equivalent of a helium nucleus, it consists of two protons and two neutrons. *Alpha particles* have a typical energy range of 4-8 MeV and are easily dissipated within a few centimeters of air (or less than 0.005mm of aluminum).

ambipolar plasma diffusion Diffusion process in which a buildup of spatial electrical charge creates electric fields (*see* ambipolar plasma potential) which cause electrons and ions to leave the plasma at the same rate.

ambipolar plasma potential Electric fields that are self-generated by the plasma and act to preserve charge neutrality through ambipolar diffusion.

amorphous Refers to material that has no crystalline structure. Glass is an example of *amorphous* material with no long-range ordering of atoms.

amplitude, scattering The scattering cross-section for particles by a potential $V(\mathbf{r})$ can be expressed in terms of *scattering amplitudes* $\sigma(\Omega) = |f(\Omega)|^2$, where it is assumed solutions exist to the Schrödinger equation $[-\hbar/2m\Delta + V(\mathbf{r})]\psi_k(\mathbf{r}) = E)\psi_k(\mathbf{r})$ whose behavior at infinity is of the form $e^{i\mathbf{k}\cdot\mathbf{r}} + f(\Omega)e^{ikr}/r$.

Andrade's equation A simplification of the log-quadratic law determining the viscosity of liquids:

$$\mu = Ae^{B/T}$$

where A and B are constants, and T is the absolute temperature of the liquid.

anemometer Any device specifically used to measure the velocity of air; often used generically for the measurement of velocity in any gas. (anemometry, e.g., hot-wire *anemometry*).

angstrom (Å) Unit of length equal to one trillionth of a meter (10^{-10} m or 1/10th of a nanometer). An *angstrom* is not an SI unit.

angular momentum A property of any revolving or rotating particle or system of particles. Classically, a particle of mass m moving with velocity v at a distance r from a point O carries a momentum relative to (or about) O defined by the vector (cross) product $\mathbf{L} = \mathbf{r} \times \mathbf{p} = m\mathbf{r} \times \mathbf{v}$.

Quantum mechanically, values of orbital *angular momentum* are quantized in units of $\hbar = h/2\pi$, while the intrinsic *angular momentum* possessed by particles (*see* spin) is quantized in units of $\frac{1}{2}\hbar$.

An azimuthal (orbital *angular momentum*) quantum number, ℓ, denotes the quantized units of orbital *angular momentum* and distinguishes the different shaped orbitals of any given energy level (radial quantum number), n. The quantum number ℓ can have any integer value from 0 to $n-1$.

angular momentum operator   An operator rule that, when applied to a state function, returns a new wave function expressible as a linear combination of eigenfunctions weighted by the corresponding angular momentum eigenvalue. The classical expression for angular momentum (*see* angular momentum) $\mathbf{L} = \mathbf{r} \times \mathbf{p}$ is re-expressed with r and p interpreted as quantum mechanical dynamic variables (operators) themselves: $\mathbf{L}_{op} = \mathbf{r}_{op} \times (-i\hbar \nabla)$.

In Cartesian coordinates:

$$\mathbf{L}_x = -i\hbar(y\partial/\partial z - z\partial/\partial y)$$
$$\mathbf{L}_y = -i\hbar(z\partial/\partial x - x\partial/\partial z)$$
$$\mathbf{L}_z = -i\hbar(x\partial/\partial y - y\partial/\partial x) .$$

In spherical polar coordinates:

$$\mathbf{L}_x = i\hbar(\sin\varphi\, \partial/\partial\theta + \cot\theta \cos\varphi\, \partial/\partial\varphi)$$
$$\mathbf{L}_y = -i\hbar(\cos\varphi\, \partial/\partial\theta - \cot\theta \sin\varphi\, \partial/\partial\varphi)$$
$$\mathbf{L}_z = -i\hbar\, \partial/\partial\varphi .$$

The application of this operator is synonymous with taking a physical measurement of the angular momentum of that state. The operator representing the square of the total orbital angular momentum

$$\mathbf{L}^2 = -\hbar^2 \left[(1/\sin\theta)\partial/\partial\theta(\sin\theta\, \partial/\partial\theta) + \left(1/\sin^2\theta\right) \partial^2/\partial\varphi^2 \right]$$

has eigenvalues of $\ell(\ell+1)\hbar^2$ where $\ell = 0, 1, 2, \ldots$ and ℓ is known as the orbital angular momentum quantum number. $\mathbf{L}_z$ can be shown to have eigenvalues of $m\hbar$ where m takes on integer values from $-\ell$ to $+\ell$. For spherically symmetric potentials, the wave function in the direction of the polar axis is arbitrary and the wave functions must be eigenfunctions of both the total angular momentum and the z-component (along the polar axis). In general, only values for $\mathbf{L}_z$ and $\mathbf{L}_2$ can be precisely specified at the same time.

angular momentum states   An eigenstate of quantum mechanical angular momentum operators. In quantum mechanics there are two types of angular momenta. The first, represented by the operator $\mathbf{L} = \mathbf{r} \times \mathbf{p}$, is the orbital angular momentum. There exists an intrinsic angular momentum, called spin, that is represented by operator $\mathbf{S}$ and whose components also obey angular momentum commutation relations $[J_i, J_j] = i\hbar\epsilon_{ijk}J_k$. Here, J_k is the kth component of an angular momentum operator and ϵ_{ijk} is the unit antisymmetric tensor. An orbital angular momentum eigenstate is an eigenstate of $\mathbf{L}^2$ and L_z, the z-component of $\mathbf{L}$. The eigenvalues are labeled by quantum numbers l and m respectively. For spin angular momentum, the labels s and m_s denote the eigenvalues corresponding to the operators $\mathbf{S}^2$ and S_z. The allowed values for l are integers and for s are half integers. Linear sums of products of orbital and spin angular momentum can be constructed to form eigenstates of total angular momentum $\mathbf{J} \equiv \mathbf{L} + \mathbf{S}$.

an-harmonic interaction   The interaction corresponding to the an-harmonics in the energy expansion.

anions   A negatively charged ion, formed by addition of electrons to atoms or molecules. In an electrolysis process, *anions* are attracted toward the positive electrode.

anisotropy   A medium is said to be *anisotropic* if a certain physical characteristic differs in magnitude in different directions. Examples of this effect are electrical *anisotropy* in crystals and polarization properties in crystals with different directions.

annealing   The process of heating a material to a temperature below the melting point, and then cooling it slowly.

annihilation The result of matter and antimatter (for example an electron and a positron, particles of identical mass but opposite charge) undergoing collision. The resulting destruction of matter gives off energy in the form of radiation. Conservation of energy and momentum prevents this radiation from being carried by a single photon and demands it be carried by a pair of photons. *See* antimatter (antiparticle), creation of matter.

annihilation diagram The Feynman diagram describing the annihilation process of a particle and its antiparticle. The diagram for $e + e- \to \gamma\gamma$ pair annihilation, for example, is constructed with two copies of the primitive quantum electrodynamics (QED) $ee\gamma$ vertex.

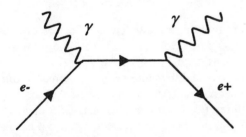

Annihilation diagram.

The external lines of incoming e^+e^- and outgoing γs represent the observable particles. The internal lines describe virtual particles involved in the process, consistent here with the conservation of energy and momentum demands for two photons in the final states.

The *annihilation diagram* for electron-positron scattering, also built with a pair of primitive $ee\gamma$ vertices, carries an internal photon line.

See Feynman diagram; quantum chromodynamics.

annihilation operator (1) The vacuum state is an eigenstate of this operator, and has the null eigenvalue. If operator a and $a^\dagger$ obey the following commutation relation $[a, a^\dagger] = 1$, then a is called an *annihilation operator* and its adjoint $a^\dagger$ is called the creation operator. If $|n>$ is an eigenstate of the number operator $N \equiv a^\dagger a$, then $a|n>$ is also an eigenstate of N, but with eigenvalue $n - 1$. *Annihilation operators* are

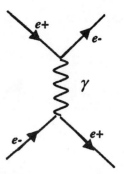

Annihilation diagram.

fundamental in field theory. Here the state $|n>$ represents a quantum state of definite occupation number n, the number of particles. The action of a on that state produces a state with one less particle, hence the label *annihilation operator*. If $n = 0$, i.e., the vacuum state, then $|0>$ is an eigenstate of a.

(2) In quantum field theory, an operator that, when acting on a state vector, decreases the eigenvalue of the number operator by one and the charge operator by z. If the expression $u_k \phi_p$ represents the state vector with energy-momentum (four-momentum) $p - k$, where $k^2 = m$, then operator u_k can describe the annihilation of a particle of mass m, charge z, and four-momentum k.

For particles obeying Fermi-Dirac statistics (fermions such as electrons and muons), the operators must satisfy "anticommutator relations" $\{u_k, u^\dagger_{k'}\} = \delta_{kk'}$ and $\{u_k, u_{k'}\} = \{u^\dagger_k, u^\dagger_{k'}\} = 0$ in order to incorporate the Pauli exclusion principle. *See* creation operators.

anomalous (magnetic) moment A correction to the gyro-magnetic ratio of a particle which accounts for the complications introduced by virtual pairs existing in the particle's own electric field. Virtual photons are continuously emitted and reabsorbed, and their presence affects the interactions with other particles, such as those measuring the gyro-magnetic ratio. The *anomalous magnetic moment* is expressed in terms of the departure of a constant g from its expected bare electron value of two: $g = 2[1 + (e^2/4\pi\hbar)1/2\pi + \cdots]$ and can be accounted for by a phenomenological term in the interaction

Hamiltonian of the form

$$\mathcal{H}\text{int} = -(e\hbar k)\left[\frac{1}{2}F_{\nu\mu}\bar{\psi}\sigma_{\nu\mu}\psi\right]$$

called the *anomalous moment* interaction. *See* gyromagnetic ratio.

anomalous plasma diffusion Particle or heat diffusion in a plasma that is larger than what was predicted from theoretical predictions of classical plasma phenomenon. Classical diffusion and neo-classical diffusion are the two well-understood diffusion theories, although neither is adequate to fully explain the experimentally observed magnitude of anomalous diffusion.

anomalous Zeeman effect Term used to describe the shifting of atomic levels in the presence of an external magnetic field. The ordinary Zeeman effect describes energy shifts that are proportional to the orbital azimuthal quantum number m. In the *anomalous Zeeman effect*, the spin azimuthal quantum number is also taken into account. The total shift is then proportional to $m + 2m_s$, where $m_s = 1/2$ for a single electron. *See* azimuthal quantum number.

anti-bonding orbital Electronic state for a system of two atoms in which the atoms repel each other as they approach. The *anti-bonding orbital* contrasts with the bonding orbital, in which chemical forces favor a bound configuration of the atoms.

anticommutation relations *See* commutation relations.

anticommutator (1) With the product of operators defined as the successive application of operators, $(AB)\psi \equiv A(B\psi)$, in general any two operators A and B will not likely commute, $AB\psi \neq BA\psi$. Operators for which $(BA)\psi = -(AB)\psi$ are said to anticommute, and the *anticommutator* $\{A, B\}$ defined by $\{A, B\} \equiv AB + BA$ vanishes. *See* commutator; anticommutation relations.

(2) The product $AB + BA$ of two operators A and B in Hilbert space. The bracket symbol $\{A, B\}_+$ is often used to denote the *anticommutator*.

anti-ferromagnetic crystals At the temperature below Neel temperature, the magnets of atoms (or ions) are anti-parallel. The net moment is zero for *anti-ferromagnetic crystals*.

antiferromagnetism A phenomenon in certain types of material that have two or more atoms with different magnetic moments. The magnetic moment of one set of atoms can align anti-parallel to the atoms of the other type. In *antiferromagnetism*, the susceptibility increases with temperature up to a certain value (*see* Nèel temperature). Above this temperature, the material is paramagnetic.

anti-linear operator An operator that has the property $Ac|\psi> = c^*A|\psi>$, where A is the *anti-linear operator*, c is a scalar, and $|\psi>$ is a vector in Hilbert space.

antimatter (antiparticle) Species of subatomic particles that have the same mass and spin as normal particles, but opposite electrical charge (and therefore magnetic moment) from their normal matter counterparts. Antineutrons differ from neutrons and magnetic moment. Positrons, the counterpart to electrons, have a positive charge and antiprotons have a negative charge. Photons are their own *antimatter* counterpart. When a particle of matter collides with a particle of *antimatter*, both particles are destroyed and their mass is converted to photons of equivalent energy. *See* annihilation; charge conjugation.

anti-stokes line In Raman scattering, if the frequency of the incident photon is w_0, the scattered photon at $w_0 + dw$ is called the *anti-stokes line*, where dw is the frequency of the absorbed phonon.

antisymmetric state A state in which an interchange of coordinates for two indistinguishable particles results in a sign change of the wave function.

antisymmetric wave function A wave function of a multiparticle system ($\psi(1, 2, \ldots, n; t)$, where each number represents all the coordinates (position and spin) of individual particles, which changes only by an overall sign under the interchange of any pair of particles. Since the Hamiltonian $\mathcal{H}$ is symmetric in these arguments, $\mathcal{H}\psi$, and therefore, $\partial\psi/\partial t$ are antisymmetric, which implies that the symmetry character of a state does not change with time. Particles described by *antisymmetric wave functions* obey Fermi-Dirac statistics and are called fermions. *See* fermion.

antisymmetrization operator An operator that projects the antisymmetric component, with respect to particle permutation or exchange, of a many-body wave function for identical particles. If P is the particle permutation operator for the special case of two particles, then the *antisymmetrization operator* can be written $A = \frac{1}{2}(1 - P)$. For a many-body system, the *antisymmetrization operator* can be expressed by the sum of many-particle permutation operators.

anti-unitary operator An operator that can be written as a product of a unitary operator and an anti-linear operator. In quantum mechanics, time reversal symmetry is associated with an *anti-unitary operator*. *See* anti-linear operator.

anyon A particle whose wave function, for a many-*anyon* system, undergoes an arbitrary phase change following the interchange of coordinates of an anyon pair. In the standard description, a fermion wave function undergoes a sign change following the interchange of coordinates. Boson wave functions are invariant under particle interchange. The former case corresponds to a phase change of value π and the latter to a modulus 2π change.

apparent viscosity For non-Newtonian fluids, if the shear stress and velocity gradient relation is written as

$$\tau = k \left|\frac{du}{dy}\right|^{n-1} \frac{du}{dy} = \eta \frac{du}{dy}$$

the quantity $\eta = k \mid du/dy \mid^{n-1}$ is called the *apparent viscosity* of the fluid.

APW method Augmented plane waves; this is one way to calculate energy band in crystal.

Archimede's law A body immersed in fluid experiences an upward force equal to the weight of the fluid displaced by the body.

arc, or plasma arc A type of electrical discharge between two electrodes; characterized by high-current density within the plasma between the electrodes.

aspect ratio Geometric term relating the width (span, b) and area, A, of a wing planform:

$$AR \equiv \frac{b^2}{A}.$$

For a rectangular wing, this reduces to $AR = b/c$.

aspirator Device utilizing the principle of entrainment around a jet to create a suction effect. Typically, the jet is water or some other liquid which effluxes into a cavity open to the atmosphere. As the jet enters an exit in the cavity, it draws surrounding air with it and generates a suction force.

associated Laguerre polynomial Symbol: L_q^p. Member of a set of orthogonal polynomials that has applications in the quantum mechanics of Coulomb systems. The *associated Laguerre polynomial*, $L_q^p(x)$, is a solution to the following second order differential equation, $x\frac{d^2}{dx^2}L_q^p(x) + (p + 1 - x)\frac{d}{dx}L_q^p(x) + (q - p)L_q^p(x) = 0$. The radial hydrogenic wave functions are related to the *associated Laguerre polynomials* for the special case $p = 2l+1, q = n+l$, where l is the angular momentum quantum number and n is a positive integer, the principal quantum number. *See* angular momentum states.

associated Legendre polynomial Symbol: P_l^m. Member of a set of orthogonal polynomials that has applications in quantum systems possessing spherical symmetry. The Legendre polynomial, $P_l^m(x)$, is a solution to the following second order differential equation, $[(1 - x^2)\frac{d}{dx}P_l^m(x)]' - (l(l+1) - \frac{m^2}{1-x^2})P_l^m(x) = 0$, where the prime signifies differentiation with respect to x. For the case $m = 0$, the *associated Legendre polynomial* is called the Legendre polynomial.

astrophysical plasmas Includes the sun and stars, the solar wind and stellar winds, large parts of the interstellar medium and the intergalactic medium, nebulae, and more. Planets, neutron stars, black holes, and some neutral hydrogen clouds are not in a plasma state. Approximately 99% of the observable universe can be described as being in a plasma state.

atmosphere, standard (US) Average values of pressure, temperature, and density of air in the Earth's atmosphere as a function of altitude. At sea level, $p = 101.3$ kPa, $T = 15.0°$ C, and $\rho = 1.225$ kg/m^3.

The *US standard atmosphere* is a defined variation in the Earth's atmospheric pressure and temperature. The hydrostatic equation

$$\frac{dp}{dz} = \rho g$$

shows that pressure varies with height for a constant density. However, as density is not constant in the Earth's atmosphere, we use the ideal gas equation to write

$$\rho = p/RT$$

which is substituted into the hydrostatic equation to give

$$\frac{dp}{dz} = -\frac{gp}{RT}.$$

Temperature is also a variable. The *US standard atmosphere* defines the variation in temperature for average conditions as follows:

Troposphere:	
$T = T_{sl} - \alpha z$	$:0 \leq z \leq 11.0$ km
Stratosphere:	
$T = T_{hi}$	$:11.0$ km $\leq z \leq 20.1$ km

where α is the *lapse rate*, T_{sl} is the average sea level temperature, and T_{hi} is the average temperature of the stratosphere (assumed constant). Thus, the temperature decreases linearly until 11 km, whereafter it is constant. (It increases again after that, but the validity of the hydrostatic relation decreases significantly.) These values are given as

	SI	US
α	6.50 K/km	18.85°R/mile
T_{sl}	288 K	518.4°R
T_{hi}	218 K	392.4°R

In the troposphere, this becomes

$$p = p_{sl} (1 - \alpha z/T_{sl})^{g/\alpha R}$$

where $p_{sl} = 101$ kPa (14.7 psi), the sea level pressure. The pressure decreases to 22.5 kPa (3.28 psi) at 11 km. In the stratosphere, temperature is a constant $T = T_{hi}$, so

$$p = p_{hi} e^{-g(z-z_{hi})/RT_{hi}}$$

where $p_{hi} = 22.5$ kPa and $z_{hi} = 11.0$ km.

atom The basic building block of neutral matter. *Atoms* are composites of a heavy, positively charged nucleus and much lighter, negatively charged electrons. The Coulomb interaction between the nucleus and electrons binds the system. The nucleus itself is a composite system of protons and neutrons held together by the so-called strong, or nuclear, forces.

atomic level States in the sub-manifold of an atomic state. *Atomic levels* are usually split by small perturbations, but the resulting energy defects of levels are much smaller than the energy defects between atomic states.

atomic spectra The characteristic radiation observed when atoms radiate in the optical frequencies. Because atoms exist in well defined, discrete quantum energy states, the emitted radiation is seen at discrete frequencies or wavelengths. With modern instruments, atomic radiation can also be measured in the ultraviolet and X-ray regions of the electromagnetic spectrum. For the hydrogen atom, the radiation spectra is predicted by the Bohr model of the atom. For many electron atoms, the Schrödinger equation must be used to predict accurate energy levels, hence spectra.

aufbau principle Derived from the German word *aufbau*, which means to build up. The *aufbau principle* in atomic theory explains how complex atoms are organized. The *aufbau principle* can be used to predict, in a qualitative way, the chemical property of an element.

Auger effect *See* autoionization.

aurora Called aurora borealis in the northern hemisphere and aurora australis in the southern hemisphere, *aurorae* are light emissions by atmospheric atoms and molecules after being excited by electrons precipitating from the Earth's magnetosphere.

autoionization The process in which excited atoms decay due to inter-electronic interactions. In a many-electron atom, we can construct approximate, mean field states that are products of bound one-electron states. They provide a qualitative description of the atom. However, because of electron–electron interaction, excited states described by the independent particle approximation are unstable and have a finite lifetime. The process in which a multi-electron atom in an excited state subsequently decays, resulting in the ejection of electrons, is called *auto-ionization*. This phenomena is also called the Auger effect.

avogadro number (N_0) The number of molecules in one mole of a substance. It is the same for all substances and has the value 6.02×10^{23}. *See* mole.

axial vector A vector quantity which retains its directional sign under space inversion $\mathbf{r} \rightarrow \mathbf{r}'$ (an inversion of the coordinates axes $x \rightarrow -x$, $y \rightarrow -y$, $z \rightarrow -z$). Polar vectors like position $\mathbf{r}$ and momentum $\mathbf{p}$ reverse sign. Angular momentum is an example of an *axial* or pseudo *vector*, since under space inversion, $\mathbf{L} = \mathbf{r} \times \mathbf{p} \rightarrow (-\mathbf{r} \times (-\mathbf{p}) = +\mathbf{L}$.

azimuthal quantum number Symbol: m. Quantum number associated with the component of angular momentum along the quantization axis. If $\mathbf{J}$ is the angular momentum operator and $|jm>$ is an angular momentum eigenstate, then $J_z|jm> = m\hbar|jm>$ and m is called the *azimuthal quantum number*. The quantization axis is usually taken, by convention, along the **z** axis. *See* angular momentum states.

B

backwater curve The increase in the surface height of a stream as it approaches a weir.

Baker-Hausdorff formula Follows from the theorem: given two operators A and B that commute with operator $AB - BA \equiv [A, B]$, the identity $\exp(A)\exp(B) = \exp(A+B)\exp(1/2[A,B])$ holds true.

ballooning mode A plasma mode which is localized in regions of unfavorable magnetic field curvature (also known as "bad curvature") that becomes unstable (grows in amplitude) when the force due to plasma pressure gradients is greater than the mean *magnetic pressure force*.

Balmer formula *See* Balmer series.

Balmer series The characteristic radiation of atomic hydrogen, whose wavelength λ follows the empirical relation $1/\lambda = R_H(1/n^2 - 1/4)$, $R_H = 1.07 \times 10^7 \, m^{-1}$ is the Rydberg constant, and n is an integer whose value is greater than 2. This is called the Balmer formula and, as an empirical relation, pre-dates the Bohr derivation by a couple of decades.

banana orbit In a toroidal geometry, the fast spiraling of a charged particle around a magnetic field line is accompanied by a slow drift motion of the particle's center around the spiral. When projected onto the poloidal plane of a toroidally confined plasma, the drift orbit has the shape of a banana. These orbits are responsible for neoclassical diffusion and for bootstrap current.

band calculation Each calculation of the energy band for a given crystal includes a complicated calculation and a suitable approximation for the exchange interaction. There are a lot of ways to calculate the band, each with a different approximation, such as LCAO, OPW, APW, etc.

band gap The results of band calculation show that electrons in crystal are arranged in energy bands. Because of some perturbations which come from long range or short range interaction in the crystal, there are some forbidden regions in these bands which are called *band gaps* or energy band *gaps*.

band theory An electron in a crystalline solid can exist only in certain values of energy. Electrons in solids are influenced by the array of positive ions. As a result, there are bands of energy, of allowed energy levels instead of single discrete energy levels, where an electron can exist. The allowed bands are separated by gaps of forbidden energy called forbidden gaps. The valence electrons in a solid are located in an energy band called the valence band. The energy band in which electrons can freely move is called the conduction band.

bare mass The mass value appearing in the Dirac equation which, however, differs from the real or physically observable particle mass (sometimes called the renormalized mass). The self-energy (interaction energy, for example, between an electron and its own electromagnetic field which is visualized as the continuous emission by the electron of virtual photons that are subsequently reabsorbed) becomes an inseparable part of a particle's observed rest mass.

barn A unit of area typically used in nuclear and high energy physics to express subatomic cross sections, equal to 10^{-24} cm^2. 1 millibarn = 10^{-27} cm^2. 1 nanobarn = 10^{-33} cm^2.

baroclinic Flow condition in which density is not a function of pressure only. Lines of constant pressure and density are not necessarily parallel.

baroclinic instability Geophysical instability of baroclinic flows that results in fluid motion slightly inclined to the horizontal. Mid-latitude disturbances favor *baroclinic instability*.

barometer Device used to measure atmospheric pressure.

barotropic Flow condition in which density is a function of pressure only. Lines of constant pressure and density are parallel.

barotropic instability Geophysical instability of barotropic flows arising from a sign change of the vorticity gradient. Occurs primarily in low-latitude regions since baroclinic instability is favored at higher latitudes.

barrier penetration A quantum wave phenomena important in nuclear, atomic, and condensed matter physics. In classical physics, a particle trajectory cannot sample regions of space where its total energy is less than the potential energy. In contrast, quantum theory does allow a finite probability for finding a particle in this region. An important application of this quantum phenomena is called *barrier penetration,* and refers to the fact that a particle has a finite probability to penetrate a potential barrier, such as the Coulomb repulsion barrier between two nuclei.

barrier, potential A potential $V(r)$ showing appreciable relative variation over a distance of the order of a wavelength, substantial enough to classically confine a particle with $E(r) < V(r)$ for some range in r. Simple illustrative examples include the idealized potential of the discontinuous square well, where the wave function must vanish at the edge of an infinitely high *potential barrier* but otherwise is partially transmitted (tunneling) by a finite barrier.

baseball coils Coils (copper or superconducting) that carry electrical current for producing magnetic fields that are shaped like the seams of a baseball, also known as yin-yang coils.

basis In crystal lattices, what is repeated is called the *basis*. A *basis* can be an atom, molecule, etc.

basis functions *Basis functions* define rows and columns of matrices in group theory. That is to say, they define what is operated on for vectors. *Basis functions* are non-unique. There are different choices of *basis functions*.

basis states A term used to describe a class of vectors in Hilbert space. In Hilbert space, any vector can be expressed as a sum over a set of complete orthonormal vectors. The members of this set are called *basis states. See also* completeness.

BCS states *BCS states* are superconductive. In *BCS states,* electrons are bonded in pairs called Cooper pairs. Because of the attractive interaction between two electrons in Cooper pairs, the total energy of a *BCS state* is lower than that of a Fermi state.

BCS theory BCS stands for the names of three physicists: Bardeen, Cooper, and Schrieffer. *BCS theory* is regarded as the basis of superconductivity theory. It predicts a criterion temperature Tc below which some material will become a superconductor.

BCS wave function The wave function which describes cooper pairs $\mathbf{K} \uparrow$ and $\mathbf{K} \downarrow$, where $\mathbf{K}$ is the wave vector. $\uparrow$ means up spin and $\downarrow$ means down spin. The electronic superconductivity and energy gaps in metals can be derived from the *BCS wave function.*

beam A concentrated, ideally unidirectional stream of particles characterized by its flux (number per unit area per unit time) and energy. In high energy experiments, typically a few MeV to TeV in energy with intensities as high as 10^{33}/cm^2/sec directed at targets of only a few mm^2 in area for the purposes of studying collisions and measuring cross sections.

beam-beam reaction Fusion reaction that occurs in neutral beam heated plasmas from the collision of two fast ions originating in the neutral beams injected into the plasma for heating purposes. Distinguished from beam-plasma, beam-wall, and thermonuclear (plasma-plasma) reactions.

beam-plasma reaction Fusion reaction that occurs in neutral beam heated plasmas from the collision of a fast beam ion with a thermal plasma ion.

beam-wall reaction   Fusion reaction that occurs in neutral beam heated plasmas from the collision of a fast beam ion with an ion embedded in the plasma vacuum wall.

Bell inequalities   Provide a test of quantum mechanics and its classical alternatives, the so-called local hidden variable theories. According to a paper published by Einstein, Podolsky, and Rosen, in which they discuss the Einstein-Podolsky-Rosen (EPR) gedankenexperiment, reality cannot be completely described by quantum mechanics. Supposedly local hidden variable theories provide such a complete description. Bell proved (1) the possible existence of hidden variable theories in the context of the EPR experiment, and (2) the statistical predictions of any hidden variable theory for the correlations of two particle systems in an entangled state obey the *Bell inequalities,* whereas the statistical predictions of quantum mechanics can violate those inequalities. Therefore an experimental distinction between the two is possible. However, due to the strict experimental requirements imposed on a test, i.e., high detection efficiencies, the strongest form of the *Bell inequalities* has never been tested. Tests of weaker forms of the *Bell inequalities,* e.g., photon experiments based on the cascade decay of atoms or parametric downconversion, have confirmed quantum mechanics.

The procedure for a test of the *Bell inequalities* is as follows: generation of an entangled singlet state between two particles, separation of the two constituents, and measurement of the correlation between the components of the entangled parameter with respect to certain directions. This can be, for instance, polarization in the case of photons or spin for atoms, etc.

Bell J.S.   Irish physicist (1923–1998) noted for his statement of the Bell inequalities. *See* Bell's inequality.

Bell's inequality   A set of relations, first laid down by John Bell, that provides constraints on the values obtained in the experimental measurement of spin correlations between particles that are separated by macroscopic distances, but which must be described by a quantum mechanical wave function. If, in a measurement, the inequality is violated, the measurement is in agreement with the predictions of the quantum theory. If the equality is satisfied, it suggests that a classical, causal, and local model is adequate to explain the outcome of the measurements. To date, experiments have confirmed that correlations are consistent with quantum theory in systems that are separated as far as tens of kilometers.

bend loss   *See* loss, minor.

Bérnard convection   Convection in a horizontal layer due to a temperature difference across the layer. Above a critical Rayleigh number of 1700, the fluid begins to move as hot fluid from the bottom of the layer rises, and cold fluid from the top of the layers descends. The instability forms regular convective Bérnard cells across the fluid layer. As the Rayleigh number increases, spatial regularity is lost and the fluid mixing becomes turbulent.

Bernoulli's equation   Simplification of the Euler equation in which the variation of flow properties along a streamline are constant such that

$$\frac{1}{2}|\mathbf{u}|^2 + \int \frac{dp}{\rho} + gz = \text{constant}$$

where u, p, and z are variable. For two points connected by a streamline, this can be written as

$$\frac{1}{2}u_1^2 + \frac{p_1}{\rho} + gz_1 = \frac{1}{2}u_2^2 + \frac{p_2}{\rho} + gz_2.$$

The flow must be steady, incompressible, and adiabatic.

Bernstein mode   Type of plasma mode that propagates perpendicular to the equilibrium magnetic field in a plasma. Bernstein waves, named after the plasma physicist Ira Bernstein, have their electric field nearly parallel to the wave propagation vector and their frequency between harmonics of the electron cyclotron frequency.

Berry's phase   Phenomena associated with the adiabatic evolution of a quantum system. According to the adiabatic theorem, the state of a quantum system that undergoes slow, or adiabatic, evolution acquires a dynamical phase factor. Under certain conditions, the state can acquire an additional pre-factor that has the form given by $\exp(i \int_C d\mathbf{R} \cdot \mathbf{A})$, where the path integral is taken in the parameter space that governs the evolution of the Hamiltonian. The resulting, non-vanishing, value of the circuit integral is called *Berry's phase*. *See also* adiabatic theorem.

beta decay The decay of a free neutron (or neutron within the nucleus of a radioactive isotope) producing a final state electron (negative beta particle). This decay is an example of a weak interaction that transforms one of its constituent's down quarks into an up quark through a process involving the emission and subsequent decay of a W boson.

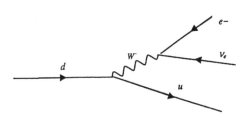

Beta decay.

beta limit Also known as the troyon limit in a tokamak, the *beta limit* is the maximum achievable ratio (beta, or beta value) of plasma pressure to magnetic pressure for a given plasma to remain stable. In a tokamak, if the beta value is too high, ballooning modes become unstable and lead to a loss of plasma confinement.

beta, or beta value Ratio of plasma kinetic pressure to magnetic field pressure. *Beta* is usually measured relative to the total local magnetic field, but in some cases it can be measured relative to components of the total field, such as the poloidal field in tokamaks.

beta particle, beta radiation High-speed charged particle emitted from the nucleus of some atoms in their radioactive decay. Positively charged *beta particles* are positrons and negatively charged *beta particles* are electrons. Because *beta particles* are harmful to living tissue (*beta particles* can cause burns), protection can be provided by thin sheets of metal.

beta plane model Simplified model accounting for the variation of Coriolis forces with latitude in geophysical flows. In the governing equations, the angular velocity of rotation Ω is taken to be a function of position such that

$$\Omega = \Omega \cos \theta + \beta y$$

where θ is the co-latitude, y is the northerly distance, and $\beta = \Omega \sin \theta / R_E$. The β-*plane model* results in approximate solutions for Rossby waves.

Bethe, Hans (1906-) American physicist. Pioneer of modern atomic and nuclear physics. *H. Bethe* was the first to provide the theoretical explanation for the Lamb shift in atoms. Professor Bethe was awarded the Nobel Prize in physics, with Enrico Fermi, for elucidating the nuclear life cycle of stars.

Bethe-log An expression that involves a sum over atomic states and that is needed for calculation of the self-energy shift in atomic levels. This shift, also called the Lamb shift, arises from electron interaction with the vacuum of the radiation field.

Bethe-Salpeter equation A relativistic covariant equation that describes two-body quantum systems in the relativistic regime. The equation is derived from quantum electrodynamics (QED) assuming the ladder approximation for the covariant two particle Green's function.

Bhabha scattering The scattering of electrons by positrons treated theoretically by H.J. Bhabha (1935). The particles are distinguishable by their charge and the process may proceed through the two mechanisms illustrated by the Feynman diagrams graphed on the next page. At left, scattering by photon exchange, at right, scattering proceeds via the annihilation diagram.

bias A potential applied in a device to produce the desired characteristic.

bilinear covariants Probability densities of the form $\bar{\psi}\Gamma\psi$, where Γ is a product of (Dirac)

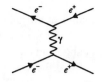

Bhabha scattering.

gamma matrices, which have definite transformation properties under Lorentz transformations. As an example, $\bar{\psi}\psi$ transforms as a scalar, $\bar{\psi}\gamma_\mu\psi$ as a vector, and $\bar{\psi}\gamma_5\gamma_\mu\psi$ as an axial vector.

binary alloy A mixture of two pure components containing a fraction x_A of component A and $x_B = 1 - x_A$ of component B. The fraction x_A that specifies the composition of the alloy can be measured as a fraction of the weight, volume, or moles of the alloy.

binding energy In crystal, the energy differences between free atoms and the crystal composed by the atoms are called *binding energy*. If the *binding energy* is larger, the crystal is more stable.

binding force The interaction between atoms, ions, or molecules in crystal. The kind of crystal depends on the kind of *binding force* in the crystal. For example, in molecular crystal, the *binding force* is the Van der Waal force.

Bingham plastic Fluid which behaves as a solid until a minimum yield stress is exceeded and subsequently behaves as a Newtonian fluid. The shear stress relation is given by

$$\tau = \tau_{\text{yield}} + \mu \frac{du}{dy}.$$

Some pastes and muds exhibit this behavior.

binomial distribution The probability distribution $W_N(N_A)$ of distributing N objects into two groups A and B containing N_A and $N_B = N - N_A$ objects, respectively, where an object belongs to group A with probability p and to group B with probability, $1 - p$. $W_N(N_A) = (N!/(N_A!N_B!))p^{N_A}(1-p)^{N_B}$.

Biot-Savart law Kinematic relation between velocity and vorticity. For two vortex filaments, it takes on the form

$$\mathbf{U}_n = \sum_{m=1}^{2} \Gamma_m \int \frac{\mathbf{R}_{mn} \times d\mathbf{L}_m}{4\pi \mid \mathbf{R}_{mn} \mid^3}, \ n = 1, 2.$$

bipolar transistor A solid state electronic device with three terminals, used in amplifiers. It controls the current between two terminals (the source and the drain) by the voltage at a third terminal called the gate. A heavily doped p-type semiconductor forms a gate. A single piece of n-type semiconductor with a source at one end and drain at the other end with a gate in the middle is an n-type field effect transistor (n-FET). In the FET, only one type of charge carrier, electrons in n-FET and holes in p-FET, determines the current and is thus known as a unipolar transistor. In the *bipolar* junction *transistor*, the positive and negative charge carriers contribute to the current.

black-body An ideal body that completely absorbs all radiant energy striking it and, therefore, appears perfectly black at all wavelengths. The radiation emitted by such a body when heated is referred to as *black-body* radiation. A perfect *black-body* has an emissivity of unity.

black-body radiation The intensity distribution of light emitted by a hot solid. The spectral distribution for a black-body in thermal equilibrium with its surroundings is a function only of its temperature, but is unsolvable using a classical interpretation of electromagnetic radiation as a continuous wave.

In a statistical mechanics treatment of the problem, Planck found (1900) that in order to fit the distribution with a functional form, an assumption had to be made that the solid radiated energy in integral multiples of $h\nu$, where h was a proportionality constant, now known as Planck's constant, equal to 6.6×10^{-27} erg/sec. This was the first introduction of energy quantization into physics. *See* Boltzmann distribution.

Blasius solution Solution for the viscous flow in a boundary layer over a flat plate. Solution is given by simplification and similarity arguments. For laminar flow, the shape of the boundary layer is given by

$$\frac{\delta}{x} = \frac{4.9}{\sqrt{\text{Re}_x}}$$

where x is the distance from the leading edge of the flat plate, and Re_x is the local Reynolds number, $\text{Re}_x \equiv \frac{Ux}{\nu}$.

Blasius theorem Relation between lift L and drag D of a body in a two-dimensional potential (irrotational) flow field given by the velocity field (u, v) such that

$$D - iL = \frac{i}{2}\rho \oint (u - iv)^2 \, dz$$

where z is the complex variable

$$z \equiv x + iy \, .$$

Bloch, F. (1905–1983). American physicist. Noted pioneer in the application of quantum theory to the physics of condensed matter.

Bloch oscillator In crystal, an electron will oscillate when it moves across the superlattice plane. This phenomenon is called *Bloch oscillator*.

Bloch wall Divides crystal into domains. In each domain, there is a different orientation of the magnetization.

bloch wave A wave function expressible as a plane wave modulated by a periodic function $u_k(\mathbf{r})$: $\psi(\mathbf{r}) = e^{i\mathbf{k}\cdot\mathbf{r}} u_k(\mathbf{r})$. Such forms are applicable to systems with a potential that is periodic in space (like that felt by an electron within a crystal lattice). Such a wave function will be an eigenfunction of the translation operator $\mathbf{r} \to \mathbf{r} + \mathbf{a}$, where $\mathbf{a}$ corresponds to the crystal lattice spacing and $u_k(\mathbf{r})$ has the same periodicity as the lattice.

blocking Effect of bodies in a flow field on the upstream and downstream flow behavior. Particularly important in stratified flows (such as in geophysical fluid dynamics) and flows in ducts (such as in wind tunnels).

blower Pump classification in which the pressure rise of the gas is approximately less than 1 atmosphere but still significant; the increase in pressure may cause a slight density change, but the working gas will most likely remain at the initial density. *Compare with* compressor.

body-centered cubic primitive vectors For a body-centered cubic primitive cell with length a, we define primitive vectors as $a\mathbf{x}$, $a\mathbf{y}$, and $a/2(\mathbf{x} + \mathbf{y} + \mathbf{z})$. On the other hand, we can regard body-centered cubes as simple cubes with basis $(0, a/2(\mathbf{x} + \mathbf{y} + \mathbf{z}))$ and primitive vectors $a\mathbf{x}, a\mathbf{y}$, and $a\mathbf{z}$.

body-centered cubic structure One of the most common metallic structures. In the *body-centered cubic structure,* atoms are arranged in a cubes, and an additional atom is located at the center of each cube.

Bohm diffusion A rapid loss of plasma particles across magnetic field lines caused by plasma microinstabilities that scales inversely with the magnetic field strength, unlike classical diffusion that scales inversely as the square of the magnetic field strength. Named after the plasma physicist David Bohm, who first proposed such scaling.

Bohr atom A model of the atom successfully developed for hydrogen by Bohr (1913). By constraining hydrogen's atomic electron to move only in one of a number of allowed circular orbits (stationary states), its energy became quantized. Transitions between stationary states required the absorption or emission of a quantum of light with frequency $\nu = \Delta E/h$, where ΔE is the energy difference between two states. Applying Newtonian mechanics, Bohr was able to derive a formula for hydrogen atom energy levels in complete agreement with the observed hydrogen spectrum. The theory failed, however, to account for the helium spectrum or the chemical bonds of molecules.

Bohr, Niels (1884–1962) Danish physicist/philosopher. The father of atomic theory and a leading figure in the development of the modern quantum theory. Bohr's Institute for Advanced Studies in Copenhagen was host to many leading physicists of the time. *Niels Bohr* also played an leading role in the development of modern nuclear physics. *See* Copenhagen interpretation.

Bohr quantization Rule that determines the allowed electron orbits in Bohr's theory of the hydrogen atom. In an early atomic theory, Bohr suggested that electrons orbit parent nuclei much like planets orbit the sun. Because electrons are electrically charged, classical physics predicts that such a system is unstable due to radiative energy loss. Bohr postulated that electrons radiate only if they "jump" between allowed prescribed orbits. These orbits are called Bohr orbits. The conditions required for the allowed angular momenta, hence orbits, is called *Bohr quantization* and is given by the formula $L = n\hbar$, where L is the allowed value of the angular momentum of a circular orbit, n is called the principal quantum number, and $\hbar$ is the Planck constant divided by 2π.

Bohr radius (a_0) (1) The radius of the electron in the hydrogen atom in its ground state, as described by the Bohr theory. In Bohr's early atomic theory, electrons orbit the nucleus on well defined radii, the smallest of which is called the first *Bohr radius*. Its value is 0.0529 nm.

(2) According to the Bohr theory of the atom (*see* Bohr atom), the radius of the circle in which the electron moves in the ground state of the hydrogen atom, $a_0 \equiv \hbar^2/m_e^2 = 0.5292$ Å. A full quantum mechanical treatment of hydrogen gives a_0 as the most probable distance between electrons and the nucleus.

Boltzmann constant (k_B) A fundamental constant which relates the energy scale to the Kelvin scale of temperature, $k_B = 1.3807 \times 10^{-23}$ joules/kelvin.

Boltzmann distribution A law of statistical mechanics that states that the probability of finding a system at temperature T with an energy E is proportional to $e^{-E/KT}$, where K is Boltzmann's constant. When applied to photons in a cavity with walls at a constant temperature T, the *Boltzmann distribution* gives Planck's distribution law of $E_k = \hbar ck/(e^{\hbar ck/KT} - 1)$.

Boltzmann factor The term, $\exp(-\varepsilon/k_B T)$, that is proportional to the probability of finding a system in a state of energy ε at absolute temperature T.

Boltzmann's constant A constant equal to the universal gas constant divided by Avogadro's number. It is approximately equal to 1.38×10^{-23} J/K and is commonly expressed by the symbol k.

Boltzmann statistics Statistics that lead to the Boltzmann distribution. *Boltzmann statistics* assume that particles are distinguishable.

Boltzmann transport equation An integro-differential equation used in the classical theory of transport processes to describe the equation of motion of the distribution function $f(\mathbf{r}, \mathbf{v}, t)$. The number of particles in the infinitesimal volume $d\mathbf{r}\,d\mathbf{v}$ of the 6-dimensional phase space of Cartesian coordinates $\mathbf{r}$ and velocity $\mathbf{v}$ is given by $f(\mathbf{r}, \mathbf{v}, t) d\mathbf{r}\,d\mathbf{v}$ and obeys the equation

$$\frac{\partial f}{\partial t} + \alpha \cdot \nabla_v f + \vec{v} \cdot \nabla_r f = \left(\frac{\partial f}{\partial t}\right)_{\text{coll.}}$$

Here, α denotes the acceleration, and $(\partial f/\partial t)_{\text{coll.}}$ is the change in the distribution function due to collisions. The integral character of the equation arises in writing the collision term in terms of two particle collisions.

bonding orbital *See* anti-bonding orbital.

bootstrap current Currents driven in toroidal devices by neo-classical processes.

Born approximation An approximation useful for calculation of the cross-section in collisions of atomic and fundamental particles. The *Born approximation* is particularly well-suited for estimates of cross-sections at sufficiently large relative collision partner velocities. In potential scattering, the *Born approximation* for the scattering amplitude is given by the expression $f(\theta) = -\frac{2\mu}{\hbar^2 q} \int_0^\infty r\,sin(qr) V(r)\,dr$, where θ is the observation angle, $\hbar$ is Planck's constant divided by 2π, μ is the reduced mass, $V(r)$ is the spherically symmetric potential, $q \equiv 2k sin(\theta/2)$, and k is the wave number for the collision. *See* cross-section.

Born-Fock theorem *See* adiabatic theorem.

Born, Max (1882–1970) German physicist. A founding father of the modern quantum theory. His name is associated with many applications of the modern quantum theory, such as the Born approximation, the Born-Oppenheimer approximation, etc. Professor Born was awarded the Nobel Prize in physics in 1954.

Born-Oppenheimer approximation An approximation scheme for solving the many- few-atom Schrödinger equation. The utility of the approximation follows from the fact that the nuclei of atoms are much heavier than electrons, and their motion can be decoupled from the electronic motion. The *Born-Oppenheimer approximation* is the cornerstone of theoretical quantum chemistry and molecular physics.

Born postulate The expression $|\psi(x, y, z, t)|^2 \, dx \, dy \, dz$ gives the probability at time t of finding the particle within the infinitesimal region of space lying between x and dx, y and dy, and z and dz. $|\psi(x, y, z, t)|^2$ is then the probability density for finding a particle in various positions in space.

Born-Von Karman boundary condition Also called the periodic boundary condition. To one dimensional crystal, it can be expressed as $U_1 = U_N + 1$, where N is the number of particles in the crystal with length L.

Bose-Einstein condensation A quantum phenomenon, first predicted and described by Einstein, in which a non-interacting gas of bosons undergoes a phase transformation at critical values of density and temperature. A Bose-Einstein condensate can be considered a macroscopic system described by a quantum state. Bose-Einstein condensates have recently been observed, about 70 years following Einstein's prediction, in dilute atomic gases that have been cooled to temperatures only about 10^{-9} Kelvin above absolute zero.

Bose-Einstein statistics Statistical treatment of an assembled collection of bosons. The distinction between particles whose wave functions are symmetric or antisymmetric leads to different behavior under a collection of particles (i.e., different statistics). Particles with integral spin are characterized by symmetric wave functions and therefore are not subject to the Pauli exclusion principal and obey Bose-Einstein statistics.

Bose, S.N. (1894–1974) Indian physicist and mathematician noted for fundamental contributions to statistical quantum physics. His name is associated with the term Bose statistics which describes the statistics obeyed by indistinguishable particles of integer spin. Such particles are also called bosons. His name is also associated with Bose-Einstein condensation. *See* Bose-Einstein condensation.

Bose statistics Quantum statistics obeyed by a collection of bosons. *Bose statistics* lead to the Bose-Einstein distribution function and, for critical values of density and temperature, predict the novel quantum phenomenon of the Bose-Einstein condensation. *See* Bose-Einstein condensation.

boson (1) A particle that has integer spin. A boson can be a fundamental particle, such as a photon, or a composite of other fundamental particles. Atoms are composites of electrons and nuclei; if the nucleus has half integer spin and the total electron spin is also half integral, then the atom as a whole must possess integer spin and can be considered a composite *boson*.

(2) Particles can be divided into two kinds, boson and fermion. The fundamental difference between the two is that the spin quanta number of bosons is integer and that of fermions is half integer. Unlike fermions, which can only be created or destroyed in particle-antiparticle pairs, *bosons* can be created and destroyed singly.

bounce frequency The average frequency of oscillation of a particle trapped in a magnetic mirror as it bounces back and forth between its turning points in regions of high magnetic field.

boundary layer (1) A thin layer of fluid, existing next to a solid surface beyond which the liquid is moving. Within the layer, the effects of viscosity are significant. The effects of viscosity often can be neglected beyond the *boundary layer*.

(2) The transition layer between the solid boundary of a body and a moving viscous fluid as required by the no-slip condition. The thickness of the *boundary layer* is usually taken to be the point at which the velocity is equal to 99% of the free-stream velocity. Other measures of *boundary layer* thickness include the displacement thickness and momentum thickness. The *boundary layer* gives rise to friction drag from viscous forces and can also lead to separation. It also is responsible for the creation of vorticity and the diffusion thereof due to viscous effects. Thus, a previously irrotational region will remain so unless it interacts with a *boundary layer*. This leads to the separation of flows into irrotational portions outside the *boundary layer* and viscous regions inside the *boundary layer*. The thickness of a boundary decreases with an increasing Reynolds number, resulting in the approximation of high speed flows as irrotational. A *boundary layer* can be laminar, but will eventually transition to turbulence given time. The *boundary layer* concept was introduced by Ludwig Prandtl in 1904 and led to the development of modern fluid dynamics. *See* boundary layer approximation.

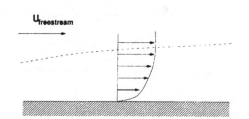

Boundary layer.

boundary layer approximation Simplification of the governing equations of motion within a thin boundary layer. If the boundary layer thickness is assumed to be small compared to the length of the body, then the variation along the direction of the boundary layer (x) is assumed to be much less than that across the boundary layer (y) or

$$\frac{\partial}{\partial x} \ll \frac{\partial}{\partial y} \quad \text{and} \quad \frac{\partial^2}{\partial x^2} \ll \frac{\partial^2}{\partial y^2}$$

where the velocity in the y-direction (v) is also assumed to be much smaller than the velocity in the x-direction (u), $v \ll u$. The continuity equation remains

$$\frac{\partial u}{\partial x} + \frac{\partial v}{\partial y} = 0$$

while the x- and y-momentum equations reduce to

$$u\frac{\partial u}{\partial x} + v\frac{\partial u}{\partial y} = -\frac{1}{\rho}\frac{\partial p}{\partial x} + \nu\frac{\partial^2 u}{\partial y^2}$$

and

$$0 = -\frac{\partial p}{\partial y}$$

respectively. This results in a tractable solution with three equations and three unknowns.

bound state An eigenstate of distinct energy that a particle occupies when its energy $E < V$ of a potential well that confines it near the force center creating the potential. The discrete energy values are forced on the system by the requirement of continuity of the wave function at the boundaries of the potential well, beyond which the wave function must diminish (or vanish). When $E > V$ everywhere, the particle is not bound but instead is free to occupy any of an infinite continuum of states.

Bourdon tube Classical mechanical device used for measuring pressure utilizing a curved tube with a flattened cross-section. When pressurized, the tube deflects outward and can be calibrated to a gauge using a mechanical linkage. *Bourdon tubes* are notable for their high accuracy.

Boussinesq approximation Simplification of the equations of motion by assuming that density changes can be neglected in certain flows due to the compressibility. While the density may vary in the flow, the variation is not due to fluid motion such as occurs in high speed flows. Thus, the continuity equation is simplified to

$$\nabla \cdot \mathbf{u} = 0$$

from its normal form.

box normalization A common wave function normalization convention. If a particle is contained in a box of unit length L, the wave function is constrained to vanish at the boundary and requires quantization of momentum. An integral of the probability density $|\psi|^2$ throughout the box is required to sum to unity and typically leads to a normalization pre-factor for the wave function given by $1/\sqrt{V}$, where V is the volume of the box. In most applications, the volume of the box is taken to have the limit as $L \to \infty$.

Boyle's law An empirical law for gases which states that at a fixed temperature, the pressure of a gas is inversely proportional to its volume, i.e., $pV =$ constant. This law is strictly valid for a classical ideal gas; real gases obey this to a good approximation at high temperatures and low pressures.

bracket, or bra-ket An expression representing the inner (or dot) product of two state vectors, $\psi_\alpha^\dagger \equiv <\alpha|\beta>$ which yields a simple scalar value. The first and last three letters of the bracket name the notational expression involving triangular brackets for the two kinds of state vectors that form the inner product.

Bragg diffraction A laboratory method that takes advantage of the wave nature of electromagnetic radiation in order to probe the structure of crystalline solids. Also called X-ray diffraction, the method was developed and applied by W.L. Bragg and his father, W.H. Bragg. The pair received the Nobel Prize for physics in 1915.

bra vector Defined by the bra-ket formalism of Dirac, which allows a concise and easy-to-use terminology for performing operations in Hilbert space. According to the bra-ket formalism, a quantum state, or a vector in Hilbert space, can be described by the ket symbol. For any ket $|a>$ there exists a bra $<a|$. This is also called a dual correspondence. If $<b|$ is a bra and $|a>$ a ket, then one can define a complex number represented by the symbol $<b|a>$, whose value is given by an inner product of the vectors $|a>$ and $|b>$.

breakeven (commercial, engineering, scientific, and extrapolated) Several definitions exist for fusion plasmas: *Commercial breakeven* is when sufficient fusion power can be converted into electric power to cover the costs of the fusion power plant at economically competitive rates; *engineering breakeven* is when sufficient electrical power can be generated from the fusion power output to supply power for the plasma reactor plus a net surplus without the economic considerations; *scientific breakeven* is when the fusion power is equal to the input power; i.e., $Q = 1$. (*See also* Lawson criterion); *extrapolated breakeven* is when scientific breakeven is projected for actual reactor fuel (e.g., deuterium and tritium) from experimental results using an alternative fuel (e.g., deuterium only) by scaling the reaction rates for the two fuels.

Breit–Wigner curve The natural line shape of the probability density of finding a decaying state at energy E. Rather than existing at a single well-defined energy, the state is broadened to a full width at half max, Γ, which is related to its lifetime by $\tau \Gamma = \hbar$. The curve of the probability density is given by

$$P(E) = \frac{\Gamma}{2\pi} \frac{1}{(E - E_0)^2 + (\Gamma/2)^2}.$$

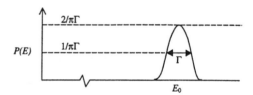

Breit–Wigner curve.

Breit–Wigner form Functional form of cross sections in the vicinity of a resonance. In resonance scattering, the presence of a metastable state, or a significant time delay, is signaled by the behavior of the scattering cross-section according to this particular functional form. Near the resonance energy E_r, the *Breit–Wigner form* for the cross-section $\sigma(E)$, is given by $\sigma(E) \sim \frac{\Gamma}{(E-E_r)^2+(\Gamma/2)^2}$, where E is the collision energy, and Γ is the lifetime of the resonance state.

Bremsstrahlung Electromagnetic radiation that is emitted by an electron as it is accelerated or decelerated while moving through the electric field of an ion.

Bremsstrahlung radiation Occurs in plasma when electrons interact ("collide") with the Coulomb fields of ions; the resulting deflection of the electrons causes them to radiate.

Brillouin–Wigner perturbation Perturbation treatment that expresses a state as a series expansion in powers of λ (the scale of the perturbation from an unperturbed Hamiltonian, $\mathcal{H} = \mathcal{H}_0 + \lambda V$) with coefficients that depend on the perturbed energy values E_n (rather than the unperturbed energies εn of the Rayleigh-Schrödinger perturbation method). An initial unperturbed eigenstate, φ_n, becomes,

$$\Psi_n = \varphi_n + \sum_{m \neq n} \varphi_m \frac{1}{E_n - E_m} \lambda \langle \varphi_m | V | \varphi_n \rangle .$$

Brillouin zone Similar to the first *Brillouin zone*, bisect all lines, among which each connects a reciprocal lattice point to one of its secondly nearest points. The region composed of all the bisections is defined as the second *Brillouin zone*. Keeping on it, we can get all *Brillouin zones* of the considered reciprocal lattice point. Each *Brillouin zone* is center symmetric to the point.

broken symmetry Property of a system whose ground state is not invariant under symmetry operations. Suppose L is the generator of some symmetry of a system described by Hamiltonian H. Then $[L, H] = 0$, and if $|a >$ is a non-degenerate eigenstate of H, it must also be an eigenstate of L. If there exists a degeneracy, $L|a >$ is generally a linear combination of states in the degenerate sub-manifold. If the ground state $|g >$ of the system has the property that $L|g > \neq c|g >$, where c is a complex number, then the symmetry corresponding to the generator of that symmetry, L, is said to be broken.

Brownian motion The disordered motion of microscopic solid particles suspended in a fluid or gas, first observed by botanist Robert Brown in 1827 as a continuous random motion and attributed to the frequent collisions the particles undergo with the surrounding molecules. The motion was qualitatively explained by Einstein's (1905) statistical treatment of the laws of motions of the molecules.

Brunt–Väisälä frequency Natural frequency, N, of vertical fluid motion in stratified flow as given by the linearized equations of motion:

$$N^2 \equiv -\frac{g}{\rho_o} \frac{d\bar{\rho}}{dz}$$

where

$$\bar{\rho}(z) = \rho - \rho' .$$

Also called buoyancy frequency.

bubble chamber A large tank filled with liquid hydrogen, with a flat window at one end and complex optical devices for observing and photographing the rows of fine bubbles formed when a high-energy particle traverses the hydrogen.

Buckingham's Pi theorem For r number of required dimensions (such as mass, length, time, and temperature), n number of dimensional variables can always be combined to form exactly $n - r$ independent dimensionless variables. Thus, for a problem whose solution requires seven variables with three total dimensions, the problem can be reduced to four dimensionless parameters. *See* dimensional analysis, Reynolds number for an example.

bulk viscosity Viscous term from the constitutive relations for a Newtonian fluid, $\lambda + \frac{2}{3}\mu$, where λ and μ are measures of the viscous properties of the fluid. This is reduced to a more usable form using the Stokes assumption.

buoyancy The vertical force on a body immersed in a fluid equal to the weight of fluid displaced. A floating body displaces its own weight in the fluid in which it is floating. *See* Archimede's law.

C

calorie (Cal) A unit of heat defined as the amount of heat required to raise the temperature of 1 gm. of water at 1 atmosphere pressure from 14.5 to 15.5 C. It is related to the unit of energy in the standard international system of units, the Joule, by 1 *calorie* = 4.184 joules. Note that the *calorie* used in food energy values is 1 kilocalorie ≡ 1000 *calories*, and is denoted by the capital symbol *Cal*.

camber Curvature of an airfoil as defined by the line equidistant between the upper and lower surfaces. Important geometric property in the generation of lift.

canonical ensemble Ensemble that describes the thermodynamic properties of a system maintained at a constant temperature T, by keeping it in contact with a heat reservoir at temperature T. The canonical distribution function gives the probability of finding the system in a non-degenerate state of energy E_i as

$$P(E_i) = \exp(-E_i/k_BT) / \sum_i \exp(-E_i/k_BT),$$

where k_B is the Boltzmann constant, and the summation is over all possible microstates of the system, denoted by the index i.

canonical partition function For a system of N particles at constant temperature T and volume V, all thermodynamic properties can be obtained from the *canonical partition function* defined as $Z(T, V, N) = \sum_i \exp(-E_i/k_BT)$, where E_i is the energy of the system of N particles in the ith microstate.

canonical variables In the Hamiltonian formulation of classical physics, conjugate variables are defined as the pair, q, $p = \frac{\partial L}{\partial \dot{q}}$, where L is the Lagrangian and q is a coordinate, or variable of the system.

capacitively coupled discharge plasma Plasma created by applying an oscillating, radio-frequency potential between two electrodes. Energy is coupled into the plasma by collisions between the electrons and the oscillating plasma sheaths. If the oscillation frequency is reduced, the discharge converts to a glow discharge.

capillarity Effect of surface tension on the shape of the free surface of a fluid, causing curvature, particularly when in contact with a solid boundary. The effect is primarily important at small length scales.

capillary waves Free surface waves due to the effect of surface tension σ which are present at very small wavelengths. The phase speed, c, of capillary waves decreases as wavelength increases,

$$c = \sqrt{\frac{k\sigma}{\rho}}$$

as opposed to surface gravity waves, whose phase speed increases with increasing wavelength.

Carnot cycle A cyclical process in which a system, for example, a gas, is expanded and compressed in four steps: (i) an isothermal (constant temperature) expansion at temperature T_h, until its entropy changes from S_c to S_h, (ii) an adiabatic (constant entropy) expansion during which the system cools to temperature T_c, followed by (iii) an isothermal compression at temperature T_c, and (iv) an adiabatic compression until the substance returns to its initial state of entropy, S_c. The Carnot cycle can be represented by a rectangle in an entropy–temperature diagram, as shown in the figure, and it is the same regardless of the working substance.

carrier A charge carrier in a conduction process: either an electron or a positive hole.

cascade A row of blades in a turbine or pump.

cascade, turbulent energy Transfer of energy in a turbulent flow from large scales to small scales through various means such as dissipation and vortex stretching. Energy fed into the turbulent flow field is primarily distributed among

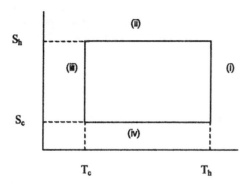

Carnot cycle.

large scale eddies. These large eddies generate smaller and smaller eddies until the eddy length scale is small enough for viscous forces to dissipate the energy. Dimensional analysis shows that the relation between the energy E, the energy dissipation ε, and wavenumber k is

$$E \propto \varepsilon^{2/3} k^{-5/3}$$

which is known as Kolmogorov's -5/3 law. *See* turbulence.

Casimir operator Named after physicist H.A. Casimir, these operators are bi-linear combinations of the group generators for a Lie group that commute with all group generators. For the covering group of rotations in three-dimensional space, there exists one *Casimir operator*, usually labeled $\mathbf{J}^2$, where $\mathbf{J}$ are the angular momentum operators. *See* angular momentum.

cation A positively charged ion, formed as a result of the removal of electrons from atoms and molecules. In an electrolysis process, *cations* will move toward negative electrodes.

Cauchy–Riemann conditions Relations between velocity potential and streamfunction in a potential flow where

$$\frac{\partial \phi}{\partial x} = \frac{\partial \Psi}{\partial y}$$
$$\frac{\partial \phi}{\partial y} = -\frac{\partial \Psi}{\partial x}$$

such that either ϕ or Ψ can be determined if the other is known.

causality The causal relationship between a wavefunction at an initial time $\psi(t_o)$ and a wavefunction at any later time $\psi(t)$ as expressed through Schrödinger's equation. This applies only to isolated systems and assumes that the dynamical state of such a system can be represented completely by its wave function at that instant. *See* complementarity.

cavitation Spontaneous vaporization of a liquid when the pressure drops below the vapor pressure. *Cavitation* commonly occurs in pumps or marine propellers where high fluid speeds are present. Excessive speed of the pump or propeller and high liquid temperatures are standard causes of *cavitation*. *Cavitation* degrades pump performance and can cause noise, vibration, and even structural damage to the device.

cavitation number Dimensionless parameter used to express the degree of cavitation (vapor formation) in a liquid:

$$Ca \equiv (p_a - p_v)/\rho U^2$$

where p_a is the atmospheric pressure and p_v is the vapor pressure of the liquid.

cellular method One method of energy band calculation in crystal. It was addressed by Wigner and Seitz. They divided a crystal into atomic cells. For a given potential, because of symmetry, the calculation reduced into a single cell. The assumption for the *cellular method* is that the normal component of the gradient of wave function will vanish at the single cell surface or at the Wigner–Seitz sphere.

Celsius temperature scale (C) Defined by setting the temperature at which water at 1 atmospheric pressure freezes at 0°C and boils at 100°C. Alternatively, the Celsius scale can be defined in terms of the Kelvin temperature T as temperature in Celsius $= T - 273.16K$.

center-of-momentum (c.o.m.) coordinates A coordinate system in which the centers of mass of interacting particles are at rest. The particles are located by position vectors ρ_{r_i} defined by the center of mass of the rest frame of the system, which, in general, moves with respect to the particles themselves.

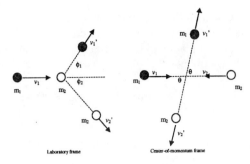

c.o.m. coordinates

In the *center-of-momentum* system, a pair of colliding particles both approach the c.o.m. head on, and then recede from the center with equal but opposite momenta:

$$\frac{\rho}{p_1} + \frac{\rho}{p_2} = \frac{\rho'}{p_1} + \frac{\rho'}{p_2} = 0$$

even if, in the laboratory frame, the target particle is at rest (as depicted above). The velocity of the c.o.m. for such a collision is $\dfrac{\rho}{v_{cm}} = \dfrac{m_1 \dfrac{\rho}{v_1}}{m_1+m_2}$. While in the laboratory frame two angles measured with respect to the line of motion of the incident particle are necessary to describe the final directions of the particles, ϕ_1 and ϕ_2, a single common angle θ suffices in the c.o.m.:

$$\tan\phi_1 = \frac{\sin\theta}{\gamma + \cos\theta} \quad \text{where} \quad \gamma = \frac{v_{cm}}{v_1'^{cm}}$$

$$= \frac{m_1 v_1}{v_1'^{cm}(m_1+m_2)}.$$

central force A force always directed toward or away from a fixed center whose magnitude is a function only of the distance from that center. In terms of spherical coordinates with an origin at the force's center,

$$\frac{\rho}{F} = \hat{r}F(r), \text{ where } r = \sqrt{x^2+y^2+z^2}$$

or in cartesian coordinates, $F_x = \frac{x}{r}F(r)$, $F_y = \frac{y}{r}F(r)$, and $F_z = \frac{z}{r}F(r)$.

centrifugal barrier A centrifugal force-like term that appears in Schrödinger's equations for central potentials that prevents particles with non-zero angular momentum from getting too close to the potential's center. The symmetry of Hamiltonians with central potentials allows the state function to be separated into radial and angular parts: $\psi(r) = f_\lambda(r)Y_{\lambda m}(\theta,\phi)$. If the radial part is written in the form $f_\lambda(r) = u_\lambda(r)/r$, the function $u_\lambda(r)$ can satisfy

$$\left(-\frac{\eta^2}{2m}\frac{d^2}{dr^2} + \frac{\eta^2}{2m}\frac{\lambda(\lambda+1)}{r^2} + V(r) - E\right)u_\lambda(r) = 0$$

a one-dimensional Schrödinger equation carrying an additional potential-like term $\eta^2\lambda(\lambda+1)/2mr^2$ which grows large as $r \to 0$.

centrigual instability Present in a circular Couette flow driven by the adverse gradient of angular momentum which results in counter-rotating toroidal vortices. Also known as the Taylor or Taylor-Couette instability.

cesium chloride structure In *cesium chloride*, the bravais lattice is a simple cube with primitive vectors $a\mathbf{x}$, $a\mathbf{y}$, and $a\mathbf{z}$ and a basis composed of a cesium positive ion and a chloride negative ion.

CFD Computational fluid dynamics.

change of state Refers to a change from one state of matter to another (i.e., solid to liquid, liquid to gas, or solid to gas).

chaos The effect of a solution on a system which is extremely sensitive to initial conditions, resulting in different outcomes from small changes in the initial conditions. Deterministic *chaos* is often used to describe the behavior of turbulent flow.

characteristic Mach number A Mach number such that

$$M' = u/a'$$

where a' is the speed of sound for $M = 1$. Thus, M' is not a sonic Mach number, but the Mach number of any velocity based on the sonic Mach number speed of sound. This merely serves as a useful reference condition and helps to simplify the governing equations. *See* Prandtl relation.

character of group representation The trace of a matrix at a representation in group theory.

charge conjugation (1) The symmetry operation associated with the interchange of the role of a particle with its antiparticle. Equivalent to reversing the sign on all electric charge and the direction of electromagnetic fields (and, therefore, magnetic moments).

(2) A unitary operator $\zeta : j_\mu(x) \to -j_\mu(x)$ which reverses the electromagnetic current and changes particles into antiparticles and vice versa.

chemical bond Term used to describe the nature of quantum mechanical forces that allows neutral atoms to bind and form stable molecules. The details of the bond, such as the binding energy, can be calculated using the methods of quantum chemistry to solve the Born-Oppenheimer problem. *See* Born-Oppenheimer approximation.

chemical equilibrium For a reaction at constant temperature and pressure, the condition of *chemical equilibrium* is defined in terms of the minimum Gibbs free energy with respect to changes in the proportions of the reactants and the products. This leads to the condition, $\sum_j v_j \mu_j = 0$, where v_j is the stoichiometric coefficient of the jth species in the reaction (negative for reactants and positive for products), and μj is the chemical potential of the jth species.

chemical potential (1) At absolute zero temperature, the *chemical potential* is equal to the Fermi energy. If the number of particles is not conserved, the *chemical potential* is zero.

(2) The *chemical potential* (μ) represents the change in the free energy of a system when the number of particles changes. It is defined as the derivative of the Gibbs free energy with respect to particle number of the jth species in the system at constant temperature and pressure, or, equivalently, as the derivative of the Helmholtz free energy at constant temperature and volume:

$$\mu_j(T, P) = \left(\frac{\partial G}{\partial N_j}\right)_{T,P} ;$$

$$\mu_j(T, V) = \left(\frac{\partial F}{\partial N_j}\right)_{T,V} .$$

Chézy relations For flow in an open channel with a constant slope and constant channel width, the velocity U and flow rate Q can be shown to obey the relations

$$U = C\sqrt{R_h \tan \theta} \quad Q = CA\sqrt{R_h \tan \theta}$$

where $C = \sqrt{8g/f}$ and is known as the Chézy coefficient; f is the friction factor and R_h is the hydraulic radius.

Child–Langmuir law Description of electron current flow in a vacuum tube when plasma conditions exist that result in the electron current scaling with the cathode–anode potential to the 3/2 power.

choked flow Condition encountered in a throat in which the mass flow rate cannot be increased any further without a change in the upstream conditions. Often encountered in high speed flows where the speed at a throat cannot exceed a Mach number of 1 (speed of sound) regardless of changes in the upstream or downstream flow field.

circularly polarized light A light beam whose electric vectors can be broken into two perpendicular elements having equal amplitudes but differing in phase by l/4 wavelength.

circulation The total amount of vorticity within a given region defined by

$$\Gamma \equiv \oint_C \mathbf{u} \cdot d\mathbf{s} .$$

Circulation is a measure of the overall rotation in a flow field and is used to determine the strength of a vortex. *See* Stokes theorem.

classical confinement Plasma confinement in which particle and energy transport occur via classical diffusion.

classical diffusion In plasma physics, diffusion due solely to the scattering of charged particles by Coulomb collisions stemming from the electric fields of the particles. In classical transport (i.e., diffusion), the characteristic step size is one gyroradius (Larmor orbit) and the characteristic time is one collision time.

classical limit Used to describe the limiting behavior of a quantum system as the Planck constant approaches the limit $\hbar \to 0$.

classical mechanics The study of physical systems that states that each can be completely specified by well-defined values of all dynamic variables (such as position and its derivatives: velocity and acceleration) at any instant of time. The system's evolution in time is then entirely determined by a set of first order differential equations, and, as a consequence, the energy of a classical system is a continuous quantity. Under *classical mechanics*, phenomena are classified as involving matter (subject to Newton's laws) or radiation (obeying Maxwell's equations).

Clausius–Clapeyron equation The change of the boiling temperature T, with a change in the pressure at which a liquid boils, is given by the *Clausius–Clapeyron equation:*

$$\frac{dP}{dT} = \frac{L}{T(v_g - v_l)}.$$

Here, L denotes the molar latent heat of vaporization, and v_g and v_l are the molar volumes in the gas and liquid phase, respectively. This equation is also referred to as the vapor pressure equation.

Clebsch–Gordon coefficients Coefficients that relate total angular momentum eigenstates with product states that are eigenstates of individual angular momentum. For example, let $|j_1 m_1 >$ be angular momentum eigenstates for operators $\mathbf{J}_1$ (i.e., its square, and z-component), and let $|j_2 m_2 >$ be the eigenstates of angular momentum $\mathbf{J}_2$. We require the components of $\mathbf{J}_1$ to commute with those of $\mathbf{J}_2$. We define $\mathbf{J} = \mathbf{J}_1 + \mathbf{J}_2$, and if states $|JM>$ are angular momentum eigenstates of $\mathbf{J}^2$ and J_z, then $|JM> = \sum <j_2 m_2 j_1 m_1 | JM> |j_1 m_1 j_2 m_2 >$, where the sum extends over all allowed values $j_1 \, j_2 \, m_1 \, m_2$. The complex numbers $<j_2 m_2 j_1 m_1 | JM>$ are called *Clebsch–Gordon coefficients*. See angular momentum states.

Clebsch–Gordon series Identity involving Wigner rotation matrices, given the Wigner matrices $D^{ja}_{m_a m_{a'}}(R)$ and $D^{jb}_{m_b m_{b'}}(R)$, where the first matrix is a representation, with respect to an angular momentum basis, of rotation R. The second rotation is a representation of the same rotation R but is defined with respect to another angular momentum basis. The matrices act on direct product states of angular momentum. For example, the first Wigner matrix operates on spin states for particle 1, whereas the second operates on the spin states for particle 2. The *Clebsch-Gordon series* relates products of these matrices with a third Wigner rotation matrix $D^{j}_{mm'}(R)$, which is a representation of the rotation R with respect to a basis given by the eigenstates of the total angular momentum (for the above example, the total spin angular momentum of particle 1 and 2).

closed system A thermodynamic system of fixed volume that does not exchange particles or energy with its environment is referred to as a *closed system*. Such a system is also called an isolated system. All other external parameters, such as electric or magnetic fields, that might affect the system also remain constant in a *closed system*.

closure See completeness.

closure relation Satisfied by any complete orthonormal set of vectors $|n>$, the relation $\sum_n |n><n| = 1$, valid when the spectrum of eigenvalues is entirely discrete, allows the expansion of any vector $|u>$ as a series of the basis kets of any observable. When the spectrum includes a continuum of eigenvalues, the relation is sometimes expressed in terms of a delta function identity:

$$\delta\begin{pmatrix} \rho \\ r \end{pmatrix} \begin{pmatrix} \rho \\ r - \rho' \\ r \end{pmatrix} = \sum_n \phi_n^*\begin{pmatrix} \rho' \\ r \end{pmatrix}$$

$$\phi_n\begin{pmatrix} \rho \\ r \end{pmatrix} + \int \phi_n^*\begin{pmatrix} \rho \\ p , r \end{pmatrix}$$

$$\phi\begin{pmatrix} \rho \\ p , r \end{pmatrix} d^3 p$$

where n enumerates the discrete eigenfunctions (the vectors above) and $\phi\begin{pmatrix} \rho \\ p , r \end{pmatrix}$ generalizes the expression to the continuous case.

cloud chamber An apparatus that can track the trajectories of atomic and sub-atomic particles in a super-saturated vapor. The tracks are a result of ionization caused by the energetic particles, followed by nucleation of cloud droplets centered at the ionization site.

cnoidal wave Periodic finite amplitude surface waves in shallow water whose shapes are given by the solution of the Korteweg-deVries equation.

c-number Fields describing single particle wave functions in the Schrödinger–Pauli representation of quantum mechanics. The representation of Dirac fields as operators acting on state vectors in occupation-number space are known as q-number fields.

Coanda effect The tendency for a flow such as a jet to attach to a wall or a flow in the same direction. The primary method is entrainment; since the flow entrains fluid from all directions, the region near the wall cannot replace fluid, and the jet is drawn towards the wall from a reduced pressure.

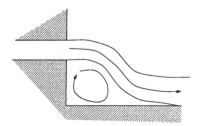

Coanda effect.

coefficient of linear expansion The fractional change in length per unit of change in temperature, assuming that the cross-sectional area does not change.

coefficient of refrigerator performance (γ) The ratio of the amount of heat extracted from the cold system per unit of work input into the cold system. For a reversible refrigerator, also called a Carnot refrigerator or an ideal refrigerator, operating between a cold temperature reservoir at absolute temperature T_c and a high temperature exhaust reservoir at absolute temperature T_h, this coefficient approaches its limiting value $\gamma = T_c/(T_h - T_c)$.

coefficient of volume expansion (α) Determines the fractional rate of change of volume with temperature, i.e.,

$$\alpha = \frac{1}{V}\left(\frac{\partial V}{\partial T}\right)_P.$$

coexistence curve The curve in a pressure–temperature phase diagram for a liquid–gas system along which two phases coexist. The *coexistence curve* separates the homogeneous, stable, one-phase system from a two-phase mixture. Similarly, a *coexistence curve* can be defined by the relevant thermodynamic variables separating the one-phase state from the two-phase state, e.g., in the temperature composition diagram for binary mixtures, or in the magnetic field vs. temperature phase diagram for magnetic systems.

coherence Property of the density matrix. *Coherences* of the off-diagonal elements of the density matrix say something about the statistical properties of a quantum system.

coherent Refers to waves or sources of radiation that are always in phase. The laser is an example of a single source of *coherent* radiation.

coherent photon The phase relationship between the photon that an atom emits with the photon that stimulated the emission. The two photons are said to be coherent. They can, when this occurs, stimulate other atoms to emit *coherent photons*.

coherent state A state in the Hilbert space of a second quantized radiation field that is an eigenstate of the annihilation operator (*see* annihilation operator) for a given mode of the radiation field.

cold atoms Atoms whose translational kinetic energy is less than about 10^{-3} K. Recent laboratory efforts have succeeded in producing atoms of temperatures on the order of 10^{-9} K. Below a critical temperature, *cold atoms* of integer angular momentum can undergo a phase transition into a Bose–Einstein condensate.

cold plasma model   Model of plasma where the plasma temperature is neglected.

Colebrook pipe friction formula   Formula to determine friction factor f in turbulent pipe flow:

$$1/\sqrt{f} = -2\log\left(\frac{\epsilon/d}{3.7} + \frac{2.51}{\mathrm{Re}_d\sqrt{f}}\right)$$

where ϵ/d is the roughness of the pipe and Re_d is the pipe Reynolds number. The *Colebrook pipe friction formula* is plotted as the Moody chart.

collisionless plasma model   Model of plasma where the density is low enough or the temperature is high enough that collisions can be neglected because the plasma time scales of interest are shorter than the particle collision times.

collision rate   The probability per unit of time that a molecule will suffer a collision. The inverse of the *collision rate* is the mean time between collisions.

color center   In crystal, a point defect, which can absorb observable light, is called the *color center* (for example, F-center). *See* absorption band.

column vectors   The components, with respect to some basis vectors, of a ket vector in Hilbert space. They can be written as a column matrix and, therefore, kets are also are also referred to as *column vectors*.

combination principle   The sum or difference of observed frequencies from the same optical spectrum often occurs as a line in the same spectrum. This observation led to the tabulation of spectral terms by Rydberg and Ritz (1905), whose pairwise differences (qualified by simple selection rules identifying those that do not occur) yield all observable frequencies.

commercial breakeven   *See* breakeven.

commutation relations   The non-commutability of operators is closely related to the Pauli exclusion principle and results in a number of important anticommutator relations. The three Pauli spin matrices anticommute, i.e., $\sigma_x\sigma_y = -\sigma_y\sigma_x = i\sigma_z$ (plus two similar relations obtained by cyclically permuting x, y, z).

Fermions wave functions must satisfy $< \psi(\mathbf{r})|\psi(\mathbf{r}')> = \delta^3(\mathbf{r} - \mathbf{r}')$, which implies that the annihilation and creation operators must satisfy $\{u_k, u_{k'}\} = \{u_k^\dagger, u_{k'}^\dagger\} = 0$ and $\{u_k, u_{k'}^\dagger\} = \delta(\mathbf{k} - \mathbf{k}')$. For photons, the annihilation and creation operators must satisfy the commutator rule $\{d_k, d_{k'}^\dagger\} = \delta(\mathbf{k} - \mathbf{k}')$. *See* commutator, exclusion principle.

commutator   Defined as the product $AB - BA$ of two operators A and B in Hilbert space. The bracket symbol $[A, B]$ is commonly used to denote the *commutator*.

commutator algebra   The set of commutation relations (*see* commutator) among a group of operators. If the commutation relations among the group elements are closed, the group constitutes a Lie group.

compatible observable operators   A set of operators that mutually commute. Given a set of quantum operators $A, B, \ldots$, that are mutually commuting i.e., $[A, B] = 0$, $[A, C] = 0$, $[A, C] = 0 \ldots$, the members of the set are called *compatible operators*. An eigenstate of one member of a set of *compatible operators* is also an eigenstate of the other members of the set.

complementarity   Since the process of observing involves an interaction between a system and some instrument, an observed state by definition is no longer isolated, and the causality between the state before and after observation is no longer governed by Schrödinger's equation. By implication, one cannot predict with certainty the final state of an observed system, but can only make predictions of a statistical nature. *See* causality.

completeness   Property of vectors in Hilbert space. Given the operator $|a><a|$ that projects onto the basis vector $|a>$, and if $\sum |a><a| = I$ where the sum extends over all basis states and

complete orthonormal basis

I is the identity operator, then the basis is said to be complete (also called closure).

complete orthonormal basis   A set of N orthogonal normed functions ϕ_n (or unit vectors $|u_n>$) with which the N-dimensions of any state vector can be expanded as a linear superposition: $\psi = \sum_n c_n \phi_n (|U> = \sum c_n |u_n>)$. The basis functions are orthogonal if $<\phi_i|\phi_j> = 0$ for $i \neq j$ and orthonormal if, additionally, they individually satisfy the normalization condition $<\phi|\phi> = 1$. If no function (vector) exists in the Hilbert (vector) space orthogonal to all N functions ϕ_n (vectors $|u_n>$) of this set, the set is said to span the space. If every function of the Hilbert space (or vector in the N-dimensional vector space) can be expanded in this way, then a set of functions ϕ_n (vectors $|un>$) is said to form a complete set.

complex phase shift   A phase shift with an imaginary component. In potential scattering theory, the S-matrix is generally taken to be unitary and the phase shift δ is considered real. However, if the potential has an imaginary component, δ will generally contain a real and imaginary part and is called a *complex phase shift*.

complex potential   A potential function that contains an imaginary part. A *complex potential* leads to complex phase shifts for the scattering solutions to the Schrödinger equation. *Complex potentials* are useful for describing loss mechanisms, such as radiative decay.

compressibility   The reciprocal of bulk modulus K.

compressible flow   Flow in which the density ρ may vary with the flow field. *Compressible flow* occurs when the Mach number is greater than 0.3. *Compressible flow* rarely occurs in liquids since the compressibility requires pressures of about 1000 atmospheres to reach sonic speeds, but *compressible flow* in gases is common where a pressure drop of 50% can create speeds approaching $M = 1$. The study of *compressible flow* is relegated to the field of gas dynamics.

compressor   Pump classification in which the pressure rise of the gas is approximately greater than 1 atmosphere or more; the large increase in pressure causes a density increase or compression of the working gas. *Compressors* are an integral part of gas turbine engines. *Compare with* blower.

Compton, A.H.   (1892-1962) American Physicist, noted for his discovery and explanation of the phenomena where the wavelength of an X-ray changes as it scatters from electrons in a metal. This phenomenon, called the Compton effect, confirmed the quantum nature of electromagnetic radiation. Along with C.T.R. Wilson, Compton was awarded the 1927 Nobel Prize for physics.

Compton effect   Discovered and explained by A.H. Compton, the *Compton effect* is a phenomena where a photon changes its wavelength as it scatters from an electron in a metal. Explanation of this effect requires the assumption that light (X-rays) be described in terms of quanta (photons). This discovery was an important experimental confirmation of wave-particle duality, first postulated by Einstein, for photons.

Compton scattering   Confirming the photon theory of light, the observation (Compton, 1923) that scattered x-rays possess a longer wavelength and correspondingly smaller frequency than the incident radiation. The shift was understood as the collision between an incident photon and a free (or weakly bound) electron. The electron gains momentum and energy, and, thus, the outgoing photon carries less energy (and therefore smaller frequency) than the incident photon. The change in wavelength $\Delta \lambda$ varies as a function of scattering angle θ and is given by Compton's formula $\Delta \lambda = 2 \frac{h}{mc} \sin^2 \frac{\theta}{2}$, where m is the rest mass of the electron.

Compton wavelength   The ratio $\lambda = \hbar/mc$, where $\hbar$ is the Planck constant divided by 2π, m is the mass of the electron, and c is the speed of light. Its value is $\lambda = 2.4 \times 10^{-10}$ cm and provides the scale of length which is important for describing the scattering of radiation on electrons.

concentration fluctuations The mean square deviation in the concentration (number of particles per unit of volume) from the average concentration in a system capable of exchanging particles with a reservoir.

condensation Compression region in an acoustic wave where the density is higher than the ambient density.

conduction A process in which there is net energy transfer through a material without movement of the material itself. For example, energy transfer could be thermal (thermal *conduction*) or electrical (electrical conductivity) in nature.

conduction band Term used to describe the set of allowed energy states, in which the electrons in a semi-conductor can occupy and produce a current. In the presence of an external electric field or an increase in temperature, electrons from the filled insulation band can be promoted into the unfilled *conduction band* and allow an electric current.

conductivity, electrical *Electrical conductivity* is defined as the ability of a material to conduct electric current. It is denoted by the symbol σ. It is also the reciprocal of resistivity.

conductor, electrical A material with a high value of electrical conductivity. Metals are generally very good *electrical conductors* because of large pools of free electrons.

conductor, thermal A substance with a high value of thermal conductivity. In general, metals are good *thermal conductors* as well. Many non-metallic materials are poor *thermal conductors*.

configurational entropy The entropy of a system that arises from the way its constituent particles are distributed in space. For example, a polymer chain has configurational entropy corresponding to the number of ways that the individual links can be arranged.

confinement time The characteristic time that plasma can be contained within a laboratory experimental device using a magnetic field, a particle's own inertia, or by other methods (e.g., electric field). The electron and ion particle *confinement time* is often distinguished from the energy *confinement time* of the plasma.

conformal mapping Method by which a complex flow pattern, by itself or around a solid body, can be mapped or transformed into a much simpler pattern allowing easier solution of the flow field. A common application is the transformation of flow around an airfoil into flow around a circular cylinder. Applicable to potential (inviscid) flow only.

conjugate momentum The differential quantities of the Langrangian with respect to the time derivative of its generalized coordinates: $P_r = \frac{\partial L}{\partial p_r^\&}(r = 1, 2, 3, \ldots, N)$. When q_r is an ordinary cartesian coordinate for a mass m and all forces it experiences are derivable from a static potential, p_r is the corresponding coordinate of the particle's momentum, $p_r = mq_r^\&$. *See* Hamiltonian; Lagrangian.

conjugate operator *See* adjoint operator.

conjugation of vectors, operators A mapping of bras to corresponding kets analogous to the complex conjugation of numbers. Any expression of vectors and operators can be conjugated by the following prescription: replace all numbers by their complex conjugate, bras by their conjugate kets (and vice versa), and operators by their Hermitian conjugates, and reverse the order of all bras, kets, and operators in every term.

connection formulae Analytic continuation rules for Wentzel-Kramer-Brillouin (WKB) functions between classical allowed and non-allowed regions. In the WKB approximation for solutions to the Schrödinger equation, *connection formulae* provide a prescription whereby a solution in a classically allowed region is analytically continued into the classical non-allowed region. *Connection formulae* are essential for determining semi-classical quantization conditions.

conservation equations Equations describing the conservation of mass, momentum, and energy in a fluid. The *conservation equations* are applicable to all flows, but typically take the form of the Navier-Stokes equations after suitable assumptions are made. In differential form, the *conservation equations* are given by continuity (mass conservation),

$$\frac{\partial \rho}{\partial t} + \nabla \cdot (\rho \vec{u}) = 0$$

momentum,

$$\frac{\partial \vec{u}}{\partial t} + \vec{u} \cdot \nabla \vec{u} = -\frac{1}{\rho} \nabla p + \vec{g} + \nu \nabla^2 \vec{u}$$

and energy

$$\frac{\partial e}{\partial t} + \vec{u} \cdot \nabla e = -\nabla \cdot \vec{q} - p(\nabla \cdot \vec{u}) + \phi .$$

conservation of flux *See* continuity equation, current density.

constants of motion Any observable C which commutes with the Hamiltonian, $[C, H] = 0$ and which does not depend explicitly on time, will have a mean value that remains constant in time $\frac{\partial}{\partial t} < C > = 0$. More generally, $[C, H] = 0$ implies $[\exp(i\xi C), H]$ so that $\frac{\partial}{\partial t} < e^{i\xi C} > = 0$, and the statistical distribution of C remains constant in time. Notice that since the Hamiltonian commutes with itself, $[H, H] = 0$, energy must be a *constant of motion*.

contact angle Formed by the interface of a liquid and solid boundary at the free surface. *See* meniscus.

continuity equation Conservation equation obeyed by solutions of the Schrödinger equation. A solution of the Schrödinger equation, $\psi(x, t)$, also obeys the following equation, $\frac{\partial \rho(x,t)}{\partial t} + \vec{\nabla} \cdot \vec{j}(x, t) = 0$, $\rho(x, t) \equiv \psi^*(x, t)\psi(x, t)$, $\vec{j}(x, t) \equiv -\frac{i\hbar}{2m}(\psi^*(x, t)\nabla \psi(x, t) - \psi(x, t)\nabla \psi^*(x, t))$, provided that the potential function in the Schrödinger equation is real. This equation is called the *continuity equation* and allows the identification of $\rho(x, t)$ as a probability density. The quantity $\vec{j}(x, t)$ is the current density, and the *continuity equation* is a mathematical statement of the fact that the rate of change of the probability in an enclosed volume is proportional to the amount of flux entering/leaving the surface enclosing that volume.

continuum hypothesis Assumption that a fluid behaves not as a group of discrete individual particles, but as a continuous distribution of matter infinitely divisible. For the hypothesis to be valid, the size of the body around which the flow is moving must be much larger than the mean free path of the molecules. The derivation of the conservation equations of motion are based on this fundamental assumption. This is characterized by the Knudsen number.

controlled thermonuclear fusion Laboratory experimental plasmas in which light nuclei are heated to high temperatures (millions of degrees) in a confined region which results in fusion reactions under controlled conditions significant enough to be able to produce energy.

control surface The surfaces of a control volume through which fluid passes.

control volume A volume fixed in space used in integral analysis of fluid motion. The volume can be variable in shape or size.

convection Transport of fluid from point to point due to the effects of temperature differences in the fluid. Natural *convection* is characterized by motion driven by buoyant forces created when the density of a fluid changes when in contact with a heated surface. Forced *convection*, in addition to a temperature differential, has a fluid motion driven by other means imposed upon the convective motion.

convective instabilities A plasma wave's amplitude increases as the wave propagates through space without necessarily growing at a fixed point in space. Compare to absolute instabilities.

converging–diverging nozzle A nozzle whose area first decreases then increases after reaching a minimum area (known as the throat). Used to accelerate a flow from subsonic velocities to sonic velocities at the throat to supersonic velocities. Subsonic flow accelerates in a converging nozzle and decelerates in a diverging nozzle, while supersonic flow exhibits the opposite behavior. The relation between the fluid velocity U, nozzle area A, and Mach number M is given by

$$\frac{dU}{U} = \frac{-dA/A}{1 - M^2}$$

which results in the conclusion that sonic flow (M=1) can only occur when $dA = 0$.

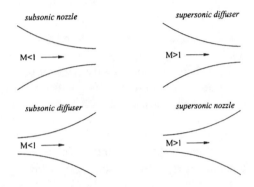

Converging–diverging nozzle and diffuser.

Cooper pair (1) A pair of electrons with opposite spins and opposite momenta that are bound together by an attractive interaction. Such a pair behaves as a boson, and the collective behavior of these pairs gives rise to superconductivity.

(2) Describes the phenomenon whereby a pair of electrons in a super-conducting medium are coupled with their spins anti-aligned and behave as a composite boson. A theoretical construct, *Cooper pairs* play a central role in the Bardeen–Cooper–Schriefer (BCS) theory of low temperature super-conductivity.

Copenhagen interpretation The standard and accepted laws and postulates of the modern quantum theory. The most controversial aspect of the *Copenhagen interpretation* involves the collapse of the wave function postulate so named because of the strong influence of N. Bohr (*see* Bohr, Niels) and the institute in Copenhagen on the development of the quantum theory.

corona The outermost part of a star's atmosphere, characterized by higher temperatures and lower densities relative to the stellar photosphere.

correlation Term used to describe the degree of deviation of the energy eigenvalues for a many-body system from that predicted by a mean field theory. For example, in the theoretical description of many electron atoms, it is useful to consider all electrons to experience a common background potential. This picture provides a good qualitative description for the system but does not predict accurate energy eigenvalues for the atom. The difference between the exact non-relativistic energy and the energy predicted by mean field theory (Hartree-Fock approximation) is called the *correlation* energy.

correlation energy *See* correlation.

correspondence principle A guiding principle, invoked by N. Bohr, that provides a connection between the predictions of the quantum theory and that of classical physics in the classical limit.

Couette flow Flow between parallel plates which is driven by the linear motion of one of the plates alone. The solution of the flow field is reduced to the solution

$$u = \frac{yU}{2b}$$

where b is the gap width between the plates and U is the speed of the moving plate. Circular *Couette flow* is the steady flow between two concentric cylinders, one of which is rotating relative to the other.

Coulomb collision Particle collisions where the Coulomb force (electrical-force attraction or repulsion) is the governing force that results in deflection of the particles away from their initial paths.

Coulomb gauge Gauge condition commonly used for calculation of properties for an atom under the influence of a radiation field. The vector potential $\vec{A}$ is arbitrary up to the gradient of a scalar function, and this fact can be exploited to insure that $\vec{\nabla} \cdot \vec{A} = 0$, the *Coulomb gauge* condition.

Coulomb scattering Atomic or nuclear scattering phenomenon in a Coulomb potential.

coupling constant A dimensionless numerical quantity characterizing the strength of an interaction. Powers of the fine-structure constant $\left(\alpha \equiv \frac{e^2}{4\pi \eta c} \approx \frac{1}{137}\right)$ appear as an expansion factor in electrodynamic perturbation theory and therefore set an order of magnitude to the intensity of electromagnetic interactions. In contrast, the strong quark-gluon interaction involves a constant for chromodynamic interactions (numerically estimated to be $\alpha_s \equiv \frac{g_s^2}{4\pi \eta c} \approx 0.1$).

covalent crystals In *covalent crystals*, the main interaction is a covalent bond. The crystals have high melting points and hardnesses. For example, diamond is a *covalent crystal* and is the hardest crystal in the world.

covariant An expression or equation that remains invariant under transformation to other Lorentz frames (guaranteed, for example, for scalar expressions).

covariant derivative A vector operator combining a simple differential operator with a term involving a vector field A_v:

$$D_v(A)u \equiv (\partial_v - ieA_v(x))\, u(x)$$

which changes gauge under a local phase transformation. Together with the self-energy terms of the vector field itself, the replacement of ordinary derivatives with this covariant form introduces interactions to the free Lagrangian and transforms it into one invariant to gauge transformations.

CPT theorem A theoretical assertion which states that the local Lagrangian of a system of quantum fields must be invariant under the simultaneous application of the three operations of charge conjugation C, space inversion P, and time inversion T. This condition is sufficient to ensure the equality of particle and antiparticle masses and lifetimes.

Canonical equations (*see* Hamiltonian) are the set of first order differential relations involving the Hamiltonian, H:

$$q_r^{\&} \equiv \frac{\partial H}{\partial p_r} \quad p_r^{\&} \equiv \frac{\partial H}{\partial q_r}$$

where r enumerates the generalized coordinates q_r of the Hamiltonian, and p_r are their conjugate momenta, from which all equations of motions (including the general laws of classical mechanics) can be derived.

creation of matter *See* pair production.

creation operators In quantum field theory, operators that, when acting on a state vector, increase the eigenvalue of the number operator by one and the charge operator by z. If the expression $u_k^\dagger \phi_p$ represents the state vector with energy-momentum (four-momentum) $p + k$, where $k^2 = m$, then the operator $u_k^\dagger$ can describe the creation of a particle of mass m, charge z, and four-momentum k.

For particles obeying Fermi–Dirac statistics (fermions such as electrons and muons), the operators must satisfy the anticommutator relations $\{u_k, u_{k'}^\dagger\} = \delta_{kk'}$ and $\{u_k, u_{k'}\} = \{u_k^\dagger, u_{k'}^\dagger\} = 0$ in order to incorporate the Pauli exclusion principle. *See* annihilation operator.

creeping flow Flows at which Re $\ll 1$. The flow is characterized by laminar viscous motion. The steady momentum equation simplifies to the form

$$\nabla p = \mu \nabla^2 \mathbf{u}$$

where the pressure and viscous forces balance.

critical exponents In the vicinity of a critical point, the divergence of many physical quantities, e.g., correlation length, specific heat and magnetic susceptibility, can be written as a power law $t^{-\alpha}$ in terms of the reduced temperature t or other parameters characterizing the critical behavior. The exponent α is called a *critical exponent*.

critical opalescence Intense scattering of light that occurs in the vicinity of a critical point due to the enhanced density fluctuations arising from the divergence of the correlation length.

critical point The point in the pressure–temperature phase diagram (point C in the figure below) above which there is no liquid-gas phase separation, i.e., the order parameter that represents the difference in the molar volume (or density) of the two coexisting phases vanishes at the critical point. The nature of the phase transition changes from first order to second order at the *critical point*. *Critical points* can be similarly defined for other systems, such as in ferromagnetic systems and binary mixtures.

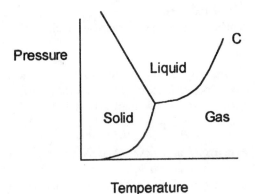

Critical point.

critical pressure The value of the pressure at the critical point.

critical state The state of a fluid in which liquid and gas phases have the same density. The temperature, pressure, and density at this state are respectively called critical temperature, critical pressure, and critical density. Above the critical temperature, a gas cannot be liquefied by increasing the pressure.

critical temperature The value of the temperature at the critical point.

cross-section The ratio of the number of particles scattered into a given solid angle with the flux of particles incident on the scattering center is the differential *cross-section*. Integrated over all solid angles, this ratio gives the total *cross-section*.

Crow instability Long-wave instability on a trailing vortex pair often observed behind aircraft. The instability is caused by the action of one vortex upon the other, resulting in a symmetric and sinusoidal instability causing the vortex pair to join, breaking up into individual vortex rings. The system is simplified by dividing into two pairs of independent modes, symmetric and antisymmetric. In the symmetric case, the filaments bend alternately towards and away from each, getting closer to each other in some regions and farther away in others. In the antisymmetric mode, the vortex filaments move with each other, preserving their separation distance. Thus, the symmetric mode is the unstable mode. The stability criteria is given by

$$\sigma = \pm \left\{ \left[1 + khK_1(kh) - \frac{1}{2}k^2h^2S(ak\delta) \right] \right.$$
$$\left[1 - khK_1(kh) - k^2h^2K_0(kh) \right.$$
$$\left. \left. + \frac{1}{2}k^2h^2S(ak\delta) \right] \right\}^{\frac{1}{2}}$$

where the symmetric mode is stable if $\sigma \in \Im$ and unstable if any part of $\sigma \in \Re$ (or $\sigma \ni \Im$). If the elliptical wing loading assumption is used, the predominant wavelength of this mode is

$$0.42b \leq \lambda \leq 8.6b$$

where b is the aircraft span.

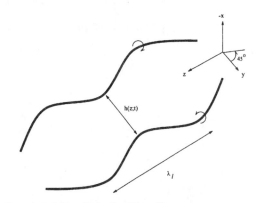

Crow instability of two slender vortices.

cryogenics The science and technology of the production of low temperatures.

crystal A material in which atoms, ions, or molecules have a regular three-dimensional repeating arrangement in space. The crystalline lattice is the arrangement of points in space at which atoms, ions, or molecules are positioned.

cubic lattice A common type of lattice. It can be divided into three groups, known as simple *cubic lattices,* body-centered *cubic lattices,* and face-centered *cubic lattices.*

cubic zinc sulfide structure Its Bravais lattice is face-centered cubic. Its basses are two different atoms located at (0,0,0) and ($a/4\mathbf{i}$, $a/4\mathbf{j}$, $a/4\mathbf{k}$), respectively (for example, GaAs and InSb).

Curie–Weiss law When temperature is below a critical temperature Tc, which is called the Curie temperature, there is a spontaneous magnetization in some crystals. When temperature is above Tc, spontaneous magnetization disappears and the crystals show paramagnetism. The paramagnetic susceptibility satisfies the *Curie-Weiss law*: $X = X_0^* C/(T - Tc)$, where X_0 is constant and C is the Curie constant.

current density A vector describing the change in the position probability density of a wave function, $P\left(\begin{smallmatrix}\rho\\r\end{smallmatrix}, t\right) = \left|\psi\left(\begin{smallmatrix}\rho\\r\end{smallmatrix}, t\right)\right|^2$, defined as a conserved probability *current density* $\begin{smallmatrix}\rho\\j\end{smallmatrix}\left(\begin{smallmatrix}\rho\\r\end{smallmatrix}, t\right) \equiv -\frac{i\eta}{2m}\left[\psi^*\nabla\psi - (\nabla\psi^*)\psi\right]$ satisfying the continuity equation

$$\frac{\partial}{\partial t}\left|\psi\left(\begin{smallmatrix}\rho\\r\end{smallmatrix}, t\right)\right|^2 + \nabla \cdot \begin{smallmatrix}\rho\\j\end{smallmatrix}\left(\begin{smallmatrix}\rho\\r\end{smallmatrix}, t\right) = 0,$$

which is an analog to classical mechanics conservation of fluid flow. *See* continuity equation.

current operator The second quantized generalization of the conserved current (*see* continuity equation) $\vec{j}(x, t) \equiv -\frac{i\hbar}{2m}(\psi^*(x, t)\nabla\psi(x, t) - \psi(x, t)\nabla\psi^*(x, t))$. In the formalism of second quantization, the function $\psi(x, t)$ is taken to represent a Heisenberg picture operator, and, thus, $\vec{j}(x, t)$ is the *current operator.*

cutoff frequency Frequency beyond which a plasma wave ceases to exist or changes its nature.

cyclic process A process in which a system undergoes various changes in state in such a way that it returns to its initial state at the end of the process.

cyclotron frequency Angular precession frequency that a particle with a magnetic moment experiences when placed in a constant magnetic field. For an electron, the *cyclotron frequency* is given by $\omega = (eB/mc)^2$, where e is the electron charge, m is the electron's mass, c is the speed of light, and B is the magnetic field strength.

cyclotron radius Radius of orbit of a charged particle about a magnetic field line. Also called gyroradius or Larmor radius.

cyclotron resonance A charged particle in a magnetic field will resonate with an electric field (perpendicular to the magnetic field) that oscillates at the particle's cyclotron frequency or harmonics of the particle's cyclotron frequency.

D

D'Alembertian operator A relativistically invariant, second order, partial differential operator in space-time. It may be written as

$$\Box = \frac{\partial^2}{\partial x_0^2} - \sum_{i=1}^{3} \frac{\partial^2}{\partial x_i^2} \;;$$

where x_0 is the time component and x_i are the spatial components of the relativistic space-time four-vector.

D'Alembert's paradox General result of potential (irrotational) flow theory that a moving body does not experience a drag force. This result, derived in the 18th century, was at odds with both intuition and observation of flow about a body in motion.

Dalgarno–Lewis method Method developed by A. Dalgarno and J.T. Lewis (1955) that occasionally enables the second order perturbative correction to the energy of a state to be evaluated exactly.

Dalitz pair A high energy gamma can convert into an electron positron pair in the electric field of a nucleus. In this situation, energy and momentum are conserved by the three particles (electron, positron, and recoil nucleus) in the final state, but because the nucleus is much more massive than the lepton pair, the energy of the gamma ray is essentially shared between the leptons. Thus a *Dalitz pair* is the pair of conversion leptons having a total energy approximately equal to the initial energy of the gamma ray.

Dalitz plot A method of graphically representing data when three particles are in the final state of a reaction $A + B$. In this case, the reaction can depend on five independent variables. Binning the data as a function of these parameters usually leads to poor statistical accuracy and a complicated presentation in which systematic trends are difficult to extract. Thus, the spectra are integrated over the relative angles of the outgoing particles, which leaves two independent variables. These may be taken as the energies of two of the particles, $E1$, $E2$, and $E3$. Energy is conserved so that $E1 + E2 + E3$ = system mass. The *Dalitz plot* is developed by placing a count in a two-dimensional diagram at a distance $E1$, $E2$, $E3$ from each side of an equilateral triangle, e.g., if all three particles have the same energy the point lies at the center of the triangle. Energy is conserved because the sum of the three distances as described above is constant. Applying momentum conservation gives a further restriction on the plot providing a boundary curve as shown in the figure below. If the final state particles have no interactions, then the distribution of points is random within the boundary curve. If a final state interaction occurs, or the reaction proceeds through sequential decay, then the density of points in various regions in the plot will be enhanced.

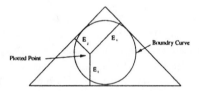

A schematic example of a Dalitz plot, which shows a plotted point.

damping The loss of energy to the surroundings, i.e., a reservoir with a large number of degrees of freedom, by means of some interaction. In the case of an atom, this can be in the form of spontaneous emission or non-radiative decay. Other forms of *damping* include the loss of photons in a cavity through partially transparent mirrors. The quantum theory of *damping* can be treated with various approaches. Among those are the reduced density operator, quantum jump, or Monte-Carlo methods.

Darcy friction factor Dimensionless measure of the skin friction τ_w (shear stress) in pipe flow

$$f \equiv \frac{8\tau_w}{\rho U_\infty^2}$$

Darcy's law Equation governing flow in a porous medium

$$\nabla p = -\frac{\mu}{K}\mathbf{u}$$

where K is the permeability of the porous medium.

Darcy–Weisbach equation Equation relating head less h_f in a pipe of length L and diameter d as a function of friction factor f

$$h_f = f\frac{L}{D}\frac{U_\infty^2}{2g}.$$

Once f has been determined for the given flow regime and geometry, the *Darcy–Weisbach equation* can be applied to a duct flow of any cross-section for both laminar and turbulent regimes.

dark matter Non-luminous matter in the universe that is postulated to exist in order to account for the motion of observed systems due to the presumed effects of gravitation. This matter could be in the form of cold matter, neutron stars, black holes, or stable elementary particles. Evidence that *dark matter* must exist in some form comes from several sources, the most compelling being the fact that the circular velocity of hydrogen clouds surrounding spiral galaxies is independent of the radius of the clouds, in direct contradiction to what is expected from a gravitationally bound system of stars (which should be proportional to 1/r). No currently proposed source of *dark matter* fully explains present observations.

dark spot trap Can be used to increase the atom density in magneto-optical traps. In standard magneto-optical traps, the radiation pressure from spontaneously emitted photons limits the achievable density. In order to increase this limit, atoms must be prevented from emitting fluorescence in the regions of highest density, i.e., the trap center.

For many trapping species in many magneto-optical traps, repumping lasers must be superimposed on the trapping lasers in order to prevent pumping of the atoms into states not accessible to the trapping lasers. By reducing or completely avoiding illumination of the trap center with the repumping laser, the atoms will be preferentially pumped into the dark states in the trap center. The fluorescence at the trapping frequency will therefore be suppressed, which significantly increases the atom density due to much lower radiation pressure.

dark state (1) A coherent effect which exist in three level systems where the three levels are connected through two coherent fields. Only the two transitions $|a\rangle \leftrightarrow |b\rangle$ and $|a\rangle \leftrightarrow |c\rangle$ are dipole-allowed. The two counterpropagating laser beams fulfill the Raman condition

$$\omega_2 - \omega_1 = \omega_{ab} - \omega_{ac},$$

where ω_{ab} and ω_{ac} are the respective transition frequencies. The interaction part of the Hamiltonian of such a system in the rotating wave approximation is given by

$$\begin{aligned}H_{\text{int}} = &-\exp(-\iota\omega_1 t)\exp \iota k_1 z \Omega_{ab}|b\rangle\langle a|\\&-\exp(-\iota\omega_1 t)\exp -\iota k_1 z \Omega_{ab}|a\rangle\langle b|\\&-\exp(-\iota\omega_2 t)\exp \iota k_2 z \Omega_{ac}|b\rangle\langle c|\\&-\exp(-\iota\omega_2 t)\exp -\iota k_2 z \Omega_{ac}|c\rangle\langle a|,\end{aligned}$$

where the Ω are the Rabi frequencies of the fields. Solution of the problem shows that it is possible to prepare the system in a superposition of the two lower states $|b\rangle$ and $|c\rangle$ so that no absorption to state $|a\rangle$ occurs, even in the presence of the fields. This is possible because of appropriate choice of the Rabi frequencies and phases of the fields. The superposition of the two states is called the *dark state,* as it is not electric dipole-connected to the upper state anymore. Due to the coherence in the system, the probability amplitudes for excitation into the excited state $|a\rangle$ interfere destructively.

This effect is also called coherent population trapping. Similar to optical pumping, this population trapping can also occur if the atom has not initially been in this dark state. By cycles of the effects of the electro-magnetic radiation and spontaneous emission, the atom is optically pumped into a dark state. An example of this effect is the electromagnetically induced transparency (EIT).

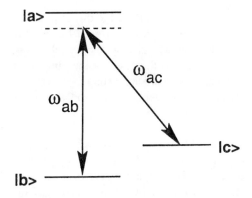

Illustration of dark states. The coherent interaction of the radiation fields with frequencies ω_{ab} and ω_{ac} which connect levels $|a\rangle$ with $|b\rangle$ and $|a\rangle$ with $|c\rangle$ optically pump the atom in a dark state which has no dipole allowed transitions to state $|a\rangle$. All atoms will be trapped in a superposition of states $|b\rangle$ and $|c\rangle$.

(2) In molecular dynamics, states that are not connected by dipole-allowed transitions to a ground state, but exist in the background and can lead to perturbations in molecular spectra. These *dark states* can be rovibrational levels of higher harmonic vibrations or combination overtones as well as other rovibronic levels in other electronic states. In the Born–Oppenheimer picture, the different motions in molecules are completely separated. Interactions do occur, however, in real molecules. The perturbations can be taken into account by the introduction of a coupling matrix element between optically active states $|\Psi_o\rangle$ connected by an optically allowed transition to the ground state and the manifold of dark background states $|\{\Psi_d\}\rangle$, which are not optically dipole-coupled to the ground state. One distinguishes several mechanisms for these interactions: intramolecular vibrational redistribution (IVR) for coupling of states within the same electronic state, intersystem crossing (ISC), and internal conversion (IC) for couplings within different electronic states of different or the same multiplicity, respectively.

IVR is particularly important in the electronic ground state of molecules, where it is the only mechanism possible, since other electronic states are much higher in energy. The possible interactions can be subdivided into the anharmonic Fermi coupling, coriolis, and centripetal couplings. The former are independent of the rotational quantum number J, whereas the latter are linear and quadratic in their J dependence.

The actual decay rate for the optically active state is then the sum of the radiative and non-radiative decays into the *dark states*. The latter become more important as the density of background states increases. The energy flow into these *dark states* is investigated in molecular dynamics.

Darwin term   Discovered by C.G. Darwin (1928) in the relativistic corrections to the Hamiltonian of a hydrogen atom, which is singular and only affects the spherically symmetric electronic states.

Davisson, Clinton   Nobel Prize winner in 1937 who, along with Thompson, observed diffraction patterns in electron scattering from crystals confirming the de Broglie hypothesis of the wave nature of matter.

Davisson–Germer experiment   Experiment by C.J. Davisson and L.H. Germer in 1927, in which electrons striking a crystal of Ni were observed to produce a diffraction pattern. This experiment unambiguously demonstrated the validity of de Broglie's concept of matter waves.

dead time   The percentage or ratio of the time period during the acquisition of experimental data when the experimental equipment is inoperative due to the collection, processing, or on-line analysis of this data.

de Boer parameter   Parameter which expresses the importance of quantum mechanical effects in a liquid, especially one of the inert elements. The parameter is approximately the square root of the ratio of the zero-point-energy to the strength of the attraction between two atoms.

de Broglie, Louis Victor   Nobel Prize winner in 1929 for proposing that matter, as well as light, has both wave and particle properties.

de Broglie wavelength   (1) The wavelength assigned to a unit of matter. It is defined by the equation $\lambda = h/p$, where h is Planck's constant and p is the momentum of the matter. An electron (the least massive elementary particle) moving at a velocity 9×10^6 cm/s (room temperature) has a wavelength $\approx 8 \times 10^{-7}$ cm.

(2) Reflects the wave-particle duality of matter. Just as light can exhibit both wave and particle character, matter particles can exhibit wave behavior. The characteristic wavelength λ that is associated with a matter particle was found by de Broglie and is given by

$$\lambda = \frac{\hbar}{mv}$$

where m is the mass of the particle and v is its speed. For slow quantum particles, this wavelength becomes very large. In fact, if the *de Broglie wavelength* and the average distance between the atoms are in the same order of magnitude, new phenomena can be observed such as Bose-Einstein condensation and the formation of degenerate Fermi gases.

de Broglie waves Profound and far reaching concept attributed to L. de Broglie (1924), stating that matter, which until then had been conceived of as particulate, would sometimes behave as a wave. Specifically, a particle of momentum p also behaves as a wave of the so-called de Broglie wavelength:

$$\lambda = h/p ,$$

where h is Planck's constant.

debye The common unit for the permanent dipole moment-present in molecules and also represents the atomic units for the dipole moment. A *debye* is defined as

1 *debye* $= 1\, ea_0 = 8.49166 \times 10^{-30}$ c m

where e is the elementary charge and a_0 is the Bohr radius.

Debye sheath The region, named after chemist Peter Debye, in front of a material surface in contact with a plasma and in the presence of electrical fields. The characteristic thickness of the sheath is the Debye length.

Debye shielding When a positive (or negative) charge is inserted into a plasma medium, it will change the local charge distribution by attracting (or repelling) electrons. The net result is an additional negative (or positive) charge density that cancels the effect of the initial charge at distances that are large compared to the characteristic Debye length (*see also* Debye sheath) associated with the shielding process.

Debye T^3 law From the Debye approximation, we can conclude that at very low temperatures, the heat capacity of crystals is linearly relative to T^3, where T is temperature. This is called the *Debye T^3 law*.

decay constant The probability that a nucleus undergoes radioactive decay in an interval δt is given by $\lambda \delta t$, where λ is the *decay constant*. This constant depends only on nuclear properties which are independent of all other physical quantities including time. Thus, the number of nuclei remaining after a given time, t, will be

$$N(t) = N(t=0) e^{-\lambda t} .$$

decay rate (A) The rate at which a state spontaneously decays to states with lower energy. The *decay rate* is also referred to as the Einstein A coefficient. The lifetime τ of this state and the *decay rate* obey the relationship

$$\tau = 1/A ,$$

where A is given in terms of ω rather than ν. For more details, *see* Einstein A coefficient. The linewidth of a transition $\Delta \nu$ is connected with the decay rate via

$$A = 2\pi \Delta \nu .$$

decay, spontaneous The decay of any system from an excited state to a state of lower energy without the application of any perturbation, with the emission of radiation or other particles. The term is especially applied to atoms, often as spontaneous emission, where it stands in sharp distinction to induced decay or emission, and alpha decay of nuclei.

decay width (Γ.) For any decaying state, system, particle, etc., with a lifetime τ, a quantity with the dimensions of energy, given by

$\Gamma = h/2\pi\tau$, where h is Planck's constant. Through the energy–time uncertainty relation, Γ expresses the uncertainty with which the energy of a state can be ascertained. *See also* resonance scattering; Breit–Wigner form; Fock–Krylov theorem.

decibel (dB) A logarithmic measure of the gain or loss in a material. The *decibel* scale is the ratio of the final intensity I and the initial intensity I_0 on a logarithmic scale. Positive numbers refer to gain, whereas negative values refer to loss. Specifically, one finds

$$G[dB] = 10 \log \frac{I}{I_0}.$$

deconfinement Elementary particles are composed of quarks which combine in such a way that the color quantum number of the entire system vanishes. If one attempts to separate a quark from this system, the interaction strength increases linearly with the distance between the quarks. This binding forms the essence of quantum chromodynamics, QCD. Thus, quarks are never free and must always be associated with at least one other quark. At sufficient densities and energies, however, quarks can exist and mix within a volume larger than that of an elementary particle, e.g., a nucleon or meson. This is called *deconfinement* and occurs only within this localized region of space at high densities and energies. Presumably, these conditions represent the situation in the very early universe. The deconfined state condenses back to elementary particles as the density decreases and the temperature cools.

decorrelation The loss of coherence in an atomic or molecular system. Such *decorrelation* originates from any damping or loss process. One distinguishes in analogy to the nuclear magnetic resonance phenomena T_1 and T_2, where the former expresses the lifetime and the latter the decorrelation time. For isolated, homogeneous systems we find the following relationship:

$$T_1 = \frac{T_2}{2}.$$

deep inelastic scattering Scattering of leptons at high energies with large momentum transfer can be calculated by perturbation techniques, since quantum chromodynamics (QCD), which is the theory of strongly interacting quarks and gluons, indicates that at high energies and small distances, the interaction between quarks is weak. In deep inelastic scattering, the lepton scatters in first order from one parton (quark) within the target. This scattering is determined by four structure functions of the parton distributions, F_1, F_2, g_1, and g_2, where the first two are spin-independent and the latter two are spin-dependent.

de-excitation The loss of excitation in a system. Mechanisms for *de-excitation* can be fluorescence, phosphorescence, and collisions, as well as other non-radiative loss processes in atomic and molecular systems. Since the radiation field is also quantized, one can view the loss of photons as a *de-excitation* process as well.

defect (**1**) An irregularity in the ordered arrangements of atoms in a crystal lattice. In general, defects are classified into two main types in crystalline solids: point *defects* and interstitial *defects*. Point *defects* correspond to a missing atom at a single lattice point. This *defect* is also referred to as a vacancy. An interstitial *defect* is an atom that is in a position other than a normal lattice point. All solids above absolute zero have a concentration of *defects* which depends on the temperature.

(**2**) *Defects* are what is different from ideal crystal. In general, there are microscopic *defects* in each real crystal. The *defects* have a great effect on the properties of crystal. There are many kinds of fundamental microscopic defects, such as dislocation, vacancy, point defect, etc.

deformation A nucleus is generally not spherical in shape, but has a nuclear density which can be defined as spheroidal (prolate or oblate). A static nuclear quadrupole moment is one indication that the nucleus is deformed. The *deformation* parameters are proportional to the difference between the radius of a spherical distribution of the same volume and the radius along the symmetry axes of the deformed system.

deformation potential (volume) Deformation can change the volume of crystals. *Volume deformation potential* is equal to energy change per volume dilation, which is the ratio of the volume change to the total volume of crystals.

deformation potentials For long wavelength acoustic phonons, atomic displacements can result in a deformation of crystals. *Deformation potentials* describe changes of the electric energy resulting from the deformation.

degeneracy When states with different quantum numbers have identical energy. One example are the Zeeman substates in an atom in the absence of a magnetic field. The *degeneracy* can be lifted due to the interaction with external fields (for instance electric or magnetic fields) or internal interactions. *Compare with* Lamb shift.

degeneracy, accidental (1) Degeneracy of states which are not related to each other by any obvious symmetry of the system. For a general system, such degeneracies are expected to occur at random, and any regularities are expected to be statistical at best. Many cases of *accidental degeneracy*, however, have historically turned out to be systematic on closer study, and are understood to arise from extra abstract symmetries of the system. Examples are the degeneracy of different angular momentum states of the hydrogen atom and the isotropic harmonic oscillator in more than one dimension.

(2) Different states with the same energy are called degenerated states. If the Hamiltonian has degenerated eigenstates with different symmetries, there are called *accidental degeneracies*.

degeneracy, Kramers A doubling of the degeneracy of every energy eigenstate, elucidated by H.A. Kramers in 1930, that necessarily arises in any system of particles with a total half-integral spin in the absence of external influences, such as a magnetic field, that break time-reversal symmetry. The theorem is of great value in understanding magnetic ions in crystals.

degeneracy, lifting of, or removal of Slight inequality of energy of a group of quantum states due to the presence of a perturbation, usually weak, that destroys or lowers the symmetry of the underlying system.

degeneracy pressure The pressure exerted by a degenerate Fermi gas, even at zero temperature, due to the occupation of non-zero momentum states as mandated by the Pauli exclusion principle. This pressure accounts for a large part of the compressibility of metals, and the stability of white dwarf stars against gravitational collapse. Synonymous with Fermi pressure.

degeneracy temperature *See* Fermi temperature.

degenerate Fermi gas A system of non-interacting fermions at a temperature much lower than the Fermi temperature.

degenerate gas (1) A gas of quantum mechanical particles (fermions or bosons) at temperatures low enough, or densities high enough, that the low-lying single particle energy levels are multiply occupied in equilibrium.

(2) A gas at temperatures and densities such that the thermal energy of a particle is comparable to the zero-point energy that it would have if confined to a box of volume equal to the volume per particle. Examples of such systems are given by the Bose condensates of the alkali gases ^{87}Rb, ^{23}Na, and ^{7}Li discovered in 1995, and, except for the fact that they are liquids, ^{4}He and ^{3}He. *See also* Bose-Einstein condensation; Fermi gas; Fermi liquid; superconductivity.

degenerate semiconductor In a compound semiconductor, after purposely adding one component, the conductivity of the semiconductor will change. Such semiconductors are called *degenerate semiconductors*.

degree of coherence Gives the coherence in a quantum system. The off-diagonal matrix elements of the density matrix give the *degree of coherence*. The *degree of coherence* can also be determined through the visibility V of interference fringes with intensity maxima I_{max} and minima I_{min} as interference in a characteristic of coherence:

$$V = \frac{I_{max} - I_{min}}{I_{max} + I_{min}}.$$

degree of freedom (1) A distribution function may depend on several variables that vary stochastically. If these variables are statistically independent, then each represents a *degree of freedom* of the distribution.

(2) The number of independent coordinates needed for the description of the microscopic state of a system is called the number of *degrees of freedom*. For example, a single point particle in three-dimensional space has three degrees of freedom; a system of N point particles has $3N$ degrees of freedom.

De Haas–Van Alphan effect In 1930, De Haas and Van Alphan measured the magnetic susceptibility x of metal Bi at a low temperature, 14.2K, and strong magnetic field. They found that x oscillated along with the change of magnetic field. This phenomenon is called the *De Haas–Van Alphan effect*.

delayed choice experiment Gedanken variant of the two-slit interference experiment with photons in which the slits and screen are replaced by two half-silvered mirrors. When only the first mirror is in place, it is possible to tell which path a photon takes. When both mirrors are in place, however, interference is observed, and the "which path" information is lost. In the *delayed choice experiment,* the decision to insert or not insert the second mirror is made after the photon has, classically speaking, passed the first mirror. Nevertheless, it is apparent that interference is observed when and only when the second mirror is in place. The experiment further confirms quantum mechanical precepts that it is not possible to assign a meaning to the notion of a trajectory to a particle in the absence of an apparatus designed to measure the trajectory.

delayed emission De-excitation of an excited nucleus usually occurs rapidly ($\leq 10^{-8}$s) after formation by gamma emission (electromagnetic interaction). Emission of protons or neutrons from a nucleus occurs on much shorter time scales due to the fact that the hadronic interaction is much stronger. Occasionally, weak decay of an unstable nucleus occurs. If this unstable nucleus then emits a nucleon delayed by the weak decay time, then *delayed emission* has occurred.

delta function A pseudo-mathematical function which provides a technique for summing of an infinite series or integrating over infinite spatial dimensions. The *delta function*, δ, is defined such that:

$$f(x) = \int_{-\infty}^{\infty} \delta(x - x') f(x') \, dx' .$$

The integral:

$$\delta(x) = (1/2\pi)^{1/2} \int_{-\infty}^{\infty} e^{-i(x-x')} \, dx'$$

forms one representation of the *delta function*. Note that, by itself, the *delta function* is not convergent, but used to find the value of a function, $f(x)$, it is well-defined if the limits are taken in an appropriate order.

delta ray A low energy electron created from the ionization of matter by an energetic charged particle passing through the material. *Delta rays,* however, have sufficient energy to further ionize the atoms of the material ($\geq$ a few ev).

delta resonance The lowest excitation of a nucleon. It has a spin/parity of $3/2^-$ and exists in four charge or isotopic spin states, $2e, e, -e,$ and $-2e$, where e is the magnitude of the electronic charge. The delta belongs to the decouplet SU(3) quark representation of the non-strange baryons.

density matrix Reflects the statistical nature of quantum mechanics. Specifically, the *density matrix,* which is sometimes also called the statistical matrix, illustrates that any knowledge about a quantum mechanical system stems from the observation of many identically prepared systems, i.e., the ensemble average. For a system in a well-defined state $|\Psi\rangle$ given by $|\Psi(\theta)\rangle = \sum_n c_n(\theta)|\psi_n\rangle$, where $|\psi_n\rangle$ forms a complete basis, the *density matrix* elements are defined as

$$\rho_{mn} = \langle \psi_m | \hat{\rho} | \psi_n \rangle ,$$

where $\hat{\rho} = |\Psi\rangle\langle\Psi|$. It follows that the individual matrix elements ρ_{mn} can also be calculated through

$$\rho_{mn} = \langle \psi_m | \Psi \rangle \langle \Psi | \psi_n \rangle = c_m c_n^* .$$

For statistical mixtures of states, the definition for the *density matrix* must be generalized to account for the uncertainties of the different admixtures of pure states:

$$\rho_{mn} = \int p(\theta) c_m(\theta) c_n^*(\theta) d\theta ,$$

where $p(\theta)$ is the probability distribution of finding the state $|\Psi(\theta)\rangle$ in the mixed state.

The *density matrix* contains information about the specific preparation of a quantum system. This is in contrast to the matrix elements $O_{nm} = \langle \psi_n | \hat{O} | \psi_m \rangle$ of an observable $\hat{O}$. O_{nm} depends only on the specific operator $\hat{O}$ and the basis set, but contains no information about the quantum state $|\Psi\rangle$ itself.

The diagonal elements ρ_{nn} are called populations, as ρ_{nn} give the populations, i.e., the probability of finding the system in state ψ_n ($\rho_{nn} = P_n$) which leads to the condition $\rho_{nn} \geq 0$. This terminology is also justified by the property of the *density matrix:*

$$\text{Tr} \rho = \sum_i \rho_{ii} = 1 .$$

The off-diagonal elements ρ_{nm} are termed the coherences, as they are measures for the coherences between states $|\Psi_n\rangle$ and $|\Psi_m\rangle$. In the case that a particular *density matrix* ρ represents a pure state, as opposed to a statistical mixture, the *density matrix* is idempotent, i.e.,

$$\rho \rho = \rho .$$

Consequently, also,

$$\text{Tr}(\rho^n) = 1 .$$

In contrast, for the *density matrix* of mixed states, we find:

$$\text{Tr}(\rho^2) \leq 1 .$$

Finally, the *density matrix* is Hermitian, i.e.,

$$\rho^\dagger = \rho \quad \text{or} \quad \rho_{mn}^* = \rho_{nm} .$$

The *density matrix* allows a straightforward calculation of expectation values $\langle \hat{O} \rangle$ for an observable $\hat{O}$:

$$\langle \hat{O} \rangle = \langle \Psi | \hat{O} | \Psi \rangle = \sum_{mn} \langle \Psi | \psi_m \rangle O_{mn} \langle \psi_n | \Psi \rangle .$$

With the help of the *density matrix*, one solves for $\langle \hat{O} \rangle$:

$$\langle \hat{O} \rangle = \sum_{mn} \hat{O}_{nm} \rho_{mn} = \sum_m [\hat{O} \hat{\rho}]_{nn} = \text{Tr}[\hat{O} \hat{\rho}] .$$

As an example, the density state for the simplest coherent state $|\Psi_{\text{coh}}\rangle$ given by

$$|\Psi_{\text{coh}}\rangle = \cos\theta |\Psi_1\rangle + \sin\theta |\Psi_2\rangle$$

yields

$$\rho = \begin{vmatrix} \cos^2\theta & \cos\theta \sin\theta \\ \cos\theta \sin\theta & \sin^2\theta \end{vmatrix} .$$

For the special case of $\theta = \pi/4$, we find:

$$\rho = \begin{vmatrix} \frac{1}{\sqrt{2}} & \frac{1}{\sqrt{2}} \\ \frac{1}{\sqrt{2}} & \frac{1}{\sqrt{2}} \end{vmatrix} .$$

In contrast, for a completely incoherent state or mixed state, where states with values of all different θ are mixed with an equal probability, we solve for the density matrix:

$$\rho = \begin{vmatrix} \frac{1}{\sqrt{2}} & 0 \\ 0 & \frac{1}{\sqrt{2}} \end{vmatrix} .$$

density of final states Represents statistically the number of possible states per momentum interval of the final particles. The particles are assumed to be non-interacting, with population density governed only by the conservation of energy and momentum.

density of modes The number of modes of the radiation field in an energy range dE. The *density of modes* is a function of the boundary conditions of the space under consideration. For free space, the *density of modes* per unit of volume and per angular frequency is given by:

$$\overline{\omega} = \frac{\omega^2}{\pi^2 c^3} .$$

For large mode volumes, the mode distribution is quasi continuous, while for small cavities, the discrete mode structure is fully apparent. This can lead to enhancement and suppression of spontaneous decay depending on the exact cavity geometry. The change in mode density originates from the boundary condition that has to be fulfilled by the different cavity modes. Specifically, for a cavity, the modes have to have vanishing electric fields on the cavity walls. The physics originating from such a modification of the mode density is explored by cavity quantum electrodynamics (CQED) and in its most basic form by the Jaynes–Cummings model.

density of states The number of states in a quantum mechanical system in a given energy range dE. One finds that

$$D(E) = \frac{dN_s}{dE},$$

where $D(E)$ is the *density of states* in an energy range between E and dE.

depolarization Scattering of nucleons from nucleons (spin 1/2 on spin 1/2 hadronic scattering) can be parameterized in terms of nine variables, but at any given scattering angle only four of these are independent due to unitarity. These parameters can be defined in different ways, one of which is to assign the production of polarization by scattering as the parameter, P, while the other parameters describe possible changes to an already polarized particle due to its scattering interactions. In general, the polarization is rotated in the collision, and in particular, the *depolarization* parameter measures the polarization after scattering along the perpendicular direction to the beam in the scattering plane if the initial beam is 100% polarized in this direction.

destruction operator (1) Abstract operator that diminishes quanta of energy or particles in Fock space by one unit. Also known as an annihilation or lowering operator in some contexts. *See also* creation operators.

(2) In quantum field theoretic calculations, the field quanta are represented in momentum space. In this space, a wave function for a quantum of the field represents a particle, and can be considered as either creating or annihilating this particle out of or into the vacuum state. The *destruction* (annihilation) *operator* is the Hermetian conjugate of the creation operator.

detailed balance The reaction matrix, U, depends on all the quantum numbers of the incoming and outgoing states. General considerations of quantum mechanics indicate that the U matrix multiplied by its Hermitian adjoint results in the identity matrix. This means that in any reaction, $A \to B$ is identical to the reversed reaction $B \to A$ with spins reversed (*detailed balance*) and with time inversion symmetry preserved.

detection efficiency loophole Due to experimental insufficiencies in tests of the Bell inequalities. As of now, the strongest form of the Bell inequalities has not been tested, since the required detection efficiencies have not been enforced. Therefore, current tests of the Bell inequalities test weaker forms that are derived by assuming that particles which are detected behave exactly the same as those that are not detected, or, in other words, that the detectors produce a fair sample of the entire ensemble of particles (fair sampling assumption). Thus, the present tests leave open a loophole. Other requirements for a definite test of the Bell inequalities are strong spatial correlation and a pure preparation of the entangled state.

determinantal wave function A wave function for a system of identical fermions consisting of an antisymmetrized product of single-particle wave functions. Also called a Slater determinant after J.C. Slater.

detuning Refers to the fact that light incident on an atomic or molecular system is not resonant with a transition in this atom/molecule. The *detuning* has the value of

$$\Delta\omega = \omega_l - \omega_0$$

where ω_0 is the resonant frequency and ω_l is the frequency of the incident light. Light is said to be red-*detuned* light when $\Delta\omega < 0$ and blue-*detuned* when $\Delta\omega > 0$.

deuteron (1) The nucleus of the hydrogen isotope deuterium consisting of a proton and a neutron.

(2) A *deuteron* is the nucleus of the isotope of hydrogen with the atomic mass number 2. It consists of a neutron bound to a proton with their intrinsic spins aligned, which gives a value of one for the total angular momentum of the bound state, *deuteron*. Since the system with anti-aligned nuclear spins is unbound, the nuclear force is spin-dependent and stronger in the $^3S_1$ state than in the $^1S_0$ state.

diabolical point For a system with a Hamiltonian parametrized by two variables, the *diabolical point* is a point in this parameter space where two energy levels are degenerate. So called because the energy surface in the vicinity of this point is a double elliptic cone, resembling an Italian toy, the diabolo. A diabolical point need not be characterized by any obvious symmetry, and is, to that extent, an accidental degeneracy.

diagonalization of matrices Used to find the eigenvectors and eigenvalues of matrices. The eigenvectors $\vec{v}_i$ and eigenvalues λ_i of a matrix M are given by the following equation:

$$\lambda_i \vec{v}_i = M \vec{v}_i .$$

If the matrix M is diagonal, i.e., $M_{ij} = 0$ for $i \neq j$, the diagonal elements M_{ii} are the eigenvalues of the matrix. Diagonalization of Hermitian matrices is of particular relevance since physical observables can be described by Hermitian matrices, i.e.,

$$\hat{O}_{ij}^* = \hat{O}_{ji} ,$$

where the corresponding matrix elements for the operator $\hat{O}$ can be written as:

$$\hat{O}_{ij} = \int d^3r \, \Psi_i^* \hat{O} \Psi_j = \langle \Psi_i | \hat{O} | \Psi_j \rangle ,$$

where the $|\Psi_i\rangle$ form a complete basis.

The matrix $\hat{O}_{ij}$ is diagonal if the $|\Psi_i\rangle$ are eigenstates of the operator $\hat{O}$. The eigenvalues are the diagonal elements. Hence the *diagonalization of a matrix* is equivalent to finding the eigenvalues of the matrix and is an important step toward finding the eigenstates of a particular problem.

diamagnetism If one material has a net negative magnetic susceptibility, it has *diamagnetism*.

diamond structure In a diamond, the Bravais lattice is a face-centered cube whose primitive vector is $a/2(\mathbf{x}+\mathbf{y}, \mathbf{y}+\mathbf{z}, \mathbf{z}+\mathbf{x})$, where a is the distance between two atoms. The lattice's bases are two carbon atoms located at $(\mathbf{0}, \mathbf{0}, \mathbf{0})$ and $a/4(\mathbf{x}, \mathbf{y}, \mathbf{z})$.

diatomic molecule A molecule made up of two atoms. Bonding can be covalent or due to van der Waals forces. *Diatomic molecules* bound by relatively weak van der Waals forces are sometimes referred to as dimers.

Dicke narrowing (motional narrowing) The narrowing of atomic or molecular transitions due to a process that increases the characteristic time an atom/molecule interacts with light. The characteristic width of Doppler broadened lines is $\Delta = 2\pi v_T/\lambda$, where v_T is the thermal speed and λ is the wavelength of the emitted or absorbed light. This width can be associated with a coherence time $1/\Delta$, in which the atom can interact with the light without interruption. Increasing this time leads to an effective narrowing of the transitions. This can be achieved for instance by means of a buffer gas: the increased number of collisions with the buffer gas leads to an increased interaction time of the species under investigation with the light and, thus, to a narrowing of the transition lines.

dielectric A nonconductor of electricity. The term *dielectric* is usually used where electric fields can exist inside a material, such as between a parallel plate capacitor.

dielectric strength The maximum electric field that can exist in a material without causing it to break down.

diesel engine A four-step cyclical engine, illustrated below. It consists of an adiabatic compression of the air and fuel mixture (i), followed by a combustion step at constant pressure (ii), and then cooled first by an adiabatic expansion (iii), with further cooling at constant volume (iv) to return the gas to the initial temperature and pressure.

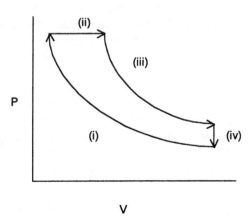

Diesel engine cycle.

difference frequency generation A nonlinear process in which radiation is generated that has an energy equivalent to the difference of the two initially present radiation fields. It is the reverse process of sum frequency generation and closely related to optically parametric down conversion. Energy and momentum conservation have to be fulfilled in the process, i.e.,

$\nu_d = \nu_1 - \nu_2$ energy conservation ,
$\vec{k}_d = \vec{k}_1 - \vec{k}_2$ momentum conservation

where ν are the frequencies and $\vec{k}$ are the wave vectors of the different radiation fields involved.

differential cross-section The nuclear cross-section per unit of energy, momentum, or angle; usually refers to the angular *differential cross-section*. The *differential cross-section* per solid angle, $\partial\Omega$, is written as:

$$\frac{\partial\sigma}{\partial\Omega} .$$

diffraction At forward angles and small momentum transfers, the scattering of high energy particles from a composite of target scattering centers, such as nucleons in a nucleus, is primarily governed by the wave nature of these projectiles. Scattering from such a system can be coherent, i.e., the incident and outgoing particle waves are identical except for a phase change, leading to a description of the scattering in terms of interfering waves. Scattering represented by this process is called diffractive scattering or *diffraction*.

diffuser A duct in which the flow is decelerated and compressed. The shape of a *diffuser* is dependent upon whether the flow is subsonic or supersonic. In subsonic flow, a *diffuser* duct has a diverging shape, while in supersonic flow, a *diffuser* duct has a converging shape. *See* converging–diverging nozzle.

diffusion The movement of a solid, liquid, or gas as a result of the random thermal motion of its atoms or molecules. *Diffusion* in solids is quite small at normal temperatures.

diffusion coefficient, diffusion length Neutrons above thermal energies lose energy by scattering from the nuclei of a material, losing energy until they are captured by a nucleus or reach thermal equilibrium with the surrounding environment. Thus, the average energy of an initial distribution of neutrons will decrease over time, and the width will increase (diffuse):

$$D = \lambda v/3 ,$$

where v/λ is the number of collisions of the neutron per unit of time, and D is the *diffusion coefficient*. The quantity,

$$L = [\lambda \Lambda/3]^{1/2} ,$$

where Λ/v is the mean-life of a thermal neutron, is the *diffusion length*. The density of thermal neutrons then obeys the equation (q_τ the number of neutrons becoming thermal per unit time),

$$\nabla^2 n - (3/\lambda\Lambda)n + 3q_\tau/\lambda v = 0 ;$$

with the boundary condition $n = 0$ on the surface of the moderator.

diffusion, plasma The loss of plasma from one region (normally the interior) to another region (normally the exterior) stemming from plasma density or pressure gradients.

diffusion, viscous Penetration of the effects of motion in a viscous fluid where the boundary layer grows outward from the surface. Near the surface, fluid parcels are accelerated by an imbalance of shear forces. As the fluid moves adjacent to the wall, it drags a portion of the neighboring fluid parcels along with it, resulting in a gradual induction of fluid moving with or retarded by the surface. In an unsteady flow, the diffusion is governed by the simplified equation

$$\frac{\partial u}{\partial t} = \nu \frac{\partial^2 u}{\partial y^2}$$

where viscous forces govern the fluid behavior.

dilatant fluid Non-Newtonian fluid in which the apparent viscosity decreases with an increasing rate of deformation. Also referred to as a shear thickening fluid.

dimensional analysis The basis of *dimensional analysis* is that any equation which expresses a physical law must be satisfied in all possible systems of units. What differentiates between one set of units and another is how the system is defined, in particular, what quantities are chosen as primary. These are the basic set of units. All other units are a combination of these and are known as secondary (these are also known as base and derived units when specifically referring to the system). In fluid mechanics, the primary dimensions are usually mass, length, time, and temperature (SI). All other physical quantities are derived from these primary dimensions.

dimensionless intensity The intensity in atomic units often used in theoretical calculations. In particular in the semiclassical theory, a dimensionless intensity can be defined which is equivalent to the number of photons n in the laser mode with volume V:

$$n = \frac{\epsilon_0 \mathcal{E}^3 V}{2\hbar \omega},$$

where ω is the angular frequency of the photons.

In the literature, the intensity is often defined as:

$$I = \frac{c}{8\pi}\mathcal{E}^2,$$

where $\mathcal{E}$ is the time averaged electric field. The standard SI unit for the intensity is W/m^2. The intensity is sometimes also referred to as the irradiance.

dimensionless parameter Any of a number of parameters characterized by value alone and which describes characteristic physical behavior of fluid flow phenomena. A *dimensionless parameter* is composed of a ratio of two quantities with the same dimensions to measure the relative effect of these quantities in a given flow (*see* Reynolds number, Mach number). Some *dimensionless parameters* of common use in fluid mechanics are listed below.

Name	Form & Ratio
Cauchy number	$Ca = U^2 \rho / \beta_s = M^2$ inertia force:compressive force
Euler number	$Eu = \Delta p / \rho U^2$ pressure force:inertia force
Froude number	$Fr = U^2 / gL$ inertia force:gravity force
Grashof number	$Gr = g\beta \Delta T L^3 / \nu^2$ buoyancy force:viscous force
Knudsen number	$K = \lambda / L$ mean free path:length scale
Mach number	$M = U/a$ velocity:sound speed
Reynolds number	$Re = UL/\nu$ inertia force:viscous force
Stokes number	$Sk = \Delta p L / \mu U$ pressure force:viscous force
Strouhal number	$St = fU/L$ vibration frequency:time-scale
Weber number	$We = \rho U^2 L / \sigma$ inertia force:surface tension force

diode An electronic device that exhibits rectifying action when a potential difference is applied between two electrodes. Current flows from one direction of the potential, called the forward direction. When the potential is reversed, the current is very small or zero.

dipolar force The attractive force between two molecules originating from the polarization of the molecules. The partially positively charged end of a molecule attracts the partially negatively charged part of the other molecule.

dipole-allowed transition *See* electric dipole-allowed transition.

dipole approximation Frequently used when the interaction between an atom and an electromagnetic wave is considered. The electromagnetic wave can be written as the resultant from a vector potential $\vec{A}$ as

$$\vec{E}(\vec{r}, t) = -\frac{1}{c}\frac{\partial}{\partial t}\vec{A}(\vec{r}, t)$$
$$\vec{B}(\vec{r}, t) = \nabla \times \vec{A}(\vec{r}, t) \ .$$

An electron subject to the vector potential $\vec{A}$ has the minimal coupling Hamiltonian:

$$H = \frac{1}{2m}\left(\vec{p} - e\vec{A}\right)^2 + eU(r, t) + \Phi(\vec{r}) \ ,$$

where $\vec{A}$ and U are the vector and scalar potentials of the field, and $\Phi(\vec{r})$ constitutes the scalar Coulomb potential. In the radiation gauge we find

$$U = 0 \quad \text{and} \quad \nabla \vec{A} = 0 \ .$$

The interaction of a two-level atom is with spherical waves that can be written with the help of the vector potential as

$$\vec{A}(\vec{r}, t) = A_2(t)\exp\left(\imath\vec{k}\vec{r}\right) + A_2(t)\exp\left(-\imath\vec{k}\vec{r}\right),$$

which gives rise to the interactions of the form

$$\langle\Psi_f|\frac{e}{m}A_2\vec{p}\exp\left(\imath\vec{k}\vec{r}\right)|\Psi_i\rangle$$

where the rotating wave approximation was assumed. In the *dipole approximation*, one assumes that the electric field of the wave ($\lambda \approx 1000$Å) does not significantly change across the dimension of the nucleus $\lambda \approx 1$Å. Mathematically it means that only the zeroth order term in the series expansion for the operator

$$\exp\left(\imath\vec{k}\vec{r}\right) = 1 + \imath\vec{k}\vec{r} + \frac{1}{2}\left(\imath\vec{k}\vec{r}\right)^2 + \cdots$$

is used. Here, $\vec{k}$ is the wave vector of the electromagnetic wave, and $\vec{r}$ is typically the extent of the nucleus, i.e., in the order of 1 Å. Therefore, the higher order terms are much smaller than the leading term and the *dipole approximation* holds. These are the electric dipole-allowed transitions (E1). Thus, using the *dipole approximation*, the interaction between states $|\Psi_f\rangle$ and $|\Psi_i\rangle$ can then be written as

$$\langle\Psi_f|\frac{e}{m}\vec{p}|\Psi_i\rangle \ ,$$

which, by means of a gauge transformation of fields and wave functions to the electric field gauge, can be shown to be equivalent to

$$\omega_{fi}^2\langle\Psi_f'|e\vec{r}|\Psi_i'\rangle$$

where ω_{fi} is the resonance frequency of the transition.

In the case that the zeroth order term has no contribution, i.e., in the case of dipole-forbidden transitions, the higher order terms can become important.

dipole field   The field of an electric dipole with dipole moment $q\vec{d}$. It is given by

$$\vec{E}(\vec{r}) = \frac{q}{4\pi\varepsilon_0}\frac{3\left(\vec{d}\vec{r}\right)\vec{r} - (\vec{r}\vec{r})\vec{d}}{r^5} \ .$$

dipole-forbidden transitions   Transitions for which the electric dipole transition moment in the dipole approximation vanishes:

$$|\langle\Psi_1|e\hat{r}|\Psi_2\rangle|^2 = \left|e\int\Psi_1^*r\Psi_2\,dr\right|^2 = 0 \ .$$

Transitions are possible due to higher order terms in the expansion of the matrix element

$$\left|\langle\Psi_1|\frac{e}{m}e\vec{p}e^{\imath\vec{k}\vec{r}}|\Psi_2\rangle\right|^2$$

which is derived considering interactions of one photon with a two-level system using the radiation gauge Hamiltonian. These transitions are much weaker than dipole-allowed transitions. The two most important types are magnetic dipole and electric quadrupole transitions. Their selection rules are:

magnetic dipole transitions:
$\Delta J = 0, \pm 1$
$\Delta L = 0$
$\Delta m = 0, \pm 1$

electric quadrupole transition:
$\Delta L = \pm 2$
$\Delta m = 0, \pm 1, \pm 2 \ .$

One also speaks of forbidden transitions in the case of intercombination lines, where the selection rule $\Delta S = 0$ is violated. This can be the case for heavy atoms, where the spin–orbit interaction is large. These transitions still have dipole characteristics, since they occur due to the admixture of other states to the bare states involved in the transitions. An example is the well known 253.7 nm transition in mercury ($^3P_1 \leftarrow {}^1S_0$).

dipole forces Result from the interaction of the induced dipole moment in an atom or molecule with an intensity gradient of the light field causing this dipole. Several models are available to describe the conservative *dipole force*. In the oscillator model, we assume a two-level system and use the rotating wave approximation (assuming that the laser frequency detuning Δ from the resonance at ω_0 is small compared to the frequency ω_0: $|\Delta| \ll \omega_0$). Thus, the force on a particle is

$$F(r) = -\nabla U_{\text{dipole}}(r) = \frac{3\pi c^2}{2\omega_0^3} \frac{\Gamma}{\Delta} \nabla I(r),$$

where ω_0 and Γ are the resonance frequency of the atom, and the linewidth of the resonance transition, and $\Delta = \omega - \omega_0$ is the detuning of the laser from the resonance; c is the speed of light. The force is conservative since it can be written as the gradient of a potential U_{dipole}. The heating of the sample due to absorption of the light by the atomic system can be measured by the scattering rate $\Gamma(r)$ of photons:

$$\Gamma(r) = \frac{3\pi c^2}{2\hbar\omega_0^3} \frac{\Gamma^2}{\Delta^2} I(r).$$

As indicated above, α is dependent on the frequency of the light field.

It is important to realize the dependence of the *dipole force* on the sign of the detuning. For red detuning, i.e., $\Delta < 0$, the force is negative. The atoms or molecules are therefore drawn to high intensities. For the case of blue detuning, i.e., $\Delta > 0$, the force is positive, and the interaction leads to a repulsion of the particles from areas with high intensity.

The potential scales with I/Δ, whereas the scattering rate, i.e., the heating, scales with I/Δ^2. Thus, large detunings lead to much smaller heating of the sample, but do require larger intensities to produce the same force.

It should be noted that for multi-level atoms, the expressions for the force and scattering rate become slightly more complicated.

The dipole trap is based on *dipole forces*.

dipole moment Associated with a charge distribution $\varrho(\vec{r})$, and given by

$$d = \int d^3r \varrho \vec{r} = -e \int d^3r \Psi_n(\vec{r})^* \vec{r} \Psi_n(\vec{r}),$$

where e is the elementary charge and we have used the relationship between the charge density ϱ and the wave function Ψ_n of a stationary electron:

$$\varrho \vec{r} = -e \Psi_n(\vec{r})^* \vec{r} \Psi_n(\vec{r}).$$

dipole operator Defined as

$$\hat{d} = -e\vec{r}$$

where e is the elementary charge.

dipole selection rule States that *electric dipole transitions* in any system take place between levels that differ by, at most, one unit of angular momentum, except in the case where both levels have zero angular momentum. Similar rules accompany magnetic dipole and higher multipole transitions.

dipole sum rule Rule that puts an upper boundary on the total absorption cross-section for any system in its ground state, under the assumption that the absorption is primarily due to dipole transitions. The rule is of value in estimating transition matrix elements, and played a historically important role in the development of quantum mechanics. Also known as the Thomas-Reiche-Kuhn rule.

dipole transition See electric dipole-allowed transition; forbidden transition.

dipole transition moment For a one-electron atom between state Ψ_n and Ψ_m, the dipole transition moment is defined as the integral

$$d = -e \int d^3r \Psi_m(\vec{r})^* \vec{r} \Psi_n(\vec{r}).$$

The value $|d|^2$ is proportional to the transition probability for an electric dipole transition between the two states Ψ_n and Ψ_m. It can be derived from the zeroth order term of the series expansion of the operator $e^{i\vec{k}\vec{r}}$, which appears in the interaction Hamiltonian. The *dipole transition moment* is derived with the help of the dipole and rotating wave approximations.

dipole traps (optical dipole traps) Allow trapping of neutral atoms and molecules. Their action is based on the dipole forces in far-detuned light. Typically, their trap depths are much lower than those of the magneto-optical traps or purely magnetic traps. They are typically below 1 mK. Therefore, atoms or molecules that are to be trapped in dipole traps must be pre-cooled with other techniques before they can be stored. However, since the trapping mechanism is based on non-resonant light, molecules as well as atoms can be trapped.

Dirac equation A quantum mechanical, relativistic wave equation which describes the interaction and motion of particles with an intrinsic spin of 1/2. The equation has the form:

$$H\psi = i\frac{\partial \psi}{\partial t},$$

where the Hamiltonian for a free particle is written as:

$$H = \gamma_4 \left(\gamma_k \frac{\partial}{\partial x_k} + m \right).$$

The γs are 4×4 matrices, the wave function, ψ, is a four-dimensional column vector, the two upper components represent the two spin states of a positive energy particle, and the lower two components represent the two spin states of the corresponding negative energy particle (anti-particle).

Dirac hole theory Theory in which the physical vacuum is regarded as obtained by filling all the negative energy single-electron states that emerge as solutions of the Dirac equation, and a positron is regarded as obtained by the removal of one of the negative energy states.

Dirac magnetic monopole Particle postulated by P.A.M. Dirac in 1931, which would act as a source of magnetic flux density **B** in the same way as an electron is a source of the electric field **E**. Thus, an infinitesimal surface enclosing a magnetic monopole would have a nonzero magnetic flux passing through it. Dirac showed that the magnetic charge g of such a particle and the electric charge e of the electron would be related by the so-called Dirac quantization condition, according to which the product ge must be an integral multiple of $hc/4\pi$, where h is Planck's constant and c is the speed of light. No magnetic monoples have been discovered to date. *See also* Dirac string.

Dirac matrix A four-dimensional matrix which is a component of the Dirac equation and which describes the operations of parity and space–time rotations of the spin degrees of freedom. There are several representations of these matrices, but one useful representation may be written in terms of the Pauli spin matrices, σ. Thus,

$$\gamma_k = \begin{pmatrix} 0 & -i\sigma_k \\ i\sigma_k & 0 \end{pmatrix};$$

and

$$\gamma_4 = \begin{pmatrix} 1 & 0 \\ 0 & -1 \end{pmatrix}.$$

See Dirac equation.

Dirac notation A nomenclature to write quantum mechanical integrals introduced by Dirac. The expectation value for an operator $\hat{A}$ for a wave function Ψ can be expressed in the Dirac notation simply as

$$\langle \Psi | \hat{A} | \Psi \rangle = \int \Psi^* A \Psi \, dr,$$

where the Schrödinger notation is used in the second part. The $\langle \Psi |$ and $|\Psi \rangle$ parts are referred to as bra and kets, respectively.

Dirac quantization condition *See* Dirac magnetic monopole.

Dirac string A convenient representation of the singularity that necessarily arises in describing a magnetic monopole in terms of a magnetic vector potential **A**. The total magnetic flux emerging from the monopole is viewed as returning to the monopole along a string of zero width anchored to the monopole. The string can wind around arbitrarily in space, but cannot be eliminated, reflecting the fact that the singularity cannot be removed by any choice of gauge.

direct band gap semiconductor In a *direct band gap semiconductor*, the conduction band edge and valence band edge are at the center of the Brillouin zone, such as GaAs, InSb, etc.

direct drive An approach to inertial confinement fusion in which the laser or particle beam energy is directly incident on a pea-sized fusion-fuel capsule resulting in compression heating from the ablation of the target surface.

direct reaction Nuclear reactions are generally described as compound or direct. Although this classification is not well-defined, a compound reaction usually occurs at low energy when a particle is absorbed by a nucleus, the incident energy is shared by at least several nuclear components, and particles are emitted to remove the excess energy. A *direct reaction* usually occurs at higher energy when an incident particle interacts with one nuclear component, directly producing the final nuclear state without the system passing through a set of intermediate states.

discharge coefficient Empirical quantity used in flow through an orifice to account for the losses encountered in non-ideal geometries from separation and other effects.

discrete spectrum A discrete set of values in quantum mechanics for the observational outcomes (the spectrum) of a physical quantity, as opposed to values that run through a continuous range. For example, the spectrum of angular momentum is wholly discrete.

dispersive wave A wave that propagates at different speeds as a function of wavelength, thus dispersing as the wave progresses in time or space.

displacement thickness In boundary layer analysis, the distance by which the wall would have to be displaced outward to maintain the identical mass flux in the flow, given by

$$\delta^* = \int_o^\infty \left(1 - \frac{u(y)}{U_\infty}\right) dy$$

where U_∞ is the free-stream velocity outside the boundary layer.

disruption, or plasma disruption Plasma instabilities (usually oscillatory modes) sometimes grow and cause abrupt temperature drops and the termination of a experimentally confined plasma. Stored energy in the plasma is rapidly dumped into the rest of the experimental system (vacuum vessel walls, magnetic coils, etc.).

dissipation The transformation of kinetic energy to internal energy due to viscous forces. It is proportional to the square of the velocity gradients and is greater in regions of high shear.

distorted wave approximation The transition matrix between two quantum mechanical states can be expressed as:

$$S_{fi} = \langle \phi_f | H_{\text{int}} | \psi_i \rangle \; ;$$

where H_{int} is the perturbing Hamiltionian that causes the transition between the states, ψ_i is a state of the complete Hamiltonian, $H = H_0 + H_{\text{int}}$ with initial boundary conditions, and ϕ_f is a state of the unperturbed Hamiltonian, H_0, with final boundary conditions. In general, ψ_i is difficult to determine and is replaced by an approximate wave function, usually found by perturbation techniques. Thus to first order when ψ is replaced by ϕ_f, one has the plane-wave Born approximation. More realistic approximations may be determined by replacing the exact Hamiltonian, H, with one which has an approximate interaction potential, but is more easily solvable, e.g., the addition of a Coulomb potential plus some central potential. Then the approximate ψ is not exactly correct but is more realistic and is distorted from the plane wave solutions, ϕ.

divergence operator The application of the *divergence operator* on a vector field gives the flux of that vector out of an infinitesimal volume per unit of volume. In Cartesian coordinates, the divergence of a vector, **A** is written:

$$\nabla \bullet A = \frac{\partial A_x}{\partial x} + \frac{\partial A_y}{\partial y} + \frac{\partial A_z}{\partial z} .$$

divergence theorem Relation between volume integral and surface integral given by

$$\int_V \nabla \cdot \mathbf{Q} dV = \int_A \mathbf{Q} \cdot d\mathbf{A}$$

where Q can be either a vector or a tensor. Also referred to as the Gauss-Ostrogradskii *divergence theorem*.

divertor, plasma divertor Component of a toroidal plasma experimental device that diverts charged ions on the outer edge of the plasma into a separate chamber where charged particles can strike a barrier and become neutral atoms.

D Meson Class of fundamental particles constructed of a charmed (anti-charmed) quark and an up or down (anti-up or anti-down) quark. The lowest representation of these mesons are the $D^{\pm}$ and the D^0, which have spin 0 and negative parity and are composed of $c\bar{d}$ or $\bar{c}d$ and $c\bar{u}$, respectively.

domain In ferroelectric materials, there are many microscopic regions. The direction of polarization is the same in one *domain;* however, in adjacent *domains*, the directions of polarization are opposite.

donor levels The levels corresponding to donors, found in the energy band gap and very close to the bottom of the conduction band.

donors In a semiconductor, pentravalent impurities which can offer electrons are called donors.

dopant *See* acceptor.

Doppler broadening The inhomogeneous broadening of a transition due to the velocity distribution of an ensemble of atoms. The broadening comes from the Doppler detuning for individual atoms, which have different velocity components with respect to the propagation direction of the light. If the ensemble of atoms exhibits a Maxwell-Boltzmann distribution for their velocities, one finds a *Doppler-broadened* line width of

$$\Delta \nu = \frac{2\nu_0}{c}\sqrt{\frac{2R \ln 2}{M}},$$

where R is the general gas constant, M is the molar mass, and λ and ν_0 are the resonance wavelength and frequency, respectively.

Doppler detuning The detuning of a transition caused by the movement of the atom relative to the source of radiation. *Doppler detuning* is sometimes called the Doppler shift.

Doppler distribution The characteristic line shape of a transition that is broadened due to the movement of the atoms. Since each atom has a different velocity and, consequently, a different Doppler shift, one speaks of an inhomogeneous distribution. For atoms with a Maxwell–Boltzmann distribution of the velocities, the distribution is given by a Gaussian profile:

$$I(\omega) = I_0 \exp\left[-\left(\frac{c(\omega - \omega_0)}{\omega_0 v_m}\right)^2\right],$$

where $v_m = \sqrt{\frac{2kT}{m}} = \sqrt{\frac{2RT}{M}}$

where ω_0 is the resonance frequency, v_m is the most likely velocity of the distribution, T is the equilibrium temperature of the atoms, and m and M are their atomic and molar masses, respectively. k and R are the Boltzmann constant and general gas constant, respectively.

However, experimentally, usually the convolution of a Gaussian (inhomogeneous) with a homogeneously broadened linewidth (collisions) is observed:

$$I(\omega) = \frac{\Gamma I_0 N c}{2 v_m \pi^{3/2} \omega_0} \int_0^\infty \frac{\exp\left[(-c/v_m)^2 (\omega_0 - \omega')^2 / \omega_0^2\right]}{(\omega - \omega')^2 + (\Gamma/2)^2} d\omega' .$$

Here, Γ is the width of the Lorentzian profile. This convoluted distribution is called the Voigt profile.

Doppler-free excitation An excitation method that circumvents the Doppler shift of the resonances due to the motion of the individual atoms so that for a given laser frequency, all atoms will be excited. Examples are two-photon spectroscopy and saturation spectroscopy.

In two-photon spectroscopy, the atom absorbs one photon out of each of two counterpropagating beams. In this way, the Doppler shift with respect to one beam is canceled by the Doppler shift occurring with respect to the second. Since there is a probability for the atom to absorb two photons out of the same beam, there will be a small pedestal underneath the Doppler-free main signal.

In saturation spectroscopy, two laser beams of different intensities — a strong pump and a weak probe derived from the same laser beam — are counterpropagated through a cell. The laser beams are both intensity-modulated with different frequencies. The laser is then tuned. Since the Doppler shifts for both beams are opposite, the probe signal will be modulated at the sum of the two modulation frequencies only when the two lasers interact with the same subclass of atoms, i.e., atoms with no movement relative to the pump and probe beam. Thus, the probe signal measured via a lock-in amplifier will be free of Doppler broadening.

Doppler limit The temperature limit in atom trapping, which was originally considered the limit for laser cooling of atoms. The limit is reached when the natural line width of the cooling transition reaches the Doppler shift associated with the movement of the atom. It is given by

$$kT_{\text{Doppler}} = \hbar\Gamma/2,$$

where k is the Boltzmann constant, $\hbar$ is Planck's constant, and Γ is the line width of the cooling transition. Experiments showed that atoms can be cooled to much lower temperatures, which is due to the internal structure, i.e., Zeeman sublevels, of the atoms. The latter cooling mechanisms are referred to as sysiphus and polarization gradient cooling.

Doppler profile *See* Doppler distribution.

Doppler shift (**1**) When either the source or the receiver is moving with respect to the reference frame in which a wave is traveling, the wavelength (frequency) in that moving frame will change. This is due to the obvious fact that the spacing between wave crests will increase or decrease due to relative motion between the frames, and is known as the *Doppler shift*. Relativistically it is expressed as:

$$\nu = \frac{\nu[1 - \beta \cos(\theta)]}{\sqrt{1 - \beta^2}},$$

where $\beta = v/c$, and θ is the angle between the wave vector and the velocity, v.

(**2**) The shift in the transition frequency of an atom or molecule that occurs when an atom is moving relative to the radiation source. The transition is red-shifted if the atom moves towards the source and blue-shifted if it moves away. The shifted resonance frequency is given by

$$\omega_D = \omega_0 + \vec{k}\vec{v} = \omega_0 \left(1 + \frac{v_z}{c}\right),$$

where ω_0 is the resonance frequency in the angular frequency of the atom, and $\vec{k}$ and $\vec{v}$ are the wave vector of the light and the velocity of the atom respectively. v_z is the atomic velocity component in the direction of light propagation.

Doppler width The broadened line width of a transition caused by the random movement of an ensemble of atoms. The resonance frequency of each atom is shifted due to the Doppler effect by a different amount corresponding to the Doppler shift for its particular velocity. Assuming a Boltzmann distribution for the velocities of the atoms with mass m at temperature T, the *Doppler width* has a value of

$$\delta\nu = 2\frac{\nu_0}{c}\sqrt{2R\ln 2/M} = \frac{2}{\lambda}\sqrt{2R\ln 2/M}$$

where c is the speed of light, R is the general gas constant, and M is the molar mass of the atom. It is apparent that the *Doppler width* is proportional to the transition frequency. Typically, the *Doppler width* is twice that of the natural line width for frequencies in the visible spectrum.

dose A measure of the exposure to nuclear irradiation. It is measured in units of 6.24×10^{12} MeV/kg (1 joule/kg) of deposited energy in the material (gray). The older unit of *dose*, the rad, is 10^{-2} gray. The gray does not include a factor for biological damage which is dependent on the type and energy of the radiation, w_R. Thus, the biological *dose* in sievert is Sv = absorbed *dose* in gray $\times w_R$. *See* gray.

double beta decay A simultaneous change of two neutrons into two protons. For a few nuclei, this may result in a lower mass nucleus, but the original nucleus is stable against single beta decay. There are 58 nuclei, all even–even (neutron number–proton number), which can result in *double beta decay*. As *double beta decay* is a second order weak process, it is extremely rare, and the lifetimes of these isotopes are $\geq 10^{19}$ years. The process is of interest, however, because it is potentially possible for neutrino-less beta decay to occur if the neutrino possesses certain properties. That is, instead of the process

$$_Z X^A \to _{Z+2} X^A + 2e^- + 2\bar{\nu} ;$$

one could have the reaction

$$_Z X^A \to _{Z+2} X^A + 2e^- .$$

This latter process violates lepton conservation, but aside from that, the latter process occurs with much higher probability than the former process. Thus, neutrino-less beta decay is a sensitive test of lepton conservation, and, in particular, of whether the emitted neutrino is a Majorana or a Dirac particle, i.e., whether the neutrino is its own anti-particle.

double escape peak In the interaction of a photon with a nucleus, the creation of electron–positron pairs is possible if the photon, has energy above two electron masses. To determine the energy of the original photon, all the deposited energy must be measured, and this includes the capture of the two annihilation photons of 0.511 MeV each, emitted when a positron at rest captures an electron. If these secondary photons escape the detector, then the measured energy of the photon is reduced by 0.511 or $2 \times 0.511 = 1.022$ Mev, depending on whether one or two photons escape. This produces a full energy peak (no escape), a single escape peak, and a *double escape peak* in the measured energy spectrum.

double resonance spectroscopy A technique often used in atomic and molecular spectroscopy. Molecular spectra usually show spectral congestion, and the multitude of lines makes their assignment difficult. At a high density of states, the lines might even overlap. Using double resonance techniques can greatly reduce this congestion, since the second resonant light provides additional selection. One distinguishes between RF/optical, microwave/optical, and optical/optical double resonance depending on the frequency range used. Other distinguishing features are the arrangement of the energy levels involved, as depicted in the figure. Usually the pump laser is fixed at a particular resonance frequency, while the other laser is tuned.

Double resonance schemes distinguished by the arrangement of the energy levels: λ-type, V-type, and step-wise.

double-slit experiment Classic experiment first performed by Thomas Young in 1801, in which light from a source falls on a screen after passage through an intervening screen with two close-by narrow slits. Under suitable conditions, a pattern of alternating dark and bright fringes (images of the slit) appears on the final screen. This experiment was the first to demonstrate convincingly the wave nature of light. The same experiment may be done (with inessential modifications) with sound, X-rays, electrons, neutrons, or any other particle, as a consequence of de Broglie's principle. *See* diffraction.

doublet A dipole in potential flow consisting of a source and sink of equal strength and infinitesimal separation between them. The streamfunction Ψ and velocity potential ϕ are given by

$$\Psi = -\frac{K \sin \theta}{r}$$

and
$$\phi = -\frac{K \cos\theta}{r}$$
where K is the strength of the *doublet*. In a superimposed uniform flow, a closed streamline is formed around the *doublet*. Doublets can be used in potential flow to simulate the flow past a body such as flow past a cylinder (*doublet in uniform flow*) or flow past a rotating cylinder (*doublet with superimposed vortex in uniform flow*).

down-conversion   A non-linear process in which, due to the non-linear interaction of a pump photon with a medium, two photons of lower energy are generated. It is often referred to as parametric *down-conversion*. Down-conversion is closely related to difference frequency generation. The generated photons are the signal (higher energy) and the idler photons. Energy and momentum have to be fulfilled in the process, i.e.,

$$\omega_p = \omega_s + \omega_i$$
$$\vec{k}_p = \vec{k}_s + \vec{k}_i ,$$

where ω and $\vec{k}$ denote the respective frequencies and wave vectors. The efficiency of the process is larger when the process is collinear, i.e., all wave vectors are either parallel or antiparallel. Generally, the process can take place only in birefringent media, because otherwise the phase-matching condition can not be met. With the exception of processes in periodically poled media, this requires that some of the three involved photons differ in polarization. One must distinguish between type-I and type-II processes. In type-I processes, the idler and signal photons have the same polarization, while for type-II processes they are perpendicular to each other. Parametric *down-conversion* processes are used to build optical parametric oscillators.

Parametric *down-conversion* can be used to produce squeezed light and entangled states between photons.

The Hamiltonian in the rotating wave approximation in the interaction picture is written as

$$H_{\text{int}} = \hbar\kappa \left(a_s^\dagger a_i^\dagger a_p + a_s a_i a_p^\dagger\right) ,$$

where κ is the coupling constant, a_s, a_i, and a_p are the annihilation operators, and $a_s^\dagger$, $a_i^\dagger$, and $a_p^\dagger$ are the creation operators at the respective frequencies. The coupling constant is among others on the second order susceptibility tensor of the non-linear material used in the non-linear process. Often the processes are studied under the parametric approximation, where the pump field is treated classically. Consequently, one also assumes that the pump field is not depleted. In this case, the Hamiltonian is written as:

$$H_{\text{int}} = \hbar\kappa\beta \left(a_s^\dagger a_i^\dagger e^{-\imath\Phi} + a_s a_i e^{\imath\Phi}\right) .$$

down quark   Fundamental hadronic particles are composed of quarks and anti-quarks. In the standard model, the quarks are arranged in three families, the least massive of which contains quarks of up and down types. Nucleons are constructed from a combination of three constituent up and *down quarks* and a sea of quark–antiquark pairs. Thus, a neutron has two *down quarks* and one up quark, while a proton has two up quarks and one *down quark*. The *down quark* has -1/3 of the electronic charge and the up quark has 2/3 of the electronic charge.

downwash   Downward flow behind a wing created as a direct result of the generation of lift. *See* trailing vortex wake.

drag   Resistive force opposed to the direction of motion. *Drag* can be generated by various forces including skin friction and pressure forces. *Drag* is primarily a viscous phenomenon (*see* D'Alembert's paradox) with boundary layers and separation as its primary causes.

drag coefficient   Non-dimensionalized drag force given by

$$C_D = \frac{D}{\frac{1}{2}\rho U_\infty^2 c^2}$$

where c is the chord length of the airfoil. Drag is used in
conjunction with lift to determine the efficiency of the airfoil.

Drell–Yan process   In nucleon–nucleon scattering, the production of lepton pairs with high transverse momentum far from a vector meson resonance is assumed to proceed by quark–antiquark annihilation. This first order process produces a virtual photon which converts into a lepton pair in the final state. Thus, the *Drell–Yan process* provides a mechanism to study the parton distributions in nuclei.

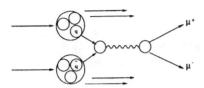

A flow diagram of the Drell–Yan process. The quark–antiquark annihilate to form a muon pair.

dressed atom   Description of an atomic or molecular system interacting with a quantized radiation field in a coupled atomic-field basis. Each energy state in this picture is expressed as an atomic excitation and a specific number of photons associated with it (*see* dressed states).

dressed states   The eigenstates for the Hamiltonian of an atomic or molecular system coupled to a quantized radiation field. The discussion is restricted here to two-level atoms with a ground state $|a\rangle$ and an excited state $|b\rangle$. For this two-level system, the Hamiltonian using the standard annihilation and creation operators can be written in the rotating wave approximation as

$$H = H_A + H_F + H_{AF}$$
$$= \hbar\omega_0|b\rangle\langle b| + \hbar\omega_F\left(c^\dagger c + \frac{1}{2}\right)$$
$$+ \hbar g\left(c^\dagger|a\rangle\langle b| + c|b\rangle\langle a|\right),$$

where $H_A = \hbar\omega_0|b\rangle\langle b|$ is the Hamiltonian of the atom with eigenstates $|b\rangle$ and $|a\rangle$ with energies $\hbar\omega_0$ and 0 respectively. $H_F = \hbar\omega_F(c^\dagger c + \frac{1}{2})$ is the Hamiltonian of the field, where $c^\dagger$ and c are the creation and annihilation operators for a photon with frequency ω_F. The term $H_{AF} = \hbar g(c^\dagger|a\rangle\langle b| + c|b\rangle\langle a|)$ is the interaction between the field and the atom, where g is the coupling constant. Without the coupling term H_{AF}, the eigenstates of the atom-field system are two infinite ladders with $|a, n\rangle$ and $|b, n\rangle$, i.e., states where the atom is in the ground state

and n photons in the field and the atom is in the excited state and n photons in the field. As depicted in the figure, the total energy of these states is given by $n\hbar\omega$ and $\hbar\omega_0 + n\hbar\omega_F$ respectively. The interaction of Hamiltonian couple states with $|a, n\rangle$ and $|b, n - 1\rangle$ leads to new eigenstates, the perturbed states or *dressed states*. The matrix element of this coupling is given as

$$v = \langle b, n - 1|H_{AF}|a, n\rangle = g\sqrt{n} = \hbar\Omega/2$$

Ω is called the Rabi frequency. The *dressed states* have the form

$$|+(n)\rangle = \sin\theta|a, n\rangle + \cos\theta|b, n - 1\rangle$$
$$|-(n)\rangle = \cos\theta|a, n\rangle - \sin\theta|b, n - 1\rangle,$$

where $\tan 2\theta = -\frac{\Omega}{\Delta}$, and $\Delta = \omega_0 - \omega_F$ is the detuning of the photons from the atomic resonance. The energy difference between these states is given by

$$\Delta E = \hbar\Omega' = \sqrt{\Delta^2 + \Omega^2},$$

which means that for the case of weak excitation, ($\Omega \approx 0$) the states go over in the unperturbed states with an energy separation equivalent to the detuning. θ takes on the value $\pi/4$ for the case of no detuning, i.e., $\Delta = 0$. Thus, we find that for the *dressed states:*

$$|+(n)\rangle = \frac{1}{\sqrt{2}}(|a, n\rangle + |b, n - 1\rangle)$$
$$|-(n)\rangle = \frac{1}{\sqrt{2}}(|a, n\rangle - |b, n - 1\rangle),$$

The *dressed state* description is valuable in understanding phenomena such as the Autler–Townes doublet in the emission of dressed three-state atoms and the Mollow spectrum of the emission of a coherently driven two-level atom.

drift chamber   A type of multiwire particle detector which uses the time that it takes an ionization charge to drift to its sense wires to interpolate the position of the track between the wires. A cross-section of a typical *drift chamber* is shown in the figure. Generally, ions drift at a velocity of ≈ 5 cm/μs, so with a typical time resolution of ≈ 1 ns, a position resolution of 100 μm can be obtained.

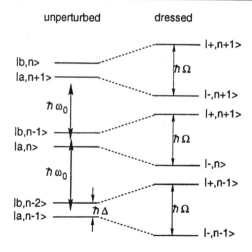

Depiction of the unperturbed and dressed states for an atom-field system.

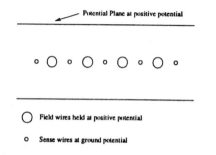

Cross-section of a drift chamber. The drift wires and foils shape the electrostatic field lines along which the ionization charge drifts.

drift motion Charged particles placed in a uniform magnetic field will have orbits that can be described as a helix of constant pitch, where the center axis of the helix is along the magnetic field line. However, if the magnetic field is not uniform, or if there are electrical fields with perpendicular components to the magnetic field, then the guiding centers of the particle orbits will drift (generally perpendicular to the magnetic field).

drift-tube accelerator A linear accelerator that uses radio-frequency electromagnetic fields. The accelerator is composed of conducting tubes separated by spatial gaps. The rf-field is imposed in the gaps between the tubes and is excluded from the interior of the conducting tubes. Thus, the particles drift, field-free, while the rf-potential polarity opposes acceleration, and are accelerated between the gaps during the other half-cycle of the rf-fields.

drift velocity The *drift velocity* of an ionization charge in a typical chamber gas is about 5cm/μs. The addition of an organic quenching gas not only provides operational stability of the wire chamber, but keeps the *drift velocity* of the ionization more or less constant, independent of the applied electric field in the wire chamber. This fortuitous circumstance makes the position vs. drift time function nearly linear in most situations.

drift waves Plasma oscillations arising in the presence of density gradients, such as at the plasma's surface.

duality, wave-particle The observation that quantum mechanical systems can exhibit wave- and particle-like behavior. The *wave-particle duality* is an independent principle of quantum mechanics and not a consequence of Heisenberg's uncertainty principle. The occurrence of wave-like behavior can be understood through the interference of indistinguishable paths of a system from one common initial state to a particular final state. Particle-like behavior occurs when this indistinguishability is destroyed and which-path information becomes available. It can be shown that the relationship

$$D^2 + V^2 < 1$$

exists, where V is the visibility of the interference fringes defined as

$$V = \frac{I_{max} - I_{min}}{I_{max} - I_{min}}$$

and D is a measure of the ability to distinguish between paths. Whether particle or wave nature is observed depends on the type of experiment performed. If the experiment aims at wave properties, those will be observed and particle features likewise.

duct flow *See* pipe flow.

dusty plasma An ionized gas containing small particles of solid matter which become electrically charged. Particles may be dielectric or conducting and typically range in size from nanometers to millimeters. *Dusty plasmas* occur in astrophysics plasmas, plasma processing discharges, and other laboratory plasmas. *Dusty plasmas* are sometimes called complex plasmas and, when strongly-coupled, plasma crystals.

dynamic pressure The pressure of a flow attributed to the flow velocity defined as $\frac{1}{2}\rho U_\infty^2$. *See* Bernoulli's equation and pressure, stagnation.

dynamic similarity When problems of similar geometry but varying dimensions have similar dimensionless solutions. *See* dimensional analysis.

dynamic Stark shift The shift in the atomic energy due to the presence of strong radiation fields. The shift can be explained with the help of the dressed state model. The ground and excited states of a two-level atom can be written as $|g, n\rangle$ and $|e, n\rangle$, where n is the number of photons. In the weak field limit, i.e., $n \approx 0$, we can neglect the photon number. For strong fields, however, the levels $|g, n\rangle$ and $|e, n\rangle$ transform into the dressed states

$$|e, n\rangle \to |+, n+1\rangle = \cos\theta_{n+1}|e, n\rangle$$
$$- \sin\theta_{n+1}|g, n+1\rangle$$
$$|g, n\rangle \to |-, n\rangle = \cos\theta_n|g, n\rangle$$
$$- \sin\theta_n|e, n-1\rangle ,$$

where $\tan 2\theta_i = \frac{\Omega_i}{\Delta}$ with Ω_i is the Rabi frequency and $\Delta = \omega - \omega_0$ is the detuning between the radiation and the atomic resonance transition. This transformation shifts the energy levels of the states $|e, n\rangle$ by $+\delta$ and $|g, n\rangle$ by $-\delta$, known as the *dynamic Stark shift*. It is also referred to as the light shift, since it depends on the Rabi frequency Ω and, hence, on the light intensity. The value of δ is given by

$$\delta = \frac{1}{2}\left(\sqrt{\Delta^2 + \Omega^2} - \Delta\right) .$$

The *dynamic Stark* shift is sometimes also called the AC Stark shift due to its analogy to the Stark shift of atomic levels in DC fields.

Dyson's equations In quantum field theory, formally exact integral equations obeyed by propagators or Green's functions in a system of interacting fields. First obtained by F.J. Dyson in 1949 in the study of *quantum electrodynamics*.

Dyson series Perturbative expansion of any Green's function or correlation function in an interacting quantum field theory as a sum of time-ordered products. First developed by F.J. Dyson in 1949.

dysprosium An element with atomic number (nuclear charge) 66 and atomic weight 162.50. The element has 7 stable isotopes. *Dysprosium* has a large thermal neutron cross-section and is used in combination with other elements in the control rods of nuclear reactors.

E

e   Symbol commonly used for the elementary charge:

$$e = 1.602176462(63) \times 10^{-19} \text{C}.$$

echo, photon   Technique analogous to spin echoes, in which the washing out of Rabi oscillations by inhomogeneous broadening in a vapor of atoms is partially reversed by a suitable pulse at the resonant frequency.

echo, spin   Ingenious technique invented in 1950 by E.L. Hahn, in which the damping of the free induction decay signal in an NMR experiment on a macroscopic sample, which arises from the inhomogeneity of the local magnetic fields experienced by the various nuclei, is reversed. In the simplest form, a so-called π-pulse of radiation at the Larmor frequency of the nuclei is applied, reversing nuclear motion in such a way as to rephase the nuclei after an interval. The echo signal provides valuable information about the interaction of the nuclear spins and by extension, the atoms with their surroundings. Many sophisticated echo protocols now exist, and the resonant echo technique is now a standard tool of analysis in many branches of physics. *See also* photon echo.

Eckert number E_c   A dimensionless parameter that appears in the non-dimensional energy equation. The *Eckert number* is given as the ratio $U^2/c_p \Delta T$, where c_p is the specific heat at constant pressure and ΔT is a characteristic temperature difference. It thus represents the ratio of kinetic to thermal energy. The *Eckert number* is the ratio of the Brinkman number to the Prandtl number. The Brinkman number represents the extent to which viscous heating is important relative to heat flow due to temperature difference. The Prandtl number is the ratio of kinematic viscosity to thermal diffusivity and represents the relative magnitudes of diffusion of momentum and heat in a fluid. For fluids with constant specific heats c_p and c_v, the *Eckert number* is related to the Mach number, Ma, by $Ec = (\gamma - 1)Ma^2$, where γ is the ratio c_p/c_v with c_v representing the specific heat at constant volume.

eddy   A loosely defined entity in a turbulent flow that is usually associated with a recognizable shape, such as a vortex, or a mushroom, and a size such as a wavelength range. *Eddies* do not exist in isolation. Smaller *eddies* usually exist with larger ones. One characteristic of turbulent flows is the continuous distribution of *eddy* sizes. The *eddy* size affects many phenomena, such as diffusion and mixing.

eddy current   Electrical current induced in a conducting material submitted to a varying magnetic field.

eddy viscosity   Turbulent flows are characterized by spatial and temporal fluctuations of the velocity components. These fluctuations are responsible for the exchange of energy and momentum among turbulence scales or eddies. This exchange results in reduction of momentum gradients similar to, yet more effective than, reduction of these gradients by molecular interactions caused by viscosity. By analogy to Newton's law of viscosity, *eddy viscosity* is used to represent the effects of momentum exchange between turbulence scales. The contribution of this exchange to the mean flow is represented by the Reynolds stress tensor written as $\rho \overline{u_i u_j}$. This term appears in the time-averaged equation of motion. Consequently, *eddy viscosity* is used to model turbulence. *Eddy viscosity* models include zero-, one-, and two-equation models. These models work well for non-separating near-parallel shear flows. In order to apply them to other flows, correction terms are usually used. *Eddy viscosity* modeling has been used to solve a variety of problems and is used in commercial fluid software packages as well. Yet, with the advancements in computing capabilities, direct numerical simulation (DNS) and large eddy simulation (LES) are becoming more common methods in numerical studies of turbulent flow fields.

edge dislocation Two-dimensional defect in a solid.

effective charge In many nuclear models, the description of the properties of a many-body quantum-mechanical state may be considered in terms of a single particle moving in some type of potential well created by other particles. However, this single particle may be assigned an *effective* mass and *charge* to better fit the observed experimental data. For example, in single-particle nuclear transitions with the emission of a gamma ray, the remaining nucleons also move about the system center-of-mass. This motion can be taken into account in a simple single-particle model by reducing the charge of this particle.

effective field Electrical field created by an effective charge.

effective mass Individual nucleons in a nucleus can be represented, in many circumstances, as though they possess the same properties as free neutrons or protons. However, there is still a residual interaction between the nucleons, and this residual interaction can, for some applications, be approximated by the insertion of an *effective mass* and charge for this particle. *See* effective charge.

effective range Angular momentum in the scattering of particles (e.g., nucleons) can be ignored if the incident energy is sufficiently low (s-waves). In this situation, information about the scattering potential is contained in the asymptotic scattering wave function, which is basically an outgoing wave, phase-shifted by the scattering potential. The s-wave phase shift can be expanded in powers of 1/kR, where R is the *effective range* and k is the momentum of the particle in units of $\hbar$. For uncharged particles this expression is

$$\text{kcot}(\delta) = -1/a + \left(k^2 R\right)/2 .$$

In this expression, a is the scattering length.

effective range formula Formula of general validity that represents quantum mechanical scattering at low energy in terms of just two parameters, the scattering length, and the effective range. While the former is often not a length characterizing the scattering potential, the latter is, especially if the potential is attractive.

efficiency of an engine (η) The ratio of the work output to the heat input in an engine. For a Carnot cycle, the efficiency η equals $1 - T_c/T_h$, where T_c denotes the temperature of the cold reservoir to which the energy exhausts heat, and T_h is the temperature of the hot reservoir from which the energy extracts heat.

effusion The flow of gas molecules through large holes.

Ehrenfest equation The equation of motion that the quantum mechanical expectation values of operators follow. In the case of the space operator $\hat{x}$ and the momentum operator $\hat{p}$, we find the following *Ehrenfest equations*:

$$\frac{d}{dt} \langle \hat{x}(t) \rangle = \left\langle \frac{\partial H}{\partial p} \right\rangle$$
$$\frac{d}{dt} \langle \hat{p}(t) \rangle = -\left\langle \frac{\partial H}{\partial x} \right\rangle ,$$

where H is the Hamiltonian of the system and $\langle \cdot \rangle$ indicates the expectation value. Those equations are equivalent to the classical equations of motions.

Ehrenfest's theorem States that the quantum mechanical expectation values follow classical equations of motion, the Ehrenfest equations.

eigenfunction *See* eigenvalue problem.

eigenstates Eigenstates of an operator $\hat{A}$ are states $|\Psi\rangle$ that obey the equation

$$\hat{A}|\Psi\rangle = c_i |\Psi\rangle ,$$

where c_i is a complex number. Any quantum mechanical system in a state $|\Xi\rangle$ can be expressed as a superposition of *eigenstates*, i.e.,

$$|\Xi\rangle = \sum_i a |\Psi_i\rangle$$

provided these states $|\Psi_i\rangle$ form a complete basis. The latter condition can be expressed as

$$\sum_i |\Psi_i\rangle\langle\Psi_i| = 1 .$$

eigenvalue problem (1) Very generally, any mathematical problem that has a solution only for a specially chosen value or set of values of some parameter. The special value is known as the eigenvalue. The special case of a linear *eigenvalue problem* is of particular importance. In its simplest form, it may be given as

$$Mx = \lambda x,$$

where M is an $n \times n$ matrix, x is a column vector of length n, and λ is a scalar, i.e., a number. The size n of the matrix M may be finite or infinite. A value of λ for which a solution to this problem exists is an eigenvalue, and the associated vector x is then an eigenvector. More generally, the linear *eigenvalue problem* may be given in terms of differential, integral, or abstract operators on a vector space in lieu of the matrix M, and the solution x may also be called an eigenfunction or mode.

Eigenvalue problems are ubiquitous in all branches of physics, arising for example, in studies of the vibrations of strings and drumheads, the modes of an electromagnetic cavity, wave motion, the onset of thermal instability in a thin layer of fluid heated from below (the Bénard problem), diverse problems of linear stability analysis, etc.

(2) The linear *eigenvalue problem* is of fundamental significance in quantum theory, in which the states of a physical system are postulated to correspond to vectors in an abstract vector space known as the Hilbert space, and all physical observables, such as energy, momentum, position, etc. are postulated to correspond to linear operators on this vector space. In this context, an eigenvector of the energy, e.g., is also known as an energy eigenfunction or energy eigenstate. A general physical state is not associated with a definite value of a physical quantity unless it is an eigenstate of that quantity. Additionally, no state may ever be a simultaneous eigenstate of certain pairs of observables, momentum and position, e.g., on account of Heisenberg's uncertainty principle.

eigenvalues Eigenvalues of an operator $\hat{A}$ are complex numbers that obey the equation

$$\hat{A}|\Psi\rangle = a|\Psi\rangle,$$

where $|\Psi\rangle$ is a state vector. When a measurement of an operator $\hat{A}$ is performed on a system in a state $|\Xi\rangle$ which can be written as a superposition of different state vectors $|\Psi_i\rangle$, which form a complete basis set of eigenvectors with respect to the operator $\hat{A}$, i.e.,

$$\Xi = \sum_i c_i |\Psi_i\rangle,$$

the measurement will result in a collapse of the wavefunction $|\Xi\rangle$ into one of the states $|\Psi_i\rangle$ with a probability $|c_i|^2$, since

$$\langle \Psi_i | \hat{A} | \Xi \rangle = |c_i|^2.$$

eigenvalue spectrum (1) The totality of eigenvalues in an eigenvalue problem. In a linear eigenvalue problem, one speaks of the *eigenvalue spectrum* of the matrix or operator in question.

(2) In quantum mechanics, the *eigenvalue spectrum* of an observable is the set of possible outcomes of measurements of that observable in a given physical setting.

eigenvector Equivalent to eigenstate.

eightfold way In a forerunner to the description of hadronic interactions in terms of quantum chromodynamics (QCD), Gell-Mann proposed that hadronic interactions, which are symmetric with respect to isospin and hypercharge, could be described by the eight-dimensional representation of an SU(3) algebra, the *eightfold way*. Thus, in addition to the three components of isospin and the hypercharge, Y, there would be four other operators forming the eight generators of an SU(3) algebra. This symmetry predicted the correct lowest order spectrum of baryons and mesons. The use of quarks and color symmetry explains more naturally the occurrence of the hadronic spectrum.

eikonal approximation Approximation in which the quantum mechanical evolution of a system is determined in terms of the classical action along the classical trajectories followed by the system. *See also* Wentzel–Kramer–Brillouin (WKB) method.

Einstein A coefficient

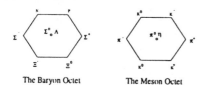

A weight diagram of the baryon and meson octets, which is the lowest representation of the SU(3) group symmetry representing these particles.

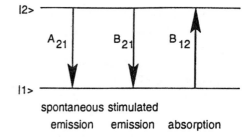

Einstein A and B coefficients.

Einstein A coefficient   Gives the probability for the spontaneous decay of an excited atom or molecule. For different types of transitions, the Einstein coefficient is given by

$$A^\nu = \frac{16\pi^3 \nu^3}{3\varepsilon_0 h c^3 g_2} S_{ed}$$

electric dipole transitions

$$A^\nu = \frac{16\pi^3 \mu_0 \nu^3}{3 h c^3 g_2} S_{md}$$

magnetic dipole transitions

$$A^\nu = \frac{8\pi^5 \nu^5}{5\varepsilon_0 h c^5 g_2} S_{eq}$$

electric quadrupole transitions

where the subscript denotes that A is given in Hz, ν is the frequency of the transition in Hz, ε is the susceptibility of the vacuum, h is Planck's constant, c is the speed of light, and S_{ed}, S_{md}, and S_{eq} are the line strengths for electric dipole, magnetic dipole, and electric quadrupole transitions.

The relationship between the lifetime τ of a state and the Einstein A coefficient is given by

$$A^\nu = \frac{1}{2\pi\tau},$$

where we assume that A is given in terms of ν. If it is given in terms of the radial frequency ω, we find

$$A^\omega = \frac{1}{\tau}.$$

Finally the Rabi frequency Ω can be calculated using the Einstein A coefficient:

$$|\Omega|^2 = \frac{\lambda^3}{4\pi^2 \hbar c} g_2 I A,$$

where g_2 is the degeneracy factor of the upper level, λ is the wavelength of the laser, and I the intensity in W/m^2.

In case a level can decay to several states, the Einstein A coefficient is given by the sum of the individual Einstein A coefficients A_i of the decays to the individual levels.

Einstein, Albert   Nobel Prize winner in 1905 for explaining the photoelectric effect. He is better known for his theories of special and general relativity. The general theory of relativity was the first fully developed field theory which provided the intellectual stimulation for modern theoretical physics.

Einstein B coefficient   Coefficient for absorption or stimulated emission of a photon from a level 1 to a level 2. If level 2 is higher in energy than level 1, the coefficient of stimulated emission B_{21} and stimulated absorption B_{12} are given by

$$B_{12} = \frac{g_2}{g_1} B_{21}$$

$$A_{21} = \frac{8\pi h \nu^3}{c^3} B_{21},$$

where A_{21} is the Einstein A coefficient for the transition from $2 \to 1$, and g_2 and g_1 are the statistical weights or degeneracy factors for level 2 and 1, respectively. ν is the transition frequency.

The Einstein B coefficient can also be expressed in terms of the oscillator strength f of the transition:

$$B_{21} = \frac{g_1}{g_2} \frac{e^2}{4m\varepsilon_0 h \nu} f,$$

where ε is the dielectric constant for the vacuum.

Einstein equation   Equation announced by Einstein in the form $E = mc^2$, where E is energy, m is mass, and c is the speed of light.

Einstein–Podolsky–Rosen experiment Introduced as a gedanken experiment by Einstein, Podolsky, and Rosen in 1935. The authors wanted to illustrate the incompleteness of quantum mechanics.

The history of the *Einstein–Podolsky–Rosen experiment* dates back to the early years of quantum mechanics. Despite its successes in predicting the outcome of experiments, many felt that quantum mechanics was an unsatisfactory theory due to its counterintuitive nature, i.e., action-at-a-distance. Among the most prominent critics were Einstein, Podolsky, and Rosen, who expressed their concerns in an article entitled "Can Quantum Mechanical Description of Physical Reality Be Considered Complete?" published in 1935 in *Phys. Rev*. They introduced the EPR gedanken experiment to demonstrate their belief that quantum mechanics was incomplete. Crucial to their discussion was the concept of entanglement between particles, since entangled states seemingly allow action-at-a-distance. The gedanken experiment involved the generation of a two particle system in an entangled state, separation of the constituents, and measurement of the correlations between the entangled quantities. The original gedanken experiment focused on an entanglement in space and momentum. The most common referenced version was introduced by D. Bohm and is based on an entanglement between two spin 1/2 particles.

Einstein, Podolsky, and Rosen left open the question of whether a complete description of reality was possible. Later such complete theories, which are classical in nature, were called local hidden variable (LHV) theories. Hidden variables were supposed to be the origin of the observed correlations, resolving the "spooky" action-at-a-distance.

For a long time, the discussions were only philosophical in nature. This changed in 1964, when J.S. Bell realized that LHV theories were at least possible. This contrasts with von Neumann's proof that it was not possible to construct LHV theories, which reproduced all the quantum mechanical predictions. Bell showed that von Neumann had been much too restrictive in his expectations for LHV theories. In addition, Bell showed that the statistical predictions of these theories showed correlations which were limited by an inequality. However, it was possible to find cases in which the statistical predictions of quantum mechanics violate this inequality. So for the first time, it was possible, at least in principle, to distinguish between quantum mechanics and its classical counterparts — the local hidden variable theories.

A test of a Bell inequality involves the measurements of correlations in an entangled state, i.e., polarization or spin components with respect to different axes. Of course, there have been tests of the Bell inequalities before. Among the more prominent tests are cascade decay and down-conversion experiments. In the former, the entanglement between a photon pair is produced by a consecutive cascade decay in an atom. In the latter, the entanglement between the polarization of two photons is generated by a non-linear process.

However, all previous tests of a Bell inequality have loopholes. Specifically, these are the detection efficiency loophole and the locality loophole. Due to low detection efficiencies in previous photon-based experiments, additional assumptions had to be introduced in order to derive a testable Bell inequality. Thus, the resulting experiments test much weaker forms of the Bell inequalities. In order to enforce the locality condition, the detector for one particle should not know the measurement orientation of the other detector. This means that within the time of the analysis and detection step at one detector, no information about its particular direction can reach the other detector. The enforcement of this condition not only requires large detector distances, but also rapid, randomized switching of the measurement directions. This was not the case in the only previous experimental attempt to enforce the locality condition.

einsteinum A transuranic element with atomic number (nuclear charge) 99. Twenty isotopes have been produced, with atomic number 252 having the longest half-life at 472 days.

EIT See electromagnetically induced transparency.

Ekman layer A boundary layer affected by rotation. *Ekman layers* develop in geophysical situations under the action of the Coriolis force.

Outside the earth's boundary layer, the flow is approximately horizontally homogeneous. In addition, the shear stresses are negligible. Consequently, the Coriolis force balances the pressure forces, i.e.,

$$f_c U_g = -\frac{1}{\rho}\frac{\partial P}{\partial y}$$

and

$$f_c V_g = \frac{1}{\rho}\frac{\partial P}{\partial x}$$

where U_g and V_g are the x and y components of the geostrophic wind (wind outside the earth's boundary layer). The parameter f_c is the Coriolis parameter and is equal to $2\omega\sin\phi$, where $\omega = 2\pi/24$ hrs $= 7.27 \times 10^{-5} s^{-1}$ and ϕ is the latitude. Based on the balance between Coriolis and pressure forces, the geostrophic wind is parallel to the isobars. Inside the earth's boundary layer, or in the *Ekman layer,* shear stresses must be considered in the equation of motion to yield

$$f_c U = -\frac{1}{\rho}\frac{\partial P}{\partial y} - \frac{\partial \overline{v'w'}}{\partial z} = f_c U_g - \frac{\partial \overline{v'w'}}{\partial z}$$

or

$$f_c(U_g - U) - -\frac{\partial \overline{v'w'}}{\partial z} = 0$$

and

$$-f_c V = -\frac{1}{\rho}\frac{\partial P}{\partial x} - \frac{\partial \overline{u'w'}}{\partial z} = -f_c V_g - \frac{\partial \overline{u'w'}}{\partial z}$$

or

$$f_c(V_g - V) + \frac{\partial \overline{u'w'}}{\partial z} = 0.$$

Thus, in the boundary layer, the balance is between pressure, Coriolis, and friction forces. The pressure forces retain the same direction and magnitude as in the outer layer. The friction force is in the opposite direction of the velocity. Because the sum of the Coriolis and friction forces must balance the pressure force, the velocity vector must change directions. In the Northern Hemisphere (where f_c is positive), the velocity vector is rotated to the left of the geostrophic wind vector. *Ekman layers* also exist in oceans, but with different boundary conditions than the atmosphere.

Ekman number   A dimensionless parameter that represents the relative importance of the viscous forces associated with fluid motions and the Coriolis force. It is written as $E = \nu/\Omega L^2$, where ν is the kinematic viscosity, Ω is the angular velocity, and L is a characteristic length. The *Ekman number* is equal to the ratio of the Rossby number (Ro $= U/\Omega L$), which is a measure of relative importance of fluid acceleration to the Coriolis acceleration, to the Reynolds number Re $= UL/\nu$, which is a measure of the relative importance of inertia to viscous forces.

elastic collision   A collision between two or more bodies in which the internal state of the bodies is left unchanged, i.e., energy is not converted to or from heat or any other internal degree of freedom.

elastic constants   Hooke's law states that for sufficiently small deformations, the strain is directly proportional to the stress, so that the strain components are linear functions of the stress components. The coefficient for strain components in each direction are called elastic compliance constants or *elastic constants.*

elasticity   The property of a material of returning of its original dimensions after a deforming stress has been removed. A material subjected to a stress produces a strain. The limit up to which stress is proportional to strain (which is called Hooke's law) is called the elastic limit. Beyond the elastic limit, the material will not return to its original condition (i.e., the stress is no longer proportional to the strain) and permanent deformation occurs.

elastic light scattering   The scattering of light in which the frequency of the scattered light is not changed. Examples of this type of process are Rayleigh scattering or resonance fluorescence.

elastic limit   The minimum stress that produces permanent change in a body.

elastic modulus   The ratio of elastic stress to elastic strain on a body. There are three types of elastic moduli depending on the types of stress applied. Young modulus refers to tensile stress, bulk modulus refers to overall pressure on the

body, and rigidity modulus refers to a shearing stress.

elastic scattering If the particles in a scattering process are the same in the final state as in the initial state, the scattering process is elastic. In an elastic process, no energy is lost in internal excitations of the colliding particles.

elastic tensor Tensor whose elements are the elastic constants.

elbow meter *See* flow meters.

electret A material containing permanent electric dipoles.

electrical conductivity The ability of a material to conduct electrical current, as measured by the current per unit of applied voltage. In a semiconductor, N-type conductivity is associated with conduction electrons, and P-type conductivity is associated with conducting holes.

electric charge The elementary charge is quantized as 1.6×10^{-19} couloumb and is the same for both positive and negative charges. This quantization remains unexplained, although it is hypothesized to be connected with the presence of magnetic monopoles.

electric current density The electrical current per unit of cross-sectional area of a conductor.

electric dipole A charge distribution which has an electric dipole moment, EDM. Although macroscopic distributions can have *electric dipole* moments, there is no experimental evidence of an *electric dipole* moment of a fundamental particle. If this were evident, then this particle would be asymmetric with respect to both parity and time reversal. Although CP violation has been observed, and consequently T violation if CPT is a good symmetry, the standard model predicts all EDMs to be much smaller than present experimental limits. Various extensions to the standard model do predict EDMs within the range of present experimental technology, and so searches for them continue. The present limit on the EDM of the neutron is $d_n \leq 10^{-25}$ e cm. *See* dipole moment.

electric dipole approximation *See* dipole approximation.

electric dipole matrix element For a one electron system this is given by:

$$d_{ij} = \langle \Psi_i | e\hat{r} | \Psi_j \rangle = e \int \Psi_i^* r \Psi_j \, dr \ .$$

For the case $i = j$, the *electric dipole matrix element* is called the electric dipole moment. For $i \neq j$, it is the electric dipole transition moment. In this case, the absolute square of the dipole matrix element is proportional to the transition probability for the electric dipole transition between states $|\Psi_i\rangle$ and $|\Psi_j\rangle$.

electric dipole moment *See* dipole moment.

electric dipole operator *See* dipole operator.

electric dipole-allowed transition A transition between states $|\Psi_1\rangle$ and $|\Psi_2\rangle$ which is electrically dipole-allowed, i.e., whose dipole matrix element,

$$\left| \langle \Psi_1 | e\hat{r} | \Psi_2 \rangle \right|^2 = \left| e \int \Psi_1 r \Psi_2 \, dr \right|^2 > 0 \ .$$

Selection rules for such a transition are given by

$$\Delta J = 0, \pm 1 \quad \text{but not} \quad J = 0 \to 0$$
$$\Delta L = 0, \pm 1 \quad \text{but not} \quad L = 0 \to 0 \ .$$

electric field A three-dimensional function (one function for each of the three coordinate directions) assigned to each point in space, such that if a charge q were placed at that point, the force on that charge would be determined by multiplying q by the value of the functions at that spatial point. Thus,

$$F_i = qE_i \ ;$$

where $i = 1, 2, 3$.

electric field operator For a standing wave with wave vector $\vec{k}$, this is given by

$$\hat{E}(z,t) = \frac{1}{\sqrt{2}} \mathcal{E}_0 \sin(kx) \left[a(t) + a^\dagger(t) \right] \ ,$$

where $a(t)$ and $a^\dagger(t)$ are the annihilation and creation operators, and $\mathcal{E}_0$ is given by

$$\mathcal{E}_0 = \sqrt{8\pi\hbar\omega/V}$$

where ω is the radial frequency of the wave and V is the mode volume.

electric multipole radiation Radiation that originates from higher order terms in the series expansion in the interaction of atoms with electromagnetic radiation. Usually, electric dipole transitions are dominant, which represents the zeroth order term in the above-mentioned expansion. If the dipole-allowed transitions are forbidden, however, *electric multipole radiation* might become possible. Most important is the electric quadrupole radiation.

electric quadrupole As described in the definition of the electric dipole moment, the potential of a charge distribution can expand in a series in increasing powers of (a/R), where a is the size of the charge distribution and R is the distance to the potential point. This defines the multipole moments of the distribution. In increasing order, the moments are the monopole, dipole, and quadrupole distributions. An example of a quadrupole distribution is two positive and two negative charges, all of the same magnitude, each placed the same distance from the origin of the coordinate system, as shown in the figure. See dipole moment.

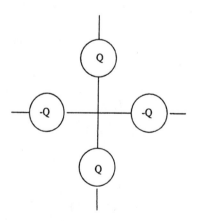

A schematic example of a quadrupole distribution of discrete charges.

electric quadrupole transition Sometimes referred to as E2 transitions. Transitions that originate from the second order terms in the series expansion of the interaction between atoms and an electromagnetic wave (see dipole approximation). These transitions occur due to linear spatial variations of the electric field across the atom. The defining quantity is the quadrupole tensor Q with elements

$$Q_{ij} = 3\int d^3r \varrho(\vec{r}) r_i r_j$$

where ϱ is the charge density.

The Einstein A coefficient in terms of the frequency ν for *electric quadrupole transitions* is given by

$$A^\nu = \frac{8\pi^5 \nu^5}{5\varepsilon_0 hc^5} S_{eq}$$

where ν is the frequency of the transition in Hz, ε is the susceptibility of the vacuum, h is Planck's constant, c is the speed of light, and S_{eq} is the line strength for the *electric quadrupole transition*.

The selection rules are given by

$$\Delta L = \pm 2$$
$$\Delta m = 0, \pm 1, \pm 2 \ .$$

electromagnetically induced transparency (EIT) A coherent effect in three or more level atoms. In a three level system, a coherent drive laser which is resonant on the transition $|a\rangle - |b\rangle$ produces a coherence between those levels so that a light field resonant on the transition $|b\rangle \to |c\rangle$ is not absorbed. One can understand this effect in the dressed atom picture. Under the influence of the drive laser, the atom is optically pumped into a state whose transition matrix element to the state $|c\rangle$ vanishes, the so-called dark state. *Electromagnetically induced transparency* is an important step in the process of lasing without inversion and wave mixing.

electromagnetic radiation Energy that is transmitted by traveling electric and magnetic fields which point in directions perpendicular to the propagation direction. The energy travels with velocity $1/\sqrt{(\mu\epsilon)}$, where ϵ is the dielectric constant and μ is the magnetic permittivity.

electromagnetic wave, or plasma electromagnetic wave (1) One of three categories of plasma waves: electromagnetic, electrostatic, and hydrodynamic (magnetohydrodynamic). Wawe motions, i.e., plasma oscillations, are inherent to plasmas due to the ion/electron species, electric/magnetic forces, pressure gradients, and gas-like properties that can lead to shock waves.

(2) Transverse waves characterized by oscillating electric and magnetic fields with two possible oscillation directions called polarizations. Their behavior can be described classically via a wave equation derived from Maxwell's equations and also quantum mechanically. For the latter picture, the waves are replaced by particles, the photons. The frequency ν and the wavelength λ of an *electromagnetic* wave obey the relationship

$$c = \lambda \nu ,$$

where c is the speed of light. Depending on the frequency and wavelength of the waves, one can divide the electromagnetic spectrum into different parts.

Name	Frequency / THz	Wavelength / nm
FM,AM radio, television	10^{-7}-10^{-3}	$3 \times 10^{12} - 3 \times 10^8$
Microwaves	10^{-3}- 0.3	$3 \times 10^8 - 10^6$
Far-infrared	0.3-6	$10^6 - 5 \times 10^4$
Mid-infrared	6 - 100	$5 \times 10^4 - 3000$
Near-infrared	100 - 385	3000-780
Visible light	385 - 790	780-380
Ultraviolet light	790 - 1500	380-200
Vacuum ultraviolet light	1500 - 3000	200-10
X-rays	3000 - 3×10^7	10-1
Gamma rays	3×10^7-3×10^9	1-10^{-1}

Within the visible light region, the human eye sees the different spectral colors at approximately the following wavelengths:

Color	Wavelength / nm
red	630
orange	610
yellow	580
green	532
blue	480

electron A fundamental particle which has a negative electronic charge, a spin of 1/2, and undergoes the electroweak interaction. It, along with its neutrino, are the leptons in the first family of the standard model.

electron affinity The decrease in energy when an electron is added to a neutral atom to form a negative ion. Second, third, and higher affinities are similarly defined as the additional decreases in energy upon the addition of successively more electrons.

electron capture Atomic electrons can weakly interact with protons in a nucleus to produce a neutron and an electron neutrino. The reaction is;

$$p + e^- \to n + \nu .$$

This reaction competes with the beta decay of a nuclear proton where a positron in addition to the neutron and neutrino are emitted.

electron configuration The arrangement of electrons in shells in an atomic energy state, often the ground state. Thus, the *electron configuration* of nitrogen in its ground state is written as $1s^2\, 2s^2\, 2p^3$, indicating that there are two electrons each in the $1s$ and $2s$ shells, and three in the $2p$ shell. *See also* electron shell.

electron cyclotron discharge cleaning Using relatively low power microwaves (at the electron cyclotron frequency) to create a weakly ionized, essentially unconfined hydrogen plasma in the plasma vacuum chamber. The ions react with impurities on the walls of the vacuum chamber and help remove the impurities from the chamber.

electron cyclotron emission Radio-frequency electromagnetic waves radiated by electrons as they orbit magnetic field lines.

electron cyclotron frequency Number of times per second that an electron orbits a magnetic field line. The frequency is completely determined by the strength of the field and the electron's charge-to-mass ratio.

electron cyclotron heating Heating of plasma at the electron cyclotron frequency. The electric field of the wave, matched to the gyrating orbits of the plasma electrons, looks like a static electric field, and thus causes a large acceleration. While accelerating, the electrons collide with other electrons and ions, which results in heating.

electron cyclotron wave Plasma waves at the electron cyclotron frequency.

electron diffraction Diffraction of electrons from matter, often crystalline. Commonly used as an analytical tool to study the structure of the diffracting matter. *See* de Broglie waves; Davisson–Germer experiment.

electronegativity A measure of the tendency of an atom to gain an electron in the formation of a chemical bond, often taken as the mean of its ionization potential and electron affinity.

electron gas The approximation describing the properties of free electrons as gas particles.

electron g-factor Additional dimensionless factor in the ratio of the magnetic moment of the electron to its spin angular momentum, over and above the value of this ratio for a classical charged spinning particle. Denoted by g. Similar g-factors can be defined for muons, protons, and many other particles. The *electron g-factor* is very close to two, and the deficit $g - 2$ is one of the most accurately measured and calculated quantities in modern physics, providing very high-precision tests of the validity of quantum electrodynamics.

electron-hole pair Excitation in metals and semiconductors, wherein an electron is excited from an occupied state below the Fermi energy to an unoccupied state above this level, leaving behind a hole.

electron-hole symmetry In atomic physics, an approximate equivalence between the energy spectrum of an atom containing n electrons in a shell, and an atom lacking n electrons (or, alternatively, containing n holes) in a shell. For example, the ratio of level separations in carbon with two electrons in the $2p$ shell is very close to the ratio in oxygen, which has four electrons (or two holes) in the $2p$ shell. A similar equivalence exists in metals between filled electron states above the *Fermi energy* and empty states below this energy.

electron shell Fundamental concept in atomic physics, according to which electrons are arranged in shells around the nucleus. A shell is characterized by a principle quantum number n, which specifies the energy on a gross scale, and an angular momentum quantum number l. Both n and l are integers, with $n \geq 1$, and $0 \leq l \leq n-1$. The values $l = 0, 1, 2, 3, \ldots$ are often indicated by letters s, p, d, f, etc. Thus, a shell with $n = 3$ and $l = 2$ is also referred to as the $3d$ shell.

A shell with numbers (n, l) can contain at most $2(2l + 1)$ electrons by virtue of Pauli's exclusion principle. This fact is responsible for the observed periodic variation of chemical properties of elements with increasing atomic numbers, and is reflected in the organization of the modern periodic table of elements. Elements with the same electron configuration in their outermost shell, i.e., differing in n but having equal numbers of electrons of the same l in their outermost shells, usually belong to the same group in this table. For example, carbon and silicon (configurations $2s^2 2p^2$ and $3s^2 3p^2$, respectively) both belong to group 4A, while copper, silver, and gold (configurations $3d^{10} 4s^1$, $4d^{10} 5s^1$, and $5d^{10} 6s^1$ respectively), all belong to group 1B.

electron volt A unit of energy equal to that obtained by a particle having one electronic charge accelerated from rest through a potential difference of 1 volt.

electroproduction A scattering process of fundamental particles in which other particles are produced via the exchange of virtual photons, as opposed to the scattering or absorption of real photons which is called photoproduction.

electrostatic confinement An approach to plasma confinement based on confining charged particles by means of electric fields rather than the magnetic fields used in magnetic confinement.

electrostatics The study of geometric fields, forces, and energies of static charge (time independent) distributions.

electrostatic wave, or plasma electrostatic wave One of three categories of plasma waves: electromagnetic, electrostatic, and hydrodynamic (magnetohydrodynamic). Wave motions, i.e., plasma oscillations, are inherent to plasmas due to the ion/electron species, electric/magnetic forces, pressure gradients, and gas-like properties that lead to shock waves. *Electrostatic waves* are longitudinal oscillations appearing in plasma due to a local perturbation of electric neutrality. For a cold, unmagnetized plasma, the frequency of electrostatic waves is at the plasma frequency.

electroweak theory The Nobel Prize was awarded to Glashow, Salam, and Weinberg in 1979 for their development of a unified theory of the weak and electromagnetic interactions. The field quanta of the *electroweak theory* are photons and three massive bosons, $W^{\pm}$ and Z^0. These interact with the quarks and leptons in a way that produces either weak or electromagnetic interaction. The theory is based on gauge fields which require massless particles. In order to explain how the bosons become massive while the photon remains massless, the introduction of another particle, the Higgs boson, is required.

element An atom of specific nuclear charge (i.e., has a given number of protons although the number of neutrons may vary). An *element* cannot be further separated by chemical means.

elementary excitation The concept, especially advanced by L.D. Landau in the 1940s, that low energy excited states of a macroscopic body, or an assembly of many interacting particles, may be understood in terms of a collection of particle-like excitations, also called quasiparticles, which do not interact with one another in the first approximation, and which possess definite single-particle properties such as energy, momentum, charge, and spin. In addition, *elementary* excitations may be distributed in energy in accordance with Bose–Einstein or Fermi–Dirac statistics, depending on the nature of the underlying system and the excitations in question. The concept proves of great value in understanding a diverse variety of matter: Fermi liquids such as ^{3}He, superfluids, superconductors, normal metals, magnets, etc.

elementary particles At one level of definition, fundamental building blocks of nature, such as electrons and protons, of which all matter is comprised. More currently, however, the concept is understood to depend on the magnitude of the energy transfers involved in any given physical setting. In matter irradiated by visible light at ordinary temperatures, for example, the protons and neutrons may be regarded as inviolate entities with definite mass, charge, and spin. In collisions at energies of around 1 GeV, however, protons and neutrons are clearly seen to have internal structure and are better viewed as composite entities. At present, the only particles which have been detected and for which there is no evidence of internal structure are the leptons (electron, muon, and taon), their respective neutrinos, quarks, photons, W and Z bosons, gluons, and the antiparticles of all of these particles.

Elitzur's theorem The assertion that in a lattice gauge theory with only local interactions, local gauge invariance may not be spontaneously broken.

Ellis–Jaffe sum rule Sum rules are essentially the moments of the parton distribution functions with respect to the Feynman variable, x. For example, the first moment of the spin dependent parton distribution function, g_1, is defined as

$$\Gamma_1^{p,n}(Q^2) \equiv \int_0^1 g_1^{p,n}(x, q^2)\, dx\,,$$

and if there is no polarization of the nucleon's strange quark sea, then Γ_1 may be evaluated to be ≈ 0.185. This is the *Ellis–Jaffe sum rule*. Experimentally, the first moment of g_1 is found to be substantially larger that this value, and this result is referred to a spin-crisis, since on face value, the nucleon's spin is not carried by the valence quarks, and a sizeable negative polarization of the strange sea is required to explain the experimental result. *See* form factor.

emission The release of energy by an atomic or molecular system in the form of electro-magnetic radiation. When the energy in a system and the photons emitted have the same energy, one speaks of resonance fluorescence. Phosphorescence is the emission to electronic states with different multiplicities. These can occur due to spin–orbit coupling in heavy atoms or the breakdown of the Born–Oppenheimer approximation in molecules. Non-radiative processes, i.e., decays of atomic levels that are not giving off radiation, are the competing mechanisms. These can be ionization (atoms and molecules), dissociation (molecules), and thermalization over a large number of degrees of freedom (molecules). The understanding of radiative and non-radiative decays and their origin in molecules is investigated in molecular dynamics.

emission, induced and spontaneous Processes by which an atom or molecule emits light while making a transition from a state of higher energy to one of lower energy. The rate for *induced emission* is proportional to the number of photons already present, while that for *spontaneous emission* is not. The total rate of emission is the sum of these two terms. *See also* Einstein A coefficient; Einstein B coefficient.

emission spectrum The frequency spectrum of the radiation which is emitted by atoms or molecules. In atoms, most frequently the emission spectrum contains only sharp lines, whereas in the case of molecules, due to the higher density of states, emission spectra can have a large number of lines and even a continuous structure. In atoms, the strength of the emitted lines is given by the electronic transition moments. In molecules, other factors, like Franck–Condon factors or Hoenl–London factors, also come into play.

end cap trap A special form of the Paul trap for atomic and molecular ions. Its advantages are its smaller size and the much higher accessibility of the trap region due to much smaller electrode sizes.

endothermic reaction That requires energy in order for the reaction to occur. In particle physics, the total incident particle masses are less than the final particle masses for an endothermic reaction.

energy band The energy levels that an electron can occupy in a solid. *See* band theory.

energy confinement time In a plasma confinement device, the energy loss time (or the *energy confinement time*) is the length of time that the confinement system's energy is degraded to its surroundings by one e-folding. *See also* confinement time.

energy conservation Fundamental physical principle stating that the total amount of energy in the universe is a constant that cannot change with time or in any physical process. The principle is intimately connected to the empirical fact of the homogeneity of time, i.e., the fact that an experiment conducted under certain conditions at one time will yield identical results if conducted under the same conditions at a later time. Other restatements of the principle are that energy cannot be created or destroyed, only transformed from one form to another, and the first law of thermodynamics.

energy density The measure of energy per unit of volume.

energy eigenstate In quantum mechanics, a state with a definite value of the energy; an eigenstate of the Hamiltonian operator. For a closed system, the physical properties of a system in an *energy eigenstate* do not change with time. Hence, such states are also called stationary states.

energy eigenvalue The value of the energy of a system in an energy eigenstate.

energy equation Describes energy interconversions that take place in a fluid. It is based on the first law of thermodynamics with consideration only of energy added by heat and work done on surroundings. In general, other forms of energy such as nuclear, chemical, radioactive, and electromagnetic are not included in fluid mechanics problems. The *energy equation* is actually the first law of thermodynamics expressed for an open system using Reynolds' transport theorem. The result can then be expressed as

$$\left\{\begin{array}{c}\text{rate of}\\\text{accumulation}\\\text{of internal}\\\text{and kinetic}\\\text{energy}\end{array}\right\} = \left\{\begin{array}{c}\text{rate of}\\\text{internal}\\\text{and kinetic}\\\text{energy in}\\\text{by convention}\end{array}\right\}$$

$$- \left\{\begin{array}{c}\text{rate of}\\\text{internal}\\\text{and kinetic}\\\text{energy out}\\\text{by convention}\end{array}\right\} + \left\{\begin{array}{c}\text{net rate}\\\text{of heat}\\\text{addition by}\\\text{conduction}\end{array}\right\}$$

$$- \left\{\begin{array}{c}\text{net rate}\\\text{of work}\\\text{done by}\\\text{surface and}\\\text{body forces}\end{array}\right\}.$$

The specific energy (energy per unit of mass) is usually considered instead of energy when writing the *energy equation*. The kinetic energy, $1/2\rho v^2$ on a per-unit-volume basis, is the energy associated with the observable fluid motion. Internal energy, means the energy associated with the random translational and internal motions of the molecules and their interactions. Note that the internal energy is thus dependent on the local temperature and density. The gravitational potential energy is included in the work term. The work term also includes work of surface forces, i.e., pressure and viscous stresses. Note that the rate of work done by surface forces can result from a velocity multiplied by a force imbalance, which contributes to the kinetic energy. It can also result from a force multiplied by rate of deformation, which contributes to the internal energy. In this case, the pressure contribution is reversible. On the other hand, the contribution by viscous stresses is irreversible and is usually referred to as viscous dissipation.

The total *energy equation* is written in index notation as

$$\frac{\partial}{\partial t}\left[\rho\left(e+\frac{1}{2}v^2\right)\right] + \partial_i\left[\rho v_i\left(e+\frac{1}{2}v^2\right)\right]$$
$$= -\partial_i q_i + \partial_i \tau_{ij} v_j - \partial_i(pv_i) + \rho v_i F_i.$$

Because the equation governing the kinetic energy can be derived independently from the momentum equation, the above equation can be divided into two equations, namely the kinetic and thermal *energy equations*. Kinetic energy is written as

$$\frac{\partial}{\partial t}\left[\rho\left(\frac{1}{2}v^2\right)\right] + \partial_i\left[\rho v_i\left(\frac{1}{2}v^2\right)\right]$$
$$= -v_i\partial_i p + v_i\partial_j \tau_{ji} + \rho v_i F_i.$$

and thermal energy is written as

$$\frac{\partial}{\partial t}(\rho e) + \partial_i(\rho v_i e) = -p\partial_i v_i + \tau_{ji}\partial_j v_i - \partial_i q_i.$$

To apply the above equations to a system one can either integrate the differential equations or consider an energy balance for the whole system.

In considering an energy balance for the whole system, one can write

$$\left\{\begin{array}{c}\text{rate of}\\\text{accumulation}\\\text{of internal,}\\\text{kinetic, and}\\\text{potential energy}\end{array}\right\} = \left\{\begin{array}{c}\text{rate of internal}\\\text{kinetic and}\\\text{potential energy}\\\text{in by convention}\end{array}\right\}$$

$$- \left\{\begin{array}{c}\text{rate of internal}\\\text{kinetic and}\\\text{potential energy out}\\\text{by convention}\end{array}\right\}$$

$$+ \left\{\begin{array}{c}\text{net rate of}\\\text{heat addition}\\\text{to system}\end{array}\right\} - \left\{\begin{array}{c}\text{net rate of}\\\text{work done}\\\text{by system}\end{array}\right\}.$$

By considering the rate of work done by the pressure with the surface terms, i.e., in and out by convection, the above equation can be rewritten as

$$\frac{\partial}{\partial t}\int_{c.\forall}\left(e+\frac{v^2}{2}+gz\right)\rho d\forall$$
$$\int_{c.s.}\left(e+\frac{v^2}{2}+gz+\frac{p}{\rho}\right)\rho\vec{V}\cdot\vec{n}dS$$
$$\dot{Q}_{\text{net in}} + \dot{W}_{\text{net in}}.$$

For one-dimensional, steady-in-the-mean flow conditions, one obtains

$$\dot{m}\left[e_{\text{out}} + \frac{p_{\text{out}}}{\rho} + \frac{v_{\text{out}}^2}{2} + gz_{\text{out}}\right.$$
$$\left. -e_{\text{in}} - \frac{p_{\text{in}}}{\rho} - \frac{v_{\text{in}}^2}{2} - gz_{\text{in}}\right] = \dot{Q}_{\text{net in}} + \dot{W}_{\text{net in}}.$$

For steady, incompressible flow with friction, the change in internal energy $\dot{m}(e_{out} - e_{in})$ and $Q_{net\ in}$ are combined as a loss term. Dividing by $\dot{m}$ on both sides and rearranging the terms, one obtains

$$\frac{P_{out}}{\rho} + \frac{v_{out}^2}{2} + gz_{out} = \frac{P_{in}}{\rho} + \frac{v_{in}^2}{2} + gz_{in} - loss + \dot{w}_{net\ in}.$$

This is one form of the *energy equation* for steady-in-the-mean flow that is often used for incompressible flow problems with friction and shaft work. It is also called the mechanical *energy equation*.

energy fluctuations The total energy of a system in equilibrium at constant temperature T fluctuates about an average value $<E>$, with a mean square fluctuation proportional to C_v and the specific heat at constant volume, $<(E- <E>)^2> = k_B T^2 C_v$.

energy gap The energy range between the bottom of the conduction band and the top of the valence band in a solid.

energy level The discrete eigenstates of the Hamiltonian of an atomic or molecular system. In more complex systems or for states with a high energy, the energy levels can overlap due to their individual natural line width such that a continuum is formed. In solid state materials, this can lead to the formation of energy bands.

energy level diagram A diagram showing the allowed energies in a single- or many-particle quantum system. So called because the energies are usually depicted by horizontal lines, with higher energies shown vertically above lower ones.

energy loss When a charged particle traverses material, it ionizes this material by the collision and knock-out of atomic electrons. These collisions absorb energy from the traversing particle causing an *energy loss*. The *energy loss* can be calculated using the Bethe–Bloch equation.

energy–momentum conservation The conservation of both energy and momentum in a physical process. The term is especially used in this form in contexts where special relativistic considerations are important. *See* energy conservation, momentum conservation.

energy shift A perturbation of the atomic or molecular structure which manifests itself in a shift of the energy levels. These shifts arise due to external fields or the interaction of other close-by energy levels. Examples of the former are Zeeman and Stark shifts due to external magnetic or electric fields. Other shifts can be induced by electro-magnetic radiation (*see* dynamic Stark shift).

energy spectrum The set of energy eigenstates of a physical system. The set of possible outcomes of a measurement of the energy; also known as the set of allowed energies.

energy-time uncertainty principle An equivalent form of the Heisenberg uncertainty principle which is written as

$$\Delta E \Delta t \geq h/2\pi,$$

where h is Planck's constant, and several complementary interpretations can be assigned to the symbols ΔE and Δt. In one interpretation, Δt is the interval between successive measurements of the energy of a system, and ΔE is the accuracy to which the conservation of energy can be determined, i.e., the uncertainty in a measurement of the system's energy. In another, Δt is the lifetime of an unstable or metastable system undergoing decay, and ΔE is the accuracy with which the energy of the system may be determined. The latter interpretation is at the heart of the notion of decay width or the width of a scattering resonance. *See also* Fock–Krylov theorem.

engineering breakeven *See* breakeven.

enrichment Refers to the increase of a nuclear isotope above its natural abundance. In particular, nuclear fuel must be enriched in the isotope of the uranium isotope with 235 nucleons in order to produce a self-sustaining nuclear fission reaction in commercial power reactors.

Various reactor designs require different enrichment factors. Enrichment must be based on some physical property of the isotopes, as chemically, all nuclear isotopes are similar. Usually, the small difference in nuclear mass between isotopes is used to enrich a sample over the natural abundance of isotope mixtures.

ensemble A collection of a large number of similarly prepared systems with the same macroscopic parameters, such as energy, volume, and number of particles. The different members of the ensemble exist in different quantum or microscopic states, such that the frequency of occurrence of a given quantum state can be taken as a measure of the probability of that particular state.

ensemble average The average over a group of particles. For an ergodic system, the *ensemble average* at a given time t is equal to the time average for a single part of the system. The particular choice of time t is not relevant.

ensemble interpretation of quantum mechanics The mostly widely accepted interpretation of quantum mechanics, which states that it is not possible to make definite predictions about the outcome of every possible measurement on a single instance of a physical system. Instead, only predictions of a statistical nature can be made, which can therefore be verified only on an ensemble of identically prepared systems. This ensemble is fully described by a wave function, or more generally, a density matrix. No finer description is possible.

entanglement A non-factorizable superposition between two or more states, i.e.,

$$|\Psi\rangle = \sum a_{i,\cdots,j} |\Psi_i\rangle \cdots |\Psi_j\rangle .$$

For a two-particle system in a spin-entangled state this reduces to

$$|\Psi\rangle = \frac{1}{\sqrt{2}} \Big(|\uparrow_1\rangle|\downarrow_2\rangle - |\downarrow_1\rangle|\uparrow_2\rangle \Big) ,$$

where $\uparrow$ and $\downarrow$ symbolize spin-up and spin-down, and the indices represent the different particles. An equal weight between the states is assumed. Such a state is called maximally entangled.

Entanglement is specific to quantum mechanical systems. In the case of photons, *entanglement* can be produced by parametric down-conversion or emission of photons in atomic cascade decays. Atomic systems can be entangled, for instance, by the consecutive passage of atoms through cavities indirectly via the interaction with the cavity or photo-dissociation of diatomic molecules. *Entanglement* is the basis of the Einstein–Podolsky–Rosen experiment and a prerequisite of any experiment in quantum information.

enthalpy (**1**) The *enthalpy h* is defined as the sum $E + pV$, where E is the internal energy and pV (product of pressure and volume) is the flow work or work done on a system by the entering fluid. From its definition, the *enthalpy* does not have a simple physical significance. Yet, one way to think about *enthalpy* is as the energy of a fluid crossing the boundary of a system. In a constant-pressure process, the heat added to a system equals the change in its *enthalpy*.

(**2**) The *enthalpy H* is the sum of $U + PV$, where U denotes the internal energy of the system, P is its pressure, and V is its volume. The change in the *enthalpy* at constant pressure is equal to the amount of heat added to the system (or removed from the system if dH is negative), provided there is no other work except mechanical work.

entrance region (entry length) When the flow in the entrance to a pipe is uniform, its central core, outside the developing boundary layer, is irrotational. However, the boundary layer will develop and grow in thickness until it fills the pipe. The region where a central irrotational core is maintained is called the *entrance region*. The region where the boundary layer has grown to completely fill the pipe is called the fully developed region in which viscous effects are dominant. In the fully developed region, the fluid velocity at any distance from the wall is constant along the flow direction. Thus, there is no flow acceleration and the viscous force must be balanced by gravity and/or pressure, i.e., work must be done on the fluid to keep it moving. In laminar pipe flow, the fully-developed flow is attained within $0.03 R_{eD}$ diameters of the entrance, where R_{eD} is the Reynolds number based on the pipe diameter, D, and average velocity. The length $0.03 R_{eD}$ diameters is known as the *entry* (or entrance) *length*. For turbulent pipe flow, the *entry length* is about 25 to 40 pipe diameters.

entropy (1) A measure of the disorder of a system. According to the second law of thermodynamics, a system will always evolve into one with higher *entropy* unless energy is expended.

(2) In thermodynamics, *entropy* S is defined by the relationship between the absolute temperature T and the internal energy U as $1/T = (\partial U/\partial S)_{V,N}$. Another definition, based on the second law of thermodynamics, gives the change in the *entropy* between the final and initial states, f and i, respectively, in terms of the integral

$$\Delta S = \int_i^f \frac{dQ_{\text{rev}}}{T}$$

where dQ_{rev} is the infinitesimal amount of heat added to the system at temperature T in a reversible process.

In statistical thermodynamics, *entropy* is defined via the Boltzmann relationship, $S = k_B \ln W$, where W is the number of possible microstates accessible to the system. Finally, *entropy* can also be defined as a measure of the amount of disorder in the system, which is seen in the information theory definition of *entropy* as $-\sum_i (p_i \ln p)_i$, where p_i denotes the probability of being in the ith state.

Eötvös experiment Published in 1890, this experiment determined the equivalence of the gravitational and inertial masses of an object. The experiment suspended two equal weights of different materials from a tortion balance. As the balance did not experience a torque, the inertial masses were measured as equal.

EPR experiment See Einstein–Podolsky–Rosen experiment.

EPR paradox (Einstein–Podolsky–Rosen paradox) Shows, according to its authors (Einstein, Podolsky, and Rosen), the incompleteness of quantum mechanics. The Einstein–Podolsky–Rosen experiment investigates the *EPR paradox*.

equation of continuity The macroscopic condition necessary to guarantee the conservation of mass leads to the *continuity equation:*

$$\frac{\partial \rho}{\partial t} + \nabla \cdot \rho \mathbf{u} = 0$$

where $\mathbf{u}$ denotes the velocity of the moving fluid and ρ denotes its density.

equations of motion There are three basic equations that govern fluid motion. These are the continuity or mass conservation equation and the momentum and energy equations. In their integral form, these equations are applied to large control volumes without a description of specific flow characteristics inside the control volume. To consider local characteristics, one needs to apply the basic principles to a fluid element, which results in the differential form of the *equations of motion*. To solve the *equations of motion*, they must be complemented by a set of proper boundary conditions, expressions for the state relation of the thermodynamic properties, and additional information about the stresses. For incompressible flow, the density, ρ, is constant, and the continuity and momentum equations can be solved separately since they would be independent of the energy equation.

equations of state (1) The relationships between pressure, volume, and temperature of substances in thermodynamic equilibrium.

(2) The intensive thermodynamic properties (internal energy, temperature, entropy, etc.) of a substance are related to each other. A change in one property may cause changes in the others. The relationships between these properties are called *equations of state* and can be given in algebraic, graphical, or tabular form. For certain idealized substances, which is the case for most gases, except under conditions of extreme pressure and temperature, the *equation of state* is written as $P = \rho RT$, where R is the gas constant. For air, $R = 287.03 \text{m}^2/\text{s}^2\text{K} = 1716.4 \text{ft}^2/\text{sec}^2\text{R}$. This equation is also known as the ideal gas law.

equilibrium An isolated system is in *equilibrium* when all macroscopic parameters describing the system remain unchanged in time.

equipartition Prediction by classical statistical mechanics that the energy of a system in thermal equilibrium is distributed in equal parts over the different degrees of freedom. Each variable with quadratic dependence in the Hamiltonian (such as the velocity of a particle) of the system has an energy of $\frac{1}{2}k_B T$, where k_B is the Boltzmann constant and T is the temperature of the system. For instance, for an ideal gas (non-interacting point-like particles) we find an energy of $E = \frac{3}{2}n_k T$, where the motion in each spatial dimension contributes $\frac{1}{2}k_B T$.

The law holds true for the classical limit in quantized systems, when the discrete energy levels can be replaced by a continuum. This means that *equipartition* does not hold for the low temperatures, since in this case only very few energy levels are populated.

equipartition of energy Whenever a momentum component occurs as a quadratic term in the classical Hamiltonian of a system, the classical limit of the thermal kinetic energy associated with that momentum will be $1/2k_B T$. Similarly, whenever the position coordinate component occurs as a quadratic term in the classical Hamiltonian of the thermal, the average potential energy associated with that coordinate will be $1/2k_B T$.

equivalence principle One of the basic assumptions of general relativity, that all physical systems cannot distinguish between an acceleration and a gravitational field.

erbium An element with atomic number (nuclear charge) 68 and atomic weight 167.26. The element has six stable isotopes.

ergodic process A process for which the ensemble average and the time average are identical.

escape peak *See* double escape peak.

eta meson An uncharged subatomic particle with spin zero and mass 547.3 Mev, which predominantly decays via the emission of neutral particles, either photons or neutral pions. It is one of the mesons of the fundamental pseudoscalar meson nonet which contains the pion, kaon, $\overline{K}$, and eta. The eta is composed of up, down, and strange quarks, mixed in quark–antiquark pairs. *See* eightfold way.

ether Before special relativity, it was expected that electromagnetic waves propagated through a medium called the *ether*. The *ether* was a massless quantity that had essentially no interaction with other matter, but permeated all space. It existed solely to support the propagation of electromagnetic waves. After relativity, the requirement of a physical medium to propagate electromagnetic waves was not needed, and the *ether* hypothesis was discarded.

Ettingshausen effect The development of a thermal gradient in a conducting material when an electric current flows across the lines of force of a magnetic field. This gradient has the opposite direction to the Hakk field.

Euclidian space A space which is flat and homogeneous. This means that the direction of the coordinate system axes and the origin is unimportant when describing physical laws in space-time.

Euler angles Two Euclidian coordinate systems having the same origin are, in general, related through a set of three rotation angles. By convention, these are generated by (1) a rotation about the z axis, (2) a rotation about the new x axis, and (3) a rotation about the new z axis. These rotations can place the (x, y, z) axes of one coordinate system along the (x, y, z) axes of the other.

Each rotation about the axes is shown in steps from 1 to 3. The Euler angles are the rotation axes.

eulerian viewpoint (eulerian description of fluid motion The Eulerian description of fluid motion gives entire flow characteristics at any

Euler–Lagrange equation

position and any time. For instance, by considering fixed coordinates x, y, and z and letting time pass, one can express a flow property such as velocity of particles moving by a certain position at any time. Mathematically, this would be given by a function $f(x, y, z, t)$. This description stands in contrast with the Langrangian description where the fluid motion is described in terms of the movement of individual particles, i.e., by following these particles. One problem with the adoption of the Eulerian viewpoint is that it focuses on specific locations in space at different times with no ability to track the history of a particle. This makes it difficult to apply laws concerned with particles such as Newton's second law. Consequently, there is a need to express the time rate of change of a particle property in the Eulerian variables. The substantial (or material) derivative provides the expression needed to formulate, in Eulerian variables, a time derivative evaluated as one follows a particle. For instance, the substantial derivative, denoted by $\frac{D}{Dt}$, is an operator that when acting on the velocity, gives the acceleration of a particle in a Eulerian description.

Euler–Lagrange equation (1) Relativistic mechanics, including relativistic quantum mechanics, is best formulated in terms of the variational principle of stationary action, where the action is the integral of the Lagrangian over space-time. Variational calculus then leads to a set of partial differential equations, *Euler–Lagrange equations,* which describe the evolution of the system with time. These equations are:

$$\frac{d}{dt}\left[\frac{\partial L}{\partial \dot{q}_i}\right] - \frac{\partial L}{\partial q_i} = 0.$$

(2) A reformulation of Newton's second law of classical mechanics. The latter describes the motion of a particle under the influence of a force F:

$$F = m\frac{d^2}{dt^2}x.$$

If the force F can be derived from a scalar or vector potential, this equation can be rewritten using the Lagrangian $L = L(x, \dot{x}, t)$:

$$\frac{d}{dt}\left(\frac{d}{dt}L\right) = \frac{\partial}{\partial x}L.$$

For classical problems, the Lagrangian L can be calculated through the relationship:

$$H(p, x) = \dot{x}p - L,$$

where p is the momentum and H is the Hamiltonian of the system.

Euler number A dimensionless number that represents the ratio of the pressure force to the inertia force and is given by $\Delta P/\rho V^2$. It is equal to one-half the pressure coefficient, cp, defined as $\Delta P/(1/2\rho V^2)$, and is usually used as a non-dimensional pressure.

Euler's equation For an element of mass dm, the linear momentum is defined as $dm\vec{V}$. In terms of linear momentum, Newton's second law for an inertial reference frame is written as

$$d\vec{F} = \frac{D}{DT}\left(dm\vec{V}\right).$$

Considering only pressure and gravity forces, neglecting viscous stresses, and dividing both sides by dm, the above equation reduces to

$$-\frac{1}{\rho}\nabla p - g\nabla z = \frac{D\vec{V}}{Dt}.$$

This equation is called *Euler's equation*. For a fluid moving as a right body with acceleration a, *Euler's equation* can be applied to write

$$-\frac{1}{\rho}\nabla p + g\nabla z = \vec{a}.$$

Also, by integrating the steady-state *Euler's equation* along a streamline between two points 1 and 2, one obtains the Bernoulli equation:

$$\frac{P_2}{\rho} + \frac{v_2^2}{2} + gz_2 = \frac{P_1}{\rho} + \frac{v_1^2}{2} + gz_1.$$

europium An element with atomic number (nuclear charge) 63 and atomic weight 151.96. The element has two stable isotopes. *Europium* is used as a red phosphor in color cathode ray tubes.

eutectic alloy The alloy whose composition presents the lowest freezing point.

evanescent wave trap   A dipole trap which is based on the trapping of atoms and molecules in the far detuned evanescent wave. Due to the exponential decay of the evanescent wave as a function of the surface distance, the *evanescent wave trap* is a two-dimensional trap.

evaporation   A mechanism by which an excited nucleus can shed energy. The basis of the *evaporation* model is a thermalized system of nucleons (something like a hot liquid drop) where the energy of a nucleon, in most cases a neutron, can fluctuate to a sufficient energy to escape the attractive potential of the other nucleons.

evaporative cooling   The cooling of an ensemble of particles that occurs through the evaporation of hotter particles from the ensemble. After the equilibration of the remaining particles, a cooler sample stays behind. An obvious example of *evaporative cooling* is the mechanism by which a cup of coffee cools down. *Evaporative cooling* has gained huge interest due to its usefulness in achieving the Bose–Einstein condensation in dilute gases. *Evaporative cooling* represents the last step in a sequence of several steps to achieve Bose–Einstein condensation: starting from a cold sample of atoms prepared in a magneto-optical trap, atoms were cooled down further using optical molasses. The cold atoms were pumped into low field seeking states and trapped magnetically. An rf-field induces transitions to high field seeking states, which are then ejected from the trap. By ramping the rf transition frequency to lower and lower frequencies, the transition is induced for atoms at positions closer to the trap center, which means that atoms with lower energies are ejected. This procedure leads to progressively lower temperatures. Elastic collisions between the remaining atoms leads to the necessary equilibrium.

Eve   The most frequently used name for the receiving party in quantum communication.

exact differential   Differential dF is called an *exact differential* if it depends only on the difference between the values of a function F between two closely spaced points and not on the path between them.

exchange energy   Part of the energy of a system of many electrons (or any other type of fermion) that depends on the total spin of the system. So called because the total spin determines the symmetry of the spatial part of the many-electron wave function under exchange of particle labels. This energy is thus largely electrostatic or Coulombic in origin, and is many times greater than the direct magnetic interaction between the spins. It underlies all phenomena such as ferromagnetism and antiferromagnetism. *See* spin-statistics theorem.

exchange force   The two-body interaction between nucleons is found to be spin dependent but parity (spatial exchange) symmetric. The nuclear force is also isospin symmetric and saturates, making nuclear matter essentially incompressible. To account for these properties, early nucleon–nucleon potentials used a combination of spin exchange (Bartlett force), space exchange (Majorana force), and isobaric exchange (Heisenberg force) operators. These are generally called exchange forces.

exchange integral   An integral giving the exchange energy in a multi-electron system. In the simplest case, the integral involves a two-particle wave function.

exchange interaction   An effective interaction between several fermions in a many-body system. It originates from the requirement of the Pauli principle that two fermions in the same spin state are repelling each other. For a many-electron system, the exchange interaction for an electron l is found to be

$$H_{int} = -\frac{e^2}{4\pi\varepsilon} \sum_{\substack{j' \\ m_{sj'}=m_{sl}}} \int d^3r \frac{1}{|r-r'|} \frac{\Psi_{j'}^*(r')\Psi_l(r')\Psi_{j'}(r)\Psi_l(r)}{\Psi_l^*(r)\Psi_l(r)},$$

where the sum is over all electrons which have the same spin state as the one under consideration. The charge density represented by H_{int} gives just the elementary charge e, integrated

excitation

over the space. This leads to the possible interpretation that the electron is under the influence of N electrons and one positive charge smeared out over the whole space, i.e., under the total influence of $N-1$ negative charges as expected.

excitation Refers to the fact that a given system is in a state of higher energy than the energetic ground state. Atomic and molecular systems can be excited by various mechanisms.

excitation function The value of a scattering cross-section as a function of incident energy. The *excitation function* maps out the strength of the interaction of a scattered particle and the target as a function of their relative energy.

exciton The electron-hole pair in an excited state.

exclusion principle Or Pauli principle, states that two-fermions cannot be in the exact same quantum state, i.e., they must differ in at least one quantum number. An alternative but equivalent statement is that the wave function of a system consisting of two fermions must be antisymmetric with respect to an exchange of the two particles. The latter fact can be expressed with the help of Slater determinants.

exothermic reaction A reaction that releases energy during a reaction. In particle physics, an *exothermic reaction* is one where the mass of the incident system is larger than that of the final system.

expansion coefficient The measure of the tendency of a material to undergo thermal expansion. A solid bar of length L_0 at temperature T_1 increases to a length L_1 when the temperature is increased to T_2. The new length L_1 is related to L_0 by the relation: $L_1 = L_0(1+\alpha(T_2-T_1))$, where α is the linear *expansion coefficient*.

expansion, thermal The change in size of a solid, liquid, or gas when its temperature changes. Normally, solids expand in size when heat is added and contract when cooled. Gases also expand when pressure is lowered.

expectation value The average value of an observable or operator $\hat{A}$ for a quantum mechanical system. It can be evaluated through the integral

$$\langle \Psi | \hat{A} | \Psi \rangle = \int \Psi \hat{A} \Psi^* d^3 r .$$

extensive air showers The result of one cosmic ray (particle) interacting with the upper atmosphere of the earth, producing cascades of secondary particles which reach the surface. Air showers as detected on the surface are mainly composed of electrons and photons from decays of the hadronic particles produced by the primary reactions; for initially energetic cosmic rays (≥ 100 TeV), air showers are spread over a large ground area. At the maximum of the shower development, there are approximately 2/3 particle per GeV of primary energy.

extensive variable A thermodynamic variable whose value is proportional to the size of the system, e.g., volume, energy, mass, entropy.

external flow Refers to flows around immersed bodies. Examples include basic flows such as flows over flat plates, and around cylinders, spheres, and airfoils. Other applied examples include flows around submarines, ships, airplanes, etc. In general, solutions to *external flow* problems are pieced together to yield an overall solution.

extinction coefficient Or linear absorption coefficient α. A measure of the absorption of light through a medium. The intensity I_0 is reduced to I

$$I = I_0 \exp(-\alpha l)$$

due to absorption after passage through a medium with thickness l with the linear absorption coefficient α. In general, the unit of α is 1/cm.

extrapolated breakeven *See* breakeven.

F

Fabry–Perot etalon A device commonly used for the spectral analysis of light. It is a multi-beam interference device consisting of two mirrors that form a cavity. The determining quantity for the achievable resolution is the finesse, which is the ratio of free spectral range and line width.

The most common *Fabry–Perot etalons* are the planar, confocal, and concentric types. The planar *Fabry–Perot etalon* has flat mirrors $R_1 = R_2 = \infty$, the confocal etalon has a mirror distance $d = R_1 = R_2$, and for the concentric case, we find $R_1 = R_2 = d/2$. $R_{1,2}$ denotes the radii of the two mirrors.

A confocal *Fabry–Perot* etalon is less susceptible to angular misalignment than a *Fabry–Perot etalon* consisting of flat mirrors; it has, however, the disadvantage that for mirrors with comparable reflectivities, the finesse is lower, since essentially two reflections on each mirror are necessary to complete a path. This leads also to a reduction in the free spectral range compared to a planar *Fabry-Perot etalon*. Another explanation for this feature is the mode degeneracy in a confocal *Fabry–Perot etalon*. Due to the mode degeneracy, an exact mode matching of the transverse profile of the light source to the etalon is not necessary.

The transmission of the *Fabry–Perot etalon* as a function of mirror distance, tilt angle, and wavelength can be evaluated using a consistent field approach. For an ideal flat *Fabry–Perot etalon* consisting of two identical, non-absorbing mirrors with reflectivity R, and a medium of index of refraction n between them, one finds a relationship of the form

$$T = \frac{I_0}{1 + a \cos \delta},$$

where I_0 is the initial intensity incident on the Fabry-Perot, $a = 4R/(1-R)^2$, and the phase $\delta = kd = 2\pi n d/\lambda$ (λ is the wavelength of the light). Examples of this curve for different finesses are shown in the figure.

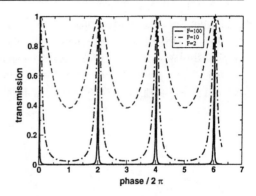

Transmission curve of a Fabry-Perot etalon for different finesses. The higher the finesse, the sharper the transmission peaks.

face-centered cube lattice A cubic crystal lattice in which atoms are also placed in the center of each face.

Fadeev equations In quantum mechanics, equations describing the collision of three bodies. Named after L.D. Fadeev.

Fadeev–Popov method Powerful method developed by L.D. Fadeev and V.N. Popov (1967) for incorporating gauge-fixing conditions into functional integrals in quantum field theory. The method guarantees that even when an explicit gauge is chosen for a calculation, certain correlation functions will be correctly found to be gauge-invariant.

Fahrenheit temperature scale (T_f) Defined by the temperature at which ice freezes, which is 32°F, and that at which water boils, to be 212°F at 1 atmospheric pressure. It can be more correctly defined in terms of the Kelvin scale of temperature T, using the relationship, $T_f = 32 + (9/5)(T - 273.15)$.

Falkhoff–Uhlenbeck formula Often used when the operator $(\vec{a}\nabla)^l$ is applied to the solid harmonics $Y_{l,m}(\vec{r})$, where $|m| = 0, \cdots, l$. It relates the solid harmonics expressed as functions of $\vec{r}$ to the solid harmonics expressed as functions of $\vec{a}$. Specifically, we have

$$(\vec{a}\nabla)^l \, r^l \, Y_{l,m}(\vec{r}) = l! \, a^l \, Y_{l,m}(\vec{a}) .$$

The solid harmonics are used to express the angular part of functions of two vectors. For example, the Legendre polynomial $P_l(\cos(\vec{a}, \vec{b}))$ for the angle between two vectors $\vec{a}$ and $\vec{b}$ can be expressed as

$$P_l\left(\cos\left(\vec{a}, \vec{b}\right)\right) = \Sigma_m (-1)^m Y_{l,m}(\vec{a}) Y_{l,-m}(\vec{b}).$$

The three simplest functions $r^l Y_{l,m}$ are given by

l	m	$r^l Y_{l,m}$
0	0	1
1	0	z
1	1	$-\frac{1}{\sqrt{2}}(x + \iota y)$

Falkner–Skan similarity solutions Prandtl treated the problem of steady, two-dimensional laminar flow along a flat plate placed longitudinally in a uniform stream. Because the velocity profiles $u(y)$ have a similar shape, Blasius was able to exactly solve Prandtl's boundary layer conditions by combining two independent variables into one similarity variable. Falkner and Skan showed that the Blasius solution is a member of a family of exact solutions of the boundary layer equations which leads to similar velocity profiles. A necessary condition for the existence of such similarity solutions is that the velocity at the outer edge of the boundary layer, $u_e(x)$ takes the form

$$u_e(x) = Cx^m.$$

By introducing a similarity variable $\eta = \frac{y}{\delta(x)}$ and dropping the x-dependence for the sake of notation, the stream function defined as

$$\varphi = \int_0^y u\, dy = u_e \delta \int_0^y \frac{u}{u_e} d\left(\frac{y}{\delta}\right)$$

can be written as

$$\varphi = u_e \delta \int_0^\eta f(\eta)\, d\eta$$

where $f(\eta) = \frac{u}{u_e}$.

Integrating the above equation yields

$$\varphi = u_e \delta f(\eta)$$

noting that

$$v = -\frac{\partial \varphi}{\partial x} = -\left(\delta' u_e + \delta u_e'\right) f + \eta u_e f' \delta'$$

$$u = \frac{\partial \varphi}{\partial y} = u_e f'$$

$$\frac{\partial u}{\partial y} = \frac{\partial^2 \varphi}{\partial y^2} = \frac{u_e f''}{\delta}$$

$$\frac{\partial u}{\partial x} = \frac{\partial^2 \varphi}{\partial x \partial y} = \frac{u_e \delta'}{\delta} \eta f'' + u_e' f'$$

and

$$\frac{\partial^2 u}{\partial y^2} = \frac{\partial^3 \varphi}{\partial y^3} = \frac{u_e f'''}{\delta^2}.$$

By substituting the momentum equation for a boundary layer, the Falkner–Skan equation is obtained

$$f''' + \frac{m+1}{2} f f'' - m f'^2 + m = 0$$

where $m = \frac{\delta^2}{\nu} u_e'$, with the boundary conditions $f(0) = f'(0) = 0$ and $f'(\infty) = 1$.

Note how the partial differential equation has been reduced to an ordinary differential equation. An exact integral of the Falkner-Skan equation has not been found. This equation is solved numerically. The *Falkner-Skan solutions* are of great significance, because, in addition to flow along a flat plate, they give flow near a stagnation point. They also show the effects of pressure gradients on the velocity profile, which are of interest for separating flows, as well as provide a good basis for approximate methods for boundary layer computation.

family (*see* generation). In the standard model, quarks and leptons are placed into families which are then placed in generations. The up and down quark *family* is associated with the electron and electron neutrino as the first generation, for example.

fanning friction factor (f) Equal to one-fourth the Darcy–Weisbach friction factor. See friction factor.

Fanno line Consider a steady compressible adiabatic (no heat transfer) flow of an ideal gas through a duct of constant cross-section where there is friction (nonisentropic flow). This flow is referred to as a Fanno flow. The basic laws governing a control volume with end sections 1 and 2 along the duct include the first law of thermodynamics, continuity, linear momentum, and the equation of state. Assume that the flow conditions at section 1 are known to give a reference point on the enthalpy–entropy or temperature-entropy diagram. The *Fanno line* is defined as the locus of all points, which represents the locus of states starting from the reference point that may be reached by changing the friction in adiabatic flow. The point of maximum entropy on the *Fanno line* corresponds to sonic conditions, while the part of the curve with higher enthalpy than at sonic conditions represents subsonic conditions. The other part corresponds to supersonic conditions. In the same manner as defining a *Fanno line,* one can define a Rayleigh line for a flow with heat transfer (adiabatic) without friction.

Fano interference The interaction of a discrete atomic or molecular level with an underlying continuum of states resulting in a line shape referred to as the Fano profile.

Fano profile The shape of a spectral line as a function of frequency of an atomic or molecular line, which originates from the coupling of a level to a background continuum of states. The spectral shape is given by

$$\sigma = \sigma_a \frac{(q+\varepsilon)^2}{1+\varepsilon^2} + \sigma_b,$$

where q is the so-called line-parameter, and the cross-sections σ_a and σ_b are due to the interaction between the continuum and the resonance line and the non-interacting part of the continuum, respectively. ε is dimensionless and defined by

$$\varepsilon = \frac{\nu - \nu_0}{c\Gamma/2},$$

where ν is the frequency and ν_0 is the resonance frequency of the transition. Γ is the line width of the transition. The maximum cross-section is given by $\sigma_{\max} = \sigma_a q^2$. Depending on the value of q, very different profile types might be observed. For $q \gg 1$ one finds a standard Lorentz profile.

Fano resonance In quantum mechanics, this is a transition to a discrete state which is embedded in a continuum close to the edge of that continuum. The transition shows up in spectra via a characteristically asymmetric shape, and was studied in detail by Ugo Fano in 1961.

Faraday effect When a plane-polarized beam of electromagnetic wave passes through a certain material in a direction parallel to the lines of a magnetic field, the plane of polarization is rotated.

Faraday rotation The polarization vector or the plane of polarization of a plane-polarized electromagnetic wave traveling along a magnetic field in a plasma experiences a rotation, which is called a *Faraday rotation*. The mechanism of the rotation is attributed to the difference in phase velocity of right and left circularly polarized waves that constitute the plane-polarized wave. This effect has been widely used to estimate densities and magnetic field orientations as well as intensities of laboratory and astrophysical plasmas.

Faraday's constant (F) The electric charge of one mole of electrons, equal to 9.648670×10^4 coulombs per mole.

far infrared The longer wavelength region of the infrared spectrum. This region is farthest from the visible region and closest to the radio-wave region.

fast wave A type of low-frequency, hydromagnetic, normal mode that exists in a magnetized plasma. These tend to propagate perpendicularly to the magnetic field. They are also called magnetosonic waves. The reason they are called fast is that their phase velocities are almost always faster than the Alfvén velocity.

f-center A lattice defect in alkali halide crystals that is usually transparent in the visible spectrum, giving a coloration.

feedback The action of returning the output of a device such as an amplifier to its input. There are two types of *feedback*: positive *feedback* and negative *feedback*. If the relative phases of the *feedback* voltage and the input signal are the same, the *feedback* is called a positive *feedback*. If the two voltages (input and output) are out of phase, then the *feedback* is negative.

fermi A unit of length equal to 10^{-13} cm, approximately the size of a nucleon.

Fermi contact interaction An interaction between the electronic and nuclear spins in an atom, so-called because it is proportional to the probability of finding the electron at the nuclear site. First discussed by E. Fermi in 1930.

Fermi-Dirac distribution The probability of occupancy of an energy level ε by a fermion at temperature T is given by the *Fermi-Dirac distribution* function:

$$f(\varepsilon) = \frac{1}{\exp\left[(\varepsilon - \mu)/k_B T\right] + 1}.$$

Fermi-Dirac statistics The statistics followed by fermions. According to the Pauli principle, two fermions must never occupy the exact same quantum state. Consequently, the total wave functions describing a system of fermions must be anti-symmetric with respect to an exchange of two fermions. The partition function Z for the *Fermi-Dirac statistics* is given by

$$Z = \sum \langle N_i \rangle = \Pi_j \left(1 + \exp\left(-\beta \left(E_j(N) - \mu\right)\right)\right),$$

where $\beta = 1/kT$ and μ is the chemical potential. The average population density $\langle N_j \rangle$ of a state with energy E_j is given by

$$f(E) = \langle N_j \rangle = \frac{2s + 1}{\exp((E_j - \mu)/kT) + 1}$$

for a fermion with spin s. For electrons $s = 1/2$, the state is either occupied or empty. $f(E)$ gives, therefore the probability of occupancy of state E_j by an electron. In solids, the chemical potential is often referred to as the Fermi level or Fermi energy, E_F. The total number of electrons in a solid is given by the total number of atoms in the lattice. Assuming the temperature of 0K, the meaning of the Fermi level becomes clear. All energy levels with $E < E_F$ are occupied by one electron, while states with $E > E_F$ are not populated. One finds $\langle N \rangle = 1/2$ for $E = E_F$.

The Maxwell–Boltzmann distribution

$$\langle N \rangle = \exp(\mu/kT) \exp(-E/kT)$$

is obtained as a limit of the *Fermi-Dirac statistics* for small population density or $E - \mu \gg kT$.

Fermi distribution Represents the probability that a particle obeying Fermi–Dirac statistics will have an energy E. This distribution has the form

$$P(E) = \frac{1}{e^{E - E_f/kT} + 1}.$$

Fermi energy (ε_F) (1) In a system of fermions, such as electrons in a metal, the energy separating the highest occupied single-particle state from the lowest unoccupied one. This definition is not sufficiently precise, however, in many contexts, such as semiconductors, and the term is used (often unwittingly) as a substitute for the *chemical potential*.

(2) The highest filled energy level at absolute zero for a fermion is called the Fermi level. All energy levels below this value are occupied, and all above this value are empty.

Fermi, Enrico A Nobel Prize winner in 1938 for his production of transuranic elements using neutron irradiation. He is known for the construction of the first controlled and self sustaining nuclear fission reactor.

Fermi function The distribution of electrons (positrons) in beta decay is calculated from the weak interaction transition matrix using plane wave functions for the electrons. However, electrons experience the Coulomb field of the nucleus, so the correct distribution must be modified by the *Fermi function,* which accounts for this effect.

Fermi gas A nucleus in some approximation can be considered as a collection of non-interacting fermions (nucleons) placed in a potential well. The potential well provides the average interactions that the nucleon experiences due to its fellow constituents. Because of Fermi statistics, the nucleons are added into phase space filling all momentum states according to the Pauli exclusion principle. Thus, nucleons fill a volume of momentum space up to a surface of radius, P_f, where P_f is the Fermi momentum, and $P_f^2/2M = T_f$ is the Fermi energy.

Fermi golden rule   Gives the transition rate for an atomic system to a group of closely lying states within an energy range $E \pm dE$ or a continuum of states. It states that the transition rate W is given by

$$W = \frac{2\pi}{\hbar} |H'|^2 \varrho_b(E),$$

where H' is the coupling energy and ϱ_b is the density of states for the continuum. One assumes that H' and ϱ_b are constant over the energy range of interest. *Fermi's golden* rule can be derived using perturbation theory.

Formula for the rate at which transitions are made between different quantum mechanical states of a system under the influence of a perturbation. The widespread applicability of the formula led E. Fermi to name it the golden rule.

Fermi liquid   A system of identical fermions which interact strongly with one another, as indicated by the word liquid. Typical examples are the electrons in a metal, the atoms in liquid ^{3}He, and the neutrons in a neutron star. The term is often used more restrictively to mean a system obeying Fermi liquid theory. *See also* Fermi gas.

Fermi liquid theory   Phenomenological theory of a Fermi liquid, developed by L.D. Landau from 1956 to 1958, with the assumption that the low lying excited states of such a system can be understood in terms of weakly interacting elementary excitations which behave almost as an ideal Fermi gas. The theory is applicable to electrons in metals, liquid ^{3}He, nuclear matter, and neutron stars.

Fermi momentum   *See* Fermi gas.

Fermi momentum, velocity, and wave vector
In an ideal Fermi gas or Fermi liquid, the momentum, velocity, and wave vector of a fermion at the Fermi surface.

fermion   (**1**) A particle with a half integer spin. It consequently obeys the Fermi-Dirac statistics. According to the Pauli principle, two *fermions* can never occupy the exact same quantum state. This has consequences concerning the symmetry of wave functions describing a system of *fermions*, i.e., the total wave function must be anti-symmetric with respect to an exchange of any two nuclei.

(**2**) Any particle, composite or elementary, with intrinsic angular momentum or spin equal to half an odd integer times $\hbar$, and thus obeying Fermi-Dirac statistics. Examples of *fermions* are electrons, neutrinos, quarks, neutrons, and ^{3}He atoms. *See also* boson; spin-statistics theorem.

Fermi pressure   *See* degeneracy pressure.

Fermi sea   In an ideal Fermi gas or Fermi liquid, the set of states below the Fermi energy.

Fermi statistics   Postulates that it is not possible for two identical fermions (particles) to have the same spatial location. Thus, the position of identical particles must be represented by a wave function antisymmetric in the exchange of any two particles.

Fermi surface   In an ideal Fermi gas of noninteracting fermions, this is a surface in momentum space that encloses the occupied states at zero temperature. The concept continues to have meaning when interactions between the fermions are important, as for instance, in liquid ^{3}He, as described by Fermi liquid theory. In this case, the surface marks a discontinuity in the probability of occupation of the single-particle momentum states. In this and all other systems with translational symmetry, the *Fermi surface* is spherical in shape.

The *Fermi surface* is of central importance in the theory of metals, but the concept requires some modification. The space in which the electrons move is no longer homogeneous on account of the crystal lattice. In other words, the system of electrons no longer has translational symmetry, and the single-electron states must be classified by their quasimomenta or Bloch wave vectors. The *Fermi surface* is now defined as the surface in quasimomentum space, separating occupied from unoccupied states (or, more precisely, as the surface of discontinuity in the occupation probability). Several consequences ensue from this consideration. First, since Bloch vectors that differ from one another by a reciprocal lattice vector are physically equivalent, the *Fermi surface* may be represented in several equivalent ways. In the repeated zone scheme, it is an infinite structure with the full periodicity of the reciprocal lattice. Or, in the reduced zone scheme, different parts of it may be translated by conveniently chosen reciprocal lattice vectors and reassembled into parts or sheets, as they are sometimes called, that are then said to lie in the first Brillouin zone, second Brillouin zone, etc. Whichever scheme is chosen, the *Fermi surface* is always closed. In general, it is not a sphere, and may be multiply connected with complicated topology. Indeed, a host of fanciful names such as the crown, the lens, and the monster have been concocted to describe the shapes encountered in various metals. The determination of *Fermi surfaces* is a large subject of its own in solid state physics, and its shape, topology, and related properties such as the Fermi velocity determine many electrical, magnetic, and optical properties of metals.

Fermi temperature (T_F) The absolute temperature corresponding to the Fermi energy, $T_F = \varepsilon_F / k_B$.

Fermi transition The weak interaction, which explains beta decay, is represented by both vector and axial vector currents. For historical reasons, the weak decay transition occurring due to the vector interaction is called a *Fermi transition*, and that due to the axial vector interaction is called Gamow-Teller transition. For allowed beta decay (first order in $(v/c)^2$) the vector transition has a spin change of zero and no parity change.

fermium A transuranic element with atomic number (nuclear charge) 100. 21 isotopes have been identified, the longest half-life at 100 days belonging to atomic number 257.

ferrimagnetism A type of magnetism in which the magnetic moments of neighboring ions tend to align antiparallel to each other.

ferrite A powdered, compressed, and sintered magnetic material.

ferroelectricity A crystalline material with a permanent spontaneous electric polarization that can be reversed by an electric field. Ammonium sulfate (NH4)2SO4 is an example of a ferroelectric material.

ferroelectric material A material in which electric dipoles can line up spontaneously by mutual interaction.

ferromagnetic material A magnetic material that has a permeability higher than the permeability of a vacuum. Typical ferromagnetic materials are iron and cobalt.

ferromagnetism The magnetism of a material caused by a domain structure. *See* domain.

Feshbach resonances Originally observed in nuclear physics, these also play an important role for ultra-cold atoms as they are prepared in magneto-optical traps and Bose–Einstein condensation. When two slow atoms collide, they usually do not stay together for very long. However, when a *Feshbach resonance* is observed, they stick together for a longer time. This can lead to a dramatic increase in the formation of molecules by photo-association in the trap or alter the properties of a Bose–Einstein condensate dramatically. Close to a *Feshbach resonance*, the atom–atom interaction is extremely sensitive to the exact shape of the potential energy curves such that small changes in, for instance, the magnetic field might switch from attractive to repulsive behavior. *Feshbach resonances* have typical signatures: the continuum wave-function shows a phase change of π over an energy range of the *Feshbach resonance*.

Feynman–Bijl formula Formula proposed by A. Bijl in 1940 and by R.P. Feynman in 1954, relating the dispersion relation for quasiparticles in superfluid ^{4}He to the spectrum of density fluctuations in the ground state, as measured by neutron scattering, for example.

Feynman diagram (1) A pictorial representation, in time order, of an interaction where lines represent particles and vertices represent interaction points. This pictorial representation was developed to help write down the perturbation series for interactions in quantum electrodynamics, where, generally, the more vertices, the higher the order of the term in the perturbation series. *Feynman diagrams* are, however, now used as a convenient exposition of any interaction, although perturbation techniques may not be so appropriate. Conservation isospin at each vertex requires the creation of an intermediate Sigma particle and the exchange of two pions. Thus, this interaction is of second order, but it may not be small due to the strength of the interaction.

(2) A system of graphs of great utility in carrying out perturbative calculations in quantum field theory. Originally invented by R.P. Feynman in 1949 for the study of quantum electrodynamics.

Feynman-Kac integral See Feynman path integral.

Feynman path integral Profound and remarkable reformulation of quantum mechanics by R.P. Feynman in 1948 (acting on P.A.M. Dirac's hint from 1933). In this formulation, the quantum mechanical amplitude necessary for a particle to make a transition from one point in space to another is given by a sum over all possible paths of a phase factor depending only on the classical action for that path in units of Planck's constant. The *path integral* is also known as a functional integral. Feynman's formulation is now part of the standard pedagogy of quantum mechanics. It has been extended in countless directions, to statistical mechanics (done by Feynman himself), to quantum field theory (again pioneered by Feynman, and profoundly developed further by J. Schwinger), to stochastic processes (notably by K. Ito, M. Kac, and N. Wiener), and to critical phenomena and the renormalization group, to name just a few, and has led to many major discoveries in all of these areas. The method is almost essential to the quantization of Yang–Mills or non-Abelian gauge theories such as that believed to underly the Glashow–Weinberg–Salam model of the electroweak interactions. Functional integration continues to be a subject of extensive research in mathematics, quantum field theory, and the semiclassical dynamics, and a practical tool lending itself to approximation and numerical calculation in these and other diverse areas of science.

Feynman, Richard Nobel Prize winner in 1965 who, with Tomonaga and Schwinger, developed the theory of quantum electrodynamics.

Feynman rules Set of rules for assigning mathematical meaning to a Feynman diagram in any quantum field theory.

Feynman scaling, Feynman variable The *Feynman variable, x*, is the ratio of the longitudinal momentum to the maximum possible longitudinal momentum. At sufficiently high energies, the invariant cross-section is almost independent of the total energy and may be represented by the product of two functions, one dependent on the transverse momentum and the other, x. Thus, the transverse momentum distribution of secondaries is independent of both the total energy and the longitudinal momentum.

Feynman variational principle Feynman developed a formulation of quantum mechanics based on the variation of the action integral over all possible paths. The classical path in space-time is the one of least action. Quantum mechanically, all paths are possible and are assigned a probability of occurrence.

fiber A flexible material of glass or transparent plastic used to transmit light.

Fick's law States that in diffusion, the flux density ($\vec{J}_n$), defined as the number of particles passing through a unit of area in a unit of time in the direction normal to the area, is proportional to the gradient of the concentration, c. The direction of flow is from a region of high concentration to low concentration.

$$\vec{J}_n = -D\nabla c .$$

The constant of proportionality, D, is the diffusion coefficient.

field amplitude The amplitude of the electric field.

field quantization The quantum mechanics of fields, as opposed to that of particles also known as second and first quantization respectively. Although field quantization is often used as a convenient tool in nonrelativistic many-body physics, it is generally regarded as an unavoidable necessity in describing quantum mechanical processes at relativistic speeds.

field-reversed configuration A plasma torus without a toroidal magnetic field generated through self-organization processes in a type of plasma confinement device called the theta pinch. Its simple machine geometry as well as physical separation of the plasma from the container are, among others, its advantages as a potential fusion reactor.

field tensor The electric and magnetic field vectors are really components of a four-dimensional, skew-symmetric tensor of second rank. This tensor is called the field tensor, and Maxwell's equations may be written in relativistically covariant form as

$$\partial_\alpha F^{\alpha\beta} = \frac{4\pi}{c} J_\beta .$$

Here, $F^{\alpha\beta}$ is the field tensor and J^β is the four-current density. The field tensor has the form

$$\begin{pmatrix} 0 & -E_x & -E_y & -E_z \\ E_x & 0 & -B_z & B_y \\ E_y & B_z & 0 & -B_x \\ E_z & -B_y & B_x & 0 \end{pmatrix} .$$

field theory Assigns a mathematical function to each point in space-time. This function is a result of sources, but is generally given physical meaning independent of the sources, so that an interaction occurs locally due to the field function at that point. The field carries both energy and momentum. The electric field is the classic example, where, although created by a charge distribution, the force on a charge is determined locally by the multiplication of that charge by the field at the charge point. The field is quantized in quantum *field theories*, where a field quantum represents a fundamental particle.

Fierz interference In the study of the weak interaction as manifested in beta decay, the most general form can contain scalar, pseudo-scalar, vector, pseudo-vector, and tensor couplings between the hadronic and leptonic currents. A particular combination of these couplings, b, can be measured experimentally. For example, b appears in the probability for emission of an electron with energy between E and $E + d$ E:

$$N(E)d(E) = \frac{A}{2\pi^3} p E q^2 (1 + b/E) .$$

Here, $q = (W_0 - E)$, where W_0 is the energy endpoint of the spectrum and p is the momentum.

For historical reasons, the term b/E is called the Fierz interference. It is found to be zero, as the possible weak interaction couplings are indeed vector and pseudo-vector.

filamentation instability A type of plasma wave instability called modulational or parametric instability, in which perturbations grow perpendicular or nearly perpendicular to the pump wave, and thus the original pump wave becomes filamented. For example, in an unmagnetized plasma, a sufficiently long and intensive electron plasma (Langmuir) wave is subject to the *filamentation* (or transverse) *instability*, and may excite ion acoustic waves that propagate predominantly perpendicular to the pump wave, as well as coupled side-band waves of electron plasma waves that propagate obliquely to the pump wave. In general, these instabilities coexist with other instabilities also driven by the same pump wave. *See also* parametric instability.

final state interaction Any reaction can be viewed in terms of an interaction which causes the reaction to happen and, perhaps, a residual component of the overall interaction acting in the background. One can view, in time sequence, a reaction occurring and then the particles in the final state interacting through the residual, background interaction. In this way, a final state interaction may numerically change the strength of a reaction or the shape of the residual spectrum.

finesse ($\mathcal{F}$) A measure of the quality of an optical resonator or a Fabry-Perot interferometer. *Finesse* is the ratio between free spectral range and the line width of a resonance. In ideal systems without absorption and ideal mirror surfaces, the *finesse* is a function of the mirror reflectivity only. In the case of a Fabry-Perot interferometer with flat mirrors, each with reflectivity R, the *finesse* $\mathcal{F}$ is given by

$$\mathcal{F} = \frac{\pi \sqrt{R}}{1-R},$$

whereas for confocal etalons, the finesse is given by

$$\mathcal{F} = \frac{\pi R}{1-R^2},$$

which reflects the fact that for a confocal etalon, the light is reflected four times between the mirrors. In practical systems, absorption and scattering reduce the achievable *finesse*.

fine structure (1) In atomic physics, a splitting of the energy levels arising from the relativistic spin-orbit interaction of orbital and spin angular momenta. The effect is so-called because atomic spectral lines are found, upon closer examination, to consist of many separate lines with spacings of only a few cm^{-1}.

(2) The splitting of spectral lines in atoms and molecules originating from the interaction between the angular momentum of the electron and its spin. The Hamiltonian for the *fine structure* interaction is given by the spin-orbit term

$$H_{\text{fine}} = \frac{1}{2m^2c^2} \frac{1}{r} \frac{dV}{dr} LS$$

$$= \frac{1}{2m^2c^2} \left(\frac{Ze^2}{4\pi\varepsilon_0} \right) \frac{1}{r^3} LS,$$

where m is the electron mass, and V is the Coulomb potential $V = -\frac{Ze^2}{4\pi\varepsilon_0 r}$.

For hydrogen-like atoms, the *fine structure* interaction leads to a splitting of transition lines into two components corresponding to $l \pm 1/2$, where l is the angular momentum. The energy shift ΔE due to the *fine structure* term can be calculated using perturbation theory. In terms of the unperturbed energy $E_n^{(0)}$, one solves for a principal quantum number n

$$\Delta E_2 = -E_n^{(0)} \frac{(Z\alpha)^2}{2nl(l+1/2)(l+1)}$$

$$\times \begin{cases} l & \text{for } j=l+1/2 \\ -l-1 & \text{for } j=l-1/2 \end{cases},$$

where $\alpha = e^2/(4\pi\varepsilon_0\hbar c)$ is the *fine structure constant*.

For $L = 0$, no splitting is observed.

fine structure constant The strength of the coupling of the electromagnetic field to charged, elementary particles is given by the fine structure constant. It has the value

$$\alpha = e^2/[4\pi\epsilon_0 \hbar c] = 1/137.036.$$

finite Larmor radius effect One of the kinetic effects in plasmas. It is well known that charged particles rotate around magnetic field lines with a radius called the Larmor radius. In fluid theories, the Larmor radius is neglected. When an inhomogeneity such as a wave with a typical scale-length, which is shorter than a Larmor radius, is present the wave property deviates from the fluid picture, and subject to the *finite Larmor radius* effect. *See also* Larmor radius.

firehose instability A type of electromagnetic low-frequency instability in magnetized plasmas driven by temperature anisotropies. Plasma particles moving along curved magnetic field lines exert centrifugal force that tends to distort the field lines just like firehoses or gardenhoses, thus triggering the instability. Its basic energy source, therefore, lies in the drift motion of plasma particles moving along a magnetic field, and thus it occurs in a magnetized anisotropic plasma in which the plasma energy parallel to the magnetic field is higher than the plasma energy perpendicular to the field combined with the magnetic field energy. Excited Alfvén waves have frequencies lower than the ion cyclotron frequency and travel predominantly parallel to the magnetic field. Being electromagnetic in nature, the *firehose instability* is not so important in low beta plasmas, and is of interest principally in space and astrophysical plasma physics.

first Brillouin zone The region in the reciprocal lattice composed of all the bisections of lines which connect a reciprocal lattice point to one of its nearest points. This is also called a Wigner-Seitz primitive cell in the reciprocal lattice.

first law of thermodynamics The statement of conservation of energy, including heat. Formally, it can be stated that the change in the internal energy of a system, ΔU, is equal to the sum of the net work done on the system, ΔW, plus the net heat input into the system, ΔQ, i.e., $\Delta U = \Delta W + \Delta Q$.

first quantization The quantization of a system of particles, so-called to distinguish it from second quantization, the quantization of a system of fields.

fission A nucleus *fissions* when it divides into two or more smaller nuclei with, perhaps, the emission of a few neutrons. A nucleus can be made to *fission* by an external reaction (scattering or capture of an another particle) or may be inherently unstable, spontaneously *fissioning* into various components. Iron forms the most stable nucleus, and nuclei heaver than iron release energy upon *fission,* while those below iron on the mass scale require energy to *fission*. A self-sustaining *fission* process can only be made to occur with a few isotopes, where the *fission* process due to neutron capture is sustained by neutron emission from previously *fissioned* nuclei.

Fitch, Val Nobel Prize winner in 1980 who, with James Cronin, discovered that nature did not conserve the product of the symmetries of charge conjugation, C, and parity, P. Because the product of CP and time reversal, T, symmetries is assumed to universally hold, this means that the symmetry of time reversal is also violated.

Fjortoft's theorem Relates to the inviscid instability of fluid flows. Rayleigh's theorem states that a necessary (but not sufficient) condition for inviscid instability is that the velocity profile, $U(y)$, has a point of inflection. *Fjortoft's theorem* is more restrictive in that it requires that if y_0 is the position of the point of inflection, then a necessary (but not sufficient) condition for inviscid instability is that

$$\frac{d^2 U}{dy^2}(U - U(y_0)) < 0$$

somewhere in the flow.

flavor Term used to identify a type of quark. Thus there are two quark *flavors* per generation and three known generations in the standard model. Quark flavors are up, down, charm, strange, top, and bottom. *See* generation, family.

Floquet's theorem Describes the solutions of Hill's differential equation with a periodic function $H(x)$. Those differential equations have the form

$$\frac{d^2 f}{dx^2} + H(x)f = 0,$$

where

$$H(x) = H(x + nd) \text{ with } n = 0, \pm 1, \pm 2 \cdots$$

The solution is given by

$$f(x) = F_1(x)\exp(\imath \mu x) + F_2(x)\exp(-\imath \mu x)$$
$$F_i(x) = F_i(x + nd)$$
$$i = 1, 2 \text{ and } n = 0, \pm 1, \pm 2, \cdots$$

Floquet's theorem enabled Bloch to formulate the solution of the Schrödinger equation for the case of a periodic potential, e.g., in a lattice. The solutions are known as Bloch waves, which are generally written as

$$\Psi(x + nd, k) = \Psi_1(x, k)\exp(\imath nkd)$$
$$+ \Psi_2(x, k)\exp(-\imath mkd)$$
$$n, m = 0, 1, 2, \cdots$$

$$\Psi\left(x, k + n\frac{2\pi}{d}\right) = \Psi(x, k),$$

where k is the quantum number of the Bloch wave. Since the Bloch waves are periodic, they are only determined within an integer multiple of $2\pi/d$. This range is called the Brillouin zone.

flow meters Devices used to measure flow rates. Examples include flow nozzles, orifices, rotameters, and Venturi and elbow meters.

flow visualization A qualitative description of an entire flow field can be obtained from *flow visualization*. Some techniques of *flow visualization* include: smoke wire visualization in air, hydrogen bubble visualization in water, particulate tracer visualization in both liquid and gases, dye injection, laser-induced fluorescence in both liquid and gases, and refractive-index-change visualizations conducted in flows with density or temperature variations. The latter techniques include shadow graph and Schlineren techniques and holographic interferometry.

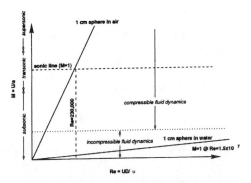

Flow map of a 1 cm sphere in air and water.

fluctuation-dissipation theorem Fundamental concept in statistical mechanics stating that the microscopic processes that underlie the relaxational or dissipative return of a macroscopic system not in equilibrium back to equilibrium are the same ones that give rise to spontaneous fluctuations in equilibrium. Originally formulated by A. Einstein and R. Smoluchowski in the study of Brownian motion in 1905 and 1906, the concept was significantly extended by H. Nyquist (1928), L. Onsager (1931), H.B. Callen and T.A. Welton (1951), and R. Kubo (1957). *See also* Onsager's reciprocal relation.

fluid A substance that deforms continuously when acted upon by a shear stress of any magnitude. This deformation is not reversible in that the *fluid* does not return to its original shape when the stress is removed. Because *fluids* deform continuously under the application of a shear stress, description of their behavior in terms of stress and deformation is not possible. The relation is between stress and the rate of the deformation. These characteristics of *fluids* stand in contrast to the response of solids to shear stresses. A solid will return to its original undeformed shape if the shear force is removed, if the magnitude of the shear and deformation are below certain limit. Moreover, for most solids, the magnitude of the shear force is proportional to the magnitude of deformation. While the distinction between a *fluid* and a solid seems simple, some substances, such as slurries, toothpaste, tar, etc. are not easily classified. They behave as a solid if the applied shear stress is small. When the stress exceeds a certain critical value, they will flow like *fluids*.

fluidization When a fluid flows upward through a granular medium, particulate *fluidization* is initiated when the upward drag becomes equal to the force of gravity and the particulates are in suspension.

fluidized beds *Fluidized bed* reactors are common in many applications. In a *fluidized bed*, solid particles move chaotically in a fluid stream. This motion causes significant mixing as well as particle–particle and particle–wall contact. *Fluidized beds* are designed to achieve effective heat and mass transfer and chemical reactions in many industrial and commercial processes.

fluorescence When a nucleus is illuminated by electromagnetic energy at a frequency corresponding to the energy of a nuclear or atomic level, the incident electromagnetic energy is absorbed and remitted as radiation or secondary particles. This is known as resonant *fluorescense* yield.

fluorine Element with atomic number (nuclear charge) 9 and atomic weight 18.9984. Only the isotope with atomic number 19 is stable. Combined with uranium as uranium hexafluoride, it is used in the gaseous diffusion process to enrich nuclear fuel for reactors.

flute instability A fluid-type electrostatic plasma instability that occurs in a magnetized inhomogeneous plasma. This instability is a special case of the gravitational instability, and is characterized by the perturbations traveling perpendicular to the magnetic field. In the case of a cylindrical plasma column in which a magnetic field exists along the axis, perturbations due to the instability grow and propagate around the surface, and make the column look like a fluted Greek column.

flux A *flux* of particles is the number of particles in a given direction per unit of cross-sectional area per unit of time.

flux conservation Assures that the magnetic flux linking an area in a non-resistive MHD plasma is conserved under the condition that the area moves with the plasma. This is the foundation of frozen field lines, i.e., field lines frozen into a plasma.

flux density (J) The amount of a quantity, such as number of particles, energy, or charge, that crosses a unit of area per unit of time.

flux quantization The fact that, irrespective of what magnetic field is actually applied, the magnetic flux passing through the hole of a ring-shaped superconductor can only be an integer multiple of the flux quantum, which in Gaussian units is given by

$$\Phi_0 = hc/2e ,$$

where h is Planck's constant, c is the speed of light, and e is the electron charge. The value of Φ_0 is 2.0679×10^{-7} gauss-cm^2.

Flux quantization was observed by B.S. Deaver and W.M. Fairbanks in 1961, and the factor of 2 in the denominator of the formula for Φ_0 provides very strong evidence that superconductivity originates in a certain association of pairs of electrons known as Cooper pairs. The phenomenon was in fact predicted on the basis of a phenomenological theory by V.L. Ginzburg and L.D. Landau in 1950, but without this factor of 2.

The flux quantum defined here is sometimes qualified with the adjective superconducting to distinguish it from the normal flux quantum, which is twice as large, i.e., hc/e. See also Aharonov–Bohm effect, Meissner effect.

flux quantum See flux quantization.

flying hot-wire anemometry One restriction of using a stationary hot-wire anemometry is its inability to measure the velocity in regions of flow reversals. *Flying hot-wire anemometry* overcomes this difficulty by moving the probe along a prescribed curve with a known velocity. The moving probe is then exposed to a relative velocity which is measured in terms of its components. The probe position and velocity and the measured relative velocity components are then used to determine the flow velocity.

Fock–Krylov theorem Theorem developed by V.A. Fock and S.N. Krylov (1947) which states that the decay of a quasi-stationary state is related to its energy distribution, in particular that the lifetime τ and the energy width Γ are related by

$$\tau = h/2\pi\Gamma ,$$

where h is Planck's constant.

Fock space In a quantum mechanical system consisting of a variable number of identical particles, this is the direct sum of spaces for 0, 1, 2, ... particles. Named after V.A. Fock. *See also* second quantization.

Fock states The states in a Fock space, often specified by giving the occupation numbers of single particle levels. The state with no particles is called the Fock vacuum. *See* number state.

Fokker–Planck equation A differential equation of an n dimensional vector P of the form

$$\frac{\partial}{\partial t} = \left[-\frac{\partial}{\partial x_j} A_j(x) + \frac{1}{2} \frac{\partial}{\partial x_i} \frac{\partial}{\partial x_j} D_{ij}(x) \right] P(x)$$

where the first term represents a drift term, while the second term is the diffusion term which leads to a broadening of P in time, provided its coefficient is positive. A is the drift vector and gives the net direction of the deterministic drift motion. D denotes the diffusion matrix. The equations of motion for the P_n are given by

$$\frac{d \langle P_i \rangle}{dt} = \langle A_i \rangle$$

$$\frac{d \langle P_i P_k \rangle}{dt} = \langle x_i A_j \rangle + \langle x_j A_i \rangle + \frac{1}{2} \langle D_{ij} + D_{ji} \rangle$$

Foldy–Wouthuysen transformation Method developed by L.L. Foldy and S.A. Wouthuysen (1950) for decomposing the Dirac equation into a pair of equations that allows the systematic evaluation of relativistic effects when these are small, and which includes the effects of electron spin in the first approximation.

forbidden band An energy band in which there can be no electrons.

forbidden transition (1) A transition between two states of an atom, molecule, or any other system that is disallowed because of symmetry. Transitions may be weakly or strictly forbidden. In atomic spectra, for instance, transitions between states with a vanishing electric dipole moment matrix element are said to be dipole-forbidden. Such transitions may still occur through higher electromagnetic multipoles. *See* dipole transition, selection rules.

(2) Transition that is not an electric dipole-allowed transition. For these transitions we have the selection rules $\Delta J = 0$ or $\Delta J = 2$. In some cases, *forbidden transitions* can be driven by two-photon processes.

force The rate of change of momentum that produces an acceleration of one meter per square second in a kilogram of mass.

forced vortex A flow field with purely tangential motion (circular streamlines). The tangential velocity increases linearly with the radius. In this flow field, the rotation and vorticity are constants. The circulation, however, is a function of the area enclosed by the contour. This flow field is similar to rigid body rotation.

force-free fields In an MHD plasma, in which gravitational field is negligible, the so-called force-free situation is realized if the magnetic field $B(x)$ satisfies $\nabla \times B(x) = f(x) B(x)$; such magnetic fields are called force-free fields.

form drag Also called pressure drag since it is generated by the pressure distribution on an object. It is called *form drag* because it strongly depends on the form or shape of the object. Similar to friction (or viscous) drag, it is due to viscosity. However, *form drag* is less sensitive to variations in Reynolds number than friction drag. This is especially true at a high Reynolds number. *Form drag* is the dominant drag component in separated flows.

form factor (or structure function) Related to a Fourier transform of the spatial distribution of the transition matrix involving the interacting particle and the target. The *form factors* for deep inelastic scattering are related to four structure functions of the parton distributions, F_1, F_2, g_1, and g_2, where the first two are spin-independent and the latter two are spin-dependent.

forward scattering Scattering between particles in which the particles continue along their initial directions of motion.

Fourier analysis Determination of the amplitude, frequency, and phase of each sinusoidal component in a given periodic signal.

Fourier transformation ($\mathcal{F}(\omega)$) The *Fourier transformation* of a function $f(t)$ is given by

$$\mathcal{F}(\omega) = \frac{1}{\sqrt{(2\pi)}} \int_{-\infty}^{\infty} f(t) \exp(-i\omega t)\, dt \,.$$

The *Fourier transformation* of $\mathcal{F}$ leads to the initial function f. The *Fourier transformation* analyzes a function f for its spectral content. Hence, aperiodic functions have a continuous frequency spectrum.

four-vector In Minkowski space describing relativistic space-time, vectors are four-dimensional corresponding to the three spatial directions and the one time dimension. A vector algebra using the Minkowski metric is defined for the scalar product between *four-vectors* so that relativistically invariant results are *produced (scalar).*

four wave mixing A non-linear frequency mixing scheme in atomic and molecular gases. *Four wave mixing* utilizes the $\Xi^{(3)}$ non-linearities, which is why it does not require an anisotropy of the medium as the $\Xi^{(2)}$ non-linearities employed in non-linear crystals. It is an effect arising from the intensity-dependent index of refraction in the medium. The interference of the coherent beams with frequencies ω_1 and ω_2 leads to an intensity variation in the medium. Due to the non-linear interaction with the atomic medium, this spatial intensity variation is accompanied by spatial variation of the index of refraction leading to a volume diffraction grating. The signal beam at frequency ω_s is scattered at this grating, forming the generated field. The signal and generated fields are therefore subject to a phase matching condition, which does not apply to the propagation of the input fields ω_1 and ω_2.

As opposed to other schemes of frequency generation by the interaction of coherent beams, such as lasing without inversion, no population transfer is occurring within the atomic or molecular system.

A special case of *four wave mixing* is the degenerate *four wave mixing,* which leads to phase conjugation. In this case, the two incoming fields with the same frequency are copropagating through the medium, which causes a stationary grating to build up. It can be shown that the generated field is the phase conjugate of the signal beam and, consequently, co-propagates with respect to it. From a viewpoint of a quantized field, one photon out of each of the pump beams is annihilated, a photon in the phase conjugate wave is generated, and the signal beam gets amplified.

fractional charge An electric charge less than that of the electron, generally by a factor expressible as a rational fraction made of small integers. The quasiparticles in the fractional quantum Hall effect, e.g., are believed to possess *fractional charges.*

fractional quantum numbers Quantum numbers associated with quasiparticles in certain systems that are simple fractions of the numbers for elementary particles. *See* fractional charge.

fractional statistics Term used to describe certain field theories in which the wave function of the many-particle system does not get multiplied by $+1$ or -1 during the exchange of any two particles, as would be the case for Bose or Fermi statistics respectively. Instead, the wave function is multiplied by a phase factor with a phase angle that is a fraction of π.

fragmentation function In a high momentum transfer reaction, a recoiling quark-parton will eventually hadronize. The *fragmentation function* represents the probability that a quark-parton of a specific type will produce a hadron in an interval dz about z, where z is the spatial direction of the recoiling hadron.

francium An element with atomic number 87. The element has 39 known isotopes, none of which are stable. The isotope with atomic mass number 223 has the longest half-life of 22 minutes. Because *francium* is the heaviest known element which chemically acts as a one-electron atom, interest has developed in using it to increase the sensitivity of atomic parity experiments.

Franck–Condon factors The overlap integrals of vibrational wave functions of different electronic states. According to the Born–Oppenheimer approximation, a molecular wave function can be written as the product of the electronic wave function Ψ_e, the vibrational wave function Ψ_v, the rotational wave function Ψ_r, and the nuclear wave function Ψ_N. In the case of an electronic transition, the transition moment is given by the integral

$$M = \left|\langle\Psi'|q\hat{r}|\Psi\rangle\right|^2 = \left|\langle\Psi'_e|q\hat{r}|\Psi_e\rangle\right|^2$$
$$\left|\langle\Psi'_v|\Psi_v\rangle\right|^2 \left|\langle\Psi'_r|\Psi_r\rangle\right|^2 \left|\langle\Psi'_N|\Psi_N\rangle\right|^2 ,$$

where $|\langle\Psi'_v|\Psi_v\rangle|^2$ is called the *Franck–Condon factor*. It constitutes the overlap integral of the vibrational wave functions in the initial and final states. Since the dipole transition moment is carried by the electronic transition, no selection rules apply between the initial and final vibrational states. However, only electronic transitions from one particular vibrational state of an electronic state to another electronic state with a different vibrational state can be excited, for these, the *Franck–Condon factor* is large. Classically, this reflects the fact that the nuclei move much slower than the electrons, and, consequently, electronic transitions are favored where the kinetic energy of the nuclei does not change. Quantum mechanically, this results in large overlap functions $|\langle\Psi'_v|\Psi_v\rangle|^2$.

The factors $|\langle\Psi'_r|\Psi_r\rangle|^2$ are called Hoenl-London factors.

Franck–Condon principle A classical principle which reflects the fact that electronic transitions in molecules occur at points on the molecular energy surface where the kinetic energy of the nuclei remains constant or at least very similar. Quantum mechanically, at these locations, the *Franck–Condon factors* are large.

Franck–Hertz experiment Experiment by J. Franck and G. Hertz in 1914, in which atoms were bombarded by low energy electrons. Franck and Hertz discovered that the electron beam current decreased sharply whenever certain thresholds in the electron energy were exceeded, The experiment demonstrated the existence of sharp atomic energy levels and provided strong support for N. Bohr's model of the atom.

Franck–Read source In a closed circular dislocation, a dislocation segment pinned at each end is called a *Franck–Read source* and can lead to the generation of a large number of concentric dislocation loops on a single slip plane.

Frank, Ilya M. Nobel Prize winner in 1958 who, with Igor Tamm, explained the Cerenkov effect.

Franson interferometer A special type of interferometer used in photon correlation measurements. A correlated pair of photons is sent through an interferometric setup as depicted in the figure. One photon is sent one way and the other photon is sent down the other path in the interferometer. By detecting coincidence counts between the two detectors on each side, the interferences between the two cases when either both photons have taken the long path or both photons have taken the short path are detected. These are second order interference effects.

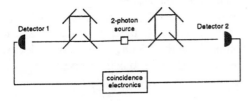

Setup of a Franson interferometer.

Fraunhofer diffraction The diffraction pattern when observed in the farfield, i.e., a large distance a from the diffracting object with a dimension d. The size of the object must be in the same order of magnitude as the wavelength λ of the light. The diffraction pattern of an object is equivalent to the spatial Fourier transformation of the diffracting object.

For instance, the diffraction pattern from a single slit with width d is given by

$$I = I_0 \frac{\sin^2 \delta}{\delta^2}$$

with $\quad \delta = \dfrac{\pi d \sin \alpha}{\lambda},$

where α is the angle from the optical axis and λ is the wavelength of the light.

free electrons Electrons detached from an atom.

free energy An energy quantity assigned to each substance, such that a reaction in a system held at constant temperature tends to proceed if it is accompanied by a decrease in *free energy*. The *free energy* is the sum of the enthalpy and entropy.

free expansion The process of expansion of a gas contained in one part of an isolated container to fill the entire container by opening a valve separating the two compartments. In this process, no heat flows into the system since it is thermally isolated, and no work is done. Thus, conservation of energy requires that the internal energy of the system remain unchanged. If the gas is ideal, there will be no temperature change; however, for a real gas, the temperature decreases in a *free expansion*.

free induction decay Term originally coined in the area of nuclear magnetic resonance to describe the decay of the induction signal of a macroscopic sample of matter containing nuclear spins which are initially tipped over from their equilibrium orientation and then undergo free precession. The term reflects the decay of the signal in a small fraction of the spins' natural lifetimes, which occurs due to inhomogeneities in the magnetic field in the sample. The term is now used in all other types of magnetic resonance, as well as in the area of resonance optics, i.e., the interaction of atoms with coherent light tuned to an atomic transition.

free particle A particle not under the influence of any external forces or fields.

free precession In magnetic resonance, the precession of the magnetic moment in a uniform static magnetic field.

free shear flows Shear flows are flows where the velocity varies principally in a direction at a right angle to the flow direction. In *free shear flows,* this variation is caused by some upstream variation or disturbance. Downstream of the disturbance, the *free shear flow* decelerates, entrains ambient fluids, and spreads. Examples of *free shear flows* include jets, wakes, mixing layers, and separated boundary layers. Viscosity has the effect of smoothing the velocity field, which causes it to become self-similar. *Free shear flows* are unstable and are characterized by large-scale structures.

free spectral range The frequency separation between adjacent transmission maxima for an optical cavity. This is an important consideration in determining the mode spacing in laser resonators or the resolution of Fabry–Perot interferometers. For a resonator with an optical path length L, one finds for the free spectral range FSR,

$$\text{FSR} = \frac{c}{L} \text{ ring resonator}$$
$$\text{FSR} = \frac{c}{2L} \text{ linear resonator}$$
$$\text{FSR} = \frac{c}{4L} \text{ confocal resonator}.$$

The value for the *free spectral range* is a consequence of the boundary conditions of the electric field amplitudes in linear resonators and the condition for constructive interference of consecutive passes in ring resonators, respectively.

free surface A surface that consists of the same fluid particles and along which the pressure is constant. In studying *free surface* flows, the shape of the *free surface* is not known initially. Rather, it is a part of the solution.

free vortex A flow field with purely tangential motion (circular streamlines). The tangential velocity is inversely proportional to the radius. Consequently, the origin is a singular point. The circulation around any contour not enclosing the origin is zero. The flow is thus irrotational.

Frenkel defect A point defect in a lattice in which an atom is transferred from the lattice site to an interstitial position.

Frenkel exciton An exciton in which the excitation is localized on or near a single atom.

Fresnel diffraction The diffraction in close proximity to the diffracting object with size d, i.e., when the wavelength of the light λ, d, and the distance from the object a are in the same order of magnitude:

$$a \approx d \approx \lambda.$$

Fresnel diffraction is in contrast to Fraunhofer diffraction, which is observed when

$$a \gg d \approx \lambda.$$

friction coefficient The ratio of the force required to move one solid surface over another surface to the total force pressing two surfaces together.

friction drag Also called viscous drag since it is generated by the shear stresses. The *friction drag* scales with the Reynolds number. It is important in flows with no separation and depends on the amount of surface area of the object that is in contact with the fluid. *See also* form drag.

friction factor A dimensionless parameter that is related to the pressure required to move a fluid at a certain rate. It is generally a function of the Reynolds number, surface roughness, and body geometry. In a pipe or duct, the relation between the *friction factor,* flow velocity, pressure drop, and geometry is given by

$$\Delta p = \frac{\rho V^2}{2} f \frac{L}{D}.$$

This relation can be written also as

$$h_L = f \frac{L}{D} \frac{V^2}{2g}$$

where f is the Darcy–Weisbach factor. For a Mach number, Ma, less than one, the *friction factor* is considered to be independent of the *Ma*.

frictionless flow A flow where viscous effects are neglected. *See also* inviscid flow.

friction velocity Defined for boundary layers as
$$u_* = (\tau_w/\rho)^{1/2}$$
where τ_w is the shear stress at the wall and ρ is the density.

Froude number A dimensionless parameter that represents the relative importance of inertial forces acting on a fluid element to the weight of the element. It is given by $V/\sqrt{g\ell}$, where V is the fluid velocity and ℓ is a characteristic length such as body length or water depth. The *Froude number* is important in flows where there is a free surface such as open channel flows, rivers, surface waves, flows around floating objects, and resulting wave generation.

frozen field lines In non-resistive MHD plasmas, the magnetic field lines are tied to the plasma so that they move and oscillate together. This state is called field lines frozen into the plasma or, simply, *frozen field lines*.

f-sum rule Relates the total amount of scattering of light, neutrons, or any other probe from any physical system, when integrated over all energies, to the number of scatterers. The rule is closely related to the dipole sum rule, and can be used in the same way to estimate the contribution of various physical mechanisms or excitations to scattering.

ft-value In allowed transitions in beta decay, the *ft-value* is a measure of the probability of the decay rate. It is proportional to the sum of the squares of the Fermi and Gammow–Teller matrix elements. In forbidden transitions, it does not give a measure of these matrix elements, but does indicate the order of forbiddenness of the decay. Since ft-value has a large range, it is usually quoted in terms of a $\log_{10}$.

fully developed flow Beyond the entrance region of the flow into a pipe or a duct, the mean flow properties do not change with downstream distance. The velocity profile is fully developed and the flow is called a fully developed flow. The entrance length, beyond which the flow is fully developed, varies between 40 to 100 diameters along the pipe and is dependent on the Reynolds number. *See also* entrance region.

fundamental vectors Vectors that can define the atomic arrangement in an entire lattice by translation.

fusion When two nuclei coalesce into a larger nucleus, nuclear *fusion* has occurred. Nuclear *fusion* is usually associated with the combination of two deuterium atoms to form a helium atom or the *fusion* of a deuterium atom with a tritium atom to form a helium atom with the release of a neutron and energy. The latter reaction is the fundamental reaction in a hydrogen bomb.

G

$g^{(2)}(\tau)$ *See* intensity correlation function.

g_a The weak interaction can be described in terms of a leptonic current interacting with a hadronic current. In general, these currents could consist of five forms — scalar, pseudoscalar, vector, pseudovector, and tensor — but the weak interaction may be described only in terms of vector and pseudovector terms. The charge involved in this interaction is called the coupling constant, and the pseudovector coupling constant is g_a.

gadolinium An element with atomic number (nuclear charge) 64 and atomic weight 157.25. The element has seven stable isotopes. *Gadolinium* has the highest thermal neutron cross-section of the known elements.

gage pressure The pressure relative to atmospheric pressure. The *gage pressure* is related to the absolute pressure by

$$P_{\text{gage}} = P_{\text{abs}} - P_{\text{atm}}.$$

The *gage pressure* is negative whenever the absolute pressure is less than the atmospheric pressure; it is then called a vacuum.

gain Growth rate of the number of photons in a laser cavity. *Gain* in a medium occurs when the rate of stimulated emission of radiation is larger than the rate of absorption, which requires that population inversion must be achieved. For a laser action, *gain* must be larger than the losses in the cavity.

gain coefficient Provides a measure of the growth rate of intensity as a function of distance in a laser gain medium. It is proportional to the population inversion and given by

$$g(\nu) = \frac{\lambda^2 A}{8\pi}\left(N_2 - \frac{g_2}{g_1}N_1\right)S(\nu),$$

where $S(\nu)$ is the Lorentzian line shape

$$S(\nu) = \frac{1}{\pi}\left[\frac{\delta\nu_o}{(\nu - \nu_o)^2 + (\delta\nu_o)^2}\right].$$

Here, A is Einstein's A coefficient, λ is the wavelength, g_1 and g_2 are degeneracies of lower and excited states, N_1 and N_2 are the number of atoms in the lower and excited states, ν and ν_o are field and atomic frequency, and $\delta\nu_o$ is the Lorentzian line width.

gain factor, photoconductivity The increase of the electrical conductivity due to illumination.

gain saturation Gain of a lasing medium decreases with an increase in photon flux in the cavity, resulting in *gain saturation*. For very large photon flux, gain approaches zero. *Gain saturation* restricts the maximum output power of a laser. In a homogeneously broadened medium, *gain saturation* causes power broadening which is given by

$$g(\nu) = \frac{g_o(\nu)}{1 + I(\nu)/I_{\text{sat}}(\nu)},$$

where $g_o(\nu)$ is unsaturated gain, $I(\nu)$ is the photon flux, and $I_{\text{sat}}(\nu)$ is the saturated photon flux at frequency ν. In a homogeneously broadened medium, *gain saturation* causes spatial hole burning, and in an inhomogeneously broadened medium, *gain saturation* causes spectral hole burning.

gain switching A technique used for generating high power laser pulses of very short duration. Using a fast pumping pulse, the inversion is raised rapidly to a value high above threshold. The rapid increase in gain does not allow photons to build up inside the cavity and, therefore, depletion is negligible. Build-up of large gain results in a short pulse of high power. Depending on the duration of the pump pulse, a laser pulse of about a nanosecond can be achieved. *Gain switching* can be achieved in any laser. Typical gain-switched lasers are diode lasers, and dye lasers.

gallium Element with atomic number (nuclear charge) 31 and atomic weight 69.72. The element has three stable isotopes. In the form

galvanometric effects Transformation of electrical current into mechanical motion.

gamma decay An excited nucleus can lose energy through the radiation of electromagnetic energy, or gamma rays. The energy of these photons is the difference between the initial and final energy levels in the nucleus.

gamma-matrices The Dirac equation for massive spin half-particles is usually written in terms of the four γ-matrices, γ^μ ($\mu = 0, \ldots, 3$), as

$$(\gamma^\mu p^\mu - m)\Psi(p) = 0$$

where p^μ is the particle four-momentum, m is the mass, and $\Psi(p)$ is the momentum-space wave function. The γ-matrices are defined as

$$\gamma^0 = \begin{pmatrix} 0 & 1 \\ 1 & 0 \end{pmatrix}, \quad \gamma^i = \begin{pmatrix} 0 & -\sigma^i \\ \sigma^i & 0 \end{pmatrix}$$

where $\mathbf{1}$ is the unit 2×2 matrix, and σ^i is the well-known Pauli spin matrix.

gamma ray A quantum of electromagnetic energy emitted from an excited nucleus as it decays electromagnetically. A *gamma ray* is a photon, but is differentiated from photons in that the source of *gamma rays* is the atomic nucleus.

gamma ray microscope A gedanken microscope first proposed by Heisenberg to measure the position of a particle. Consider the following schematic diagram:

M is the microscope, L is a lens, and P is the particle positioned along the x-axis. The particle is irradiated with gamma rays of wavelength λ. The microscope can only resolve the particle position x to precision Δx given by $\Delta x = \frac{\lambda}{\sin a}$, where a is the half-angle subtended by the lens. A particle entering the microscope imparts a recoil momentum to the particle with uncertainty in the x-direction Δp_x, given by $\Delta p_x = \frac{h}{\lambda} \sin$ (a), where h is Planck's constant. When combining, we obtain $\Delta p_x \Delta x \approx h$, which is consistent with Heisenberg's uncertainty relation.

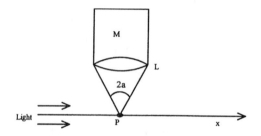

Schematic of gamma ray microscope.

Gamow factor In the alpha decay of heavy elements, the alpha particle must penetrate a Couloumb barrier. The approximate transparency for s-waves through a very high (or thick) barrier is called the *Gamow factor*, G. It is written as:

$$G \approx e^{-\pi(2Zz/137\beta)}.$$

Gamow–Teller selection rules for beta decay In the process of beta decay of a nucleus in which the nucleus emits a beta particle and a neutrino with their spins parallel, the selection rules for the emission process are known as the *Gamow–Teller selection rules*. In terms of changes in the angular momentum quantum numbers in units of $\hbar$ (ΔI) of the nuclear state due to beta decay, for allowed transitions:

$$\Delta I = \pm 1, \text{ or } 0, \text{ no change of parity}.$$

Gamow–Teller transition *See* Fermi transition.

Gamow theory (alpha decay) A theory proposed in 1928 by G. Gamow, and independently by R.W. Gurney and E.U. Condon, to describe the decay of a nucleus into an alpha particle (an He nucleus with charge 2e) plus a daughter nucleus. Gamow assumed that alpha particles exist for a short time before emission inside the nucleus. He further assumed that the potential energy $V(r)$ of the alpha particle is such that it is negative and constant in the nucleus of radius R ($r < R$) and falls off as

$$V(r) = \frac{2Z_1 e^2}{(4\pi \epsilon_0)r}, \quad r > R.$$

$Z_1 e$ is the charge carried by the daughter nucleus and ϵ_0 is the permittivity of free space.

gas State (or phase) of matter in which the molecules are relatively far apart (spacing is of an order of magnitude larger than the molecular diameter) and are practically unrestricted by intermolecular forces. Consequently, a *gas* can easily change its volume and shape. This is in contrast to solids, where both volume and shape are maintained. In the solid state, the molecules are relatively close (spacing is of the same order of magnitude as a molecular diameter) and are subject to large intermolecular forces.

gas constant (R) The constant of proportionality R, in the ideal gas law, $PV = nRT$, where P, V, and T denote the pressure, volume, and absolute temperature of n moles of an ideal gas. The value of $R = 8.31$ J/(mol.K) is a universal constant, and is equal to the product of the Boltzmann constant k_B and the Avogadro number. *See* ideal gas law.

gas dynamics The study of compressible flows, since compressible effects are more important in gas flow.

gaseous diffusion The name given to a process that is used to increase the percentage of the uranium isotope 235 to about 3% from the natural abundance of the uranium isotopes, which include ^{234}U at .0055%, ^{235}U at 0.72 %, and ^{238}U at 99.27%. Nuclear fuel composed of 3.2% ^{235}U can be used in the power producing reactors (light-water reactors) in the United States. The separation occurs based on the very small mass difference between the isotopes, which results in a slight difference in the diffusion rates.

gas lasers Lasers with gaseous gain medium. Most *gas lasers* are excited by electron collisions in various types of gas discharge which have narrow absorption bands. Common *gas lasers* are He-Ne, argon, carbon dioxide. *See* He-Ne lasers.

gauge bosons A quantum of a gauge field.

gauge field A field which has to be introduced into a theory so that gauge invariance is preserved at all points in space and time (locally). For example, consider a charged particle with wave function Ψ which transforms to Ψ' under a local gauge transformation

$$\Psi'(\mathbf{r},t) = e^{iq\Lambda(\mathbf{r},t)}\Psi(\mathbf{r}, t)$$

where $\Lambda(\mathbf{r}, t)$ is an arbitrary scalar function and q is a parameter. For the theory to be gauge invariant with respect to the above transformation, Ψ, and Ψ' must describe the same physics. If one takes $\Lambda(\mathbf{r}, t)$ as an arbitrary scalar function such that a transformation of the scalar (ϕ) and vector (**A**) electromagnetic potentials $\phi \longrightarrow \phi - \frac{\partial \Lambda}{\partial t}$ and $\mathbf{A} \longrightarrow \mathbf{A} + \nabla \Lambda$ leaves the electric and magnetic fields invariant, then ϕ and **A** are the *gauge fields* that have to be introduced into the theory for Ψ to preserve the above local gauge transformation.

gauge field theories The concept of gauge invariance may be generalized to include a theory built up by requiring invariance under a set of local phase transformations. These transformations can be based on non-Abelian groups. Yang and Mills studied the generalized theory of these fields in 1954.

gauge invariance (of the electromagnetic field) In describing the quantum interaction of an electron with an electromagnetic field, one often chooses a specific gauge to perform calculations. For an example of a gauge condition, consider the electric field $\mathbf{E}(\mathbf{r},t)$ and the magnetic field $\mathbf{B}(\mathbf{r},t)$. They can both be obtained from scalar and vector potentials $\phi(\mathbf{r},t)$ and $\mathbf{A}(\mathbf{r},t)$, respectively, by

$$\mathbf{E}(\mathbf{r},t) = -\nabla\phi(\mathbf{r},t) - \frac{\partial}{\partial t}\mathbf{A}(\mathbf{r},t)$$
$$\mathbf{B}(\mathbf{r},t) = \nabla \times \mathbf{A}(\mathbf{r},t).$$

The potentials are not completely defined by the above equations because **E** and **B** are unaltered by the substitutions $\mathbf{A} \longrightarrow \mathbf{A}+\nabla\chi$, $\phi \longrightarrow \phi - \frac{\partial}{\partial t}\chi$, where χ is any scalar. This property of the invariance of **E** and **B** under such transformations is known as *gauge invariance of the electromagnetic field*. A particular gauge which is normally chosen is called the Coulomb gauge, in which $\nabla \cdot \mathbf{A} = 0$. It should be noted that in this gauge, ∇ and **A** commute.

gauge transformation In classical electromagnetic theory, a *gauge transformation* is one that changes the vector and scalar potentials, leaving the electric and magnetic fields unchanged. This transformation is associated with conservation of electric charge. The symmetry is introduced in quantum mechanics by introducing a phase change in the wave function, where the phase change can be global or local (dependent on position). A *gauge transformation* is dependent on the interaction of a long range field and the conservation of a quantity such as electric charge.

gauss A unit of magnetic field. It is equal to 10^{-4} tesla (MKS unit), which results from the Biot-Savart law given below:

$$B = \frac{\mu_0}{4\pi} \int \frac{\mathbf{I} \times \mathbf{r}}{r^2} dl .$$

Gaussian beam A very important class of beam-like solutions of Maxwell's equation for an electromagnetic field. It retains its functional form as it propagates in free space. The field of a *Gaussian beam* propagating in the z-direction is proportional to

$$\exp\left[-ik\left[z + \left(\frac{x^2 + y^2}{2q(z)}\right)\right]\right]$$

where $q(z)$ is a complex beam parameter given by the ABCD law for *Gaussian beams*. In the paraxial approximation, the electric and magnetic fields of a *Gaussian beam* are transverse to the direction of propagation, and are therefore denoted by TEM$_{lm}$ mode. The electric field of a TEM$_{lm}$ mode is proportional to

$$H_l\left(\frac{\sqrt{2}\,x}{w(z)}\right) H_m\left(\frac{\sqrt{2}\,y}{w(z)}\right)$$
$$\exp\left[-\left(\frac{(x^2 + y^2)}{w(z)^2}\right)\right]$$
$$\exp\left[-i\left(\frac{k(x^2 + y^2)}{2R(z)}\right) + i(1 + l + m)\phi\right]$$

where $H_l(x)$ stands for the Hermite polynomial of order l with argument x. Here, the spot size $w(z)$, radius of curvature of the spherical wavefront of the Gaussian beam $R(z)$, and longitudinal phase factor ϕ are

$$w^2(z) = w_0^2[1 + (z/z_s)]$$
$$R(z) = z\left[1 + (z_s/z)^2\right],$$
$$\phi(z) = \tan^{-1}(z/z_s),$$
$$z_s = \pi w_0^2/\lambda .$$

The beam waist w_0 is spot-size at $z = 0$, λ is the wavelength, and z_s is the Rayleigh range. The lowest order Gaussian mode TEM$_{00}$ is used in many applications because of its circular cross-section and Gaussian intensity profile

$$\exp\left[-\left(\frac{2(x^2 + y^2)}{w(z)^2}\right)\right].$$

The intensity profile of a higher order gaussian mode is obtained by squaring the electric field. *Gaussian beams* are very directional. Laser is an example of a *Gaussian beam*.

Gaussian error In the limit of large numbers, a binomial probability distribution has a Gaussian form represented by

$$P(x) = \sqrt{(2/\pi)}e^{-(x-x_0)^2/2\sigma^2} .$$

where x_0 is the mean value of the distribution and σ represents the $1/e$ width. This distribution is called the normal distribution, and σ is the *Gaussian error*.

Gaussian line shape The absorption spectrum of light with a *Gaussian line shape* is seen in inhomogeneous broadening. One of the mechanisms resulting in *Gaussian line shape* is Doppler broadening. *See* inhomogeneous broadening.

Gaussian (probability) distribution Also called normal distribution. It is a probability distribution of a continuous variable $x(t)$ of the form

$$\frac{1}{\sigma\sqrt{2\pi}} \exp\left[\frac{-(x - x_o)^2}{2\sigma^2}\right],$$

where x_o is the mean and σ is the standard deviation. σ^2 is called variance.

Gaussian random processes Involves the Gaussian probability distribution, which is determined by two parameters, mean and variance.

For a *Gaussian random process*, all higher order correlations can be expressed in terms of second order correlations

$$\langle x(t_1) x(t_2) \cdots x(t_n) \rangle = \sum_{\text{all possible pairs}} \langle x(t_1) x(t_2) \rangle \cdots \langle x(t_{n-1}) x(t_n) \rangle.$$

Gaussian statistics Statistics of random variables which can be described by Gaussian random processes. *See* Gaussian random processes.

Gaussian white noise A delta correlated Gaussian random process with mean zero ($\langle \eta(t) \rangle = 0$) and variance $\langle \eta(t)\eta(t') \rangle = \delta(t-t')$, where $\delta(t-t')$ is the Dirac delta function.

Gauss–Markov process A random process $x(t)$ which is Gaussian and Makovian. It satisfies the linear differential equation

$$\frac{dx(t)}{dt} = A(t)x(t) + B(t)q(t),$$

where $q(t)$ is Gaussian white noise with zero mean ($\langle q(t) \rangle = 0$) and delta correlated variance ($\langle q(t)q(t') \rangle = \delta(t-t')$; $A(t)$ and $B(t)$ are functions of time. For time independent coefficients $A(t)$ and $B(t)$, the mean and variance of the random process decay exponentially, and the power spectrum is Lorentzian. This special case is known as the Orenstein–Uhlenbeck process. *See also* Markov process; Gaussian random processes.

Gauss's Law Gauss's law is a combination of Coulomb's law giving the force between electostatic charges, and the law of superposition, which states that the force law is linearly additive, so the total force is obtained by adding all the charges. In integral form in MKS units, the law is

$$\int_{\text{surface}} \mathbf{E} \cdot d\mathbf{A} = \text{Enclosed Charge} / \epsilon.$$

Here, ϵ is the dielectric constant and **E** is the electric field. The enclosed charge is that contained within the surface integral.

Gay-Lussac's law In 1808, J.L. Gay-Lussac discovered that when two gases combine to form a third, the volumes are in the ratio of simple integers. This law helped to confirm the atomic nature of matter.

1. Generator of translations in space: For a wave function $\Psi(\mathbf{r}, t)$ that satisfies Schrödinger's equation and that can be expanded in a Taylor series in **r**, it demonstrated that

$$\Psi(\mathbf{r} + \mathbf{r}_0, t) = e^{i\widehat{p} \cdot \mathbf{r}_0 / \hbar} \Psi(\mathbf{r}, t)$$

where $\mathbf{r}_0$ is any constant displacement, $\hbar$ is Planck's constant, and $\widehat{p} = -i\hbar\nabla$ is the momentum operator. $\widehat{p}/\hbar$ is called the generator of translations in space. ($\mathbf{r}, t$) is a point in space-time.

2. Generator of translations in time: For a wave function $\Psi(\mathbf{r}, t)$ that satisfies Schrödinger's equation, it is shown that

$$\Psi(\mathbf{r}, t + t_0) = e^{-iHt_0/\hbar} \Psi(\mathbf{r}, t)$$

where H is the Hamiltonian, t_0 is any constant time, and $\hbar$ is Planck's constant. $-\frac{H}{\hbar}$ is called the generator of translations in time.

GDH (Gerasimov–Drell–Hearn) sum rule A prediction of the first moment, Γ_1, of the spin-dependent parton distribution function, g_1, at $Q^2 = 0$. It relates the spin-dependent scattering cross-section of circularly polarized photons on longitudinally polarized nucleons to the anomalous magnetic moment of the nucleon.

$$\lim_{Q^2 \to 0} \frac{M^2}{Q^2} \Gamma_1 = \lim_{Q^2 \to 0} \frac{2M^2}{Q^2} \int_0^1 g_1(x, Q^2) dx$$
$$= -\kappa_N^2 / 4.$$

Here, κ is the anomalous moment of the nucleon, i.e., for a proton, the magnetic moment is defined as $\mu_p = (1 + \kappa_p)\mu_B$, where μ_B is the nuclear magneton. *See* gyromagnetic ratio.

Geiger counter A particle detector which is sensitive to the passage of ionizing radiation. A *Geiger counter* is constructed by inserting a thin wire along the axis of a cylindrical tube filled with a mixture of a noble gas (He, Ar, etc.) with a small amount of a quenching gas. When a voltage is applied between the cylinder and the wire, electrons from the primary ionization of a passing charged particle are accelerated in the high electric field near the wire surface. These electrons knock-out other atomic electrons from the gas causing an avalanche and creating an electronic signal.

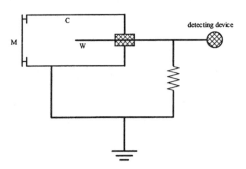

Geiger counter.

Geiger, H. (the experiment of H. Geiger, E. Marsden, and E. Rutherford) In 1906, H. Geiger, E. Marsden, and E. Rutherford carried out a series of experiments on the scattering of alpha particles by metallic foils of various thicknesses. They found that most of the alpha particles are deflected through very small angles ($< 1°$), but some are deflected through large angles. These measurements helped to establish that all the positive charge of an atom is concentrated at the center of the atom in the nucleus of very small dimensions.

Geiger–Nuttall law In 1911, Geiger and Nuttall noticed that the higher the released energy in α decay, the shorter the half-life. Although variations occur, smooth curves can at least be drawn for nuclei having the same (Z). The explanation of this rule was an early achievement of quantum mechanics and nuclear structure.

Gell-Mann, Murry Nobel Prize winner in 1969 who exploited the symmetries of the known elementary particles to classify them in a proposed scheme, the eightfold way.

Gell-Mann–Nishijima relation Gell-Mann and Nishijima proposed that in order to account for the weak decay of the kaon and the lambda particles, a quantum number called strangeness, S, which was conserved in the strong interactions, could be defined. This quantum number is related to the charge, baryon number, and the third component of isospin by

$$Q/e = B/s + S/2 + I_3 .$$

Gell-Mann–Okubo mass formula Using the static quark model, a relation between the masses of the pseudoscalar mesons can be obtained. This relation is:

$$4M_k - \pi = 3\eta_8 .$$

The prediction for the mass of η_8 is 613 MeV compared to the known $\eta(550)$ and $\eta'(960)$. If one assumes that the physical mass eigenstates are admixtures of the singlet and octet representations of the pseudoscalar mesons, the mixing angle can be calculated to be

$$\tan^2(\theta) = 0.2 ;$$

from the Gell-Mann–Okubo mass formula.

generalized Ohm's law One of four basic equations of magnetohydrodynamics, which describes the relationship between the time derivative of a current in an MHD fluid and various forces acting on the current. In the limit of a stationary, inhomogeneous, non-magnetized plasma, this law reduces to the usual Ohm's law. See also magnetohydrodynamics.

generalized oscillator strength In discussing inelastic electron scattering by a one electron atom, one defines a *generalized oscillator strength*, $\mathcal{F}_{qq'}$, as

$$\mathcal{F}_{qq'}(\Delta) = (E_q - E_{q'}) \frac{2}{\Delta^2} |\mathcal{S}_{qq'}(\Delta)|^2$$

where $E_{q'}$ and E_q are the energies of the atom before and after the scattering process. Δ is the magnitude of the wave vector difference $\vec{\Delta} = (\mathbf{k}_{q'} - \mathbf{k}_q)$ between the initial and final states of the one-electron atom. $\mathcal{S}_{qq'}(\Delta)$ is the inelastic form factor defined as

$$\mathcal{S}_{qq'}(\Delta) = \int \Psi_q^* e^{i\vec{\Delta}\cdot\mathbf{r}} \Psi_{q'} d\mathbf{r}$$

where $\Psi_{q'}$ and Ψ_q are the electron wave functions before and after the scattering, respectively.

The *generalized oscillator strength* is often used to obtain the differential scattering cross-section.

generalized Rabi frequency Consider the interaction of an intense optical field $\widetilde{\mathbf{E}}(t) = \mathbf{E}e^{-i\omega t} +$ (complex conjugate) with a two-level atom. The field is assumed to be nearly resonant with the allowed transition between the ground state $|a\rangle$ and the upper state $|b\rangle$ of the atom. The solution for the atomic wave function $\Psi(\mathbf{r}, t)$ in the presence of the applied field is given as

$$\Psi(\mathbf{r},t) = C_a(t)u_a(\mathbf{r})e^{-i\omega_a t} + C_b(t)u_b(\mathbf{r})e^{-i\omega_b t}$$

where $u_a(\mathbf{r})e^{-i\omega_a t}$ represents the wave function of the atomic ground state, $|a\rangle$, and $u_b(\mathbf{r})e^{-i\omega_b t}$ is the wave function of the excited state, $|b\rangle$. The solutions for $C_a(t)$ and $C_b(t)$ are of the form

$$C_a(t) \propto e^{-i\lambda t}$$
$$C_b(t) \propto \lambda e^{-i(\lambda+\Delta)t}$$

where λ is the characteristic frequency. λ can have two possible values $\lambda_\pm$ given by

$$\lambda_\pm = -\frac{1}{2}\Delta \pm \frac{1}{2}\Omega'$$

Ω' is the generalized Rabi frequency defined as

$$\Omega' = \left(|\Omega|^2 + \Delta^2\right)^{\frac{1}{2}}$$

where $\Omega = 2\mu_{ba}E/\hbar$ denotes the complex Rabi frequency, and $\Delta = \omega - \omega_{ba}$ represents the detuning, ω_{ba} being the transition frequency of the atom; E is the magnitude of the electric field vector $\mathbf{E}$.

general relativity Special relativity requires that the velocity of light in a vacuum is constant in inertial reference (non-accelerating) frames. The extension of special relativity to non-inertial frames is called *general relativity* and is based on the equivalence principle which states that gravitational mass and inertial mass are equivalent. Thus, gravitating bodies change the structure of space-time so that a gravitational attraction is explained through the curvature of space. For example, general relativity predicts the curvature of light in a gravitational field.

generating function Provides a way of calculating probability distribution and moments of a distribution. The *generating function* for photon counting distribution $P(m, T)$ is defined as

$$G(s, T) = \sum_{m=0}^{\infty} s^m P(m, T)$$
$$= \left\langle \exp\left[-s\eta \int_0^T I(t)\,dt\right] \right\rangle$$

where η represents the efficiency of the detector and $P(m, T)$ is the probability of detecting m photons in the counting interval $[0 - T]$. The photon counting distribution and the factorial moments can be obtained from the *generating function* as

$$P(m, T) = \frac{(-1)^m}{m!} \left[\frac{d^m}{ds^m} G(s, T)\right]_{s=1},$$

$$\langle m^{(\ell)} \rangle = \sum_{m=1}^{\infty} m(m-1) \cdots (m-\ell+1) P(m, T)$$
$$= (-1)^r \left[\frac{d^\ell}{ds^\ell} G(s, T)\right]_{s=0}.$$

The *generating function* $G(1, T)$ gives the probability of detecting zero photons in the counting interval $[0, T]$.

generation It is possible to order quarks and leptons into sets of mutually corresponding particles in a symmetry based on SU(2) × U(1). These are called *generations*, and there are three *generations* now accepted in the so-called standard model. The figure below shows the *generations*, which are coupled by weak decays between the quarks. Experiments counting neutrino flavors strongly suggest that there are only three possible *generations*.

$$\left[\begin{pmatrix} \nu_\mu \\ \mu^- \end{pmatrix} \begin{pmatrix} c \\ s \end{pmatrix} \right]$$
$$\left[\begin{pmatrix} \nu_\mu \\ \mu^- \end{pmatrix} \begin{pmatrix} c \\ s \end{pmatrix} \right]$$
$$\left[\begin{pmatrix} \nu_\mu \\ \mu^- \end{pmatrix} \begin{pmatrix} c \\ s \end{pmatrix} \right].$$

Generations →

In this figure, the family doublets are all left-handed (left helicity). In principle, there are also corresponding right-handed singlets which are sterile (have no interactions) but are usually ignored. The diagram indicates that, moving left to right, the *generations* become more massive.

geometric phase For a time-dependent Hamiltonian, $H(t)$, the eigenfunctions, $\Psi_n(t)$, and eigenvalues, $E_n(t)$, of the Schrödinger equation satisfy:

$$H(t)\Psi_n(t) = E_n(t)\Psi_n(t).$$

If $H(t)$ changes gradually, then according to the adiabatic theorem, a particle starting in an initial nth eigenstate remains in this state apart from an additional phase factor which appears in the state vector, $\Psi'_n(t)$, given as

$$\Psi'_n(t) = \Psi_n(t) e^{-\frac{i}{\hbar} \int_0^t E_n(t')dt'} e^{i\gamma_n(t)}$$

$\gamma_n(t)$ is called the *geometric phase*.

geometric phase of light Due to its vector nature, an electromagnetic wave, during its propagation through material medium, may acquire a phase in addition to the dynamic phase. This additional contribution to the phase may be reflected in the polarization state of the field. This phase, which depends only on the geometry of the path followed is called the geometric phase. This is also called the Panchratnam phase, as it was first discussed by Panchratnam (*see Panchratnam, S., Generalized theory of interference, and its applications, Proc. Ind. Acad. Sci., A44, 247, 1956*). This is a special case of topological phase discussed by Berry (Berry, M.V., *Proc. R. Soc. London*, A392, 45, 1984).

geometric probability distribution Also known as the Bose–Einstein distribution. Has a form

$$P(m) = \frac{\langle n \rangle^m}{(1 + \langle n \rangle)^{m+1}},$$

where $\langle n \rangle$ is the mean of the distribution. The generating function for this distribution is given by

$$G(s) = \frac{1}{1 + s\langle n \rangle},$$

and factorial moments are

$$\langle n^m \rangle = m! \langle n \rangle.$$

geometric similarity The prediction of prototype flow characteristics and associated effects from model flow observation is called similitude. For correct similitude studies, three types of similarities exist namely, geometric, kinematic, and dynamic similarities. The *geometric similarity* requires that the shapes of both model and prototype be the same. This requirement is satisfied by making sure that all lengths of the model and prototype have the same ratio, and that all corresponding angles are equal.

gerade wave function Molecular wave functions $\Psi(\mathbf{r}, t)$ that are even under a parity change, i.e.,

$$\Psi(\mathbf{r},t) = \Psi(-\mathbf{r},t)$$

are said to be *gerade*. Note that those that are odd are ungerade.

germanium Element with atomic number 32 and atomic weight 72.59. *Germanium* has five stable isotopes. It is extensively used in the electronic industry since, doped with other elements, it is one component of semiconductor devices. It is also used in infrared and gamma photon spectroscopy as a detector.

Germer (experiment of Davisson and Germer) In this famous experiment, a beam of 54 eV electrons irradiated a crystal of nickel normally ($\theta = 0°$). The angular distribution of the number of scattered electrons from the crystal was measured. Davisson and Germer found that the distribution falls from a maximum at $\theta = 0°$ to a minimum near 35°, then rises to a peak near 50°. The peak could only be explained by constructive interference of electron waves scattered by the regular lattice of the crystal. This was one of the most important experiments to confirm de Broglie's hypothesis.

g-factor The ratio of the number of Bohr magnetons to the units of h of angular momentum.

g-factor, Landé In the interaction of a one-electron atom with a weak uniform external magnetic field $\mathbf{B}_{\text{ext}}$, the Zeeman correction to the energy in first-order perturbation theory, E_Z^1, is given as

$$E_Z^1 = \frac{e}{2m}\mathbf{B}_{\text{ext}} \cdot \langle \mathbf{L}+2\mathbf{S}\rangle$$
$$= \left[1 + \frac{j(j+1) - l(l+1) + 3/4}{2j(j+1)}\right]\langle \mathbf{J}\rangle$$

where $\mathbf{L}$ is the orbital angular momentum and $\mathbf{S}$ is the spin angular momentum of the electron. The factor in square brackets in the above equation is called the *Landé g-factor*. j is the total angular momentum number and l is the orbital angular momentum quantum number.

Gibbs–Duhem relation Gives the variation in chemical potential μ in terms of the variation in temperature T and pressure P:

$$d\mu = -s\,dT + v\,dP,$$

where s and v denote the molar entropy and molar volume, respectively.

Gibbs energy (thermodynamic potential) (G) Defined by $G = H - TS$, where H is the enthalpy, T is the temperature, and S is the entropy.

Gibbs free energy (G) Defined as $G = U + PV - TS$, where U and S denote the internal energy and entropy, respectively, at temperature T, pressure P, and volume V. The physical significance of G is that it is minimal for a system in equilibrium with a temperature and a pressure reservoir.

Gibbs phase rule The number, f, of thermodynamic variables, such as temperature, pressure, and composition of a multi-component mixture in different phases, that can be independently varied in a system with N_c components and N_p phases is given by the *Gibbs phase rule* as $f = N_c - N_p + 2$.

GIM current Weak processes can be discussed in terms of a hadronic current interacting with a weak or leptonic current. In order to explain why strangeness-changing neutral currents are not observed, Glashow, Iliopoulos, and Maiani (GIM) proposed that there must be a charmed quark that decays into a strange and a d quark, with amplitude mixed through the Cabibbo angle, θ_c. Thus, the weak eigenstates of the first two generations are

$$\begin{pmatrix} u \\ d\cos(\theta_c) + s\sin(\theta_c) \end{pmatrix}$$

for the first generation and

$$\begin{pmatrix} c \\ s\cos(\theta_c) - d\sin(\theta_c) \end{pmatrix}$$

for the second.

Ginsburg–Landau theory of superconductivity Seven years before the BCS (Bardeen-Cooper-Schrieffer) *theory of superconductivity* was developed, Ginsburg and Landau proposed a phenomenological quantum *theory of superconductivity* based on the theory of second-order phase transitions. They sought a theory that was applicable near the critical temperature. At that temperature, the density of superconducting electrons is sufficiently small so that this number can be used as an expansion parameter. Their approach was to construct an expression for the free energy at temperature T as an expansion in the number density of superconducting electrons, n_s. The number density is proportional to the modulus square of the wave function (also called the order parameter). For the general case of an inhomogeneous superconductor in a uniform magnetic field, the Gibbs free energy is expanded in terms of the order parameter. Solutions for the wave function and the magnetic vector potential are then found which minimize the Gibbs free energy. Two equations are found; these constitute the *Ginsburg–Landau theory*.

GLAG theory The theory of type II superconductors as developed by Ginzburg, Landau, Abrikosov, and Gorkov.

Glaser, Donald A. Nobel Prize winner in 1960 for the development of the bubble chamber which could identify and track penetrating elementary particles. Bubble chambers were used for many years as the primary detector at most particle accelerators.

Glashow, Sheldon L. Nobel Prize winner in 1979 who, with Abdus Salam and Steven Wein-

glass A solid at amorphous state, containing mainly SiO_2, and transparent in the visible spectrum.

glass-laser media In many solid state lasers for gain medium, glass is used as a host material. Glass is doped with rare earth ions such as Nd, Yb, and Er. Glass, having a lower melting point than crystals, is easier to fabricate and cheaper to construct. In comparison with crystals, glass has a much lower thermal conductivity and worse thermomechanical and thermo-optical properties. Examples of glass-doped lasers are Nd:glass, Yb:Er:glass, fiber lasers.

Glauber displacement operator Coherent states, $|\alpha\rangle$, of the electromagnetic field were first used to describe quantum optical coherence, due largely to the work of R. Glauber in the 1960s. Mathematically, in addition to being an eigenstate of the photon annihilation operator with eigenvalue α, coherent states can also be constructed by the action of the *Glauber displacement operator*, $D(\alpha)$, on the Fock vacuum, $|0\rangle$, i.e.,

$$|\alpha\rangle = D(\alpha)|0\rangle$$

where

$$D(\alpha) = \exp\left(\alpha a^\dagger - \alpha^* a\right)$$

and α^* is the complex conjugate of the eigenvalue α. It should be pointed out that the coherent state is an example of a minimum uncertainty state.

Glauber's photodetection theory In 1970, R.J. Glauber showed how a simple use of perturbation theory gives a description of photodetection by atoms. Using the approximate electric dipole interaction Hamiltonian, $-\mu\cdot\mathbf{E}$, where μ is the electric dipole moment of the atom absorbing a photon from the quantized electromagnetic field $\mathbf{E}$, the transition probability $P(t)$ for the atom absorbing a photon at position $\mathbf{r}$ in some time interval from t_0 to t was calculated in its simplest form to be

$$P(t) \propto \int_{t_0}^{t} \langle i| E_i^- (\mathbf{r}, t') E_j^+ (\mathbf{r}, t') |i\rangle \, dt'$$

where $E_i^{-(+)}(\mathbf{r}, t')$ is the negative (positive) frequency part of the ith component of the electric field; $|i\rangle$ is the initial state of the electromagnetic field at time t_0.

Glauber–Sudarshan P-representation
(**1**) A diagonal coherent state ($|\alpha\rangle$) representation of the density operator

$$\hat{\rho} = \int P(\alpha)|\alpha\rangle\langle\alpha| d^2\alpha \, .$$

$P(\alpha)$ behaves as a quasi-probability density in phase space. For a classical state, it is a positive and well-behaved function which cannot be more singular than a delta function, like a classical probability density. Negative or singular behavior of $P(\alpha)$ provides a signature of a nonclassical state.

(**2**) The representation of a traceable operator, such as the density operator $\hat{\rho}$, in terms of coherent state projectors $|\alpha\rangle\langle\alpha|$ such that

$$\hat{\rho} = \int \phi(\alpha) |\alpha\rangle\langle\alpha| d^2\alpha$$

in which $\phi(\alpha)$ is some real function of the complex number α (eigenvalues of the annihilation operator with respect to the coherent states) and $d^2\alpha = d(\mathrm{Re}\,\alpha)d(\mathrm{Im}\,\alpha)$.

global modes Low-frequency hydromagnetic waves that arise together with slow Alfvén waves in magnetically confined high temperature plasmas due to nonlocal effects such as the geometry of torus during Alfvén wave heating schemes. The discrete frequencies of *global modes,* which are lower than those of the slow Alfvén waves, are determined by the boundary condition on the torus.

global symmetry One independent of space-time, i.e., the system is symmetric under the same symmetry operation at each space-time point.

glow discharge A mode of electrical conduction in relatively low pressure (10^2–10^3 pascals) ionized gases. This well-known discharge phenomenon, which is usually demonstrated in a cylindrical glass tube, is accompanied by the emission of diffuse lights with various characteristics. Typically, currents of the order of tens or hundreds of milliamperes flow, while the potential drop across the discharge region may be of the order of 100 volts.

Starting from the side of the anode, the *glow discharge* may be divided into several characteristic regions: the anode dark space, the positive column with striations comprised by successive luminous and dark regions, the Faraday dark space, the negative glow, the Crookes or Hittorf dark space, and the cathode glow. The *glow discharge* is generated as the potential is increased and the so-called Townsend region is passed. Meanwhile, as the current is increased, the *glow discharge* is transformed into the abnormal glow, and then, after a spark into the arc, discharges abruptly.

The phenomenon of *glow discharge* has been applied to the so-called voltage regulators or voltage reference tubes, which maintain relatively constant potential differences across themselves, even though the currents are changed appreciably. Thus, these devices become useful when constant reference potentials are needed.

Despite the fact that this phenomenon has been known for many years, because of the complexities of this phenomenon and difficulties involved in its measurements, many of the details of *glow discharge* remain unresolved.

glueball   A quantum of energy composed of gluons in a colorless arrangement, representing an excitation of the gluonic field.

gluon   (1) A quantum of the strong interaction. There are eight *gluons* which couple to the colored quarks in a way so that as the distance between quarks increases, the interaction strength linearly increases. As with quarks, individual *gluons* do not appear as isolated particles, but may be bound together in colorless objects, glueballs, or quark-*gluon* composites, hermaphrodites.

(2) It is believed that a property of elementary particles called color, like electric charge, gives rise to a massless gauge field of spin one. The quanta of this field are called *gluons* and they are responsible for the binding of quarks in the protons and neutrons of the nucleus.

Godfried sum rule   From the deep inelastic scattering of leptons from neutrons and protons, the electromagnetic structure functions can, in principle, be obtained. The simple quark-parton model provides the relation

$$\frac{F_2^{eP} - F_2^{eN}}{x} = (1/3)\left[u_v(x) - d_v(x)\right] ;$$

where the v subscript denotes valence. Thus, integration over the Feynman variable, x, yields

$$\int_0^1 \lim \frac{dx}{x} \left[F_2^{eP} - F_2^{eN}\right] = 1/3 .$$

This sum rule simply counts the number of valence quarks (partons) in a nucleon.

gold   Element with atomic number (nuclear charge) 79 and atomic weight 196.9665. The isotope with atomic mass number 197 is the only stable isotope. Although not abundant, *gold* is widely used. Because of its resistance to chemical attack and its excellent electrical and heat conduction, *gold* wire and plating is used extensively in the electronic industry.

Goldberger–Treiman relation   Connects the electromagnetic axial current of the quarks in the first generation with their weak interaction. The electromagnetic vector current is conserved, and the weak vector current is also assumed to be conserved:

$$\partial_\alpha J_V^\alpha = 0 .$$

However, the axial current is not conserved and can be written as

$$\partial_\alpha J_A^\alpha = \frac{f_\pi}{\sqrt{2}} \phi(x) .$$

Here, ϕ is the pion field and $f_\pi = 0.946\ m_\pi^3$. The partially conserved axial vector current, PCAC, in the limit of zero momentum transfer yields

$$\sqrt{2} M g_A m_\pi^2 = f_\pi G ;$$

where M is the nucleon mass, g_A is the weak axial vector coupling constant, and G is the πNN coupling constant.

Goldstone boson The particle required to break the chiral symmetry of the vacuum. These can be scalar or pseudoscalar. *See* Goldstone theorem.

Goldstone theorem In the limit of vanishing quark mass, chiral symmetry is built into the Dirac Lagrangian of fermion fields, i.e., the spin direction along the direction of motion is preserved. This can only be an approximate symmetry of nature and must be spontaneously broken. The *Goldstone theorem* states that if a theory has a global symmetry which is spontaneously broken, there must be a massless boson which breaks the symmetry of the vacuum. In strong interactions, this particle is associated with the pseudoscalar pion, which, while not massless, is light, and the mass is associated with explicit chiral symmetry breaking terms in the Lagrangian.

Goos–Hänchen shift In a fiber, if a wave striking a core-cladding surface has an angle of incidence larger than the critical angle, it will experience total internal reflection. In this process, the wave has an additional phase shift which can be considered an effective shift in the axial position of the wave. The corresponding shift in axial position is called the *Goos–Hänchen shift*. *See Snyder, A.L. and Lore, J.D., Appl. Opt., 15, 236, 1976.*

Gortler vortice (Gortler number) Centrifugal forces have an effect on the stability characteristics of flows with curved streamlines, such as the Couette flow, flows in curved pipes of channels, and boundary layers along concave walls. The instability of these flows is determined by Rayleigh's criteria which states that a necessary and sufficient condition for stability in axisymmetric disturbances is that the square of the circulation does not decrease anywhere, i.e.,

$$\frac{d}{dr}\left|\Omega r^2\right| < 0.$$

Physically, and for the boundary layer along a curve, the Rayleigh criterion can be explained as follows. For this flow, a pressure gradient across the boundary layer $\left(\frac{\partial P}{\partial y} = -\rho \frac{U^2}{R}\right)$ exists and balances the centrifugal force acting normal to the streamlines. If a fluid particle is displaced from an altitude y to $y + \Delta y$ away from the wall by some disturbance, and because the angular momentum about the center of curvature is conserved, the particle will have a smaller angular momentum than its surroundings. The centrifugal force is therefore smaller than what is needed to balance the pressure gradient. Consequently, it would move to larger values of y. Similarly, if the particle was displaced towards the wall, it would be subjected to a pressure gradient that is smaller than the centrifugal force. Consequently, it would move closer to the wall. This instability results in a pattern of counterrotating vortices known as *Gortler vortices*. The above argument does not take into consideration the viscosity, which has a stabilizing effect. The relative effect of the centrifugal force to that of the viscous force is measured by a parameter called the *Gortler number* given by

$$G = \frac{U\theta}{\nu}\left(\frac{\theta}{R}\right)^{1/2}$$

where U is the free stream velocity, θ is the momentum thickness, R is the radius of curvature, and ν is the kinematic viscosity. This number is similar to the Taylor number for the Couette flow and the Dean number for the curved channels. As for the boundary layers over a convex surface, the arguments presented above can be used to show that the curvature is stabilizing.

Goudsmit and Uhlenbeck spin hypothesis In 1925, Goudsmit and Uhlenbeck proposed that electrons possess an intrinsic magnetic moment, μ_s. It was suggested that electrons exhibit spin angular momentum, **S**, in addition to orbital angular momentum. The eigenvalues of **S** were calculated as $\pm\frac{\hbar}{2}$ along some chosen axis, which is normally denoted as the z-axis.

G-parity A *G-parity* operation is a charge conjugation (particle to antiparticle transformation) operation followed by a rotation of π in isotopic spin space about the 1-axis. *G-parity* is a conserved quantum number in all hadronic reactions.

grad-*B* drift In an inhomogeneous magnetic field **B** with non-zero gradient B, the guiding center of a charged particle attains a drift velocity $\mathbf{v} = \pm v_\perp r_L \mathbf{B} \infty B / 2B^2$, where $v_\perp$ is the velocity of the particle perpendicular to the magnetic field and r_L is the Larmor radius. The $\pm$ stands for the sign of the charge. This drift is caused by the difference in r_L in an inhomogeneous magnetic field.

grain boundaries In polycrystalline solids, limits of crystalline structure.

grand canonical distribution Gives the probability of finding a system with N_s particles in a state of energy E_s in equilibrium with a temperature and particle reservoir as

$$P(E_s, N_s) = \exp((\mu N_s - E_s)/k_B T) / \sum_s \exp((\mu N_s - E_s)/k_B T)$$

where kB is the Boltzmann constant, and the summation is over all possible states of the system, denoted by the index s.

grand canonical ensemble Consider a quantum system $\mathcal{S}$, consisting of an ensemble of $\mathcal{N}$ subsystems ($i = 1, \ldots, \mathcal{N}$), each characterized by a wave function $\Psi^{(i)}$, in thermal contact with a reservoir $\mathcal{R}$. If the number of particles in $\mathcal{S}$ is allowed to fluctuate while the total number of particles in the entire system is constant and energy is conserved, the system is called a *grand canonical ensemble*.

grand partition function For a system at constant temperature T and chemical potential μ, all thermodynamic properties can be obtained from the *grand partition function* defined as

$$\mathcal{Z}(T, \mu) = \sum_s \exp((\mu N_s - E_s)/k_B T)$$

where N_s is the number of particles in a state of energy E_s.

grand unification Physics attempts to explain natural phenomenon in terms of a set of fundamental axioms. It is the general goal of physics to reduce this set to its simplest, or most fundamental form. Thus there are continuing attempts to derive the four forces of nature (strong, electromagnetic, weak, and gravitational) from a common set of postulates. Although not yet possible, the electromagnetic and weak forces have at least been unified, and there is some understanding of how one might also include the strong interaction. Although string theory may provide the mathematical basis to unify all forces, this is far from settled at the moment.

graphite An amorphous form of carbon. Because of the low neutron cross-section of carbon and the abundance and ease of production, it was originally used as a moderator in nuclear reactors.

Grashof number A dimensionless parameter that measures the relative effect of buoyant to viscous forces. It generally arises in consideration of free convection heat transfer. It is given by

$$Gr = \frac{\beta g \Delta t L^3}{\nu^2}$$

where $\beta \left(= -\frac{1}{\rho} \frac{\partial \rho}{\partial T} \right)$ is the coefficient of thermal expansion, g is the gravitational acceleration, L is a characteristic length, ΔT is a characteristic temperature difference, and ν is the kinematic viscosity. The *Grashof number* has also been used in mass transport where it is given by

$$Gr = \frac{\varsigma g \Delta x L^3}{\nu^2}$$

where $\varsigma \left(= -\frac{1}{\rho} \frac{\partial \rho}{\partial x} \right)$ is a quantity that is similar to β and represents how much density varies with composition, and Δx is a characteristic concentration difference. The other terms are the same as in the previous equation.

grating equation A periodic arrangement of optical elements such as apertures, obstacles, and scatterers which can produce periodic variation in the phase or amplitude of incident light is called grating. A Fraunhofer diffraction pattern, generated by a grating with a slit separation of a, is given by the equation

$$a \sin \theta_n = n \lambda,$$

where n is the order of the principle maxima, λ is the wavelength, and θ_n is the angle of diffraction.

gravitation The weakest of the four fundamental forces known in nature. Because it is mediated by the massless gravitons and is manifested in observations between massive objects, it is an important force at astronomical distances. The quantum of the gravitational field is the graviton.

gravitational drift A charged particle with mass m and charge q located in a magnetic field **B** under the influence of a general force **F** is subject to a guiding center drift $\mathbf{v} = \mathbf{F} \infty \mathbf{B}/qB^2$. Therefore, in a gravitational field **g**, the guiding center drift of a charged particle becomes $\mathbf{v} = m\mathbf{g} \infty \mathbf{B}/qB^2$. Typically, this *gravitational drift* is negligible. However, in a curved magnetic field, where particles feel an effective gravitational force due to centrifugal force, it may become important. *See also* gravitational instability.

gravitational instability A fluid-type instability that also occurs in plasmas. This is frequently called the Rayleigh-Taylor instability. The hydrodynamic gravitational instability occurs when a light fluid supports a heavy fluid, developing relatively large amplitude ripples at the boundary separating the two fluids. Those ripples tend to grow at the expense of potential energy in the gravitational field. In plasmas, this instability develops when the plasma is inhomogeneous, having a density gradient or a sharp boundary, and is further perturbed by a force of nonelectromagnetic origin. Typically, a *gravitational instability* in a plasma is driven by a nongravitational force such as the centrifugal force of a rotating plasma, and thus has nothing to do with real gravity.

gravitational red shift In the presence of gravity, the frequency of the radiation field is lowered or shifted toward red light. This frequency shift has been measured using the Mössbauer effect.

gravitational wave (1) According to Einstein's theory of general relativity, gravitational waves can be produced by the acceleration of mass. Unlike an electromagnetic wave, which can be produced by oscillations of only one charge, at least two masses are required to generate a *gravitational wave*. Gravitational waves propagate with the speed of light from the point of source and represent a time-dependent distortion of the local space and time coordinates. They follow an inverse square law similar to electromagnetic waves. The effects of *gravitational waves* are very small and very difficult to measure. Possible detectable events include the collision of astronomical objects and the collapse of a large astronomical object. One way to measure the disturbance generated by a *gravitational wave* is by using a Michelson interferometer. Massive laser interferometers such as the laser interferometer gravitational wave observatory (LIGO) have been built to detect *gravitational waves*. Similar efforts have been made in other parts of the world to detect *gravitational waves*. These detectors can detect fluctuations of a few parts in 10^{21}. Efforts are being made to further improve the system so that fluctuations of a few parts in 10^{23} may be detected. Squeezed light and measurements from the space also have been proposed for detecting the *gravitational wave*.

(**2**) The theory of general relativity predicts that the gravitational field will propagate as a wave similar to the propagation of the electromagnetic field, with the exception that the electromagnetic field is a vector and the gravitational field is a tensor of rank two.

graviton The field quanta of the gravitational interaction. It represents a field tensor of rank two, and thus carries an intrinsic angular momentum of $2\hbar$.

gravity waves Waves for which the dominant restoring force is gravity. As the wind blows over the water surface, the pressure and stress deform the water surface, and small rounded waves with V-shaped troughs develop. These waves are called capillary waves, and the dominant restoring force is the surface tension. As the capillary waves gain energy, they increase in height and length. When their wavelength exceeds 1.75 cm, they take on the shape of sine functions, and gravity becomes the most dominant restoring force.

Gravity plays a dominant role in restoring waves with periods between 0.5 sec and 1 minute. These are mostly wind-generated waves which have a peak near a period of about 10 sec. Gravity and surface tension are the principal restoring forces for waves with periods between 0.1 and 0.5 sec. For waves with periods between 1 minute and 1 hour, Coriolis and gravity forces are dominant. In addition to forming at obvious density discontinuities, *gravity waves* form along density interfaces beneath the ocean surface and they are referred to as internal *gravity waves*.

gray (*See* dose.) The unit of exposure to ionizing radiation (dose) and equal to the deposition of 6.24×10^{12} Mev/kg of matter (1 joule/kg).

gray solitons A soliton with a rapid intensity dip which does not decrease all the way to zero. It is similar to a dark soliton with continuous background. It undergoes a phase shift of π at time $t = 0$ but not as abruptly as the dark soliton. It is described by

$$\psi(z,t') = \frac{A}{|B|}\left[1 - B^2 \text{sech}^2\left(\frac{|A|t'}{t_o}\right)\right]^{1/2}$$
$$\exp\left\{i\phi\left(|A|t'/t_o\right) - i\frac{\pi}{2}\frac{|A|^2}{B^2}\frac{z}{z_o}\right\}$$

where

$$\phi(x) = \sin^{-1}\left[\frac{-B\tanh(x)}{\left[1 - B^2 \text{sech}^2(x)\right]^{1/2}}\right],$$

$$t' = t - t_o \frac{\pi|A|\left(1-B^2\right)^{1/2}}{2B}\frac{z}{z_o}.$$

Here, t_o is the pulse duration, z_o ($= \pi t_o^2/2$ (GVD)) is called the characteristic length, GVD is the group velocity dispersion, and B is the blackness parameter. Solitons with $|B| < 1$ are called *gray solitons*. See Tomilson, W.J., Stolen, R.H., and Shank, C.V., *J. Opt. Soc. Am.* **1**, 139, 1984.

Green's function The solution to a linear field equation of physics for a point source. Since the field obeys a linear equation, the solution due to an extended source can be constructed from a superposition of solutions of a point source multiplied by the strength of the source at that point.

Green's function, Schrödinger equation The time-independent *Schrödinger equation* describing the wave function, Ψ, of a particle moving in a potential V can be written as

$$\left(\nabla^2 + k^2\right)\Psi = Q$$

where $k \equiv \frac{\sqrt{2mE}}{\hbar}$ and $Q \equiv \frac{2m}{\hbar^2}V\Psi$. The integral form of the *Schrödinger equation* can be written as

$$\Psi(\mathbf{r}) = \Psi_0(\mathbf{r}+)\int G(\mathbf{r}-\mathbf{r}_0)Q(\mathbf{r}_0)d^3r_0$$

where the *Green's function* $\mathbf{G}(\mathbf{r})$ is defined as

$$\mathbf{G}(\mathbf{r}) = -\frac{e^{ikr}}{4\pi r}$$

and $\Psi_0(\mathbf{r})$ satisfies the free-particle *Schrödinger equation*.

Gross–Llewellyn–Smith sum rule One expects quark distributions in a nucleon to be a function of the Feynman variable, x. These structure functions can be used in integrals over the variable, x, to check quark-parton models of nuclei. Thus, the number of quarks minus anti-quarks is obtained from the sum rule

$$N(q) - N\left(\overline{(}q)\right) = 1/2$$

$$\int_0^1 \left[(u(x) + d(x)) - \left[(\overline{u}(x) + \overline{d}(x))\right]\right] dx.$$

Here, $u(x)$, and $d(x)$ represent the (up, down) quark distributions as a function of the Feyman variable, x.

Grotrian diagram Diagram which shows the energy level structure of an atom and the allowed transitions between the various energy levels. The allowed transitions are usually obtained from specific selection rules.

Grotrian, W. In 1928, *W. Grotrian* suggested that it should be possible to induce transitions among the excited states of the hydrogen atom by using radio waves. This suggestion formed the basis of a very important experiment by Lamb and Retherford in which microwave techniques were used to stimulate radio-frequency transitions between the $2s_{\frac{1}{2}}$ and $2p_{\frac{1}{2}}$ levels of hydrogen. Lamb and Retherford found that the $2s_{\frac{1}{2}}$ level lies above the $2p_{\frac{1}{2}}$ level by an amount of about 1000 MHz. This energy difference is called the Lamb shift.

ground state (1) The state of lowest energy. For example, the solutions for the energies, E_n, and corresponding states, Ψ_n, of the one-dimentional Schrödinger equation for a spinless particle of mass m moving in an infinite square well potential $V(x)$ of width a, i.e.,

$$V(x) = \begin{cases} 0, & \text{if } 0 \leq x \leq a \\ \infty, & \text{otherwise} \end{cases}$$

are

$$\Psi_n = \sqrt{\frac{2}{a}} \sin\left(\frac{n\pi}{a}x\right)$$

$$E_n = \frac{n^2\pi^2\hbar^2}{2ma^2}, \quad n = 1, 2, 3, \ldots.$$

The lowest energy occurs when the quantum number $n = 1$, and therefore the ground state Ψ_1 is

$$\Psi_1 = \sqrt{\frac{2}{a}} \sin\left(\frac{\pi}{a}x\right)$$

(2) The lowest energy level for a particle in a given system.

group delay The time delay τ ($= z/v_g$) of a wave packet traveling a distance z with a group velocity v_g.

group delay dispersion (GDD) Measures the pulse broadening per unit of bandwidth of the pulse. For a pulse centered at frequency ω_o after traveling a distance z, the GDD is given by

$$z\left(\frac{d^2k}{d\omega^2}\right)_{\omega=\omega_0}$$

A small value of GDD is preferred in short pulse propagation and fiber optics communications.

group velocity The velocity of a wave packet. The *group velocity* of a wave packet with the envelope centered at frequency ω_o is given by

$$v_g = 1 / \left(\frac{dk}{d\omega}\right)_{\omega=\omega_0},$$

where k is the wave vector.

group velocity dispersion (GVD) Measures pulse broadening per unit of length of the medium per unit bandwidth of the pulse. It is given by

$$GVD = \left[\frac{d(1/v_g)}{d\omega}\right]_{\omega=\omega_0} = \left[\frac{d^2k}{d^2\omega}\right]_{\omega=\omega_0}$$

group velocity of a free particle The general solution of the one-dimensional Schrödinger equation for a particle traveling in free space (zero potential), $\Psi(x, t)$, takes the form

$$\Psi(x, t) = \frac{1}{\sqrt{2\pi}} \int_{-\infty}^{\infty} \phi(k) \exp(kx - \omega t) \, dk$$

where (x, t) is a space-time point, k is the wave vector, and $\omega = \frac{\hbar k^2}{2m}$ is the angular frequency. $\phi(k)$ is a narrowly-peaked amplitude whose spread in k-space represents the spread in momentum of the particle. If we assume $\phi(k)$ is peaked about k_0, then the group velocity, v_g, of such a wave packet $\Psi(x, t)$ is defined as

$$v_g = \left.\frac{d\omega}{dk}\right|_{k=k_0}$$

Physically, it represents the speed of the envelope.

guiding center The instantaneous center of a charged particle gyrating around a uniform magnetic field with a radius called the Larmor radius. Therefore, the orbit of a charged particle moving in a uniform magnetic field may be described as the sum of the circular gyration due to the magnetic field and a drift of the *guiding center*.

guiding-center approximation Used to describe some properties of a plasma in a slightly non-uniform magnetic field; based on Taylor expansion of the magnetic field with the use of the position of the particle relative to its guiding center. *See also* guiding center.

guiding center plasma To describe some plasmas, it is sufficient to follow the motion of the guiding centers of charged particles only. Such plasmas are called *guiding center plasmas*. *See also* guiding center.

Guoy effect In 1890, Guoy discovered that any beam with a finite cross-section acquires a phase shift of π radians when it passes through the region where the beam focuses. He measured the phase shift using interference pattern on planes, which were before and after the focusing region. For a Gaussian beam the Guoy phase shift is given by

$$\phi(z) = \tan^{-1}(z/z_s) \,,$$

where z is the distance from the beam waist and z_s is Rayleigh range.

Gupta–Bleuler formalism In calculating the expectation value of the Hamiltonian describing the electromagnetic field in the Lorentz gauge with respect to some chosen photon Fock state $|\psi\rangle$, it is found that negative expectation values are possible for some of the states. Such states are, therefore, nonphysical. To deal with the unwanted negative values, Gupta and Bleuler demanded that physical states must satisfy the requirement that $\partial_\mu A^{(+)\mu} |\psi\rangle = 0$, where $\partial_\mu = \left(\frac{1}{c}\frac{\partial}{\partial t}, \frac{\partial}{\partial x}, \frac{\partial}{\partial y}, \frac{\partial}{\partial z}\right)$ are derivatives with respect to time t and space coordinates x, y, and z, c is the speed of light *in vacuo*, and $A^{(+)\mu}$ are components of the positive frequency part of the four-vector potential. This condition gives positive expectation values of the Hamiltonian for physical states satisfying the Gupta-Bleuler condition.

g_v The vector coupling constant. See g_a.

gyrofrequency See cyclotron frequency.

gyromagnetic ratio The magnetic moment of an elementary particle may be written as

$$\mu = g\mu_B \mathbf{S} \,;$$

where $\mathbf{S}$ is the spin operator, μ_B is the classical moment $\frac{e\hbar}{2mc}$, and g is the Lande g-factor. The *gyromagnetic ratio* is $g\mu_B$.

gyroradius See Larmor radius.

gyroscope, laser Gyroscopes have been used for detecting angular motion of a system. A *laser gyroscope* is made of a ring laser in which two monochromatic light beams are propagating in opposite directions, one clockwise and the other counterclockwise. Superposition of these two waves forms an interference pattern. The angular motion of the gyroscope produces a difference in the time required to complete the cycle by the two beams. This time difference gives rise to phase difference, which are detected by interferometeric measurements.

gyrotron A device for generating high-power microwaves, in which a strong axial magnetic field in a cavity resonator is utilized to bunch an electron beam. The bunched electron beams then act as the efficient sources of microwave radiation.

H

hadron From the Greek word *hadros*, which means strong. Particle that interacts through the strong interaction. *Hadrons* have a complex internal structure in terms of quarks. They can be divided into two groups: the *baryons*, which have half-integer spin (*see* intrinsic angular momentum), and the mesons, which have zero of integer spin. For example, the proton and the neutron are baryons; the pion is a meson.

hadronic force(s) The strong forces between hadrons. For instance, the attractive force that keeps protons and neutrons together inside nuclei is a *hadronic force*. The *hadronic force* is a residual interaction of the (fundamental) strong force which acts between quarks via gluon exchange. Since hadrons are colorless multiquark systems, the residual force between them is much weaker than the one between quarks (inside hadrons). An analogy for the *hadronic force* is the interatomic force that holds, e.g., the H_2 molecule together. This residual electromagnetic interaction is much weaker than the Coulomb force that holds protons and electrons together inside one hydrogen atom. *See also* hadron.

Hagen–Poiseuille flow For a steady, incompressible fully developed flow in a straight pipe or duct section, the influx and outflux of momentum are equal. Thus, the flow is in equilibrium under the action of static pressure, gravity, and viscous stresses. For a Newtonian fluid and when the flow is laminar, the viscous stress is proportional to the velocity gradient $\frac{du}{dy}$. One can, thus, use the equilibrium of forces to relate the velocity gradient to the pressure drop in the direction of the flow. For a circular pipe with radius r_o, this relationship is given by:

$$-\mu \frac{du}{dr} = -\frac{r}{2}\frac{d}{dx}(p + \rho g h).$$

By applying the no-slip boundary condition, the above equation can be integrated to give

$$u(r) = \frac{1}{4\mu}\frac{d(p + \rho g h)}{dx}\left(r^2 - r_0^2\right).$$

The velocity profile is therefore parabolic for steady laminar flow in a circular pipe. The volume flow rate is obtained by integrating $u(r)$ over a cross-section to give

$$Q = -\frac{\pi r_0^4}{8\mu}\frac{d(p + \rho g h)}{dx}.$$

For a horizontal pipe with a pressure drop Δp over a section of length L, the pressure drop and flow rate are thus related by

$$\Delta p = \frac{8\mu Q L}{\pi r_0^4}.$$

The relations for the velocity profile and flow rate in a duct of height D and width W in terms of the pressure drop are

$$u(y) = \frac{D^2}{8\mu}\left(\frac{d(p + \rho g h)}{dx}\right)\left[\left(\frac{2y}{D}\right)^2 - 1\right]$$

and

$$Q = \frac{WD^3}{12\mu}\left(\frac{d(p + \rho g h)}{dx}\right).$$

The differential governing equations for both pipe and duct flows could have also been obtained from the Navier-Stokes equations with the elimination of different terms based on the assumptions made above.

Hahn, Otto German physical chemist (1879-1968). Based upon research conducted with Lise Meitner and Fritz Strassmann, he discovered fission of uranium in 1938, for which he received the Nobel Prize in chemistry in 1944.

hairpin vortices Observed in boundary layers. They have an axis in the transverse direction with legs near the wall and a head and neck that extend away from the wall. They have been associated with intense Reynolds stress producing events.

Haken representation A representation for describing atomic states, introduced by Haken and his co-workers, based on a characteristic function in a normal order. The characteristic function in this representation is defined as

$$\chi_N\left(\xi, \xi^*, \eta\right) = \text{trace}\left(\hat{\rho} \exp\left(-i\xi\hat{\sigma}_+\right)\right.$$
$$\left.\exp\left(-i\eta\hat{\sigma}_z\right) \exp\left(-i\xi\hat{\sigma}_-\right)\right)$$

and the quasi-probability distribution is given by

$$p\left(v, v^*, m\right) =$$
$$\frac{1}{(2\pi)^3} \int d^2\xi \int d\eta \, \chi_N \, \exp\left(-i\xi^*v\right)$$
$$\exp\left(-i\xi v^*\right) \exp(-i\eta m).$$

(For more detail, *see* the papers by Haken, H., Risken, and Weidlich, *Z. Physik*, 206, 355, 1967; Haken, H., in *Light and Matter I c,* edited by Genzel, L., Handbuch der Physik vol. XXV/2C, pp. 64-65, Springer-Verlag, Berlin, 1970.)

half-life   The time after which a sample of radioactive nuclei will have reduced to one-half of its initial number. The law of radioactive decay is

$$N(t) = N(0)e^{-\lambda t}$$

with $N(t)$ representing the number of nuclei present at any time t, $N(0)$ representing the initial number of nuclei, and λ denoting the disintegration constant. Setting $N = N(0)/2$ in the above equation and solving for the corresponding time (namely, the *half-life $t_{1/2}$*), one obtains

$$t_{1/2} = \frac{\ln 2}{\lambda}.$$

half-life (excited state)   The time it takes half the atoms in a large sample to make a transition from some excited state.

half-wave plate   Optical device generally made from uniaxial crystals which are birefringent. It is used for introducing a relative phase difference of π radians between two coherent waves with polarizations parallel and perpendicular to the symmetry axis of the crystal. Thickness d of the plate is determined from

$$d\,|n_1 - n_2| = (2m+1)\lambda/2$$

where λ is the wavelength of the light, and n_1 and n_2 are the refractive index of waves with polarizations parallel and perpendicular to the symmetry axis.

Hall coefficient   The constant of proportionality in the relation of the transverse electric field to the product of current density and magnetic flux density.

Hall constant   The transverse electric field divided by the product of the current density and the magnetic field strength.

Hall effect   The development of an electric field between the two faces of a current-carrying material whose faces are perpendicular to a magnetic field.

Hall mobility   The product of the conductivity and the Hall constant which gives a measure of the mobility of the electrons or holes in a conductor.

halo, nuclear   In some nuclei which contain a much larger number of neutrons than protons (or vice versa), some neutrons (protons) are only very loosely bound to the nucleus. Their effective radius R is much larger than what is obtained from the formula $R = r_0 A^{1/3}$ (where A is the total number of neutrons and protons, and r_0 is a constant approximately equal to 1.2×10^{-15} m), which gives correct estimates for the radius of most atomic nuclei. That is, for such nuclei, neutrons (protons) have a considerable probability to be found outside this radius. This phenomenon is known as *nuclear halo*. Scattering experiments with beams of such nuclei are performed to investigate the properties of these so-called halo nuclei.

Hamiltonian   (1) A function first introduced by Sir William Rowan Hamilton (1805-1865) in the context of classical mechanics. Any suitable set of (independent) coordinates which specifies the position of every part of a system is said to be a set of generalized coordinates. Considering, for simplicity, a system described by one generalized coordinate q and the associated generalized momentum p, the *Hamiltonian* is defined as

$$H = p\dot{q} - L(q, \dot{q})$$

where $\dot{q}$ (the generalized velocity) is equal to dq/dt, and L is the Lagrangian. In the above expression, the generalized velocity must be expressed in terms of the generalized momentum; that is, the *Hamiltonian* is an explicit function of p and q (and, in some cases, the time t). In many cases of interest, the *Hamiltonian* can be equal to the total energy of the system. Namely,

$$H = \frac{p^2}{2m} + V(q)$$

where the first term on the right is the non-relativistic kinetic energy, and V is the potential energy function. This is the classical *Hamiltonian*, or Hamilton's function. The *Hamiltonian* operator in quantum mechanics is obtained by replacing the coordinates and momenta with the appropriate quantum mechanical operators.

(2) The total energy (kinetic plus potential) of a quantum mechanical system expressed in operator form. For example, in a one-dimensional system consisting of a single particle moving in a potential, the *Hamiltonian* operator $\hat{H}$ is given as

$$\hat{H} = -\frac{\hbar^2}{2m}\frac{d^2}{dx^2} + V(x)$$

where $\hbar = \frac{h}{2\pi}$ and h is Planck's constant, m is the mass of the particle, and V is its potential energy.

(i) of a charged particle in an electromagnetic field: The problem is finding an appropriate Lagrangian L such that Lagrange's equation of motion

$$\frac{d}{dt}\left(\frac{\partial L}{\partial \dot{q}_i}\right) - \frac{\partial L}{\partial q_i} = 0, \quad i = 1, 2, \ldots$$

with q_i as generalized coordinates. For a particle of charge q, mass m, and velocity $\mathbf{v}$ in an electromagnetic field described by electric and magnetic fields $\mathcal{E}(\mathbf{r},t)$ and $\mathcal{B}(\mathbf{r},t)$ respectively, the Lorentz force $\mathbf{F}$ exerted on the particle is given as

$$\mathbf{F} = q\left(\mathcal{E} + \mathbf{v} \times \mathcal{B}\right)$$
$$= q\left(-\nabla\phi - \frac{\partial \mathbf{A}}{\partial t} + \mathbf{v} \times (\nabla \times \mathbf{A})\right)$$

where ϕ and $\mathbf{A}$ are the scalar and vector potentials respectively. If we take L as

$$L = \frac{1}{2}mv^2 - q\phi + q\mathbf{v}\cdot\mathbf{A}$$

we obtain, in Cartesian coordinates, $\mathbf{F} = m\ddot{\mathbf{r}}$. The *Hamiltonian*, H, is then obtained from

$$H = \sum_{i=1}^{3} p_i \dot{q}_i - L$$
$$= \frac{1}{2m}(\mathbf{p} - q\mathbf{A})^2 + q\phi.$$

(ii) for hydrogen in the presence of a constant magnetic field: The non-relativistic Schrödinger equation for the hydrogen atom in the presence of a constant magnetic field is

$$H = -\frac{\hbar^2}{2m}\nabla^2 - \frac{e^2}{4\pi\varepsilon_0 r} - \frac{i\hbar e}{m}\mathbf{A}\cdot\nabla + \frac{e^2}{2m}\mathbf{A}^2$$

where m and e are the electron's mass and charge respectively, and $\mathbf{A}$ is the vector potential.

(iii) for many-electron atoms: A many-electron atom consists of a nucleus of charge Ze (Z being the atomic number of the atom), and N electrons each of charge $-e$. For an infinitely heavy nucleus and considering only the attractive Coulomb interactions between the electrons and the nucleus and the coulomb repulsions between the electrons, we write the *Hamiltonian* of the N-electron atom (ion) in the absence of external fields as

$$H = \sum_{i=1}^{N}\left(-\frac{\hbar^2}{2m}\nabla_{r_i}^2 - \frac{Ze^2}{(4\pi\varepsilon_0)r_i}\right)$$
$$+ \sum_{i<j=1}^{N}\frac{e^2}{(4\pi\varepsilon_0)r_{ij}}$$

where $\mathbf{r}_i$ denotes the relative coordinate of electron i with respect to the nucleus, and $r_{ij} = |\mathbf{r}_i - \mathbf{r}_j|$. The last summation is a summation over all pairs of electrons.

(iv) for a rigid rotator: Consider two particles, each of mass m attached to the ends of a massless rigid rod of length a. The system is free to rotate in three dimensions about the center, which is kept at a fixed position. This system represents the rigid rotator. For masses moving with speed v, the total energy of the system $H = 2\left(\frac{1}{2}mv^2\right) = mv^2$. The magnitude of the orbital angular momentum $|\mathbf{L}| = L = 2\left[\frac{a}{2}mv\right] = amv$. Therefore, the *Hamiltonian* $H = \frac{L^2}{ma^2}$. The eigenvalues E_l of this

Hamiltonian are proportional to the eigenvalues of L^2 which is $\hbar^2 l(l+1)$. Therefore $E_l = \frac{\hbar^2}{ma^2} l(l+1)$.

hamiltonian for an atom-field interaction

Consists of three terms, which correspond to free atom ($\hat{H}_A$), free field ($\hat{H}_F$), and atom-field interaction ($\hat{H}_I$). The field is described in terms of a harmonic oscillator, and for most of the problems considered in quantum optics, the atom-field interaction can be approximated by dipole contribution.

$$\hat{H} = \hat{H}_A + \hat{H}_F + \hat{H}_I,$$

$$\hat{H}_A = \sum_j E_j |j\rangle\langle j|,$$

$$\hat{H}_F = \sum_k \hbar\omega_k \left(\hat{a}_k^\dagger \hat{a}_k + 1/2\right)$$

$$\hat{H}_I = \sum_{ijk} [e |i\rangle\langle i|\mathbf{r}|j\rangle\langle j|] \cdot \left[\varepsilon_k \left(\frac{\hbar\omega_k}{2\epsilon_o V}\right)^{1/2} \left(\hat{a}_k^\dagger + \hat{a}_k\right)\right].$$

Here, E_j is the energy of the atomic level $|j\rangle$, $\hat{a}_k^\dagger$ and $\hat{a}_k$ are creation and annihilation operators describing the field, ε_k is the polarization vector of the kth mode, and V is volume. A special case of this Hamiltonian is the Jaynes–Cummings model, for which a single two-level atom is considered.

Hanbury–Brown–Twiss experiment Experiments performed by Hanbury, Brown, and Twiss in the 1950s to measure the correlation between two partially correlated optical intensities. These were the first measurements of the second order intensity correlation function and the results were published in *Nature*, 177, 27, 1957, *Proc. Roy. Soc.*, 142, 300, 1957, and *Proc. Roy. Soc.*, 143, 241, 1958. In the original experiment, light with in frequency of 435.8 Hz from a mercury lamp was split into two parts at a beam splitter, and the intensity of each beam was detected by separate detectors. One of the detectors was mounted on a sliding track to introduce path difference. Outputs from the two detectors were correlated to measure the intensity correlation function.

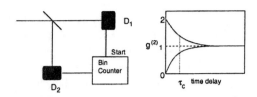

Principle of a Hanbury–Brown–Twiss experiment to measure the photon statistics of a stream of photons and qualitative plot of the second order correlation function $g^{(2)}$ for thermal and non-classical light. The characteristic time τ_c is called the coherence time and is a measure for the temporal coherence of the light source.

handedness A property of a particle associated with the direction of its spin (*see* intrinsic angular momentum) relative to the direction of its momentum. For instance, the spin of the neutrino is always antiparallel to its momentum, while the spin of the antineutrino is always parallel to its momentum. This gives the neutrino a definite *handedness*, with the same meaning attributed to the *handedness* of a screw. Within this analogy, the neutrino can be compared to a left-handed screw, while the antineutrino can be compared to a right-handed screw. The *handedness* is also expressed in terms of helicity, with the neutrino and the antineutrino having helicity -1 and $+1$, respectively. *See also* helicity.

Hanle effect Atomic coherence measurement performed by Hanle in 1924. He showed that when atoms are placed in a weak magnetic field applied in the z-direction and an electric pulse propagating in z-direction with polarization in the x-direction is applied, the scattered light could have a polarization in the y-direction.

Hanle laser In a *Hanle laser* the active medium is made of three level atoms of V type configuration. Upper levels are prepared in coherent superposition and decay to a lower state by emitting radiation with a different polarization. The coherently prepared atoms generate correlated spontaneous emission resulting in reduction in noise and coherent light source.

hard core (of the nuclear force) Very strong repulsive component of the nuclear force experienced by nucleons (protons and neutrons) at relative distances of less than approximately 0.5 fm (1 fm = 10^{-15} m).

hardness Property of a solid determined by its ability to abrade or indent another solid.

hard sphere interaction The interaction of particles of finite size modeled with an interparticle interaction that is infinitely repulsive if the separation between the centers of the two particles becomes less than the diameter of the particle, i.e., the spheres representing the particles are completely impenetrable. This is also known as the excluded volume interaction.

hard superconductor A superconductor that requires a strong magnetic field to destroy superconductivity.

harmonic generation A monochromatic light of frequency ω passing through a non-linear crystal generates non-linear polarization which is proportional to higher powers of the electric field. This effectively can generate higher order harmonics such as 2ω, 3ω, etc. Second *harmonic generation* is most commonly used in nonlinear optics. *Harmonic generation* requires phase matching and energy conservation within uncertainty limits.

harmonic oscillator A particle acted on by a linear restoring force that is, a force proportional to the distance of the particle from its equilibrium position and opposite the direction of the displacement. In the presence of such a force, a particle performs harmonic oscillations around its equilibrium position. A particle attached to a spring is one example.

harmonic oscillator (linear) A prototype for systems exhibiting small vibrations about an equilibrium point. The Hamiltonian operator $\widehat{H}$ for such a system is

$$\widehat{H} = -\frac{\hbar^2}{2m}\frac{d^2}{dx^2} + \frac{1}{2}kx^2$$

where k is a force constant. The eigenvalues of the Hamiltonian, E_n, consist of an infinite sequence of non-degenerate discrete levels as

$$E_n = \left(n + \frac{1}{2}\right)\hbar\omega, \qquad n = 0, 1, 2, \ldots$$

where $\hbar = \frac{h}{2\pi}$ and h is Planck's constant, and ω is the angular frequency.

harmonic oscillator potential The potential energy function associated with a linear restoring force; *see also* harmonic oscillator. For a simple harmonic oscillator, the potential energy function has the quadratic form

$$V(x) = \frac{1}{2}kx^2$$

with k denoting the force constant and x denoting the displacement from the equilibrium position.

harmonic oscillator wave function The solution of the Schrödinger equation in the presence of a harmonic oscillator potential. *See* harmonic oscillator potential, harmonic oscillator.

Harris instability A type of microinstability in plasmas driven by temperature anisotropes. Its basic energy source lies in the thermal energy of the particle gyration, and thus it occurs in a magnetized anisotropic plasma in which the temperature perpendicular to the magnetic field is higher than the temperature parallel to the field. Excited electrostatic waves have frequencies around the electron (or ion) cyclotron frequency or its harmonics and travel obliquely to the magnetic field. They are observed mainly in laboratory plasmas.

Hartree equation A single-particle Schrödinger equation is called the *Hartree equation*.

Hartree–Fock method The Hartree method extended by properly antisymmetrizing the many-fermion wave function. *See* Hartree method theory.

Hartree method theory A method to self-consistently derive the best single-particle potentials and wave functions for a system of many interacting fermions. The method is based upon the assumption that the mutual interactions among the particles lead to an average potential felt by each particle. The *Hartree method* and the closely related Hartree–Fock method allow accurate predictions of atomic energy levels and wave functions. This method also works reasonably well for the shell model of the atomic nucleus.

harvard classification   A classification arising out of the study of the absorption spectra of stars. Stars were classified according to the strength of the hydrogen lines in their spectra. Letters of the alphabet were used to identify the classes, with class A corresponding to the stars having the strongest hydrogen lines, class B the next strongest, etc.

H center   A lattice defect (hole center) in alkali crystal.

head   A term used to express different quantities with the dimension of length. For instance, it is sometimes convenient to express the pressure in terms of a height of a column of fluid rather than in terms of a force per unit area. This height is then referred to as pressure *head*. In a similar manner, one can define a velocity *head* $\left(\frac{V^2}{2g}\right)$. The sum of the elevation or geodetic *head*, velocity *head*, and pressure *head* is then referred to as the total *head*. In a similar manner, it is also common to refer to the energy term, $\left(\frac{\dot{W}}{mg}\right)$, associated with a pump or a turbine as a pump *head* or a turbine *head*, and to the energy loss per unit weight of a fluid as a *head* loss.

head loss   *See* head.

heat   The energy transferred to or from a system due to a thermal interaction with another system. *Heat* is more rigorously defined in terms of the first law of thermodynamics as $\Delta Q = \Delta U - \Delta W$, where ΔU is the change in the internal energy and $-\Delta W$ is the net work done by the system.

heat capacity   The amount of energy transferred to a system that raises its temperature by one degree.

heat conduction   The process by which thermal energy is transported directly through a material across a temperature gradient from one place to another without a bulk transport of particles.

heat engine   A device for the conversion of heat into work.

heat exchanger   A device used to remove heat from a hot object by transferring the thermal energy to a large reservoir.

heat of fusion   The latent heat that is removed per unit of mass (or per unit mole) from a substance undergoing a phase transformation from a liquid to a solid.

heat of reaction   The change in the enthalpy per unit of chemical reaction in the vicinity of the equilibrium state.

heat of vaporization   The latent heat that is added per unit of mass (or per unit mole) to a substance undergoing a phase transformation from a liquid to a gas.

heat pump   A device that extracts heat from a cold reservoir and pumps it to an enclosure at a lower temperature with the input of work.

heat reservoir   A large system whose temperature remains essentially unchanged when heat flows into or out of it.

heavy bosons   The $W^{\pm}$ (mass of 80 GeV/c^2) and the Z^0 (mass of 91 GeV/c^2) bosons. These bosons are understood to be the carriers of the weak interaction.

heavy ions   Charged particles resulting from adding charges to or removing charges from (heavier) atoms, a process known as ionization. *See also* ion, ionization chamber.

heavy meson   A strongly interacting particle (*see also* hadron) with zero or integer spin. Heavy typically refers to the mass of a meson relative to the lightest meson, the pion (π), which has a mass of approximately 140 MeV/c^2.

heavy-water reactor   A nuclear reactor using heavy water (D_2O) as a moderator instead of ordinary (light) water (H_2O). Light water reactors have difficulties reaching critical conditions because of the large probability of neutron absorption by hydrogen (or a large neutron absorption cross-section). Reaching critical conditions is therefore facilitated when hydrogen is replaced with deuterium, which has a smaller neutron capture cross-section. *Compare with light-water reactor.*

Heisenberg–Langevin equations In quantum treatment of an atom-field system, damping is introduced by coupling the system of interest with a large reservoir. It is assumed that the reservoir has a large degree of freedom, and therefore the system does not affect the reservoir significantly. Generally, the evolution of the reservoir is not of much interest. The evolution of the system is obtained by adding together the reservoir operators, which results in equations of motion for the system operators. These equations for the system operators have a form similar to the classical Lengevin equation with damping and noise operator terms. These equations are called Heisenberg–Langevin equations. For example, a Heisenberg–Langevin equation for an atomic system correctly describes spontaneous emission.

Heisenberg uncertainty principle This principle was formulated by W. Heisenberg in 1927. The essence of the principle is that certain pairs of variables describing for example, a particle, cannot be determined with arbitrary precision. These pairs of variables are often called complementary variables and some examples are energy (E), time (t), position (**r**), and momentum (**p**). The following are the mathematical forms of the commutation relations:

$$\Delta E \Delta t \gtrsim \hbar$$
$$\Delta x \Delta p_x \gtrsim \hbar, \quad \Delta y \Delta p_y \gtrsim \hbar, \quad \Delta z \Delta p_z \gtrsim \hbar$$

where $\mathbf{r} = \mathbf{i}x + \mathbf{j}y + \mathbf{k}z$ is the particle position in terms of Cartesian components (x, y, z); **i**, **j**, and **k** are unit vectors along the x-, y-, and z-axes. $\Delta x = [\langle (x - \langle x \rangle)^2 \rangle]^{\frac{1}{2}}$ and $\Delta p_x = [\langle (p_x - \langle p_x \rangle)^2 \rangle]^{\frac{1}{2}}$ and $\langle x \rangle, \langle p_x \rangle$ are expectation values of the position and momentum, respectively. Similar definitions apply to the y and z components.

Heitler–London method In this method for obtaining the molecular wave functions, the orbitals of the separated atoms are used as the trial wave functions in any variational method used to obtain the approximate molecular wave functions.

helicity A property of a particle associated with the component of its spin (*see* intrinsic angular momentum) along the direction of the particle motion. *See also* handedness.

helicon modes Also called the whistler mode, the *helicon mode* is an electromagnetic right-hand circularly polarized mode present in a magnetic field; the mode propagates predominantly parallel to the magnetic field, and has a frequency somewhere between the proton (or ion) cyclotron frequency and the electron cyclotron frequency. At frequencies lower than the proton (or ion) cyclotron frequency, the *helicon mode* branch of the dispersion relation is connected to the magnetosonic mode branch. These modes are of extreme importance in the study of ionospheric phenomena and condensed matter. They are also heavily used in plasma processing.

helium (**1**) The second lightest chemical element. The *helium* atom contains two electrons. The nucleus of *helium* consists of two protons and two neutrons, and is known as the α-particle. Due to its closed-shell atomic structure, *helium* does not form chemical bonds with any other element and is therefore known as one of the noble gases.

(**2**) A two-electron atom with atomic number $Z = 2$. Assuming that the nucleus of this atom is at rest, its Hamiltonian can be written as

$$\widehat{H} = \left(-\frac{\hbar^2}{2m}\nabla_1^2 - \frac{1}{4\pi\varepsilon_0}\frac{2e^2}{r_1}\right)$$
$$+ \left(-\frac{\hbar^2}{2m}\nabla_2^2 - \frac{1}{4\pi\varepsilon_0}\frac{2e^2}{r_2}\right)$$
$$+ \frac{1}{4\pi\varepsilon_0}\frac{e^2}{|\mathbf{r}_1 - \mathbf{r}_2|}$$

which consists of the sum of two hydrogenic (with nuclear charge $2e$) Hamiltonians, one for electron 1 and one for electron 2, with the symbols representing their usual meanings. The final term describes the repulsion energy of the two electrons. The spatial eigenfunctions of *he-*

lium can be either space-symmetric (para states) or space-antisymmetric (ortho states). If the effect of the total spin of the two electrons is included in the total wave function, one finds that the para state is coupled to the spin singlet state ($S = 0$ and $M_S = 0$) and the ortho state is coupled to one of three spin states (spin triplet, $S = 1, M_S = -1, 0, 1$).

Helmholtz equation For a function $\psi(\mathbf{r})$, the inhomogeneous *Helmholtz equation* takes the form

$$\nabla^2 \psi(\mathbf{r}) + k^2 \psi(\mathbf{r}) = -\rho(\mathbf{r})$$

where k is a constant and $\rho(\mathbf{r})$ is a scalar function of the spatial coordinate $\mathbf{r}$. The homogeneous form of the equation is

$$\nabla^2 \psi(\mathbf{r}) + k^2 \psi(\mathbf{r}) = 0.$$

In quantum mechanics, we often write the time-independent Schrödinger equation as an inhomogeneous *Helmholtz equation*,

$$\nabla^2 \psi(\mathbf{r}) + k^2 \psi(\mathbf{r}) = Q$$

where $k = \frac{\sqrt{2mE}}{\hbar}$ and $Q = \frac{2m}{\hbar^2} V \psi(\mathbf{r})$; m is the particle mass, V is the potential energy, E is the total energy, and $\hbar = \frac{h}{2\pi}$ where h is Planck's constant.

Helmholtz free energy (F) Defined as $F = U - TS$, where U and S denote the internal energy and entropy, respectively, at temperature T and volume V. The physical significance of F is that it is a minimum for a system of constant volume in equilibrium with a temperature reservoir.

Helmholtz theorem A theorem that describes the rate of change of vorticity and is stated as follows. If there exists a potential for all forces acting on a non-viscous fluid, no fluid particle can have a rotation if it did not originally rotate, fluid particles always belong to the same vortex line, and vortex filaments must be either closed tubes or end on the boundaries of the fluid. The *Helmholtz theorem* applies for incompressible and homogeneous flow and is proven by taking the curl of Euler's equation

$$\nabla x \left(\frac{D\vec{v}}{Dt} \right) = \nabla x \left(\vec{f} - \frac{\nabla p}{\rho} \right).$$

Using vector identities and manipulating the equations, one arrives at the following equation:

$$\frac{D\vec{\omega}}{Dt} = \vec{\omega} \cdot \nabla \vec{v} - \vec{\omega} \nabla \cdot \vec{v}$$

where $\vec{\omega}$ is the vorticity vector. For incompressible flow, $\nabla \cdot \vec{v} = 0$, and one is left with

$$\frac{D\vec{\omega}}{Dt} = \vec{\omega} \cdot \vec{\nabla} \vec{v}.$$

The term on the right represents the action of velocity variations on the vorticity. Using this equation, one can show that the changes in length and direction of a line joining two elements on a vortex line (line drawn in the direction of local vorticity) are exactly equal to the changes of the corresponding vorticity vector. Therefore, fluid elements on a certain vortex line will always remain there. The above equation also shows that if the vorticity is zero, then

$$\frac{D\vec{\omega}}{Dt} = 0$$

i.e., if a fluid element has no vorticity at some instant, it can never gain any vorticity. In other words, under the action of potential forces, all motions of an inviscid incompressible fluid set up from a state of rest or uniform motion are permanently irrotational.

He-Ne lasers One of the most commonly used gas lasers. The first *He-Ne laser* was constructed in 1960 and operated at wavelength 1.15 μm by A. Javan, W.R. Bennett, Jr., and D.R. Harriott. It was the first gas laser and first continuous wave (cw) laser constructed. Laser action is achieved from the transition of Ne atoms. Ne atoms are excited by collision with He atoms which, being nearly resonant, facilitate the pumping process. Typically, the ratio of He to Ne varies from 5:1 to 10:1, and total gas pressure in the tube is about 1 torr. It can operate at various wavelengths such as 632.8 nm (red), 543 nm (green), 1.15 μm (infrared), and 3.39 μm (infrared). In an He-Ne laser, lasing of a red line is achieved by using mirrors which are highly reflecting for a 632.8 nm wavelength but not for wavelengths corresponding to other transition lines, thus suppressing other modes.

Hermite-Gaussian modes *See* Gaussian beam and TEM modes.

Hermite polynomials The one-dimensional Schrödinger equation for a particle of mass m in a linear harmonic oscillator potential, $V = \frac{1}{2}kx^2$, admits solutions of the form $\Psi_n(\xi)$, such that

$$\Psi_n(\xi) = e^{-\frac{\xi^2}{2}} H_n(\xi), \qquad n = 1, 2, 3, \ldots$$

where $\xi = \left(\frac{mk}{\hbar^2}\right)^{\frac{1}{4}} x$ for a particle of mass m and force constant k. $H(\xi)$ are *Hermite polynomials* defined as

$$H_n(\xi) = (-1)^n e^{\xi^2} \frac{d^n e^{-\xi^2}}{d\xi^n}.$$

Hermitian operators In quantum mechanics, observable quantities are represented by *Hermitian operators*. A *Hermitian operator* $\widehat{T}$ satisfies $\widehat{T}^\dagger = \widehat{T}$, where $\widehat{T}^\dagger$ is the Hermitian conjugate of $\widehat{T}$. *Hermitian operators* have very useful properties, including real eigenvalues, their eigenvectors belonging to distinct eigenvalues are orthogonal, and their eigenvectors span the space.

heterodyne detection A technique used for reducing background noise. In *heterodyne detection*, a signal beam is superposed at a beam splitter with a coherent beam of a local oscillator of different frequency and constant relative phase. It is used for demodulating FM and AM signals.

heteronuclear/heteropolar molecule A molecule whose atoms are not identical. *Compare with* homonuclear/homopolar molecule.

heterostructure lasers Semiconductor lasers which are made of heterojunctions. A heterojunction is made of layers of two different types of materials which are doped by p and n types of atoms. For example, layers of n and p type GaAs and AlGaAs can form a heterojunction. The layered structure gives rise to a larger energy band gap than a homojunction, and the active region has greater confinement of electrons and holes. A larger refractive index of GaAs also helps in laser action by confining the electromagnetic radiation and current in the active region. These lasers have a low threshold current and can operate at room temperature. Also, because of their small size, these lasers are suitable for many practical applications such as fiber-optics communications.

hidden variables The theory of *hidden variables* is based on the premise that the wave function of a system does not provide all the information about a quantum system. Additional information is needed through variables called *hidden variables* to describe the system completely. It is believed that the indeterminism arising in quantum mechanics is due to our lack of knowledge of such variables.

Higgs particle A boson whose rest mass is expected to be of the order of 1 TeV (10^{12} eV). The Higgs boson has not been observed. Theoretical calculations show that the Higgs boson should be produced in head-on collisions of protons with energies of approximately 20 TeV. The existence of the Higgs boson is predicted by the electroweak theory, which is part of the standard model. According to the electroweak theory, the electromagnetic and weak interactions should be considered different manifestations of the fundamental electroweak interaction. The fact that the electromagnetic interaction is mediated by the massless photon, whereas the weak interaction is mediated by the $W^\pm$ and the Z^0 bosons, which have masses of about 100 GeV/c^2, is explained in terms of a spontaneously broken symmetry. The symmetry-breaking mechanism is provided by the *Higgs particle*. Thus, its discovery would be of enormous relevance as it would give the strongest support to the electroweak theory.

higher order correlation Generalization of the second order correlation function, which can be defined as

$$g^{(m,n)}(\mathbf{r}_1, \mathbf{r}_2, \ldots, \mathbf{r}_{m+n}; t_1, t_2 \cdots, t_{m+n}) = \frac{\left\langle \mathbf{E}^{(-)}(\mathbf{r}_1,t_1)\cdots \mathbf{E}^{(-)}(\mathbf{r}_m,t_m) \mathbf{E}^{(+)}(\mathbf{r}_{m+1},t_{m+1})\cdots \mathbf{E}^{(+)}(\mathbf{r}_{m+n},t_{m+n}) \right\rangle}{\left[\left\langle \mathbf{E}^{(-)}(\mathbf{r}_1,t_1) \mathbf{E}^{(+)}(\mathbf{r}_1,t_1) \right\rangle \cdots \left\langle \mathbf{E}^{(-)}(\mathbf{r}_{n+m},t_{n+m}) \mathbf{E}^{(+)}(\mathbf{r}_{n+m},t_{n+m}) \right\rangle \right]^{1/2}}.$$

Here, $\mathbf{E}^{(-)}(\mathbf{r}_1, t_1)$ and $\mathbf{E}^{(+)}(\mathbf{r}_1, t_1)$ are negative and positive frequency parts of the electric field. For a quantum system, $\mathbf{E}^{(-)}(\mathbf{r}_1, t_1)$ and $\mathbf{E}^{(+)}(\mathbf{r}_1, t_1)$ become operators.

higher order Gaussian modes *See* Gaussian beam and TEM modes.

high powered unstable lasers Unstable resonators are sometimes used to generate high power laser beams. These lasers must achieve very high gain in a short distance to compensate for all the losses. The advantages of unstable resonators are larger mode volume, efficient power extraction, collimated beam output with low diffraction, and better far-field patters. These lasers are easier to align than lasers with stable resonators. Mirrors of these lasers must be cooled to avoid damage due to high power.

high pressured gas laser Has a gaseous gain medium at very high pressures (> 50 torr). These lasers are also called transversely excited atmospheric pressure laser (TEA lasers). High pressure increases the gain but may lead to current in the gas, resulting in non-uniform excitation. Special design of the laser is required to eliminate this problem. One example of such a laser is the CO_2 laser.

Hilbert space The wave function describing a particle belongs to a function of space or set of functions called *Hilbert space*. The space has special properties such as completeness, and a well-defined inner product exists. Completeness means that there is a special set of vectors called a basis, such that every vector in the *Hilbert space* can be written as a linear combination of the members of this set. An inner product $\langle \phi | \psi \rangle$ is defined between any two vectors or wave functions of the *Hilbert space*, ϕ

and ψ, as

$$\langle \phi | \psi \rangle = \int_{-\infty}^{\infty} \phi^*(x) \psi(x) \, dx < \infty.$$

hohlraum A small metallic (typically gold) chamber used for converting high power laser beams into (soft) X-rays with efficiencies up to 50%. In addition to X-rays, interaction between the laser beams and a *hohlraum* produces an energetic, rapidly expanding plasma. This mechanism has also been applied to inertial confinement fusion as well as astrophysics.

hole A mobile vacancy in the electronic valence structure of a material.

hole burning (spatial) Multimode oscillations observed in a homogeneously broadened media are due to *spatial hole burning*. A laser in a cavity forms a standing wave. At the nodes of the standing wave of a laser, with an intensity smaller than of other points, the inversion keeps growing, and gain saturation is much lower than in other regions. Thus, the spatial variation in inversion gives rise to spatial *hole burning* in the gain curve. Due to the spatial variation of the inversion, another mode may oscillate in the cavity resulting in multimode oscillation.

hole burning (spectral) With an increase in intensity, gain in a medium saturates. In a homogeneously broadened medium, line shape cannot change. Therefore, the shape of the gain curve is restricted by the gain of the central mode. In an inhomogeneously broadened medium, the gain can saturate differently at different frequencies. Due to the selective saturation, the frequencies resonant with the cavity modes can saturate more than it the nonresonant frequencies. This results in a spectral hole in the gain curve. This gives rise to multimode oscillation in a laser.

hole state A vacancy left by a particle which has undergone a transition to a different energy level. For instance, an electron removed from its site leaves a vacancy at that site, namely a hole. The hole left by the electron behaves like a positive charge carrier.

hologram Recording which contains holographic images. *See also* holography.

holography A three-dimensional imaging technique based on interferometeric techniques. In this technique, an interference pattern due to coherent superposition of the object wave and a reference wave is recorded in a plate or film medium. The recording, which is called a hologram, contains both phase and amplitude information of the object wave. The image is reconstructed by illuminating the recorded plate/film by a reference beam similar to the one used for recording and sending the reference beam from the same direction as the original reference beam. The wave diffracted from the plate reconstructs the image which looks like the original object. *Holography* was invented by Dannis Gabor in 1948.

homenergic flow A flow where the enthalpy is constant. *See* enthalpy.

homentropic flow A flow where every fluid particle has the same value of entropy. *See* entropy.

homodyne detection Superposing a signal beam with a coherent beam of a local oscillator of the same frequency and its constant relative phase at a beam splitter. The *homodyne detection* technique has been used to detect squeezed light and enhance antibunching.

homogeneous broadening Broadening mechanism which is identical for all absorbing or emitting atoms. Examples of *homogeneous broadening* are radiative broadening and collisional broadening, which have a Lorentzian line shape of the form

$$\frac{\delta v_o/\pi}{\left[(\omega - \omega_o)^2 + (\delta v_o)^2\right]}.$$

Here, δv_o is the homogeneous line width.

homogeneous turbulence Situation in which the average properties of the turbulent fluctuations are independent of the position in the fluid. In general, all turbulent flows are inhomogeneous. Yet, the assumption of *homogeneous turbulence* in the theoretical treatment of turbulent flows gives a better understanding of certain details that are the same in both homogeneous and inhomogeneous flows, e.g., the turbulent energy transfer processes. Experimentally initiating homogeneous turbulent motion is extremely difficult. Even if this problem is overcome, maintaining the energy in such flows is difficult. As such, *homogeneous turbulence* is usually produced by placing grids in a flow, which renders the flow inhomogeneous, yet stationary, in one direction.

homojunction laser Simplest form of a semiconductor laser in which the active region is a p-n junction depletion region. Doping p and n type atoms in the same type of material creates the p-n junction in this laser. The threshold current density of this laser is very high at room temperature, therefore it is operated at a low temperature such as liquid nitrogen temperature.

homonuclear/homopolar molecule A molecule whose atoms are identical, such as the hydrogen molecule, H_2. *Compare with* heteronuclear/heteropolar molecule.

Hooke's law For small displacements, the size of the deformation is proportional to the deforming force.

hopping Microscopic motion of electrons in the presence of both lattice potential and the external field. This motion consists of individual steps in which the electron hops from one localized state to the next.

horseshoe vortex When a boundary layer encounters a surface-mounted obstacle, the adverse-pressure gradient causes a separation in which the near-wall vorticity of the boundary layer is reorganized into a vortex that has the shape of a horseshoe. This vortex is composed of two streamwise legs of vorticity, each leg having a vorticity of opposite sense. *Horseshoe vortices* are observed in many flows, including the flows around wing-fuselage junctures on aircraft, ship, and submarine appendages, bridge-piers, and turbomachinery blade-rotor junctures. *Horseshoe vortices* are undesirable in many of these flows. For instance, they increase flow losses, they are a source of noise generation, and they cause scouring of stream beds around piers. The characteristics of the *horseshoe vortex* and its effects are dependent on the nature of the incoming boundary layer.

hot wire (hot film) anemometry A technique to measure fluid velocity. The principle of operation is based on the fact that the rate of cooling of a heated wire by a flow is dependent on the velocity. Hot wire probes are usually made out of a thin (5μm in diameter) short platinum or tungsten wire through which a current of electricity is passed. The current causes the wire to heat up. In one method of operation, the current through the wire is kept constant. The velocity is then obtained by measuring the voltage across the wire, which depends on the resistance, and thus the temperature, of the wire. In a more common method of operation, a feedback circuit is used to maintain the wire at a constant temperature. The current and voltage needed to do this are then related to fluid velocity. Hot wires are usually used in gas flows while hot films are used in liquids. In hot films, the heated element consists of a thin metallic film on the surface of a wedge-shaped probe. Single hot wires are used to measure a velocity component. To measure more than one component, different configurations are used. Because hot wires and hot films are not absolute instruments, they always require calibration of the voltage with the fluid velocity. Hot wires and hot films have several advantages over other velocity measurement techniques. For instance, in comparison with other techniques, they have a short time response and can thus pick up rapid fluctuations in velocity, which is necessary for measurements in turbulent flows. Moreover, they are small enough to give local measurements instead of average values over comparatively large regions, as in the case of Pitot tube. On the other hand, their inability to fulfill the requirement of calibration over a certain velocity range and their intrusive character are two main shortcomings of hot wires and hot films in comparison with other techniques such as laser doppler anemometry.

Hubble's law This term is encountered in the context of nuclear astrophysics. *Hubble's law* states that the velocity of the recession of an object with respect to the earth is given by

$$v = Hd$$

with d representing the distance from the earth. H is the Hubble constant, which has a value of approximately 2×10^{-18} s^{-1}.

Hund–Mulliken (molecular orbital) method Electronic wave functions for molecular systems containing several electrons are constructed from one-electron molecular orbitals. This is the *molecular-orbital method*. For example, in determining the wave functions of the hydrogen molecule H$_2$, we use the wave functions of H$_2^+$ as a starting point.

Hund's rules A set of rules that have been established empirically to determine the ground state configuration of an atom. Use the Russell–Saunders notation to denote a particular state of the electrons in an atom, i.e., $^{2S+1}L_J$, where J is the total angular momentum quantum number and S represents the total spin of the electrons. L takes on the code letters S, P, D, ... for values of $L = 0, 1, 2, \ldots$. For example, the ground state for hydrogen is $^2S_{\frac{1}{2}}$. *Hund's rules* are as follows: (1) the state with the largest possible value of S has the lowest energy; the energy of the other states increases with decreasing S, and (2) for a given value of S, the state having the maximum possible value of L has the lowest energy.

hybrid frequency In a magnetized plasma, there are two types of *hybrid frequencies* that characterize electrostatic plasma waves propagating perpendicularly to the magnetic field. One is the lower-*hybrid frequency* and the other is the upper-*hybrid frequency*, both of which contain in their expressions either plasma or cyclotron frequency of protons (or ions) and electrons.

hybridization Phenomenon which occurs when atomic orbitals are combined to produce a molecular orbital of lower energy than the energy of the individual orbitals.

hybrid modes Modes of cylindrical dielectric wave guides such as optical fibers, in which both axial electric and magnetic field components are finite. *Hybrid modes* are classified into two groups: HE modes in which the axial electric field is significant compared to transverse electric components and EH modes in which the axial magnetic field is significant compared to transverse magnetic field component. These modes are further characterized by two integers corresponding to radial and azimuthal variations.

hydraulic diameter A term used to account for the shape as well as the size of a conduit. It is defined as four times the ratio of the cross-sectional area to the wetted perimeter of the cross-section. Wetted perimeter means the portion of the perimeter where there is contact between the fluid and the solid boundary.

hydraulic jump A phenomenon that takes place when a supercritical flow (Froude number $FR > 1$) undergoes a transition to a subcritical flow (Froude number $FR < 1$). *Hydraulic jumps* occur downstream of overflow structures such as spillways or underflow structures such as sluice gates where the velocities are very high. They are also observed in sinks or bathtubs when the tap water comes down at certain rates. *Hydraulic jumps* can be used as dissipaters of energy to prevent problems arising from high speeds, such as the scouring of channel bottoms. They can be used in water and sewage treatment designs to enhance chemicals mixing with the flow. Depending on the upstream Froude number, the *hydraulic jump* can assume an undular water surface ($FR < 2.5$) or a rough water surface with intermittent jets from the bottom. The hydraulic bore, also known as surge, formed by rapidly releasing water into a channel or by abruptly lowering a downstream gate is one form of a *hydraulic jump* usually referred to as a translating *hydraulic jump*.

hydraulic radius A term used to take into consideration the shape, as well as the size, of a conduit. It is defined as the ratio of the cross-sectional area to the wetted perimeter of the cross-section. Wetted perimeter means the portion of the perimeter where there is contact between the fluid and the solid boundary. For a circular pipe flowing full, the *hydraulic radius* is equal to one-fourth the diameter and is, therefore, not equal to the radius of the pipe. For a circular pipe flowing half-full, the *hydraulic radius* is equal to one-half the diameter of the pipe.

hydraulics A term originally used to describe applied and experimental aspects of fluid behavior (mainly water) and develop empirical formulas for practical problems. It stands in contrast to hydrodynamics, a term that was used to describe theoretical and mathematical aspects of idealized or frictionless fluid behavior. The introduction of the concept of boundary layers, and interest in new fields such as aerodynamics at the beginning of the twentieth century led to the synthesis of both approaches to what is known today as the science of fluid mechanics.

hydrodynamics *See* hydraulics.

hydrodynamic stability A field of study that deals with the prediction of whether a flow pattern is stable and with its transition to turbulence. It involves the linear stability theory, which examines the amplification rates of small disturbances, the subsequent non-linear stages of the transition where the growing instabilities interact with each other, and the different mechanisms for breakdown to turbulence.

hydrogen Chemical element with the lightest atomic weight. The nucleus of *hydrogen* consists of one proton. Thus, the atom contains one proton and one electron. Hydrogenic isotopes are deuterium and tritium, consisting of one proton and one neutron and one proton and two neutrons, respectively.

hydrogen atom Considered the most important two-particle system in quantum physics. It consists of a relatively heavy nucleus of mass M containing one proton of charge e together with an electron orbiting around it with charge $-e$ and mass m. The spherically symmetric potential energy $V(\mathbf{r})$ of the electron in the electric field of the nucleus is obtained from Coulomb's

hydrogen atom

law and is given as

$$V(\mathbf{r}) = \frac{-e^2}{4\pi\varepsilon_0}\frac{1}{|\mathbf{r}|}$$

(i) *ground state of:* The state of lowest energy (ground state) is described by the spatial wave function (for an infinitely heavy nucleus), $\psi_{100}(\mathbf{r})$, as

$$\psi_{100}(\mathbf{r}) = \frac{1}{\sqrt{\pi a^3}}\exp(-|\mathbf{r}|/a)$$

where a is the Bohr radius defined as

$$a = \frac{4\pi\varepsilon_0\hbar^2}{me^2} \approx 0.529 \times 10^{-10} \text{ m}$$

ε_0 is the permittivity of free space, $\hbar = \frac{h}{2\pi}$ where h is Planck's constant, and m is the mass of the electron. The subscripts on the function ψ denote the values of the principal, orbital angular momentum, and magnetic quantum numbers n, l, and m_l, respectively, with $n = 1, l = 0$, and $m_l = 0$. The energy of the ground state E_1 is -13.6 electron-volts.

(ii) *Schrödinger equation for:* The time-independent *Schrödinger equation* for the hydrogen atom in the center-of-mass system of coordinates in which the effect of the finite mass of the nucleus is taken into account is given as:

$$\left[\frac{-\hbar^2}{2\mu}\nabla^2 - \frac{e^2}{4\pi\varepsilon_0 r}\right]\psi(\mathbf{r}) = E\psi(\mathbf{r})$$

where μ is the reduced mass defined by $\mu = \frac{Mm}{M+m}$, $\mathbf{r} = \mathbf{r}_1 - \mathbf{r}_2$ is the relative displacement between the proton position $\mathbf{r}_1$ and the electron $\mathbf{r}_2$, and $r = |\mathbf{r}_1 - \mathbf{r}_2|$.

(iii) *allowed energies of:* The *allowed energies*, E_n, of the bare (unperturbed) *hydrogen atom* in the non-relativistic approximation and for a nucleus with mass taken as infinite form a discrete spectrum which depends only on the principal quantum number n. E_n is defined by the following formula

$$E_n = -\left[\frac{m}{2\hbar^2}\left(\frac{e^2}{4\pi\varepsilon_0}\right)^2\right]\frac{1}{n^2}$$

where m is the mass of the electron, e is the magnitude of the electronic charge, ε_0 is the permittivity of free space, and $\hbar = \frac{h}{2\pi}$ where h is Planck's constant.

(iv) *wave functions for:* The spatial *wave functions*, $\psi_{nlm}(\mathbf{r})$, for hydrogen are labeled by three quantum numbers n, l, and m_l. They are defined (up to some normalization constant) by

$$\psi_{nlm_l}(\mathbf{r}) = R_{nl}(r)Y_l^{m_l}(\theta,\phi)$$

where (r, θ, ϕ) are spherical polar coordinates. $R_{nl}(r)$ is the radial part of the wave function defined as

$$R_{nl}(r) = \frac{1}{r}\rho^{l+1}e^{-\rho}v(\rho)$$

where $\rho = \frac{r}{an}$ and a is the Bohr radius of the hydrogen atom. $v(\rho)$ is a polynomial of degree $i_{max} = n - l - 1$ in ρ such that

$$v(\rho) = \sum_{i=0}^{i_{max}} a_i \rho^i .$$

The coefficients in the expansion for $v(\rho)$ are determined from the following recursion relation

$$a_{i+1} = \left\{\frac{2(i+l+1)-A}{(i+1)(i+2l+2)}\right\}a_i$$

where A is a constant given by $A = \frac{me^2}{2\pi\varepsilon_0\hbar^2\left(\frac{\sqrt{-2mE_n}}{\hbar}\right)}$. The angular parts $Y_l^m(\theta,\phi)$ are called the spherical harmonics and are defined as

$$Y_l^m(\theta,\phi) =$$
$$\gamma\sqrt{\frac{(2l+1)(l-|m|)!}{4\pi(l+|m|)!}}e^{im\phi}P_l^m(\cos\theta)$$

where $P_l^m(\cos\theta)$ is an associated Legendre polynomial and $\gamma = (-1)^m$ for $m \geq 0$ and $\gamma = 1$ for $m \leq 0$.

(v) *spectrum of:* When an excited *hydrogen atom* with energy E_i decays to a lower energy level E_f, a photon is emitted with energy E_ν and frequency ν. In terms of the principal quantum numbers n_i and n_f of the initial and final energy levels, respectively, the energy of the emitted photon is determined from the following

$$E_\nu = E_i - E_f = -13.6\text{eV}\left(\frac{1}{n_i^2} - \frac{1}{n_f^2}\right) = h\nu$$

where eV stands for electron volts. For transitions to the ground state ($n_f = 1$), the emitted photons have frequencies in the ultraviolet region of the spectrum and this constitutes the Lyman series. For transitions to the first excited state, $n_f = 2$, the frequencies fall in the visible region and are called the Balmer series. Transitions to the $n_f = 3$ produce photons in the infrared called the Paschen series.

(vi) radius of: The effective radius of the electron in the ground state is denoted by a and is called the Bohr radius. It is defined as

$$a = \frac{4\pi\varepsilon_0 \hbar^2}{me^2}.$$

(vii) isotopes of: Atoms with the same atomic number as hydrogen ($Z = 1$) but with different masses are the *isotopes of hydrogen*. The neutral isotopes are deuterium and tritium. The nucleus of deuterium (deuteron) contains one proton and one neutron, while that of tritium (triton) contains one proton and two neutrons. Other isotopes, called light isotopes, are positronium and muonium. In positronium, the nucleus is a positron and in muonium it is a positive muon.

(viii) gravitational energy shift: In the *hydrogen atom*, in addition to the electromagnetic interaction of the proton and the electron, there is also a gravitational interaction. The perturbation to the unperturbed Hamiltonian, H', due to this is $H' = -\frac{GmM}{r}$, where M is the mass of the proton and G is the gravitational constant. For the first order in perturbation theory, the energy shift of the ground state (1s), E_{1s}^1 is $-\frac{GmM}{a_\mu}$, where

$$a_\mu = \frac{4\pi\varepsilon_0 \hbar^2}{\mu e^2}$$

and $\mu = \frac{mM}{m+M}$ is the reduced mass. All other symbols have their usual meanings.

(ix) molecule: This is a diatomic molecule composed of two *hydrogen atoms* with chemical formulae of H_2 bonded together covalently. It contains two electrons moving in the electromagnetic field of two protons. The equilibrium separation of the two protons is about 0.749 Å. The approximate ground state wave function Φ_T of molecular hydrogen can be written as a linear combination of two functions, Φ_M^{cov} and Φ_M^{ion}, as

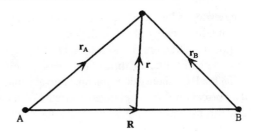

A coordinate system for the hydrogen molecular ion H_2^+.

$$\Phi_T = (1-\lambda)\Phi_M^{cov} + (1+\lambda)\Phi_M^{ion}$$

where λ is a parameter to be determined after minimizing the energy E, i.e., $\frac{\partial E}{\partial \lambda} = 0$. Φ_M^{cov} is a wave function describing the situation in which one electron is associated with one nucleus. For large separation of the protons, it reduces to a wave function describing two isolated *hydrogen atoms*. Φ_M^{ion} corresponds to the situation in which both electrons are bound to one nucleus. In the limit of large internuclear distance, this function describes a proton and a negative hydrogen ion.

(x) molecular ion: Considered the simplest of all molecules. Its chemical formula is H_2^+ and it is composed of two protons and one electron. The following gives a coordinate system for calculating the wave functions of the moving electron:

The Schrödinger equation for the electronic motion is:

$$\left[-\frac{\hbar^2}{2m}\nabla_r^2 - \frac{1}{4\pi\varepsilon_0}\frac{e^2}{|r_A|} - \frac{1}{4\pi\varepsilon_0}\frac{e^2}{|r_B|} + \frac{1}{4\pi\varepsilon_0}\frac{e^2}{|R|} - E\right]\Phi = 0$$

where Φ is the wave function. In the limit of large internuclear separation, we can approximate the ground state wave function in terms of superpositions of the ground state wave functions of the isolated *hydrogen atoms*, i.e.,

$$\Phi_g(\mathbf{R};\mathbf{r}) = \frac{1}{\sqrt{2}}[\Psi_{1s}(r_A) + \Psi_{1s}(r_B)]$$

$$\Phi_u(\mathbf{R};\mathbf{r}) = \frac{1}{\sqrt{2}}[\Psi_{1s}(r_A) - \Psi_{1s}(r_B)]$$

where $\Phi_g(\mathbf{R};\mathbf{r})$ is symmetric (gerade) and is called the bonding orbital. It represents the state of lowest energy. $\Phi_u(\mathbf{R};\mathbf{r})$ is antisymmetric (ungerade) and is called the antibonding orbital. The binding energy of H_2^+ is 2.79 eV.

(xi) fine structure of: In the study of the hydrogen atom, the calculation of the energy levels using the Bohr theory does not take into account the correction due to relativistic effects of the electron motion or magnetic effects due to the orbital motion of the electron. Such corrections to the energies constitute fine structure correction, and for first order in perturbation theory, depend on the principal and total angular momentum quantum numbers n and j respectively. The energy levels E_{nj} are given by

$$E_{nj} = \frac{-13.6\text{eV}}{n^2} \frac{\alpha^2}{n^2} \left(\frac{n}{j+\frac{1}{2}} - \frac{3}{4} \right)$$

where α is the fine structure constant defined as

$$\alpha = \frac{e^2}{4\pi\varepsilon_0 \hbar c} \approx \frac{1}{137.036}$$

(xii) relativistic correction to the spectrum: In the non-relativistic approximation, the contribution to the Bohr Hamiltonian of the kinetic energy of the electron in the atom is $-\frac{\hbar^2}{2m}\nabla^2$, which classically is $\frac{1}{2}mv^2$ with v denoting the electron velocity. The relativistic kinetic energy is

$$\text{K.E}_r = \frac{mc^2}{\sqrt{1-\left(\frac{v}{c}\right)^2}} - mc^2$$

where the first term is the total relativistic kinetic energy and the second term is the rest energy. In terms of the relativistic momentum p, where

$$p = \frac{mv}{\sqrt{1-\left(\frac{v}{c}\right)^2}}$$

$$\text{K.E}_r = \sqrt{p^2c^2 - m^2c^4} - mc^2$$

$$\approx \frac{p^2}{2m} - \frac{p^4}{8m^3c^2} + \cdots .$$

The lowest-order relativistic contribution to the Hamiltonian is thus

$$H_r^1 = \frac{-\hat{p}^4}{8m^3c^2}$$

where $\hat{p} \to -i\hbar\nabla$. For the first order in perturbation theory, the correction to the Bohr energy, E_r^1, is

$$E_r^1 = \langle H_r^1 \rangle = \frac{-E_n^2}{8mc^2} \left[\frac{4n}{l+\frac{1}{2}} - 3 \right]$$

where E_n is the Bohr energy corresponding to the principal quantum number n; l is the orbital angular momentum quantum number.

(xiii) correction to the spectrum due to spin-orbit coupling: The electron in hydrogen moves relative to the nucleus, which is positively charged. Due to this relative motion, the electron moves in a magnetic field $\mathbf{B}$ generated by the relative motion of the nucleus. This magnetic field interacts with the electron, giving it an additional energy, $H_{so} = -\boldsymbol{\mu} \cdot \mathbf{B}$, where $\boldsymbol{\mu}$ is the magnetic moment of the electron. In the rest frame of the electron, the magnetic field generated by the moving proton is

$$\mathbf{B} = \frac{1}{4\pi\varepsilon_0} \frac{e}{mc^2r^3} \mathbf{L}$$

where $\mathbf{L}$ is the total orbital angular momentum of the proton. The magnetic moment of the electron $\boldsymbol{\mu}$ with spin angular momentum $\mathbf{S}$ is $\boldsymbol{\mu} = -\frac{e}{m}\mathbf{S}$. After making a correction due to the fact that the electron's rest frame is non-inertial (Thomas precession), the spin-orbit Hamiltonian then works out to be

$$H_{so} = -\boldsymbol{\mu} \cdot \mathbf{B} = \frac{1}{8\pi\varepsilon_0} \frac{e^2}{m^2c^2r^3} \mathbf{S} \cdot \mathbf{L} .$$

This Hamiltonian gives a first order correction E_{so}^1 to the Bohr energy E_n as

$$E_{so}^1 = \frac{E_n^2}{mc^2} \left[\frac{n\left[j(j+1) - l(l+1) - \frac{3}{4}\right]}{l\left(l+\frac{1}{2}\right)(l+1)} \right] .$$

(xiv) correction to the spectrum due to spin-spin coupling (hyperfine splitting): The proton in the atom is a magnetic dipole with magnetic dipole moment $\boldsymbol{\mu}_p = \frac{ge}{2m_p}\mathbf{S}_p$, where $\mathbf{S}_p$ is the spin angular momentum of the proton, g is its gyromagnetic ratio, and m_p is its mass. The proton sets up a magnetic field $\mathbf{B}$ due to its magnetic dipole moment given as

$$\mathbf{B} = \frac{\mu_0}{4\pi r^3}\left[3\left(\boldsymbol{\mu}_p \cdot \hat{\mathbf{r}}\right)\hat{\mathbf{r}} - \boldsymbol{\mu}_p\right] + \frac{2\mu_0}{3}\boldsymbol{\mu}_p \delta(\mathbf{r}) .$$

The Hamiltonian of the electron H'_{hf} in this magnetic field is

$$H'_{\text{hf}} = \frac{\mu_0 g e^2}{8\pi m_p m_e} \frac{\left[3\left(\mathbf{S}_p \cdot \hat{\mathbf{r}}\right)\left(\mathbf{S}_e \cdot \hat{\mathbf{r}}\right) - \mathbf{S}_p \cdot \mathbf{S}_e\right]}{r^3}$$
$$+ \frac{\mu_0 g e^2}{3 m_p m_e} \left(\mathbf{S}_p \cdot \mathbf{S}_e\right) \delta\left(\mathbf{r}\right).$$

If we take the simplest case for which the orbital angular momentum of the electron is zero (e.g., the ground state), the correction, E'_{hf}, due to this spin–spin interaction is

$$E'_{\text{hf}} = \frac{\mu_0 g e^2}{3\pi m_p m_e a^3} \langle \mathbf{S}_p \cdot \mathbf{S}_e \rangle .$$

In the triplet and singlet states where the total spins are one and zero, respectively, we obtain

$$E'_{\text{hf}} = \frac{4 g \hbar^4}{3 m_p m_e^2 c^2 a^4} \times \begin{cases} \frac{1}{4} & \text{(triplet)} \\ -\frac{3}{4} & \text{(singlet)} \end{cases}.$$

hydrology A field that involves the study of the water of the earth, its precipitation, its movement over the surface and below the surface, its evaporation and transpiration, and its reaction with its environment.

hydromagnetic equilibrium Achieved in an magnetized plasma if all the forces are balanced. They can be analyzed by the MHD equations. The study of *hydromagnetic equilibrium* is important particularly for the magnetic confinement of laboratory plasmas.

hydromagnetics At frequencies well below the proton (or ion) cyclotron frequency, the electric component of a plasma wave in a magnetic field is not important, and the magnetic component dominates physical processes. The physics of such a plasma are called *hydromagnetics,* and are usually well described by the MHD (magnetohydrodynamic) equations. Although hydromagnetic processes tend to proceed slowly as their typical saturation levels are quite high, they can play dominating roles in many cases. *Hydromagnetic* has been widely applied to various problems in the fields of nuclear fusion, space physics, and astrophysics. *See also* magnetohydrodynamics (MHD).

hydromagnetic wave Low frequency electromagnetic ion oscillations in a magnetized plasma, in which electric field components, compared with their magnetic counterparts, no longer play an important role. Among the most important *hydromagnetic waves* are Alfvén waves and magnetosonic waves. The former propagate along a magnetic field, and the latter often propagate perpendicularly to it. These waves with relatively large scale-lengths are often quite significant in nuclear fusion, space physics, and astrophysics.

hydrometer A device based on the principle of buoyancy which is used to determine the specific gravity of a liquid. It is a device with tiny metal spheres placed at its bottom. It has a stem of constant cross-sectional area, A_s, that protrudes through the free surface when placed in the liquid. It is calibrated that when it floats in distilled water, the submerged volume is V_0. When floating in another liquid, the stem will sit higher or lower by a distance Δh than the position in distilled water. The relationship between Δh and specific gravity S of the liquid is then written as:

$$\Delta h = \frac{V_0}{A_s} \frac{S-1}{S} .$$

This relationship allows for direct reading (through calibration) of the specific gravity of the liquid.

hydrostatic force The force on a fully or partially submerged body that results from the hydrostatic pressure distribution.

hydrostatic pressure The pressure in a fluid at rest. It increases linearly with depth.

hydrostatics A field that involves the study of fluids under static conditions.

Hylleraas trial functions In calculating the ground state energy of two-electron atoms, the Rayleigh–Ritz variational method is often employed. This method requires trial functions for the approximate ground state wave functions.

Hylleraas used coordinates

$$s = r_1 + r_2, \quad 0 \leq s \leq \infty$$
$$t = r_1 - r_2, \quad -\infty \leq t \leq \infty$$
$$u = r_{12}, \quad 0 \leq u \leq \infty$$

where $\mathbf{r}_1$ and $\mathbf{r}_2$ are position coordinates of the electrons with respect to the nucleus, and $r_{12} = |\mathbf{r}_1 - \mathbf{r}_2|$. Using these coordinates, Hylleraas constructed trial functions of the type

$$\phi(s, t, u) = e^{-ks} \sum_{l,m,n}^{N} c_{l,2m,n} s^l t^{2m} u^n$$

where k and $c_{l,2m,n}$ are variational parameters.

hyperbolic secant pulse Also known as optical solitons. A *hyperbolic secant pulse* can pass through a medium without changing its shape. It has a shape, as given by

$$\frac{2}{T} \operatorname{sech}\left[\frac{(t - t_o)}{T}\right],$$

where at time $t = t_o$ the pulse has maximum, and T is the effective width of the pulse. If a group of inhomogeneously broadened atoms with resonant frequency ω_o is applied with a secant pulse of frequency ω then each atom completes a whole cycle of excitation and returns back to the ground state, resulting in no absorption of the energy. This gives rise to the phenomenon of self-induced transparency. *See* soliton.

hypercharge A property of a particle defined as

$$Y = A + S$$

where A is the *baryon number* (equal to 1 for baryons and 0 for mesons), and S is the strangeness. The strangeness is a conserved quantity in all strong interactions. Strange particles have the strange quark among their constituents.

hyperfine interaction The interaction of nuclear magnetic moments with any electromagnetic fields present in the environment. This interaction is responsible for the hyperfine splitting of atomic energy levels. *See also* hyperfine structure.

hyperfine structure This term refers to all effects on the atomic energy levels originating from the coupling of nuclear spins and moments to their environment. This includes interactions internal to the atom, such as coupling between electronic and nuclear angular momenta.

hyperfragment Formed by the capture of strange particles (such as Λ^0) by nuclei. *Hyperfragments* are unstable and decay into either nucleons and pions or nucleons only. *See also* hypernucleus.

hypernucleus Nuclei where some of the nucleons are replaced by hyperons. *See* hyperon.

hyperon A strange baryon, namely, a baryon with non-zero strangeness. An example is the lambda particle (Λ^0) with a mass of about 1116 MeV/c^2, which decays into a proton and a negatively charged pion.

hyperonic atoms These are special atoms containing a nucleus of charge Ze and a negative hyperon of charge $-e$ which is a particle (of the baryon family) with half-integer spin and interacts via the strong force.

hypersonic flow Flow with a speed that is much larger than the ambient speed of sound. The onset of *hypersonic flow* characteristics is usually gradual and varies with body geometry, flow speed, and properties of the ambient atmosphere. In a *hypersonic flow*, large variations in the temperature exist, to the extent that changes in the chemical composition of the medium become important in characterizing the flow. One flow where supersonic effects are achieved is re-entry of a the space shuttle into the earth's atmosphere.

I

ideal flow An inviscid and incompressible flow.

ideal fluid A fluid that is assumed to have zero viscosity. In such a fluid, there are no frictional effects between the fluid particles, and thus there is no boundary layer. The motion of an *ideal fluid* is analogous to the motion of a solid body on a frictionless surface.

ideal gas A gas of non-interacting point particles.

ideal gas law The equation of state for an ideal gas, $PV = nRT$. Here, P, V, and T denote the pressure, volume, and temperature of n moles of an ideal gas, and R is the universal gas constant.

idealized squeezed state Obtained by squeezing the vacuum state and then displacing the state:
$$|\alpha, \xi\rangle = \hat{D}(\alpha)\hat{S}(\xi)|0, 0\rangle ,$$
where
$$\hat{D}(\alpha) = \exp\left[\alpha\hat{a}^\dagger - \alpha^*\hat{a}\right]$$
is the displacement operator and
$$\hat{S}(\xi) = \exp\left[\frac{1}{2}\left\{\xi^*\left(\hat{a}\right)^2 - \xi\left(\hat{a}^\dagger\right)^2\right\}\right]$$
is the squeezing operator. α and ξ are known as displacement and squeezing parameters, respectively. *See* squeezed state.

identical particles Particles that cannot be distinguished by any intrinsic property. This is one of the basis tenets of quantum mechanics. Therefore, all electrons are identical. Because trajectories are well-defined in classical physics, it is possible, in principle, to distinguish between individual classical particles. *Identical particles* can be interchanged without any change in the physical system. In a two-particle system, if one particle is in state $\psi_\alpha(\mathbf{r})$ and the other in $\psi_\beta(\mathbf{r})$, where $\mathbf{r}$ denotes position, the composite wave function $\psi(\mathbf{r}_1, \mathbf{r}_2)$ is given by
$$\psi_\pm(\mathbf{r}_1, \mathbf{r}_2) = N\left[\psi_\alpha(\mathbf{r}_1)\psi_\beta(\mathbf{r}_2) \pm \psi_\beta(\mathbf{r}_1)\psi_\alpha(\mathbf{r}_2)\right]$$
where N is a normalization factor. Thus, there are two kinds of *identical particles,* one which uses the plus sign and the other which uses the minus sign. For particles with integer spin (bosons) the plus sign is used, and for particles with half-integer spin the minus sign is used.

identical-particle symmetry For particles that cannot be distinguished from one another, certain symmetry conditions are naturally imposed on the wave functions describing identical particles. If such particles are truly identical, then the probability density should be symmetries under the interchange. For two particles with wave function $\Psi(q_1, q_2)$, where q denotes both position and spin
$$|\Psi(q_1, q_2)|^2 = |\Psi(q_2, q_1)|^2 .$$
The two solutions ψ^S and ψ^A satisfy the following symmetry requirements:
$$\psi^S(q_1, q_2) = \psi^S(q_2, q_1)$$
$$\psi^A(q_1, q_2) = -\psi^A(q_2, q_1)$$
and are called symmetric and antisymmetric solutions respectively.

idler photon In parametric down-conversion, one of the down-converted photons is called the idler photon, and the other is called the signal photon.

impact Trajectory of a charged particle in a non-head-on collision.

impact parameter Distance from the scattering center (the target) perpendicular to the initial direction of motion of the incident particle (the projectile). That is, in an absence of forces, the incident particle would travel along a straight line which passes a distance equal to the *impact parameter* from the scattering center.

impulse approximation An approximation used to simplify the analysis of nuclear reactions, in particular the scattering of nucleons off

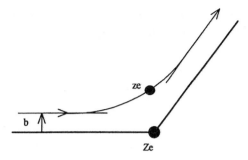

Impact.

nuclei. When the incident energy is sufficiently high, one may treat the process as a single scattering of the incident nucleon from one nucleon in the nucleus; i.e., one keeps only the first term in the multiple-scattering series.

impulse momentum principle Derived from Newton's second law, which states that the sum of external forces on a body is equal to the rate of change of momentum of that body. Since acceleration is the rate of change of velocity, one can rewrite Newton's second law as

$$\vec{F}\, dt = m\, d\vec{V}.$$

The impulse is the integral of the left-hand side. The right-hand side represents the change in momentum when integrated. The equality of these two quantities is the *impulse momentum principle*.

impurity An atom that is foreign to the material in which it exists.

impurity band An energy level caused by the presence of impurity atoms.

incident energy The energy of the incoming particle in a collision process.

incident flux The number of particles incident upon the target area per unit of time.

inclusive reaction A reaction for which the measured cross-section includes all possible final states. This is in contrast to an *exclusive* reaction, where one final state is singled out.

incoherence Two beams are in *incoherence* when they cannot form an interference pattern. Visibility of the fringe pattern in this case is zero.

Incoherence is also measured in terms of mutual coherence function defined as

$$\mathcal{V}(\mathbf{r}_1, \mathbf{r}_2, \tau) = \frac{\mathcal{I}(\mathbf{r}_1, \mathbf{r}_2, \tau)}{[\mathcal{I}(\mathbf{r}_1, \mathbf{r}_1, 0)]^{1/2} [\mathcal{I}(\mathbf{r}_2, \mathbf{r}_2, 0)]^{1/2}}$$

where the two-time field correlation $\mathcal{I}(\mathbf{r}_1, \mathbf{r}_2, \tau)$ is a function defined as

$$\mathcal{I}(\mathbf{r}_1, \mathbf{r}_2, \tau) = \langle E^*(\mathbf{r}_1, t)\, E(\mathbf{r}_1, t + \tau) \rangle.$$

For complete coherence $\mathcal{V}(\mathbf{r}_1, \mathbf{r}_2, \tau) = 1$, and for complete *incoherence* $\mathcal{V}(\mathbf{r}_1, \mathbf{r}_2, \tau) = 0$.

incoming (particle, wave) The projectile particle in a collision process. In the context of a quantum scattering process, the *incoming particle* is also referred to as the *incoming wave*.

incompatible observables The commutator $[A, B]$ between two operators, A and B, is defined as

$$[A, B] = AB - BA.$$

Operators A and B are incompatible or noncommuting if $[A, B] \neq 0$. Physically, this means that there will be an uncertainty principle for such a pair and, as such, they do not have a complete set of common eigenvectors.

incompressible flow Flow in which, following a fluid element, the density variations are negligible. Liquid flows are usually treated as incompressible because they have a large bulk modulus, i.e., even large pressures can only produce negligible density variations. Gas flows can be treated as *incompressible flows* when the exerted pressure variations are small enough not to produce significant density variations.

independent-particle model A model of the nucleus (also known as the shell model), based upon the assumption that each nucleon experiences an average field produced by all the other nucleons. Furthermore, each nucleon occupies a quantized energy state with a well-defined angular momentum.

index matching Matching the refractive index of the crystal and its surrounding medium in nonlinear optics to minimize Fresnel losses.

index of refraction The ratio of the velocity of a wave in a vacuum to that in a specified material.

indirect band gap semiconductor In an *indirect band gap semiconductor*, the conduction band edge and valence band edge are not at the center of the Brillouin zone, such as Ge, Si, etc.

indistinguishability This principle is the same as the identical particle concept. Essentially, identical particles cannot be distinguished based on their physical properties. For example, if we consider two electrons and identify the degrees of freedom as the spatial coordinates $r_{1,2}$ and z-components of spin $S_{z1,2}$, both electrons have charge $-e$, mass m, and spin $\frac{1}{2}$. Intrinsically, therefore, they have identical labels and cannot be distinguished in principle.

induced drag The increase in drag due to the finite length of the wing. The finite wing and the tip and starting vertices constitute a large vortex ring inside which there is a downward velocity (usually referred to as downwash velocity) induced by the vortices. Prandtl showed that this induced velocity is constant if the distribution of lift is elliptical over the wing. This induced velocity component caused the effective angle of attack of the wing to decrease. One effect of this reduction is that the lift vector, which must be perpendicular to the effective approach velocity, must be rotated. Consequently, the lift vector has two components, one of which is the true lift, which is perpendicular to the velocity vector without the induced velocity component, and the other is parallel to the velocity vector. The parallel component is in the direction of drag and is accordingly called *induced drag*. The induced drag is related to the lift as

$$C_{Di} = \frac{C_L^2}{\pi B^2/A} = \frac{C_L^2 C}{\pi B}$$

where B is the span of the airfoil and C is the mean chord length. The ratio B/C is the aspect ratio. *Induced drag* is critical during takeoff and landing when the lift coefficient is especially large.

induced fission The term fission is used to describe the splitting of the nucleus. This can occur either spontaneously or, for example, by irradiation with thermal neutrons *(induced fission)*. The *induced fission* of ^{235}U by thermal neutrons can be represented as

$$n_{thermal} + {}^{235}U \rightarrow \left[{}^{236}U\right] \rightarrow X + Y$$

where X and Y are fragments from the fission process.

induced reaction A process that is stimulated by an external agent. For instance, a photo-induced *reaction* is a reaction induced by exposure to light.

induced scattering Also called *induced scattering* off ions or non-linear Landau damping. One of major processes in weakly turbulent plasmas, in which a plasma wave is coupled with another through background particles such as ions. Such a coupling may be realized between two waves with frequencies ω_1, and ω_2 and wavenumbers k_1, and k_2, respectively, when the following resonance condition between the beat of the two waves and the particle velocity v, $\omega_1 - \omega_2 = (k_1 - k_2)v$ is satisfied. A plasma wave that is unstable to *induced scattering* eventually decays to other waves, making the plasma more turbulent.

inelastic cross-section For collisions, the ratio of the number of events of this type per unit of time and per of unit scatterer to the flux of the incident particles is defined as the *inelastic cross-section*.

inelasticity parameter This term is typically used in the analysis of particle scattering. If inelastic scattering occurs, there is a loss of flux from the incident (elastic) channel. The extent to which flux is removed from the elastic channel is expressed in terms of the *inelasticity parameter* η (this notation is customary, e.g., in the analysis of two-nucleon scattering data). The *inelasticity parameter* is equal to one for energies below the inelastic threshold, namely the energy threshold above which particle production becomes energetically possible.

inelastic scattering A scattering process where the initial kinetic energy is not conserved.

Namely, the kinetic energy of the initial-state particles is not the same as the kinetic energy of the final-state particles. When additional kinetic energy is produced in the final state, the reaction is said to be exoergic. On the other hand, when kinetic energy is absorbed in the process, the reaction is called endoergic. In the context of relativistic kinematics, produced (absorbed) kinetic energy corresponds to reduced (increased) rest mass energy.

inequivalent electrons Electron configurations in many-electron atoms are of two types: those which describe equivalent electrons and those which describe *inequivalent electrons*. A configuration which describes *inequivalent electrons* is such that the assignment of the orbital quantum numbers (n, l) are different for electrons outside closed subshells. Some examples are the He configuration $1s2s$ and the C configurations $1s^2 2s^2 2p3s$ and $1s^2 2s^2 2p3p$.

inert core Nucleons in the nucleus which are inactive, except for providing binding energy to the valence (or outermost) nucleons.

inertial confinement One of the techniques used to confine the plasma for the purpose of achieving nuclear fusion. In this scheme, a very high particle density (about 5×10^{25} particles/cm^3) is combined with a very short confinement time, usually 10^{-11} to 10^{-9} s. Under these conditions, due to their inertia, the particles will not be able to move appreciably from their initial positions.

inertial confinement fusion (ICF) To achieve nuclear fusion, excessively high temperatures that cannot be supported by ordinary metal containers become necessary. For the confinement of such high-temperature fusion fuel, the inertial confinement and magnetic confinement have been the most thoroughly tested reliable schemes. In the scheme of *inertial confinement fusion,* a pellet of fusion fuel such as solid DT (deuterium-tritium) is targeted by intensive beams of laser light or particles; therefore, this scheme is also called the pellet fusion.

The rapid implosion of a high-density spherical pellet accompanied by shock waves leads to the production of a heated core that fuses before it can explode. As in the case of other fusion schemes, one goal of this scheme is to satisfy the Lawson criteria, attaining a sufficiently large value of $n\tau > 10^{20}$ sm^{-3}, where n is the particle density and τ is the average confinement time of the plasma. To achieve such a goal, extremely large compression factors of the order of 10,000 are required. In addition, the confinement time τ in the inertial fusion is set by the time taken for the plasma to expand freely. Therefore, extremely high pulsed powers of hundreds of terawatts [TW] must be focused down and delivered to small sizes of approximately one millimeter in ultra-short times of nanoseconds or less. For this reason, a large array of high-power infrared (CO_2 or Nd-glass) laser beams or particle (electrons or ions) beams have been employed.

More specifically, in the case of laser fusion the pellet is irradiated by arrayed laser beams from all directions. The light energy is absorbed by parametric processes at the critical layer, where the laser frequency equals the plasma frequency, and a plasma shell is heated. Subsequently, the shell expands, and its momentum outward is used to compress the central core inward, triggering nuclear fusion inside the core.

Gekko XII at Osaka University.

The figure shows portions of the Gekko XII laser fusion facility being operated at the Institute of Laser Engineering, of the Osaka University in Japan, showing several of 13 laser amplifier trains. This system, which already had successfully compressed hollow shell targets to 1000 times their initial solid density in 1992, is designed to deliver more than 55 TW of optical power. (The Nova laser at the Lawrence Livermore National Laboratory, USA has also achieved high densities with fusion targets.) The employment of array would help increase the uniformity of the irradiation over the surface of pellets. The presence of even a small inhomogeneity would cause a major reduction in strongly desired compression through some hydrodynamic instabilities such as the Rayleigh-Taylor instability.

In the laser fusion scheme thus far, high temperatures as well as substantial values of $n\tau$, which are comparable to those in magnetic confinement fusion schemes, have been obtained. *See also* fusion.

inertial frame A non-accelerated frame. All coordinate systems which move with uniform relative velocity to each other are inertial. Interactions are as follows:

(i) Coulomb: The electrostatic interaction between charged particles is expressed by Coulomb's law, which is an example of an inverse-square law. For particles with charge q_1 and q_2 and separation r, the magnitude of the force, F, is given as

$$F = \frac{q_1 q_2}{4\pi\varepsilon_0 r^2}.$$

(ii) electromagnetic: The interaction of a charged particle (of charge q and mass m) with an electromagnetic field is given by the Lorentz force law for the force **F** acting on the particle:

$$\mathbf{F} = q\,(\mathbf{E} + \mathbf{v} \times \mathbf{B})$$

where **E** and **B** are electric and magnetic field vectors and **v** is the velocity of the particle. It should be noted that this force cannot be written as the gradient of a scalar potential energy function. The classical Hamiltonian describing this situation is

$$H = \frac{1}{2m}(\mathbf{p} - q\mathbf{A})^2 + q\phi$$

where **A** is the vector potential and **p** is the generalized momentum of the particle; ϕ is the scalar potential satisfying $\mathbf{E} = -\nabla\phi - \frac{\partial \mathbf{A}}{\partial t}$. The Schrödinger equation describing the interacting system is

$$i\hbar\frac{\partial \Psi}{\partial t} = \left[\frac{1}{2m}(-i\hbar\nabla - q\mathbf{A})^2 + q\phi\right]\Psi$$

where Ψ is the wave function of the particle.

(iii) electron–phonon: The fundamental interaction involved in the formation of Cooper pairs in superconductors. Cooper discovered that electrons can form bound pairs by exchanging phonons in the crystal lattice. The resultant force between the electrons is attractive. The electrons which interact in this way have energies close to the Fermi surface and have equal and opposite momenta.

(iv) weak: The force which is responsible for β decay of nuclei. By analogy with electromagnetism, we can set up a current–current interaction taking place in nuclei which is similar to the interaction between currents in electromagnetism. Consider the following diagram:

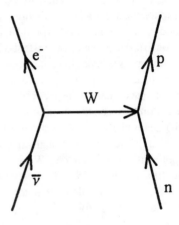

The weak interaction shown in terms of a antineutrino ($\bar{\nu}$) current and a neutron (n) current.

The n–p line constitutes the weak nucleon current and the $\bar{\nu}$–e^- (antineutrino–electron) line is the weak lepton current. Along the n–p line a neutron changes into a proton, while the $\bar{\nu}$–e^- line, represents the emission of an antineutrino–electron pair. These interactions take place by means of the exchange of intermediate particles called W bosons, which are the force-carrying particles.

(v) electroweak: Constructed by unifying the weak and electromagnetic interactions into a single theory. The unified theory deals with the interactions of quarks (the constituents of protons and neutrons) and leptons (electrons, antineutrinos, etc.). The interaction between quark and leptonic currents is mediated by some combination of four distinct quanta: W^+, W^0, W^-, and B^0. The coupling of the electromagnetic currents of quarks and leptons selects a particular combination of the above four particles. Some combinations, like B^0 and W^0, can produce a massless particle called a photon. Other combinations produce massive particles through a process called spontaneous symmetry breaking.

(vi) strong: Interaction which takes place between the nucleons in the nucleus. The interacting currents are carried by quarks in the neutrons and protons, and the intermediate (force-carrying) particles are called gluons.

(vii) Van der Waals: Dutch physicist J.D. van der Waals proposed that atoms with closed shells, e.g., He and H, can be liquified at sufficiently low temperatures. The force of attraction responsible for such liquifaction is called the London or van der Waals force. The force arises due to the fact that at any instant of time, a neutral atom possesses an instantaneous electric dipole moment. This generates an electric field which polarizes a nearby atom. The result is a net attraction between the atoms.

(viii) Lennard-Jones: When atoms of the noble gases approach close enough they experience both attractive (van der Waals) and repulsive forces. In 1925, J.E. Lennard-Jones suggested a phenomenological potential, V_{LJ}, which described these forces as

$$V_{LJ} = \alpha \left[\left(\frac{\varepsilon}{R}\right)^n - \left(\frac{\varepsilon}{R}\right)^6 \right]$$

where R is the internuclear separation. α and ε are parameters to be determined depending on the type of atoms involved. Frequently, $n = 12$ and ε is the value of R such that $V_{LJ}(R) = 0$.

inertial sublayer   *See* inner region.

inertial subrange   For the universal equilibrium to exist, the energy-containing scales and the dissipating small scales must be widely separated. In between, there exists a range of scales that are not directly affected by the energy maintenance or dissipation mechanisms. This range of scales is referred to as the *inertial subrange*.

inexact differential   The differential dF is called an *inexact differential* if its value depends on the path between two closely spaced points and thus cannot be written as the difference in the value of a function at the two points.

infinite nuclear matter   An idealized infinite system of nucleons with uniform density. If the number of neutrons and protons is the same, the matter is called symmetric nuclear matter. Nuclear matter density approximates the central density of a heavy nucleus. This system is often used in place of a finite nucleus in order to facilitate theoretical studies. A finite nucleus has a large surface region where the density drops to zero with increasing distance from the center.

infinite square well (1D)   If a particle moves through a potential $V(x)$ where x is the particle position along the x-axis such that

$$V(x) = \begin{cases} 0, & \text{if } 0 \leq x \leq a \\ \infty & \text{otherwise} \end{cases}$$

then this potential is called an infinite potential of width a. A particle in this potential is completely free, except at $x = 0$ and $x = a$, where an infinite force prevents it from escaping from inside the well.

inhomogeneous broadening   Broadening mechanism which causes different atoms to absorb or emit light at different resonant frequencies. The most common *inhomogeneous broadening* is Doppler broadening, which results from different atoms having different velocities. The line shape for Doppler broadening is a Gaussian line shape given by

$$\frac{1}{\sqrt{\pi}\,\delta} \exp\left[-\frac{(\omega - \omega_0)^2}{\delta^2} \right],$$

where $2\delta\sqrt{ln2}$ is the FWHM, or Gaussian line width.

inhomogeneous lifetime   The inverse of the inhomogeneous line width gives the inhomogeneous lifetime.

inhomogeneous line width   *See* inhomogeneous broadening.

inner product   In ordinary three-dimensional space we are familiar with the notion of the cross product and the dot product. The generalization of the dot product to n-dimensional vector spaces leads to what is known as an *inner product*. The *inner product* of two vectors $|a\rangle$ and $|b\rangle$ is a complex number and has the following properties:

$$\langle b | a \rangle = \langle a | b \rangle^*$$
$$\langle a | a \rangle \geq 0, \quad \langle a | a \rangle = 0$$
$$\implies |a\rangle = |0\rangle$$
$$\langle a | (\beta |b\rangle + \gamma |c\rangle) = \beta \langle a | b \rangle + \gamma \langle a | c \rangle .$$

It is useful to note that a vector space with an *inner product* is called an *inner product* space.

inner region The region in a wall-bounded turbulent flow where the sum of viscous and Reynolds' stresses is equal to the shear stress. This region is subdivided into three regions. The first is the viscous sublayer region which is bounded by $0 < y^+ < 5$, where y^+ is a dimensionless coordinate that is equal to (yu_τ/ν). In this region, the viscous stresses are much larger than the turbulent stresses. The velocity profile is linear. The second region is the inertial sublayer region which is bounded by $30 < y^+ < 10^4$. This region is basically a turbulent core region where viscous shearing is insignificant in comparison with the Reynolds' stresses. The velocity profile is logarithmic. In between the viscous and inertial sublayers, $5 < y^+ < 30$, there is an overlapping region where neither the viscous nor the Reynolds' stresses can be neglected. This region is referred to as a buffer region.

instability A system property where any disturbance will grow so that the system will never return to its original state. In fluid mechanics, a transition from one type of flow to another takes place as a result of *instability* of the first flow. Examples of *instability* include the *instability* caused by heating a fluid layer from below Benard convection. In this case, the *instability* is the cause of motion. A similar type of *instability* is surface tension *instability*. However, in most cases, *instability* causes transition from one type of motion to another. Examples include *instability* of the flow between concentric cylinders rotating at different flow rates (Couette flow) and shear flow *instability*. A shear flow is a flow where the velocity varies principally in a normal direction to the flow direction. Examples of shear flows include wakes, mixing layers, boundary layer, and pipe flows.

insulator A material that has high electrical resistance.

integral length (time) scale Turbulent flows are usually characterized by their scale content. The *integral length (time) scale* is based on the second order velocity correlation function. For velocity measurements in turbulent flows, the correlation function approaches zero over a certain separation in distance (time). The zero value for the correlation implies that the velocities are no longer correlated over this separation. In general, this distance (time) is considered to represent the largest scale of length (time) dimensions in the turbulent flow. Because it is mathematically obtained by integrating the correlation function, it is usually referred to as the *integral length (time) scale*.

intensity Also called irradiance. *Intensity* is the average energy passing through a unit of area per unit of time. It is proportional to the square of electric field amplitude and given by the average value of the pointing vector:

$$I = \epsilon_o c^2 \langle \mathbf{E} \times \mathbf{B} \rangle .$$

intensity correlation function Also called the second order *intensity correlation function*, or the intensity auto correlation function, or $g^{(2)}(\tau)$. It provides a measure of the correlation of the intensity of a beam at two times and/or two points in space:

$$g^{(2)}(\mathbf{r}_1, t_1 : \mathbf{r}_2, t_2) = \frac{\langle I(\mathbf{r}_1, t_1) I(\mathbf{r}_2, t_2) \rangle}{\langle I(\mathbf{r}_1, t_1) \rangle \langle I(\mathbf{r}_2, t_2) \rangle} .$$

For a stationary field, the *intensity correlation function* depends only on the time difference $(t_2 - t_1 = \tau)$, and if identical space points $(\mathbf{r}_1 = \mathbf{r}_2 = \mathbf{r})$ are chosen, we can write:

$$g^{(2)}(\mathbf{r}, \tau) = \frac{\langle I(\mathbf{r}, 0) I(\mathbf{r}, \tau) \rangle}{\langle I(\mathbf{r}, 0) \rangle \langle I(\mathbf{r}, 0) \rangle} .$$

Often, when only one detector is used, the space argument is suppressed in the expression for a two-time *intensity correlation function*.

intensity cross correlation function Provides a measure of the correlation of the intensity of the two beams at different times and different points in space:

$$g_{12}^{(2)}(\mathbf{r}_1, t_1 : \mathbf{r}_2, t_2) = \frac{\langle I_1(r_1, t_1) I_2(r_2, t_2) \rangle}{\langle I_1(r_1, t_1) \rangle \langle I_2(r_2, t_2) \rangle}.$$

Generally, in heterodyne and homodyne detection, cross correlation between the two beams at the output ports are studied.

intensity fluctuations Difference between instantaneous intensity and the average value of the intensity.

intensity interferometer Measures the second order intensity correlation function. See Hanbury–Brown–Twiss experiment.

intensity operator In a quantum picture, the intensity of an electromagnetic field becomes an operator as

$$\hat{I}(\mathbf{r}, t) = 2\epsilon_o c \left\langle \hat{\mathbf{E}}^{(-)} \hat{\mathbf{E}}^{(+)} \right\rangle,$$

where $\hat{\mathbf{E}}^{(+)}$ and $\hat{\mathbf{E}}^{(-)}$ are positive and negative frequency components of the electric field operator and are given by

$$\hat{\mathbf{E}}^{(+)} = \sum_{k\lambda} i \left(\frac{\hbar\omega}{2\epsilon_o V}\right)^{1/2} \boldsymbol{\varepsilon}_{k\lambda} \hat{a}_{k\lambda} \exp[i(\mathbf{k}\cdot\mathbf{r} - \omega t]$$

Here, $\boldsymbol{\varepsilon}_{k\lambda}$ is the polarization vector, and $\hat{a}_{k\lambda}$ is the annihilation operator for a photon with wave vector $\mathbf{k}$ and polarization λ. $\hat{\mathbf{E}}^{(-)}$ is the hermitian conjugate of $\hat{\mathbf{E}}^{(+)}$.

intensive variable A thermodynamic variable whose value is independent of the size of the system. Temperature, density, specific heat, and pressure are examples of *intensive variables*.

interacting boson model (IBM) This model was first introduced by Arima and Iachello in 1975, based on earlier ideas by Iachello and Feshbach, and was originally derived from symmetry considerations. It is an alternative to the collective model of the nucleus. The model is based upon the assumption that pairs of like nucleons couple to integer spins (zero or two). Thus, pairs of nucleons would behave like bosons.

interaction Reciprocal action. For instance, particles that are interacting have an influence or effect upon one another.

interaction Hamiltonian for atom-field system See Hamiltonian for atom-field system; Jaynes–Cummings model.

interaction representation Used when the Hamiltonian H can be split into two parts as

$$H = H_0 + V$$

where V is taken to be a perturbation which generally can be time-dependent. H_0 is normally time-independent and its eigenfunctions and eigenvalues are known. If $\psi(t)$ is the Schrödinger wave function and O is an operator with no intrinsic time dependence, then in the *interaction representation,* the wave function and operator are, respectively,

$$\tilde{\psi}(t) = e^{iH_0(t-t_0)} \psi(t) \quad \text{and}$$
$$\tilde{A}(t) = e^{iH_0(t-t_0)} A e^{-iH_0(t-t_0)}.$$

They then satisfy the following equations of motion:

$$i\hbar \frac{\partial}{\partial t} \tilde{\psi}(t) = \tilde{V} \tilde{\psi}(t)$$
$$i\hbar \frac{\partial}{\partial t} \tilde{O} = [\tilde{O}, H_0].$$

interchange instability When a light fluid supports a heavy fluid, the system is stabilized by interchanging the two. This is the *interchange instability,* which is also called the Rayleigh–Taylor instability. In the case of a plasma, which is supported against gravity by a magnetic field, if the magnetic field intensity decreases, with distance outward from the boundary, a similar instability can be triggered; consequently, the plasma will exchange places with the magnetic field.

interchange operator When acting on a many-particle wave function for identical particles, this operator results in a wave function with position and spin variables interchanged for two particles. Thus, if the *interchange operator* is denoted as P_{ij}, and the wave function of the N-particle system is $\psi(q_1, \ldots, q_i, \ldots, q_j, \ldots,$

q_N), where q_i denotes the position and spin variables describing particle i, then

$$P_{ij}\psi(q_1,\ldots,q_i,\ldots,q_j,\ldots,q_N)$$
$$= \psi(q_1,\ldots,q_j,\ldots,q_i,\ldots,q_N).$$

intercombination lines Radiative transitions between singlet and triplet spin states produce spectra with lines called *intercombination lines*.

intermediate boson The particle which mediates a particular kind of interaction in the framework of gauge field theories. For instance, the weak interaction is understood as being mediated by the W and Z bosons, whereas the electromagnetic interaction is mediated by the photon.

intermediate coupling In the study of many-electron atoms (ions), the Hamiltonian for the system containing N electrons is

$$H = \sum_{i=1}^{N}\left(\frac{-\hbar^2}{2m}\nabla_{r_i}^2 - \frac{Ze^2}{(4\pi\varepsilon_0)r_i}\right)$$
$$+ \sum_{i<j=1}^{N}\frac{e^2}{(4\pi\varepsilon_0)r_{ij}}$$

where the atomic number of the atom is taken as Z, and $\mathbf{r}_i$ is the relative coordinate of the electron with respect to the nucleus. In atomic units ($m = \hbar = e = 4\pi\varepsilon_0 = 1$), the Hamiltonian can be written more simply as

$$H = \sum_{i=1}^{N}\left(-\frac{1}{2}\nabla_{r_i}^2 - \frac{Z}{r_i}\right) + \sum_{i<j=1}^{N}\frac{1}{r_{ij}}.$$

In the central-field approximation, the Hamiltonian is approximated as H_c and is given as

$$H_c = \sum_{i=1}^{N}\left(-\frac{1}{2}\nabla_{r_i}^2 + V(r_i)\right)$$

where $V(r_i)$ is an average potential experienced by each electron, taking into account the attraction of the nucleus and the electron–electron repulsions. The corrections to H that arise from making the central-field approximation and from spin–orbit coupling when both are roughly of the same magnitude are known as *intermediate coupling*.

intermediate energy Term used in nuclear physics for laboratory kinetic energies in the range from approximately 300 MeV to 1000 MeV. Energies within this range are typically used to study nucleon–nucleus interaction. At lower energies, multiple collisions (due to the longer transit time of the projectile nucleon through the nucleus) may complicate the analysis of the reaction. At much higher energies, increased production of particles (especially π-mesons) has to be taken into account.

intermediate state An experimentally unobserved configuration of a system undergoing a transition from its initial state to a final state.

intermittency Concept used to describe intermittent behavior in flow fields. For instance, it is used to describe small turbulent bursts in an otherwise laminar flow. It has also been used to characterize the flow transition from a laminar state to a turbulent state. Other examples include the turbulent bursts in the outer layer of the boundary layer, the variations in the small scales in turbulent flows, and changes in energy levels of turbulent scales. Based on this concept of *intermittency*, an *intermittency* factor is usually defined to quantify these variations. The *intermittency* factor can be defined as the fraction of time the burst or turbulent event takes out of the total time of observation.

internal conversion A nucleus in an energetically excited state usually decays by emission of γ-rays (electromagnetic radiation). In the process of *internal conversion*, however, an electron is ejected from the atom instead of a photon.

internal energy The total energy content of a system. A thermodynamic equilibrium state of a system is characterized by a function of state called the *internal energy*, which is constant for an isolated system. The state of equilibrium of a system with fixed volume and entropy is given by the minimization of the *internal energy*.

internal flows Flows bounded by stationary or moving solid surfaces.

internal waves *Internal waves* occur within the ocean, yet they are, in principle, similar to surface waves. *Internal waves* occur at the interface between wave layers of different densities. Interfaces where *internal waves* form involve only small density differences between the layers, in comparison to the much larger density difference between air and water. As a result, *internal waves* can be much higher than surface waves and they generally move much slower.

international system of units Set of units adopted by the "Conférence générale des poids et mesures", based on meters, kilograms, seconds, amperes, kelvin, candela, and moles.

interval rule A prescription used to calculate the energy difference between two closely-spaced energy levels.

intrinsic angular momentum Also referred to as spin. A property of a particle which reveals itself in interactions with magnetic fields. Nuclear spins were predicted by Pauli in 1924 and originally associated with the orbital motion of nuclei. The spin of the electron was discovered in 1925 by Uhlenbeck and Goudsmit as an *intrinsic angular momentum* equal in magnitude to $\frac{1}{2}\hbar$, with $\hbar = \frac{h}{2\pi}$, where h is Planck's constant. Just like a small current loop or orbiting charge, the spin generates a magnetic moment which is proportional to the spin itself and interacts with an externally applied magnetic field. Thus, the spin becomes observable only in the presence of a magnetic field.

intrinsic carrier concentration The equilibrium density of electrons and holes in a semiconductor.

intrinsic conductivity Conductivity of a material due to the intrinsic carriers.

intrinsic parity A property (quantum number) associated with the behavior of a system under space reflection, namely reflection of the coordinate system. Conservation of parity means that the phenomenon under consideration is invariant upon reflection. Strong and electromagnetic interactions conserve parity, whereas the weak interaction does not. When a particle is absorbed or emitted during a nuclear or electromagnetic process, an *intrinsic parity* is assigned to the particle so as to re-establish conservation of parity.

invariance A quantity or physical phenomenon is invariant with respect to a given operation when it remains unchanged upon performing that operation. For instance, a property or phenomenon is rotationally invariant if it remains unchanged after a rotation of the coordinate system has been performed.

invariant mass The mass of a particle or system of particles which is the same regardless of the frame of reference. *See also* Lorentz invariance.

inverse beta decay The process whereby an electron and a proton combine to produce neutrons and neutrinos, i.e., $e^- + p^+ \rightarrow n + \nu$. This process takes place in stars of extremely high density.

inverse bremsstrhalung Process in which electromagnetic waves are absorbed rather than emitted; this is the resistive damping of the electromagnetic wave that occurs due to enhanced electron–ion collisions. This mechanism, in which the electromagnetic energy is converted into the thermal energy of a plasma, is a fundamental process in laser fusion and is particularly effective for short-wavelength waves.

inverse poisson transform Inverse poisson transform ($\mathcal{P}^{-1}$) and poisson transform ($\mathcal{P}$) are defined as

$$P(n) = \mathcal{P}[f(x)] = \int_0^\infty dx\, f(x)\, \frac{x^n e^{-x}}{n!},$$

$$f(x) = \mathcal{P}^{-1}[p(n)] = \mathcal{L}^{-1} \sum_0^\infty \frac{(-1)^n}{n!} P(n) s^n,$$

where $\mathcal{L}^{-1}$ is the inverse Laplace transform.

inversion Also called population *inversion*. Consider a laser gain medium in which laser action occurs between two levels, upper and lower.

If N_1 and N_2 are the number of atoms in the lower and upper levels and g_1 and g_2 are degeneracies of the lower and the upper levels, then *inversion* occurs when $N_2 > (g_2/g_1)N_1$. Inversion allows the medium to be a gain medium, which is one of the most important requirements for laser action.

inversion clamping In a laser operating above the threshold, an increase in the field intensity causes the inversion to saturate to a value of inversion at the threshold. Thus, even though the laser is operating much above the threshold, the inversion is held at the threshold value. This phenomenon is called *inversion clamping* or inversion pinning.

inversion frequency The frequency of oscillation of certain molecules between two stable configurations. For example, in the ammonia molecule NH_3, the nitrogen atom can be on either side of the hydrogen atoms which form a plane. Because these two configurations are ground states, the N atom oscillates between these two positions with a frequency (= 23870 MHz) called the *inversion frequency*.

inviscid flow Flow in which viscous effects are relatively small compared with effects from other forces such as pressure and gravity. Other terms used for *inviscid flow* include nonviscous or frictionless flow. Inviscid flows are modeled by letting the viscosity be zero, which eliminates all the viscous terms. The assumption of *inviscid flow* gives an excellent prediction of the flow and the associated lift on streamlined bodies such as airfoils and hydrofoils where boundary layers are very thin.

iodine lasers Iodine is used as a gain medium, and pumping is achieved by chemical or photodissociation of the molecules. It operates at a wavelength of 1253.73 nm in the far infrared region. Typical power output is 10 mW.

ion A charged atom obtained by either adding charges to or removing charges from the neutral atom. Adding electron(s) produces a negative *ion*, while removing electron(s) produces a positive *ion*.

ion acoustic double layer An electrical discharge phenomenon first studied by I. Langmuir in 1929. As amplitudes of ion acoustic waves propagating in a plasma are increased due to a plasma instability, for example, it becomes possible for them to form localized non-linear waveforms. In the absence of dissipation, they form either ion acoustic solitons or double layers. Double layers have a step-like or shock-like potential structure with a size of a few Debye lengths. However, unlike electrostatic shocks, double layers have nonzero internal currents that maintain their structure, and they tend to accelerate some particles streaming through them effectively, while reflecting the rest. In fact, in the process of auroral arc formation, a series of double layers positioned along the auroral field lines are thought to accelerate electrons streaming downward. Furthermore, double layers may be divided into strong double layers and weak double layers depending on their amplitudes. *See also* ion acoustic wave, ion acoustic shock wave.

ion acoustic shock wave As amplitudes of ion acoustic waves propagating in a plasma are increased, it becomes possible for them to form localized non-linear waveforms. In a streaming plasma, they tend to form shock waves characterized by a step-like potential structure with a size of a few Debye lengths or longer. Their energy is supplied by the ions streaming through them, while electrons are accelerated by the shock waves. Therefore, the ion streaming energy determines the magnitude of the shocks. *See also* ion acoustic wave.

ion acoustic solitons As amplitudes of ion acoustic waves propagating in a plasma increase, it becomes possible for them to form localized non-linear waveforms. One such wave is a soliton, which is a stable isolated pulse. Properties of non-linear ion acoustic waves are described by the Korteweg-de Vries (K-dV) equation, solutions of which correspond to ion acoustic solitons, i.e.,

$$A \operatorname{sech}^2 \sqrt{\frac{\alpha A}{12\beta}} \left(x - \frac{\alpha A}{3} t \right),$$

where A is the amplitude, α, β are constants, t is the time, and x is the position. Hence, as the amplitude increases the speed increases, while the width shrinks. Solitons are related to shock waves through a quasi-potential called the Sagdeev potential.

ion acoustic wave The only normal mode of ions allowed in nonmagnetized plasmas, *ion acoustic waves* are essentially driven by thermal motions of both electrons and ions. In fact, their phase and group velocities are given by the ion acoustic or sound speed $c_s = \{(KT_e + 3KT_i)/m_i\}^{1/2}$, where K is the Boltzman constant, T_e, and T_i are the electron and ion temperatures, and m_i is the ion mass. With the use of the ion acoustic speed, the dispersion relation of the *ion acoustic wave* with frequency ω and wave number k is given by $\omega = kc_s$. There are two damping mechanisms for *ion acoustic waves;* one is Landau damping and the other is the non-linear Landau damping that occurs after trapping of particles inside the electrostatic wave potential of relatively large *ion acoustic waves*. *Ion acoustic waves* are heavily damped if $T_e < T_i$, so that such waves usually propagate only in plasmas with $T_i \ll T_e$. Various nonlinear states of *ion acoustic waves* have been the subjects of intensive research in plasma physics for many years. As they are amplified, these waves may form solitons, double layers, and shock waves. *See also* ion wave.

ion beam instabilities There are several instabilities driven by an ion beam, which, in a magnetized plasma, usually propagates along an external magnetic field. Electrostatic instabilities are the ion acoustic instability driven by the relative drift between the electrons and the beam ions and the ion–ion drift instability. The former generates principally field-aligned waves, and the latter generates either field-aligned or oblique waves. Among electromagnetic instabilities are the ion–ion resonant and nonresonant instabilities; the former excite right-hand circularly polarized waves, and the latter excite left-hand circularly polarized (Alfvén) waves at relatively low drift speeds, i.e., the fire-hose instability and right-hand circularly polarized waves at higher speeds. Whistler waves can also be generated. Production of these right-hand circularly polarized waves can be enhanced by increased drift speed as well as increased perpendicular temperature of the beam.

ion cyclotron resonance *See* cyclotron resonance.

ion cyclotron resonance heating (ICRH) Has been utilized to heat plasmas by electromagnetic waves. For this scheme, an electromagnetic ion cyclotron wave is launched from an external source into a plasma with a frequency ω, which is lower than the local ion cyclotron frequency Ω_i of the target plasma. As the wave propagates into a decreasing magnetic field, it will eventually heat the target plasma efficiently through cyclotron acceleration when the local resonance condition $\omega = \Omega_i$ is satisfied. This heating scheme is frequently used in several fusion devices such as tokamaks.

ion cyclotron wave When magnetized, plasmas can support electrostatic *ion cyclotron waves* that propagate nearly perpendicular to the external magnetic field. The dispersion relation is given by $\omega^2 = \Omega_i^2 + k^2 c_s^2$, where ω is the frequency, k is the wave number of the wave, Ω_i is the ion cyclotron frequency, and c_s^2 is the ion acoustic speed. Experimentally, *ion cyclotron waves* were first observed by Motley and D'Angelo in a device called a Q-machine.

On the other hand, electromagnetic *ion cyclotron waves* propagate predominantly along the magnetic field, and are left-hand polarized. These waves are frequently used to heat ions in plasma confinement devices, i.e., ion cyclotron resonance heating (ICRH). *See also* ion wave.

ionic bonding The bonding in structures that results from the net attraction between oppositely charged species. For example, in compounds of the alkalis and a halogen atom (e.g., sodium chloride, NaCl), the chlorine atom detaches an electron from the sodium atom, forming Na^+ and Cl^- ions which together can form a stable configuration or crystal structure. The variation of the energy of the ($Na^+ + Cl^-$) system, $E_s(R)$, relative to the sum of the energies of the isolated neutral atoms is given as

$$E_s(R) = E_s(\infty) - \frac{1}{R} + Ae^{-hR}$$

where R is the internuclear separation and A and h are constants.

ionic conduction Electrical conduction due to ions.

ion implantation A method of introducing impurity atoms in a material by bombarding it with ions.

ionization The process whereby an electron is completely removed from an atom. The minimum energy required to do this is called the *ionization* energy.

ionization chamber One of the oldest instruments used in nuclear physics to detect charged particles. It consists of a chamber filled with a substance, usually in the form of a gas, which becomes ionized from the passage of charged particles. Ions can then be collected with the help of an electric field.

ionization energy The energy required to produce ions from neutral atoms or molecules.

ion Landau damping *See* Landau damping.

ion plasma frequency *See* plasma frequency.

ion sound wave *See* ion acoustic wave, ion wave.

ion wave Even in an essentially collision-free plasma, ions can collectively move and oscillate and form *ion waves* due to the electromagnetic force. In a nonmagnetized plasma, the only allowed type of *ion wave* is the ion acoustic (or sound) wave. However, when magnetized, a plasma can allow a variety of *ion waves*. Among the electrostatic waves are the ion acoustic (or sound) waves along the magnetic field and the ion cyclotron waves traveling nearly perpendicular to the field. Among the electromagnetic waves are the Alfvén waves along the magnetic field and the magnetosonic waves propagating perpendicular to the field. Compared to electron plasma waves, *ion waves* have lower frequencies and longer wavelengths as well as higher energy.

irreversible processes A process occurring in an isolated system so that the system cannot be taken back to its initial state by simply imposing or removing the constraints that led to the process. The entropy of a system always increases in an *irreversible process*.

irrotational flow A flow with the vorticity at every point equal to zero. *Irrotational flow* is a pre-requisite for existence of the velocity potential.

isentropic flow A flow where the entropy of a fluid particle in a continuous inviscid flow remains constant as the particle travels along with the flow, i.e., $Ds/Dt = 0$.

isentropic process A process at constant entropy. Note that an adiabatic process is isentropic, but it refers to the specific case of no heat transfer.

Ising model A model for ferromagnetism in which only the interactions between nearest neighbor spins is taken into account.

isobaric process A process at constant pressure.

isobaric spin Also called isotopic spin. *See* isospin.

isobars All nuclei (or nuclides) with the same mass number. The mass number (A) is the number of protons and neutrons.

isochoric process A process in which the volume of a system does not change, or a process in which no mechanical work is done on the system.

isoelectronic Pertains to atoms that have the same number of electrons.

isoelectronic defects Impurity introducing a bound state in a solid.

isolator Used to block off the back-reflected radiation. *Isolators* allow electromagnetic waves to propagate only in one direction. This is achieved by rotating the polarization of reflected light by 90° with respect to the incident light, and therefore the reflected light is blocked by the polarizer. Rotation of the polarization is generally achieved by using Faraday rotation in magneto-optical material. Optical *isolators* are very common in optical communications systems.

isomer (1) One of two or more nuclides that have the same atomic and mass numbers but differ in other properties.
(2) A nucleus which has the same proton and neutron number as in other nucleus, but which has a different state of excitation.

isomer (nuclear) An excited state of a nucleus which has a measurable mean life. The radioactive decay of such a state is said to occur in an isomeric transition and the phenomenon is known as nuclear isomerism.

isoscalar particle A particle with isospin equal to zero.

isospin A property (or quantum number) which distinguishes a proton from a neutron. With respect to the nuclear force, a proton and a neutron behave in essentially the same way. In contrast protons and neutrons interact differently with a Coulomb field. With an isospin of $\frac{1}{2}$ assigned to the nucleon, the two nucleons are then distinguishable through the third component of the isospin being $+\frac{1}{2}$ for the proton and $-\frac{1}{2}$ for the neutron.

isothermal bulk modulus (β_T) A measure of the resistance to volume change without deformation or change in shape in a thermodynamic system in a process at constant temperature. It is the inverse of the isothermal compressibility.

$$\beta_T = -V \left(\frac{\partial P}{\partial V} \right)_T .$$

isothermal compressibility (κ_T) The fractional decrease in volume with increase in pressure while the temperature remains constant during the compression.

$$\kappa_T = -\frac{1}{V} \left(\frac{\partial V}{\partial P} \right)_T .$$

isothermal process A process at constant temperature.

isotone One of two or more nuclides that have the same number of neutrons in their nuclei but differ in the number of protons.

isotope One of two or more nuclides that have the same atomic number but different numbers of neutrons so that they have different masses. The mass is indicated by a left exponent on the symbol of the element (i.e., ^{14}C).

isotope effect The correction to the energy levels of a bound-state system due to the finite mass of the nucleus.

isotope effect (superconductivity) Early in the development of the theory of superconductivity, it was found that different isotopes of the same superconducting metal have different critical temperatures, T_c, such that

$$T_c M^a = \text{constant}$$

where M is the mass of the isotope and $a \approx 0.5$ for most metals. This effect made it clear that the lattice of ions in a metal is an active participant in creating the superconducting state.

isotropic Independent of direction, or spherically symmetric.

isotropic turbulence Implies that there is no mean shear and that all mean values of quantities such as turbulence intensity, auto- and cross-correlations, spectra, and higher order correlation functions of the flow variables are independent of the translation or rotation of the axes of reference. These conditions are not typical in real flows. On the other hand, assumptions of *isotropic* and homogeneous *turbulence* have led to understanding of many aspects of turbulent flows.

isotropy Having identical properties in all directions.

isovector particle A particle with isospin equal to one and, thus, three possible charge states corresponding to the three possible values (0, ±1) of the third component of the isospin vector.

ITER Originally proposed at a summit meeting between the USA and the USSR in 1985, the purpose of the international thermonuclear experimental reactor [ITER] project is to build a toroidal device called a tokamak for magnetic confinement fusion to specifically demonstrate thermonuclear ignition and study the physics of burning plasma. The initial phase of this project was jointly funded by four parties: Japan, the European community, the Russian Federation and the United States. In July of 1992, ITER engineering design activities [ITER EDA] were established to provide a fully integrated engineering design as well as technical data for future decisions on the construction of the ITER. To meet the objectives, the linear dimensions of ITER will be 2–3 times bigger than the largest existing tokamaks. According to the 1998 design, the major parameters of the ITER are as follows: total fusion power of 1.5G W, a plasma inductive burn time of 1000 s, a plasma major radius of 8.1 m, a plasma minor radius of 2.8 m, a toroidal magnetic field at the plasma center of 5.7 T, and an auxiliary heating power by neutral beam injection of 100 MW.

J

Jacobi coordinates In describing the dynamics of many-particle systems, we are often faced with the task of choosing an appropriate set of coordinates. For example, in the two-body problem, the motion relative to the center of mass is described by the one-body Schrödinger equation:

$$i\hbar \frac{\partial \Psi(\mathbf{r}, t)}{\partial t} = \left[\frac{-\hbar^2}{2\mu} \nabla_r^2 + V(\mathbf{r}) \right] \Psi(\mathbf{r}, t)$$

$\mu = \frac{m_1 m_2}{m_1 + m_2}$ is the reduced mass for particles of mass m_1 and m_2, and $\mathbf{r} = \mathbf{r}_1 - \mathbf{r}_2$ are the relative position vectors of particles 1 and 2. Suitable sets of center-of-mass coordinates can be similarly constructed for systems containing any number of particles. For example, consider the three-body problem

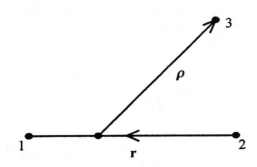

A set of Jacobi coordinates for a three-body system.

We first consider particles 1 and 2 as a sub-system with relative coordinate $\mathbf{r}$ and center of mass μ. The motion of the center-of-mass of this sub-system relative to the third particle is described through the second position vector ρ. The Schrödinger equation for this system then reads:

$$i\hbar \frac{\partial \Psi(\mathbf{r}, \rho, t)}{\partial t} = \left[\frac{-\hbar^2}{2\mu} \nabla_r^2 + \frac{-\hbar^2}{2\mu'} \nabla_\rho^2 + V(\mathbf{r}, \rho) \right] \Psi(\mathbf{r}, \rho, t)$$

where $\mu' = \frac{(m_1 m_2) m_3}{m_1 + m_2 + m_3}$, with m_3 representing the mass of particle 3. Coordinate systems of this kind where the kinetic energy is separable are called *Jacobi coordinates*.

Jahn Teller effect (rule) A non-linear molecule in a symmetric configuration with an orbitally degenerate ground state is unstable. The molecule will seek a less symmetric configuration with an orbitally nondegenerate ground state. Although this rule was introduced to describe molecules, it has applications to impurities and defects in solids. An impurity ion can move from a symmetric position in a crystal to a position of lower symmetry to lower its energy. A free hole in an alkali halide crystal (such as KCI) can be trapped by a halogen ion and becomes immobile; it moves only by hopping to another site if thermally activated.

Jansky, K. Astronomers have always searched for ways of studying celestial objects like comets, stars, and galaxies. One of the most widely used methods of studying objects in the sky is through the electromagnetic radiation reaching us from these objects. Because of the absorption of electromagnetic radiation propagating from outer space to us, we can only use limited bands (ranges of frequencies). One band was discovered in 1931 by *K. Jansky*. He discovered radio waves coming from the Milky Way. This discovery was very ground-breaking as it opened up a new field called radioastronomy, through which new discoveries about the universe such as pulsars, quasars and the universal radiation at 3 K have been made.

Jaynes–Cummings model (1) Describes dynamics of a two-level atom interacting with a single mode of radiation field in a lossless cavity. This model is perhaps the simplest solvable model that describes the fundamental physics of radiation–matter interaction. This somewhat idealized model has been realized in the laboratory by using Rydberg atoms interacting with the radiation field in a high-Q microwave cavity. The Hamiltonian for the *Jaynes–Cummings model* in the rotating-wave approximation is given by

$$\hat{H} = \frac{1}{2}\hbar\omega_0\hat{\sigma}_3 + \hbar\omega\left(\hat{a}^\dagger\hat{a} + 1/2\right) + \hbar\lambda\left[\hat{\sigma}_+\hat{a} + \hat{a}^\dagger\hat{\sigma}_-\right].$$

Here, the Pauli matrices $\hat{\sigma}_+$, $\hat{\sigma}_-$, and $\hat{\sigma}_3$ represent the raising, lowering, and inversion operators for the atom, ω_0 is the transition frequency for the atom, and ω is the field frequency. Operators $\hat{a}^\dagger$ and $\hat{a}$ are the creation and annihilation operators of the field-satisfying boson commutation relations.

(2) The simplest model in cavity quantum electrodynamics. In the *Jaynes-Cummings model,* one assumes that a two-level atom with upper level $|a\rangle$ and lower level $|b\rangle$ interacts with only one mode of the quantized electromagnetic field. Furthermore, this mode is assumed to be resonant with the atomic transition frequency. The Hamilton operator in the rotating wave approximation for this problem is given by

$$H = \omega_0 b^\dagger b + \frac{1}{2}\hbar\omega_0\sigma_z + \hbar g\left(b\sigma^+ + b^\dagger\sigma^-\right).$$

Here g is the coupling constant, ω_0 is the resonant transition frequency of the atoms, and σ^+, σ^-, and σ_z are the well-known Pauli spin matrices. This reflects the possibility of interpreting a two-level system as a spin 1/2 system with spin up when the population is in the upper state and spin down for a population of the lower state.

The first two terms of the Hamiltonian describing the energy eigenstates of the photons and the two-level atom commute with the second part describing the interaction of the system. This results in the possibility of writing the eigenstates for the Hamiltonian as a combination of the eigenstates of the atom and field.

The eigenstates and eigenvalues for such a system are given by

$$|\Psi^+\rangle = \frac{1}{2}\left(|n,a\rangle + |n+1,b\rangle\right)$$
$$|\Psi^-\rangle = \frac{1}{2}\left(|n,a\rangle - |n+1,b\rangle\right)$$

where n is the number of photons in the field. The eigenvalues for these states are $\pm\Omega\hbar$, where

$$\Omega^2 = \sqrt{\Delta^2 + 4g^2(n+1)}$$

is called the Rabi frequency. A possible detuning of the quantized cavity field with the atomic resonance Δ is also taken into account here. Assuming that the atom is initially in the excited state and the field has n photons, one can calculate the probability of finding the atom in the excited state and the atom in a state with n photons at time t to

$$P_{n,a}(t) = \cos^2(\Omega t).$$

One sees oscillatory behavior in time, which is called the Rabi oscillations or Rabi mutations. In case the radiation field is in a coherent superposition, quantum effects like recurrence phenomena can be observed.

Of greatest interest is the strong coupling limit where the coupling g is stronger than the dissipation processes of the cavity and the spontaneous decays of the atomic levels.

The *Jaynes-Cummings model* is the basis for the micromaser experiments, where a single atom interacts with a high-Q cavity. The two-level characteristics of the atom are approximated by exciting the atom into a Rydberg state before entering the cavity. The interaction time can be determined by using velocity selective excitation into the Rydberg states. Pure quantum phenomena such as quantum collapse and revival can be observed.

Jeans instability A plasma under the influence of a gravitational force is unstable due to the *Jeans instability,* for which waves longer than the Jeans length grow exponentially. This phenomenon is analogous to ordinary plasma waves propagating without being Landau-damped, provided that their wavelengths are sufficiently long.

Jeans, Sir J. *Sir J. Jeans,* together with Lord Rayleigh, derived a spectral distribution function to describe black-body radiation. Their theory was called the Rayleigh–Jeans theory and could only explain the long-wavelength behavior of the spectrum. They derived a spectral function $\rho(\lambda, T)$, where λ is wavelength and T is temperature, for the radiation emitted from an enclosed cavity (black-body) using the laws of classical physics. They modeled the thermal waves in the cavity as standing waves (modes) of wavelength λ. They calculated the number of modes per unit of volume in the wavelength range $\lambda \to \lambda + d\lambda$, $n(\lambda)$, as $\frac{8\pi}{\lambda^4} d\lambda$. This was then multiplied by the average energy in the mode, $\bar{\varepsilon}$, to give the spectral density

$$\rho(\lambda, T) = \frac{8\pi}{\lambda^4} \bar{\varepsilon}.$$

Rayleigh and Jeans surmised that the standing waves are caused by constant absorption and emission of radiation of frequency ν by classical linear harmonic oscillators in the walls of the cavity. They assumed that the energy of each oscillator can take any value from 0 to ∞, which turned out to be an erroneous assumption. The average energy of a collection of such oscillators was calculated, using classical statistical mechanics, to be $k_B T$, where k_B is the Boltzmann constant. Thus, they predicted the black-body distribution to be

$$\rho(\lambda, T) = \frac{8\pi}{\lambda^4} k_B T.$$

This is the Rayleigh–Jeans law. It agrees only in the long-wavelength limit and diverges for $\lambda \to 0$.

jellium A model in which the positive charges of the ions in a metal are uniformly spread (like jelly) in the volume occupied by the ions. It is the closest realization of the Thomson atom.

jellium model Used in the study of the correlation effects in an electron gas. The basic premise is that the atoms in the lattice are replaced with a uniform background of positive charge.

jet Efflux of fluid from an orifice, either two- or three-dimensional. In the former case, the *jet* is emitted from a slit in a wall. In the latter case, the *jet* exits through a hole of finite size. *Jets* expand by spreading and combining with surrounding fluid through entrainment. A *jet* may either be laminar or turbulent.

JET The Joint European Torus *(JET)* located at Abingdon in Oxfordshire, England is a toroidal tokamak-type device for magnetic confinement fusion jointly operated by 15 European nations. The *JET* project was set up in 1978, and there are approximately 350 scientists, engineers, and administrators supported by a similar number of contractors. Even though the project was officially terminated in 1999, the *JET* facilities have still been in operation since then. This device, being the largest of its kind in the world as well as the first to achieve the break even condition (input power = output power), is of approximately 15 meters in diameter and 12 meters high. The central portion of the device is a toroidal vacuum vessel of major radius 2.96 meters with a D-shaped cross-section of 2.5 meters by 4.2 meters; the toroidal magnetic field at the plasma center is 3.45T, and the plasma currents are 3.2–4.8 MA. It also has an additional heating power of over 25MW. It is presently the only device in the world which is capable of handling as its fuel the deuterium–tritium [DT] mixtures used in a future fusion power station.

jet instability From linear stability theory, jets are unstable above a Reynolds number of four, similar to Kelvin–Helmholtz instability. The resulting jet motion consists of vortical structures which roll up with surrounding fluid and dissipate downstream.

jet pump Similar in design to an aspirator, except both working fluids are usually of the same phase.

jets in nuclear reactions Back-to-back streams of hadrons produced in nuclear reactions. Jets are usually observed when quarks and antiquarks (free for just a very short time) fly apart. This can be observed, for example, through the reaction $e^+ + e^- \to \gamma \to q + \bar{q} \to$ hadrons. When the quarks reach a separation of about 10^{-15} m, their mutual strong interaction is so intense that new quark-antiquark pairs are produced and combine into mesons and baryons, which emerge in two (and sometimes three) back-to-back jets.

j–j coupling A possible coupling scheme for spins and angular momenta of the individual nucleons in a nucleus. In the $j--j$ scheme, (as opposed to the LS scheme), first the intrinsic spin and orbital angular momentum of each nucleon are added together to yield the total angular momentum of a single nucleon. Then the angular momenta of the individual nucleons are summed up to give the total angular momentum of the nucleus.

j-meson/resonance Also known as the Ψ meson. Particle discovered in 1974, which confirmed the existence of the fourth quark (the charm quark).

Johnson noise Noise in an electric circuit arising due to thermal energy of the charge carrier. Noise power P generated in the circuit due to the *Johnson noise* depends on the temperature T and frequency band $\Delta\nu$ considered, but is independent of the circuit elements.

$$P = \frac{h\nu\Delta\nu}{\exp[-h\nu/kT] - 1},$$

where k is the Boltzmann constant. For $kT \gg h\nu$, the noise power can be approximated to be $kT\Delta\nu$. This noise can be reduced by cooling the components generating the noise. It is also called Nyquist noise.

Jones calculus Introduced by R. Clark Jones to describe the evolution of a polarization state when it passes through various optical elements. In the Jones matrix formulation, the polarization of a plane wave is represented by a pair of complex electric field components E_1 and E_2, along two mutually orthogonal directions transverse to the direction of propagation, written as a column matrix with ($0 \leq \beta \leq \pi/2$):

$$\frac{1}{\sqrt{E_1^2 + E_2^2}} \begin{bmatrix} E_1 \\ e^{i\delta} E_2 \end{bmatrix} = \begin{bmatrix} \cos\beta \\ e^{i\delta} \sin\beta \end{bmatrix},$$

$$\beta = \tan^{-1}\left(\frac{E_2}{E_1}\right).$$

The action of various polarizing elements is then described by complex 2×2 matrices which act on the column matrix representing the polarization state. For example, the Jones matrix for a quarter wave plate whose fast axis is horizontal is given by

$$M = e^{i\pi/4} \begin{bmatrix} 1 & 0 \\ 0 & i \end{bmatrix}.$$

These matrices are derived in paraxial approximations.

Jones matrix 2×2 matrix which describes the effect of an optical element on the polarization of light. The polarization of the light can be described with a two-dimensional Jones vector. Horizontal and vertical polarization can be described as two vectors

$$\begin{pmatrix} 1 \\ 0 \end{pmatrix} \text{ (horizontal) and } \begin{pmatrix} 0 \\ 1 \end{pmatrix} \text{ (vertical)}.$$

An ideal polarizer (without loss) at an angle θ with respect to the horizontal has the *Jones matrix*,

$$\begin{pmatrix} \cos^2\theta & \sin\theta\cos\theta \\ \sin\theta\cos\theta & \sin^2\theta \end{pmatrix}.$$

For a linear retarder, which introduces a phase-shift of δ to one polarization direction and is aligned so that the optic axis makes an angle θ with respect to the horizontal, we find a *Jones matrix* given by

$$\begin{pmatrix} \cos^2\theta + \sin^2\theta \exp(-\imath\delta) & \cos\theta\sin\theta(1 - \exp(-\imath\delta)) \\ \cos\theta\sin\theta(1 - \exp(-\imath\delta)) & \sin^2\theta + \cos^2\theta \exp(-\imath\delta) \end{pmatrix}.$$

The special cases for a $\lambda/2$-plate and a $\lambda/4$ plate are easily calculated using $\delta = \pi$ and $\delta = \pi/2$ respectively.

Jones vector Used to represent the polarization of an electromagnetic wave. It can also be used to represent any vector in a two-dimensional space. These vectors can be expressed as a superposition of two basis vectors. The coefficients for the two vectors can be written as the components of a two-dimensional vector, which is called a *Jones vector*. Vertical and horizontal polarization can then be represented as

$$\begin{pmatrix} 1 \\ 0 \end{pmatrix} \text{ (horizontal) and } \begin{pmatrix} 0 \\ 1 \end{pmatrix} \text{ (vertical)}.$$

Any operation on this vector can then be expressed as a 2×2 matrix, which is the Jones matrix.

An alternative basis for describing the polarization properties is via left and right circular polarized light. These can be written as

$$\frac{1}{\sqrt{2}} \begin{pmatrix} 1 \\ -i \end{pmatrix} \text{ (lefthand circular) and}$$

$$\frac{1}{\sqrt{2}} \begin{pmatrix} 1 \\ i \end{pmatrix} \text{ (righthand circular)}.$$

Jones zones Volumes in <u>k</u> space (reciprocal lattice) bounded by planes which are perpendicular bisectors of reciprocal lattice vectors (as in the case of the Brillouin zones). These planes correspond to strong Bragg reflection for x-rays. Strong x-ray scattering suggests strong Bragg reflection for electron waves and the presence of large Fourier coefficients $V(G)$ for the potential which the electron sees, where $\underline{G}$ is the reciprocal lattice vector involved. This means that if the *Jones zone* is nearly filled with electrons, those electrons near the zone boundary within an energy interval of approximately $1/2|V(G)|$ will lower their energy by approximately $|V(\underline{G})|$, or $|V|$ for short, each below the free electron energy. The net energy reduction for the electron gas is approximately $1/2 N(E_f)|V|^2$, where $N(E_f)$ is the electron density of states at the Fermi energy E_f which gives a binding energy of $3/4|V|^2/E_f$ per electron. This method can be applied even to a covalent crystal such as diamond, silicon, or germanium. Direct lattice is a face-centered cubic with cube side a, and has two atoms per unit cell separated by the vector $\tau = (1, 1, 1)a/4$. The Fourier coefficient $V(\underline{G})$ of the crystal is that of a monoatomic crystal $V_0(\underline{G})$ multiplied by the structure factor $(1 + \exp(-i\underline{G} \bullet \underline{\tau}))$, which we call $S(\underline{G})$. Since reciprocal space is a body-centered cubic lattice with side $(2/a)2\pi$, we see that the eight reciprocal lattice vectors $(2\pi/a)(\pm 1, \pm 1, \pm 1)$ give $|S|^2 = 2$ and will define a *Jones zone* which can accommodate $(9/8)N$ states for each spin direction (and not N as we always have for Brillouin zones). Here, N is the number of unit cells (Bravais) of direct lattice. A larger *Jones zone* can be constructed from the twelve reciprocal lattice vectors of the type $4\pi/a(\pm 1, \pm 1, 0)$, which can accommodate all the valence electrons of the crystal ($8N$). Such ideas might explain the stability of certain metals and alloys. *See* nearly free electrons.

Jönsson, C. The wave behavior of electrons was demonstrated in 1961 by *C. Jönsson* in an electron diffraction experiment.

Jordan, P. Two equivalent formulations of quantum mechanics were put forward at about the same time between 1924–1926. The first formulation, called wave mechanics, was developed by E. Schrödinger. The other is matrix mechanics, which was developed by W. Heisenberg, M. Born, and P. Jordan.

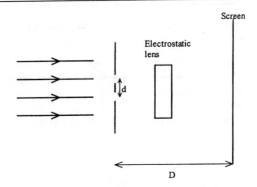

Jönsson used 40 keV electrons. The slits were made in a copper foil and were very small ~ 0.5 microns wide and the slit separation ~ 2 microns. Interference fringes were observed on a screen at a distance of 0.4 m from the slits. Since the fringe separation was very small, an electrostatic lens was used to magnify the fringes.

Josephson, B.D. In 1962, B.D. Josephson published a paper predicting two fascinating effects of superconducting tunnel junctions. The first effect was that a tunnel junction should be able to sustain a zero-voltage superconducting dc current. The second effect was that if the current exceeds its critical value, the junction begins to generate high-frequency electromagnetic waves.

Josephson effect (1) (i) DC effect: In a Josephson junction, an insulating oxide layer is sandwiched between two superconductors.

In each superconductor, electrons condense into Cooper pairs, which tunnel through the insulating layer. We define a wave function, also called an order parameter, for each superconductor. In superconductor 1, the order parameter is written as

$$\Psi_1(x, t) = n_s^{\frac{1}{2}} e^{-i\phi_1}$$
$$\phi_1 = \phi_{s1} + \omega t$$

where ϕ_{s1} is the phase of the time-independent part of the order parameter. Similarly, for su-

Josephson effect

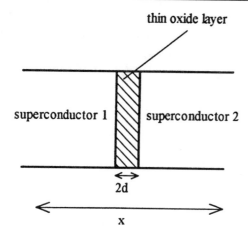

Josephson junction made from two superconductors separated by a thin oxide layer.

perconductor 2,

$$\Psi_2(x,t) = n_s^{\frac{1}{2}} e^{-i\phi_2}$$
$$\phi_2 = \phi_{s2} + \omega t$$

n_s is the number density of Cooper pairs in the left and right superconductors, which is assumed to be the same. Using the familiar expressions for current in terms of the wave functions $\Psi_{1,2}$ that are used in studying tunnelling in potential barriers, we obtain the current J as

$$J = J_0 \sin \theta$$

where $\theta = \phi_1 - \phi_2$. Thus a DC current flows across the barrier if there is a phase gradient.

(ii) AC effect: If a voltage V is applied across the junction, there is a change in the energy of the Cooper pairs, resulting in a change in the phase of the time-dependent part of the order parameter. We obtain

$$\phi_1 = \phi_{s1} + \left(\omega + \frac{eV}{\hbar}\right) t$$

and

$$\phi_2 = \phi_{s2} + \left(\omega - \frac{eV}{\hbar}\right) t .$$

Thus, we have applied a potential of $\frac{V}{2}$ to superconductor 1 and $\frac{-V}{2}$ to superconductor 2. The current in this case is time-dependent, since $\theta = \phi_1 - \phi_2 = (\phi_{s1} - \phi_{s2}) + \frac{2eV}{\hbar}t$. Due to the nature of the current this case is called the *Josephson AC effect*.

(2) A Josephson junction can be made of two good superconductors separated by a thin layer of 10 Å of an insulator, and a normal (nonsuperconducting metal) or weaker superconductor. A current of Cooper pairs (bound electron pairs) would flow across the junction even if there is no potential difference (voltage) between the two good superconductors. If a DC voltage V_0 is applied, an oscillating pair current of angular frequency $|qV_0/h|$ results where q is the charge on the Cooper pair (twice e, the electron charge) and h is Planck's constant divided by 2π. If, in addition to V_0, we add an oscillatory voltage $v \sin \omega t$, we find that the pair current J is given by

$$J \approx \sin[\delta_0 + (qV_0 t/h) + (qv/h\omega) \sin \omega t] ,$$

where δ_0 is a constant. This formula predicts that when $\omega = (qV_0/hn)$, where n is an integer, there will be a DC current component present. Two or more Josephson junctions can be connected in parallel in a magnetic field, and their current displays interference effects similar to those of diffraction slits in optics.

Josephson radiation If a DC current greater than the critical current flows through a Josephson junction, it causes a voltage V(t) to appear

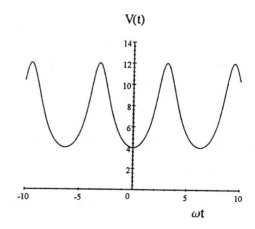

Variation of the voltage V(t) across a Josephson junction versus ωt.

across the junction which oscillates with time. This causes the emission of electromagnetic radiation of frequency ω, such that the average voltage across the junction, $\overline{V}$, is given as

$$2e\overline{V} = \hbar\omega .$$

The first experimental observation of *Josephson radiation* was reported in 1964 by I.K. Yanson, V.M. Svistunov, and I.M. Dmitrenko. The English translation of this paper appears in *Sov. Phys. JETP*, 21, 650, 1965.

Josephson vortices Consider the following Josephson junction in a magnetic field H_0:

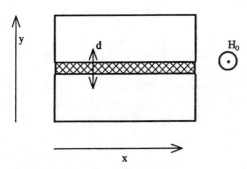

Josephson junction in a magnetic field H_0.

If the junction is placed in a magnetic field H_0 directed along the z-axis, a screening supercurrent is generated at the outer surfaces of each slab. Such current is constrained to flow within a thin layer. The magnetic field at x can be shown to be proportional to $\frac{d\phi}{dx}$, where ϕ is the phase difference between the superconductors. The differential equation which describes ϕ (Ferrell–Prange equation) is

$$\frac{d^2\phi}{dx^2} = \frac{1}{\lambda_J^2} \sin\phi$$

where λ_J is the Josephson penetration depth and gives a measure of penetration of the magnetic field into the junction. In a weak magnetic field, the above equations give solutions for the phase difference ϕ and magnetic field H as

$$\phi(x) = \phi(0) \exp(-x/\lambda_J)$$
$$H(x) = H_0 \exp(-x/\lambda_J) .$$

If the external field increases beyond a certain critical value which is characteristic of the junction, the magnetic field penetrates into the junction in the form of a soliton or vortex. This is called a *Josephson vortex*.

Joukowski airfoil See Zhukhovski airfoil.

joule Unit of energy in the standard international system of units.

Joule effect (Joule magnetostriction) Change in the length of a ferromagnetic rod in the direction of the magnetic field when magnetized. See magnetostriction.

Joule heating The electrical energy dissipated per second as heat in a resistor of resistance R ohms and carrying a current of I amperes is equal to I^2R watts.

Joule–Thompson effect A process in which a gas at high pressure moves through a porous plug into a region of lower pressure in a thermally insulated container. The process conserves enthalpy and leads to a change in temperature.

***j*-symbols** Symbols used in the context of angular momentum algebra in quantum mechanics. For example, the symbol $< j_1 j_2 m_1 m_2 | J M >$ indicates the coupling of the two angular momenta j_1 and j_2 to a total angular momentum J. In this framework, m_1, m_2, and M are the magnetic quantum numbers associated with the component of their respective angular momenta along a pre-chosen direction.

JT-60 In September 1996, the breakeven plasma condition (input power = output power) was first achieved by *JT-60*, which proved the feasibility of a fusion reactor based on the tokamak scheme. Located in Naka, Japan, and operated by Japan Atomic Energy Research Institute[JAERI], *JT-60*, a toroidal device for magnetic confinement fusion, is one of the largest tokamak machines in the world. JT-60U, the upgraded version of *JT-60* had a negative-ion based neutral beam injector installed in 1996, and the divertor transformed from open into W-shaped semi-closed in 1997. The major parameters of *JT-60* are as follows: a plasma major radius of 3.3 m, a plasma minor of radius 0.8 m, a plasma current of 4.5M A, a toroidal magnetic field at the plasma center of 4.4 T, and an auxiliary heating power by neutral beam injection of 30 MW.

JT-60U at JAERI.

jump conditions Variation in Mach number and other flow variables across a shock wave. For a normal shock wave, a variation in Mach number across a shock is only a function of the upstream Mach number as

$$M_2^2 = \frac{(\gamma - 1)M_1^2 + 2}{2\gamma M_1^2 - (\gamma - 1)}$$

where γ is the ratio of specific heats. For $M_1 = 1$, $M_2 = 2$; this is the weak wave limit where the wave is a sound wave. For $M_1 \infty$, $M_2 = \sqrt{(\gamma - 1)/2\gamma}$; this is the infinite limit which shows that there is a lower limit which the subsonic flow can attain. For air, $\gamma = 1.4$; this becomes $M_2 = 0.378$. Thus, the Mach number (but not the velocity) can go no lower than this limit. The jump in density and velocity is related by the continuity equation

$$\frac{\rho_2}{\rho_1} = \frac{u_1}{u_2} = \frac{(\gamma + 1)M_1^2}{(\gamma + 1)M_1^2 + 2}$$

while momentum yields the jump in pressure

$$\frac{p_2}{p_1} = 1 + \frac{2\gamma}{\gamma + 1}\left(M_1^2 - 1\right).$$

These can be combined with the ideal gas equation to obtain

$$\frac{T_2}{T_1} = \frac{a_2^2}{a_1^2} = \frac{\left[2\gamma M_1^2 - (\gamma - 1)\right]\left[(\gamma - 1)M_1^2 + 2\right]}{(\gamma + 1)^2 M_1^2}.$$

Since the flow is adiabatic, the stagnation or total temperature across a shock wave is constant. Thus,

$$\frac{T_{02}}{T_{01}} = 1.$$

The above relations show that pressure, density, and temperature (hence, speed of sound) all increase across a shock wave, while the Mach number and total pressure decrease across a shock.

junction (i) p–n: Formed when a semiconductor doped with impurities (acceptors) is deposited on another semiconductor doped with impurities (donors). It should be noted that a semiconductor doped with donors is called an n-type semiconductor, and those doped with acceptors are called p-type semiconductors. A semiconductor doped with acceptors possesses holes in its valence band. For example, suppose a small percentage of atoms in pure silicon are replaced by acceptors like gallium or aluminium. Gallium and aluminium each have three valence electrons occupying energy levels just above the valence band of pure silicon ($\sim$ 0.06 eV). It is energetically favorable for an electron from a neighboring silicon atom to become trapped at the acceptor atom, forming an Al^- or Ga^- ion. This electron originates from the valence band and leaves a vacancy or hole in this band. Such holes can carry a current which dominates the intrinsic current of the host. Donor impurities in silicon have five valence electrons. Each of the electrons can form a covalent bond with one of the four valence electrons in a silicon atom. This leaves an extra unpaired electron that is loosely bound to the donor atom. The energy levels of this extra electron lie close to the conduction band of silicon ($\sim$ 0.05 eV below) and can thus be excited to the conduction band and added to the number of charge carriers. Some uses of the p–n *junction* are in making solar cells, rectifiers, and light-emitting diodes.

(ii) p–n–p: Type of *junction* is often used as an amplifier in transistors. It consists of an n-type semiconductor sandwiched between two p-type semiconductors. Small changes in the applied voltage cause changes in the emitter current. For $V_{in} \ll V_E$, the change in the collector current is given by

$$\Delta I_C = \eta \Delta I_E$$

where η is a measure of the fraction of the emitter current reaching the collector, and ΔI_E is the change in the emitter current due to a change in V_{in} (ΔV_{in}). The resulting amplification is then given by $\frac{\Delta V_{out}}{\Delta V_{in}}$ and can be in excess of 100.

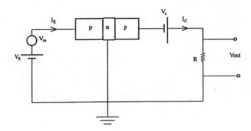

p-n-p junction as an amplifier.

K

Kadomtsev instability   One of the screw (or current convective) instabilities that occurs when an electric current flows through a magnetized fully ionized plasma having screw-shaped density perturbations. As a result of the instability, spiral clouds of protons (or ions) and electrons are generated and move along the field lines in the opposite directions, creating a charge separation and, thus, an electrostatic instability.

Kadomtsev–Nedospasov instability   One of the screw (or current convective) instabilities that occurs when an electric current flows through a magnetized partially ionized plasma having screw-shaped density perturbations. Therefore, this is also called the screw instability in a partially ionized plasma. This instability is triggered when the parallel drift speed of electrons exceeds a threshold velocity that depends, among other factors, on the collision frequency between electrons and neutrals.

kaon   A meson with a rest mass equal to approximately 494 MeV/c^2. The *kaon* has a lifetime of 1.24×10^{-8} s and decays (mostly) into muons and neutrinos. The *kaon* is a strange particle, namely it has the strange quark among its constituents.

Kármán constant   From the law of the wall, the constant k in the equation describing the overlap layer

$$f(y^+) = \frac{1}{k} \ln \frac{(y^+)}{+A}$$

where $y^+ \equiv yu/\nu$ and $u^* \equiv \sqrt{\tau_o}\rho$. A varies depending on the geometry. Observations show that $k \approx 0.41$.

Kármán momentum integral   Approximate solution for an arbitrary boundary layer for both laminar and turbulent flows. The equation is derived from the momentum equation and is given by

$$\frac{d}{dx}\left(U^2\theta\right) + \delta^* U \frac{dU}{dx} = \frac{\tau_o}{\rho}$$

where θ is the momentum thickness and δ^* is the displacement thickness.

Kármán–Tsien rule   Compressibility correction for pressure distribution on a surface at a high subsonic Mach number in terms of the incompressible pressure coefficient, C_{p_o}:

$$C_p = \frac{C_{p_o}}{\sqrt{1-M_\infty^2} + \left(\frac{M_\infty^2}{1+\sqrt{1-M_\infty^2}}\right)\frac{C_{p_o}}{2}}.$$

Kármán vortex street   Periodic vortex wake behind a circular cylinder at moderate Reynolds numbers, $80 < \text{Re} < 200$. The wake is characterized by regular vortical structures shed from opposite sides of the cylinder at a Strouhal number of 0.2. The motion becomes chaotic, but the street is still prevalent until a Reynolds number of approximately 5000.

kayser (1k)   A traditional spectroscopic unit. Today the inverse centimeter (cm^{-1}) has replaced the kayser as the unit for the wave number: 1 cm^{-1} = 1 k.

K-capture   Process in which the nucleus of an atom captures one of the atomic K-electrons (electrons of the innermost shell) and emits a neutrino. The general electron capture reaction can be written as

$$^A_Z X + e^- \rightarrow ^A_{Z-1} X + \nu_e$$

where X is a nucleus with Z protons and A nucleons, and ν_e is an electron neutrino.

Kelvin–Helmholtz instability   Instability formed at the interface between two parallel flows of different velocities. The shear resulting from the discontinuous velocity rolls up into a periodic row of vortices.

Kelvin scale of temperature (K)   Defined by choosing the unit of temperature so that the triple point of water, the temperature at which water, ice, and water vapor coexist, is exactly 273.16 K.

Kelvin's circulation theorem   The circulation around a closed loop in an inviscid barotropic flow remains constant over time, such that

$$\frac{D\Gamma}{Dt} = 0$$

which means that circulation does not decay. For flows with viscosity, circulation decays due to viscous dissipation such that

$$\frac{D\Gamma}{Dt} < 0.$$

Kelvin wedge   Envelope of surface wave disturbances emitted at successive times from a moving point on the surface of water. In deep water, the wedge has a half-angle of 19.5°.

Kennard packet   In quantum mechanics a particle is described by a wave function so that its position and momentum cannot be specified simultaneously. A *Kennard packet* is the wave packet describing the particle state that resembles a classical particle state as closely as possible. The root-mean-square deviations (Δx and Δp) of position and momentum from their respective mean values are chosen to be as small as possible. Their product is assumed to be equal to one half of Planck's constant divided by 2π.

Kerr effect   Birefringence caused in an optically isotropic material by a transverse electric field. The amount of birefringence induced by the electric field of strength E is proportional to the square of the electric field:

$$|n_o - n_e| \propto E^2,$$

where n_o and n_e are the ordinary and extraordinary indices of refraction respectively.

Some materials can exhibit an intensity-dependent index of refraction of the form

$$n = n_0 + n_2 E^2 = n_0 + n_2' I.$$

These Kerr media have a potential use in quantum non-demolition measurements. As depicted in the figure, they can be brought into one arm of a Mach–Zehnder interferometer. Depending on the intensity in the signal beam, the index of refraction in the Kerr medium will change, and the probe beam will undergo a phase shift, which will result in a shift of the interference fringes without affecting the signal beam itself.

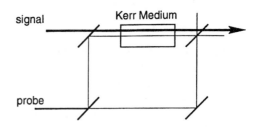

Use of a Kerr medium for quantum non-demolition measurements.

Kerr effect, electro-optical   Effect obtained if an electric field $\underline{E}$ is applied to an isotropic medium or a cubic crystal. The index of refraction for light polarized in the direction of the field $n_\parallel$ differs from that for light polarized perpendicular to the field $n^\perp$ by a term which is quadratic in the field. The medium becomes birefringent with ordinary and extraordinary rays as obtains in uniaxial crystals.

Kerr effect, magneto-optical   Deals with changes in the reflection of light from the surfaces of magnetized media. Magnetization introduces off-diagonal elements in the dielectric tensor which are linear in the components of the magnetization $\underline{M}$. For example, the reflected wave becomes elliptically polarized for a normally incident linearly polarized wave when $\underline{M}$ is also normal to the surface.

ket vector   A state vector as an element of the Hilbert space representing quantum states of a system. The name *ket vector* is used in the following example: the momentum eigenstate with momentum eigenvalue p is denoted by a *ket vector* $|p\rangle$. The name was invented by P.A.M. Dirac from the word bracket. Consequently, the Hermitian conjugate of the *ket vector* is called the bra vector.

kinematics   The study of motion in its time development.

kinematic viscosity Absolute viscosity divided by the fluid density,

$$\nu \equiv \frac{\mu}{\mu}.$$

The quantity is useful as it tends to quantify how rapidly a fluid will diffuse velocity gradients in a flow field.

kinetic energy A form of energy associated with motion. Every moving particle has kinetic energy. The kinetic energy of a non-relativistic particle with mass m and speed v is equal to $\frac{1}{2}mv^2$.

kinetic plasma instabilities A plasma and its behavior are describable via a set of velocity distribution functions combined with a kinetic equation such as the Vlasov equation, particularly when the plasma is indescribable via a set of fluid equations. If the distribution function of a plasma is non-Maxwellian, it is frequently unstable the to *kinetic plasma instabilities,* and its properties can be analyzed by the Vlasov equation.

kinetic theory A model to describe the macroscopic thermodynamic properties of a system of particles by incorporating the interactions of all the particles in the system. In principle, *kinetic theory* works for both equilibrium and non-equilibrium systems.

kink instability A hydromagnetic plasma instability. A current flowing in a plasma may be unstable due to two types of instabilities — electrostatic and electromagnetic. The former is the two-stream instability and the latter is the *kink instability* that is elucidated here. If a current column is present in a plasma, it generates a poloidal field around itself. Assume that, due to a perturbation, the current is slightly "kinked", then the field intensifies more on the inside of the kink than on the outside. Therefore, the magnetic field pressure increases on the inside of the kink, further pushing the kink outward, leading to *kink instability* and an eventual disruption. This instability may be stabilized by adding magnetic shear.

kink mode Helical or "kinked" hydromagnetic modes generated by the kink instabilities.

K–KR A method introduced by Korringa in 1947 and Kohn and Rostoker in 1954 for calculating energy bands in solids by formulating the problem as a scattering problem. Korringa used the scattering matrix method while Kohn and Rostoker used the Green's function method. The crystal potential is assumed to be an array of spherically symmetric nonoverlapping wells (muffin tin type).

Klein–Gordon equation A manifestly covariant equation which describes a fully relativistic free particle. The equation reads:

$$\left(\Box + m^2\right)\psi(x) = 0$$

with $\Box$ denoting the covariant derivative

$$\Box = \frac{\partial^2}{\partial t^2} - \nabla^2$$

and m symbolizing the mass of the particle. $\psi(x)$ is the particle wave function. In this framework, x is the four-component vector $(\mathbf{r}, t)$.

Klein–Nishina formula A formula for the differential cross-section for the scattering of a photon off an electron at rest (Compton scattering). The formula reads

$$\frac{d\sigma}{d\Omega} = \frac{\alpha^2}{4m^2}\frac{k'^2}{k^2}\left[\frac{k'}{k} + \frac{k}{k'} + 4(\epsilon \cdot \epsilon')^2 - 2\right]$$

where α is the fine-structure constant, k and k' are the initial and final momenta of the photon, ϵ and ϵ' are the photon's initial and final polarization vectors. This formula was derived by O. Klein and Y. Nishina in 1929.

Klein paradox Suppose that an electron described by the Dirac equation is moving in a space under a potential field. The space is separated by a potential step which is greater than twice the rest mass energy of the electron (approximately 1 Mev). In one side of the space the potential is high and hence the electron possesses a positive energy, while in the other side the potential is so low that the electron energy is negative. In between, the space is filled by an intermediate potential height where the Dirac equation has no solution. This region would work as an insurmountable barrier separating the positive and the negative energy states if the electron were a classical entity. However, the electron can penetrate the barrier by quantum mechanical tunneling. Accordingly, the negative energy states inherent in the Dirac equation seem to be a serious problem. This is the *Klein paradox*. To solve the paradox, Dirac postulated that the negative energy state of a vacuum is completely filled by electrons so that the Pauli exclusion principle prohibits the invasion of the positive energy electron into the negative energy region.

Knight shift A shift in the magnetic resonance frequency of nuclei when their environment changes from diamagnetic to paramagnetic. The shift is almost always toward higher frequency. The resonance frequency of Cu^{63}, for example, is higher in metals than in a diamagnetic salt such as CuCl. In metals, the magnetic field which polarizes the nuclei also polarizes the electron gas (Pauli susceptibility). The magnetic moments of the electrons interact with the nuclear magnetic moments through the contact interaction (Fermi, hyperfine) and tend to align the nuclear moments further, which is equivalent to increasing the original magnetic field B, which the nuclei see by $\Delta \underline{B}$. The *Knight shift* is measured by $\Delta \underline{B}/B$. It is absent if the electron wave function vanishes at the nucleus.

See the first two articles by Pake, G.E. and Knight W.D., in *Solid State Physics*, Vol. 2, Academic Press, New York 1956.

knock-on This term is encountered most often in the context of nucleon–nucleus scattering. An important part of this process is the exchange mechanism, where the two interacting nucleons are interchanged. This is necessary because the two nucleons are indistinguishable. This process is known as knock-on exchange.

knock-out reactions A reaction where the projectile (typically a proton) knocks out a nucleon (or a cluster of nucleons) and gets captured in one of the nuclear shells. Typical *knock-out reactions* are (p, n) or (p, α), where a proton comes in and a neutron or an α particle is knocked out.

Knudsen number Ratio of the molecular mean free path λ to a length scale in flow l,

$$Kn = \frac{\lambda}{l}.$$

For the continuum hypothesis to be considered valid, $Kn \ll 1$. From kinetic theory, it can also be shown that

$$Kn = 1.26\sqrt{\gamma}\frac{M}{\text{Re}}.$$

Kohn effect (anomalies) In a lattice vibration of wave vector $\underline{q}$ and angular frequency ω, the Coulomb interaction of the ions in the metal is screened by the electron gas. The derivative of the electron dielectric function is singular at $q = 2k_f$, the diameter of the Fermi surface. This leads to a sharp change in the ω vs. $\underline{q}$ curve, and $d\omega/dq$ becomes infinite when $\underline{q}$ or $|\underline{q} + \underline{G}| = 2k_f$, where $\underline{G}$ is a reciprocal lattice vector. Such mild kinks have been observed by careful neutron scattering experiments.

Kolmogorov length scale Also known as a Kolmogorov microscale. Length scale η of turbulent diffusion is given by dimensional analysis

$$\eta \sim \left(\frac{\nu^3}{\varepsilon}\right)^{1/4}$$

where ϵ is the turbulent dissipation rate as given by

$$\varepsilon \sim \frac{u^3}{l}.$$

The length scale is typically on the order of a millimeter or less.

Kolmogorov's law Also known as Kolmogorov's $-5/3$ law. Scaling argument that shows that the energy spectrum of isotropic turbulence varies as the wave number (inverse wavelength) to the $-5/3$ power in a given range.

Kondo's theory In 1963, Kondo explained the long standing resistivity minimum due to dilute magnetic impurities in metals. He assumed an exchange interaction between the electron spin and the impurity spin of the form $-2J\underline{S}\cdot\underline{s}$, where J is a negative constant, $\underline{S}$ is the impurity spin and, $\underline{s}$ is the electron spin, and calculated the electron scattering beyond the first (Born) approximation. He found that the resistivity rises with decreasing temperature below $\sim 10K$ and explained the occurrence of the minimum.

Korteweg–de Vries (KdV) equation (1) To explain the solitary wave traveling in the windings of a channel first observed by J.S. Russel in 1834, sixty years later in 1895, D.J. Korteweg and G. de Vries derived the following hydrodynamic equation for the motion of waves in shallow waters: $u_t + \alpha u u_x + \beta u_{xxx} = 0$, where u is the displacement, α, and β are constants, t is the time and x is the position. This equation, called the *Korteweg-de Vries equation,* is also valid for ion waves propagating in a plasma.

It was later found that the *KdV equation* indeed has some soliton solutions, one of which is given by

$$A\mathrm{sech}^2 \sqrt{\frac{\alpha A}{12\beta}} \left(x - \frac{\alpha A}{3}t \right),$$

where A denotes the amplitude. In plasmas, properties of non-linear ion-acoustic waves are described by the *Korteweg-de Vries (KdV) equation,* solutions of which correspond to ion acoustic solitons described by the above equation.

The *KdV equation* has been extended to cases of two-dimensional planar solitons (the two-dimensional *KdV equation* or the Kadomtsev–Petviashvili equation) as well as three-dimensional cylindrical solitons (cylindrical *KdV equation*).

(2) Non-linear equation describing the motion of finite amplitude waves in shallow water where $10 < \lambda/H < 20$, the solution of which gives rise to cnoidal and soliton waves.

Kramer–Kronig relations (1) Integral relations between the real and imaginary parts of the dielectric function $\varepsilon_1 + \varepsilon_2$, namely,

$$\varepsilon_1(\omega) = 1 + P \int_{-\infty}^{\infty} \frac{\varepsilon_2(\omega')}{\omega' - \omega} \frac{d\omega'}{\pi}$$

$$\varepsilon_2(\omega) = -P \int_{-\infty}^{\infty} \frac{\varepsilon_1(\omega') - 1}{\omega' - \omega} \frac{d\omega'}{\pi}$$

where P is the principal part. Similar relations hold for the index of refraction $n + i\kappa$ and many other linear response coefficients. *See* linear response theory.

(2) Reflects the strong relationship between absorption and dispersion. Any medium with a wave-length dependent index of refraction must also be absorbing. Specifically, the *Kramer–Kronig relations* are

$$\chi'(\nu) = \frac{2}{\pi} \int_0^\infty \frac{s\,\chi''(s)}{s^2 - \nu^2} ds$$

$$\chi''(\nu) = \frac{2}{\pi} \int_0^\infty \frac{\nu\,\chi'(s)}{\nu^2 - s^2} ds,$$

where ν is the frequency and χ' and χ'' are the real (dispersion) and imaginary (absorption) parts of the susceptibility. In other words, if either the, real or imaginary part of the susceptibility is known, the other can be calculated.

Kramer's degeneracy In an external electric field, the states of a system consisting of an odd number of electrons are at least two-fold degenerate. The degeneracy is lifted by a magnetic field.

Kronig–Penney model A model for a one dimensional crystal in which the crystal potential is represented by an array of Dirac delta functions located at the lattice sites. The problem is soluble in closed form and can be extended to more than one atom per unit cell. It has the interesting feature that the energy gap at a Bragg reflection remains finite at high energy. It is interesting to compare the exact results of this model with the results of other energy band calculations methods.

Kruskal–Schwarzschild instability A plasma instability which is analogous to the classical Rayleigh–Taylor instability. Instead of a light fluid supporting a heavy fluid, in the *Kruskal–Schwarzschild* instability, a plasma is supported against gravity by a magnetic field. Against a gravitational field, a plasma can never be supported by a uniform magnetic field alone in a stable manner, and thus it becomes unstable due to the *Kruskal–Schwarzschild* instability, developing ripples at the boundary.

Kruskal-Shafranov condition Ensures that if the smallest value of the so-called safety factor is greater than unity in a cylindrical plasma, which approximates a tokamak plasma, the plasma will be stable against the $m = 1$ internal kink mode with toroidal mode number $n = 1$. This mode corresponds to a rigid displacement of the entire plasma.

Kurie plot A method of analyzing the energy spectrum of electrons emitted in β-decay, which is the decay of a nucleus through the emission of electrons (β^- decay) or positrons (β^+ decay). The method consists of plotting the number of electrons emitted vs. the energy.

Kutta condition Rule stating that for flow over a two-dimensional wing with a sharp trailing edge, circulation of sufficient magnitude is developed to locate the rear stagnation point to the trailing edge.

Kutta–Zhukovski lift theorem The lift, L, of an airfoil or other aerodynamic body is proportional to the free-stream velocity, U, and circulation, Γ, about the body:

$$L = \rho b U \Gamma$$

where b is the airfoil span. The lift due to this circulation is sometimes called circulation lift. This equation is the basis for much of modern aerodynamics.

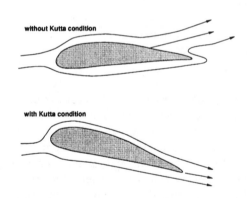

Kutta condition.

L

laboratory frame   Referred to most often in the context of particle scattering. It is the frame of reference in which the target is at rest.

Lagrangian   First introduced by J.L. Lagrange (1736–1813) in the context of classical mechanics. Given any set of independent coordinates which are suitable to specify the position of each part of a system (generalized coordinates), the *Lagrangian,* or Lagrange's function, is defined as
$$L = T - V$$
with T representing the kinetic energy and V denoting the potential energy.

Lagrangian flow description   Method of analyzing fluid flow by following the history of individual particles, as opposed to the Eulerian flow description. This requires keeping track of the motion in time and space of each and every particle and is therefore used only when necessary.

Laitone rule   Compressibility correction for pressure distribution on a surface at a high subsonic Mach number in terms of the incompressible pressure coefficient, C_{p_o}:
$$C_p = \frac{C_{p_o}}{\sqrt{1-M_\infty^2} + \left[M_\infty^2\left(1+\frac{\gamma-1}{2}\right)M_\infty^2/2\sqrt{1-M_\infty^2}\right]C_{p_o}}.$$

lambda particle   All baryons with strangeness equal to -1. One example is the Λ^0 particle, with a mass of about 1116 MeV/c^2, which decays into a proton and a negatively charged pion.

lambda scheme   Specific form of energy level diagram that resembles the Greek letter Λ. It consists of three levels labeled $|a>$, $|b>$ and $|c>$. Decays between $|a>$ and $|b>$ as well as between $|a>$ and $|c>$ are electrically dipole allowed. $|b>$ and $|c>$ can be hyperfine or finestructure levels in atoms, but can also be rovibrational levels in molecules.

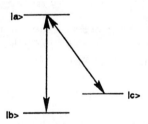

Lambda scheme.

Lambert's law   Gives the luminous intensity I of a light source as a function of the angle θ:
$$I(\theta) = I_0 \cos\theta \,.$$
Many light sources radiate according to *Lambert's law.*

Lamb–Oseen vortex   Vortex satisfying the Navier-Stokes equation given by the tangential (circumferential) velocity field
$$u_\theta = \frac{\Gamma}{2\pi r}\left(1 - e^{-r^2/4\nu t}\right)$$
where Γ is the circulation of the vortex. *See* vortex.

Lamb shift   (**1**) Energy difference between, e.g., the $2P_{1/2}$ and the $2S_{1/2}$ levels in the spectrum of hydrogen. The difference, discovered by W.E. Lamb and R.C. Retherford in 1947, is 4.4×10^{-6} eV and is due to vacuum fluctuations. To label levels, we have used the spectroscopic notation nL_j, where n is the principal quantum number, s is the total spin, j is the total angular momentum, and L refers to the orbital angular momentum. An S-state has zero orbital angular momentum, while a P-state has an orbital angular momentum equal to 1.

(**2**) Is responsible for the lift in degeneracy of the $s_{1/2}$ and $p_{1/2}$ levels in hydrogen, which is predicted by the Dirac equation. Its origin is the necessary radiative correction due to a lowering of the Coulomb potential close to the nucleus by vacuum fluctuations. Since the s-electron is more often close to the nucleus the effect is largest for s-states. One finds the Lamb shift to be

$$\Delta E = \begin{cases} \alpha^5 mc^2 \frac{1}{4n^3} f(n) & \text{for } l = 0 \\ \alpha^5 mc^2 \frac{1}{4n^3} \left(f(n,l) \pm \frac{1}{\pi(J+1/2)(l+1/2)} \right) & \text{for } l \neq 0 \end{cases}$$

where α is the fine structure constant, m the electron mass, c the speed of light, n the principal quantum number and $j = n \pm 1/2$ and $12.7 < f(n) 13.2$, and $f(n,l) < 0.05$ are numerical factors dependent on n and l, respectively.

The value of the Lamb shift in hydrogen for the $2s_{1/2}$ and $2p_{1/2}$ level is 1057.864 MHz. The three major contributions to this value are the electron mass renormalization (1017 MHz), vacuum polarization (−27 MHz) and anomalous magnetic moment (68 MHz).

laminar flow Regime of viscous flow in which the fluid follows well-defined layers (laminae). No macroscopic mixing takes place, but microscopic diffusion is possible. *Laminar flow* occurs for low Reynolds numbers.

Landau damping Damping of longitudinal waves in a plasma caused by a transfer of energy from the wave to those charged particles with velocities nearly the same as the phase velocity of the wave (resonant particles).

Landau diamagnetism In 1930, L. Landau calculated the diamagnetic contribution of the electron gas in a metal to the magnetic susceptibility and found it to be $-\chi_p/3$, where χ_p is the paramagnetic Pauli susceptibility of the electron gas. $\chi_p = 3n\mu_B^2/(2E_f)$, where μ_B is the electron magnetic moment due to its spin (Bohr magneton), n is the electron density, and E_f is the Fermi energy.

Landau levels A solution of Schrödinger's equation for a charged particle, such as an electron, of charge e and mass m in a magnetic field $B(0, 0, 1)$ can easily be obtained by assuming the vector potential (o, Bx, o). The wave function is the product of a plane wave in the z- and y-directions and a harmonic oscillator wave function in the x-direction with a frequency equal to the cyclotron frequency $\omega = (eB/mc)$, where c is the speed of light. The energy levels are given by $E_n = (\hbar^2 k_z^2/2m) + (n+1/2)\hbar\omega$, with $n = 0, 1, 2, \ldots$, and are known as *Landau levels*. For semiconductors in a magnetic field, we can obtain *Landau levels* by using the effective mass approximation method. These levels lie near the bottom of the conduction band E_c, with energies $E = E_c + E_n$, and near the top of the valence band E_v, with energies $E = E_v - E_n$. In both cases we assume parabolic bands, and m is replaced by the effective mass $|m^*|$ for that band. The optical transitions for this system take place by transitions between levels with the same n (and the same k_z). This is an example of a magneto-optical phenomenon.

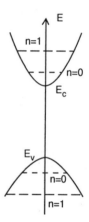

Landau levels.

Landau–Zener model Has several prominent applications in atomic physics. In general, it can be applied in the case of time varying potential energy curves, which form avoided crossings. Specifically, it has applications in atomic and molecular collisions, pulsed excitation which chirped pulses, the field ionization of Rydberg atoms, etc.

The common ground of these cases is that potential energy curves are changing as a function of a parameter q. This parameter could be the nuclear distance in the case of collisions, Stark shifts due to laser pulses or increasing electric fields. The potential energy curves of states can come closer due to these effects and form due to an interaction matrix element V_{ab} avoided crossings as depicted in the figure below.

The *Landau–Zener model* treats the time evolution of such a system. For the Landau-Zener model to be valid, we assume a linear varia-

Landé g-factor (spectroscopic splitting factor)

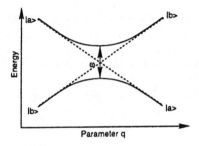

Two potential energy curves form an avoided crossing. Depending on the slew rate the transitions will be undergone adiabatically (solid curves) or diabatically (dashed curves).

tion of the parameter q with time. Interesting is whether the system will cross the avoided crossing diabatically or adiabatically. In case of an adiabatic evolution of the system, the population will follow the solid curves, and a population transfer will occur. This is in contrast to a diabatic evolution, where the system will follow the dashed curves.

Critical in the evaluation whether or not the system is evolving adiabatically is the ratio of the interaction $|V_{ab}|$ and the slew rate of the potentials dE/dt. The critical slew rate S_c is given by

$$S_c = \frac{|V_{ab}|^2}{\frac{dE}{dt}} = \frac{\omega^2}{\frac{dE}{dt}}$$

where
$$\frac{dE}{dt} = \frac{dE}{dq}\frac{dq}{dt},$$

and ω is the minimum energy separation at the avoided crossing.

If the actual slew rate S is much larger than S_c the evolution will be diabatically, i.e., along the dashed lines. When the actual slew rate is much smaller than S_x the states will follow the solid curves, i.e., adiabatically. This becomes also clear from the Landau-Zener probability for a diabatic jump along the dashed lines, which is valid for the interesting case of intermediate evolution:

$$P = \exp\left(-\pi \frac{\omega^2}{2\frac{dE}{dt}}\right).$$

For large slew rates $P \to 0$ and for very small slew rates $P \to 1$ corresponding to what was said earlier.

Landé g-factor Proportionality factor between the magnetic moment $\vec{\mu}$ of an orbiting charge and its total angular momentum. In the case of a pure orbital angular momentum, the relation is

$$\vec{\mu}_L = -\frac{g_L \mu_B \vec{L}}{\hbar}$$

with $g_L=1$. For a pure spin angular momentum, the corresponding relation is

$$\vec{\mu}_S = -\frac{g_S \mu_B \vec{S}}{\hbar}$$

with $g_S = 2.00232$. In the expressions above, μ_B is the Bohr magneton, which has a value of 5.59×10^{-5} eV/tesla, and is defined as

$$\mu_B = \frac{e\hbar}{2m_e}$$

where $\hbar = h/2\pi$ (h is the Planck's constant, equal to 6.626×10^{-34} Js), m_e is the electron mass, and e is the magnitude of the (negative) electron charge. In nuclear physics, magnetic moments are expressed in terms of nuclear magnetons, defined in the above equation, with the mass of the proton instead of the mass of the electron. The nuclear magneton is about 2000 times smaller than the Bohr magneton. The equations given above for $\vec{\mu}_L$ and $\vec{\mu}_S$ apply to the proton if the negative sign is suppressed (due to the positive charge of the proton), and the Bohr magneton is replaced with the nuclear magneton.

Landé g-factor (spectroscopic splitting factor) The total angular momentum of an atom of one or more electrons $\hbar \underline{J}$ is the sum of the orbital angular momentum $\hbar \underline{L}$ and the spin angular momentum $\hbar \underline{S}$. The magnetic moment μ (to a good approximation) is equal to $(e\hbar/2mc)(\underline{L}+2\underline{S})$. In the Russel–Saunders coupling, both $\underline{L}$ and $\underline{S}$ can precess around $\underline{J}$, and the average of $\underline{\mu}$ is given by

$$\langle \underline{\mu} \rangle = \frac{e\hbar}{2mc} g \underline{J},$$

where g is known as the *Landé g-factor* which is given by

$$g = \frac{3}{2} + \frac{S(S+1) - L(L+1)}{2J(J+1)}.$$

Landé obtained this result before the development of quantum mechanics and the Wigner–Eckart theorem. The average of μ is understood to be $\overline{\mu}$ and equals $-\mu_B g J$, where $|e\hbar/(2mc)|$ is the Bohr magneton μ_B.

Lander interval rule Gives the energy separation of two adjacent hyperfine levels in LS-coupling. The energy separation ΔE between the levels E_{J+1} and E_J is given by

$$\Delta E = E_{J+1} - E_J = a(J+1)$$

where a is a constant and is called the interval factor. The *Lande interval* rule can be used to check whether LS-coupling is valid since otherwise the interval rule is violated.

Langevin–Debye formula If a permanent electric dipole of moment p can assume any orientation in an electric field $\overline{E}$, then classical statistical mechanics states that the average of the cosine of the angle which p makes with the field is given by the Langevin function $\cos x - 1/x$, where $x = \beta p E$, $1/\beta$ is the thermal energy kT, k is the Boltzmann constant, and T is the absolute temperature. For a small value of x it reduces to $\beta p E/3$, and the electric susceptibility is $np^2\beta/3$, which is the *Langevin–Debye formula*; here, n is the number of dipoles per unit of volume. For magnetic dipoles, despite the fact that the orientations are restricted by the quantization of the angular momentum, the formula applies for weak fields. The general result, however, is given by a Brillouin function. *See* paramagnetism.

Langevin equation Is an equation of the form

$$\frac{d}{dt}x(t) = -\beta x(t) + g(t),$$

where $g(t)$ is a randomly varying stationary, Gaussian-shaped random process with a mean value of zero. Brownian motion can be expressed by a Lagrangian equation. The force $g(t)$ is here the random force of all the particles surrounding the sample particle, whose motion is being predicted.

Langmuir probe Insulated wire with an exposed tip in which the voltage is varied in order to measure the electron density, temperature, and electric potential in plasmas.

Laplace transform $F(p)$ of a function $f(t)$ is given by

$$F(p) = \int_0^\infty f(t)e^{-pt}dt.$$

lapse rate Rate at which temperature decreases in the Earth's atmosphere. *See* atmosphere, standard.

large aspect ratio expansion Approximation used in the theory of toroidal plasmas in which the major radius is taken to be much larger than the minor radius.

Larmor frequency (1) Term encountered in the context of the interaction of an atom with an external magnetic field **B**. A particle of charge e and mass m will precess in a magnetic field with the *Larmor frequency*:

$$\omega = \frac{eB}{2m}.$$

(2) Frequency of gyration of charged particles in a magnetic field. The *Larmor frequency* in radians per second is given by the charge of the particle times the magnetic field strength divided by the mass of the particle.

(3) A homogeneous magnetic field with strength b produces no force of the spin, but rather results in a precession of the spin around the axis of the magnetic field. The characteristic frequency of this precession is called the Larmor frequency. It is given by

$$\omega_L = \frac{\mu}{\hbar}B,$$

where μ is the magnetic moment.

Larmor orbit Nearly circular orbit followed during the gyration of charged particles in a magnetic field.

Larmor radius Radius of the orbit of a charged particle as it gyrates in a magnetic field. This radius is given by the velocity of the particle divided by its Larmor (or cyclotron) frequency.

laser (maser) (1) A device which amplifies light (microwaves and electromagnetic waves in general) by stimulated emission. The basic element of the device is an active medium with (at least) two energy levels, E_1 and E_2, with N_1 and N_2 particles in these states which are connected by a radiative transition. Assume that by some means, such as pumping or separation, N_2 is made larger than N_1 (population inversion), then a radiation of frequency $\omega = (E_2-E_1)/\hbar$ would be amplified by stimulated emission. The radiation must be contained in a cavity as in *masers*, or between two reflecting mirrors as in *lasers*, so that the process continues. Active media can be gases, liquids, or solids such as *p–n* junctions and ruby crystals.

(2) Is the acronym for light amplification by stimulated emission of radiation. The *laser* has quickly evolved to the most important tool in atom physics and quantum optics. It also has a wide range of applications ranging from such fields as applied optics, material processing, printing, medicine, and more.

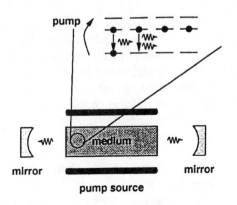

Laser diagram.

Three ingredients are crucial to a laser: (1) a laser medium, in which the light amplification is achieved (2) an energy source that pumps the medium and leads to a population inversion in the medium and (3) a cavity, which is in general formed by mirrors in order to provide feedback, such that photons that are spontaneously emitted into the cavity are amplified (see figure).

Most lasers work with three or four level schemes, since otherwise the necessary condition of population inversion cannot be achieved. An exception is lasing without inversion. Pump sources can include currents, electron collisions, electrical discharges, flash lamps or other lasers.

The cavity is usually formed by two or more mirrors forming a standing wave or running wave resonator. One mirror has often a lower reflectivity than the others and acts as the output coupler for the radiation. Other output coupling schemes are polarizing beamsplitters or output coupling via frustrated total internal reflection.

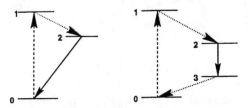

Three (left) and four (right) level schemes. The pump transitions are indicated by dashed lines, the lasing transistors by bold lines and fast relaxation processes by dotted lines. In order for the laser to operate, a higher population in the upper lasing level than in the lower lasing level is required. This is termed population inversion.

Important considerations in the design of a cavity is its stability, i.e., whether or not a propagated beam gets magnified as one roundtrip through the cavity is completed. If so, eventually the beam will leave the cavity and one speaks of an unstable cavity and of a stable cavity otherwise. The stability analysis of a cavity can be performed using the ABCD matrix technique or more sophisticated approaches that take diffraction into account as for instance the Fox-Li algorithms.

For cavities consisting of two mirrors the g parameter is helpful in determining the stability. It is given by

$$g_{1,2} = 1 - \frac{d}{R_{1,2}},$$

where the indices stand for the two mirrors and d and R are the distance between the two mirrors and the radius of the mirrors, respectively. It is found, that under the condition

$$0 \leq g_1 g_2 \leq 1$$

a stable cavity is formed. The figure below depicts the range of stability. The g parameter lets

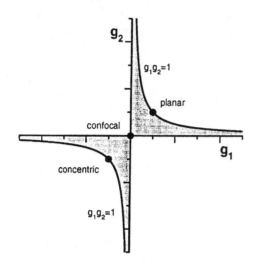

Stability diagram for an optical resonator. Also depicted are special cavity configurations and their location in the stability diagram.

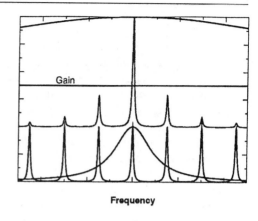

Single mode output from a laser by placing additional optical elements inside a cavity. The curve on top shows a gain profile which convoluted by the cavity modes and an additional optical element produces a gain curve shown in the middle. Only a single mode has a gain larger than the losses in the cavity.

one also calculate the parameters of the Gaussian beam inside the cavity, i.e., Rayleigh range z_R and the distance of the waist from the mirrors $z_{1,2}$. One finds

$$z_r = \frac{g_1 g_2 (1 - g_1 g_2)}{(g_1 + g_2 - 2 g_1 g_2)^2} d^2$$

$$z_{1,2} = \frac{g_{2,1}(1 - g_{1,2})}{g_1 + g_2 - 2 g_1 g_2} d$$

The linewidth of a laser is given by the convolution of the cavity mode structure, gain curve of the lasing medium and transmission curves of optical elements additionally placed in the cavity. If the gain curves include several discrete cavity modes, the laser will generally lase on multiple modes. The exact mode structure depends on mode competition. By introducing other cavity elements with wavelength dependent transmission profiles, like filters, Fabry-Perot etalons, prisms, gratings, or birefringent filters single mode operation can be achieved. If the laser medium has a gain bandwidth smaller than the separation of two cavity modes, single mode output of the laser is achieved. However, the cavity must be stabilized such that a mode coincides with the gain maximum.

The different laser types can be divided into different classes depending on their characteristics, such as operating mode (pulsed or continuous wave), frequency (tunable, fixed frequency), medium type (solid state, semiconductor, gas, liquid).

Several techniques can be used for the generation of laser pulses. The particular choice of technique depends on the required time scale. In the nanosecond regime and below Q-switching by choppers, rotating mirrors, or acousto- and electro-optic modulators in the cavity can be achieved. Q-switching works by rapidly switching the feedback of the cavity, i.e., the lifetime of the photons in the cavity. In this way the stimulated emission, i.e., the amplification can be controlled.

Most common gas laser types are excimer and CO_2 lasers as well as the HeNe laser. Solid state lasers of the biggest importance are the Nd:YAG and the widely tunable Ti:Sapphire laser. Dye lasers, i.e., inorganic dyes dissolved in organic solvents, are widely tunable and can cover the visible part of the electromagnetic spectrum as well as parts of the UV and IR regions. Increasingly important are the semiconductor diode lasers. They combine high efficiency with a compact and rugged design making them extremely interesting in the communication and mass product industries. In the future fiber lasers, based on fibers doped with a lasing medium (mostly rare earth atoms like Nd, Yb, Er, etc.), will also gain importance due to their high power capabilities, compactness, and reliability. In the Free-electron laser, in which the radiation given off by accelerated electrons is used, the wavelength range extends further into the VUV as well as the longer wavelengths.

laser cooling   Is the reduction of the temperature of atoms in the gas or bulk phase by means of laser radiation. Most often the cooling is associated with a reduction in the speed of the atoms and a narrowing of their velocity distribution.

Laser cooling can be performed by irradiating the atoms with light red-detuned from the atomic resonance. Each absorption process transfers a momentum kick to the atoms. This is followed by spontaneous emission. The latter has no net-effect since it occurs randomly in a 4π radian. Due to the red-detuning of the laser beam, the atom is more likely to absorb from a laser beam which is counterpropagating with the atom leading to a slowing of the atom. In order to keep the decelerating atom on resonance with the laser, the atom or the laser frequencies must be tuned. The former can be achieved with a spatially varying magnetic field (*see* Zeeman slower), the latter by sweeping the frequency of the lasers in synchronous with the loss in velocity. *See also* magneto-optical trap.

laser fluctuations   Are fluctuations in phase and amplitude of a laser. Intensity and phase fluctuations stem from spontaneous emission. The photons in a laser follow the Poisson statistics and scale with the square root of the photon number. The phase undergoes a random walk which is also termed the phase diffusion. Phase locking allows the locking of the phases of two lasers with respect to each other.

laser fusion   A process in which intense lasers are used to implode a pellet containing thermonuclear fuel. The power delivered by the lasers causes the surface material of the pellet to ablate, which compresses and then heats the material in the center of the pellet to produce nuclear fusion reactions.

laser induced fluorescence (LIF)   Is an important tool in spectroscopy of atoms and molecules. After excitation of a single transition from state $|a> \rightarrow |b>$ with a narrow linewidth laser, the system decays spontaneously to lower levels. The emitted fluorescence is spectrally analyzed. The selective emission of single levels facilitates a high degree of simplification in the spectra, which enables us to draw conclusions about the transition strengths. The requirement of the selective excitations is a narrow linewidth laser and that the Doppler linewidth of the different transitions is smaller than the separation between lines.

laser wakefield accelerator   Particle accelerator that uses an intense short pulse of laser light to excite plasma oscillations that are used to accelerate charged particles to high energy.

latent heat (L)   The heat absorbed or given off from a system undergoing a first order phase transition. It is related to the molar change in entropy of the two phases, $\Delta s = s_I - s_{II}$, by $L = T \Delta s$.

lattice conductivity   The contribution of lattice vibrations to the thermal conductivity of the crystal. The thermal conductivity K is given by $K \sim \frac{1}{3} Cvl$, by a simple argument attributed to Debye, where C is the specific heat, v is an average velocity for lattice waves, and l is a mean free path which is proportional to $1/T$ at high temperature T. Peierls pointed out the importance of three lattice wave processes for which the conservation of $\underline{k}$ brings in a reciprocal lattice vector $\underline{G}$ (umklapp processes).

lattice, crystal lattices   Perfect crystals are periodic structures, and it is this periodicity which makes their study easier. A *lattice* is a mathematical set of points defined by the vectors $\underline{r} = n_1 \underline{a}_1 + n_2 \underline{a}_2 + n_3 \underline{a}_3$, where n_1, n_2, and n_3 are integers, and the vectors a_1, a_2, and a_3 are linearly independent, but their choice is not unique. A crystal structure results when the atoms are assigned positions in this *lattice* (such an assignment is denoted by a basis). When one atom is assigned per *lattice* site, the crystal has a Bravais *lattice*. The three cubic *lattices,* simple cubic, body-centered cubic, and face-centered cubic, are all Bravais *lattices*. The physical properties of a crystal, such as the electron density and the potential $V(\underline{r})$ which an electron sees, are periodic functions with the periodicity of its *lattice*, and it is convenient to describe such properties in terms of a Fourier series. For this purpose, we introduce reciprocal *lattice* which is spanned by the vectors,

$$\underline{G} = m_1 \underline{b}_1 + \underline{b}_2 + m_3 \underline{b}_3,$$

where m_1, m_2, and m_3 are integers, and $\underline{b}_1 = 2\pi \underline{a}_2 \times \underline{a}_3 / v_c$, $\underline{b}_2 = 2\pi \underline{a}_3 \times \underline{a}_1 / v_c$, and $\underline{b}_3 = 2\pi \underline{a}_1 \times \underline{a}_2 / v_c$, where v_c is the volume of the unit cell in the direct lattice, namely, $\underline{a}_1 \cdot \underline{a}_2 \times \underline{a}_3$. The volume of the unit cell in reciprocal space is $8\pi 3/v_c$. Thus, the potential $V(r)$ can be written as

$$V(\underline{r}) = \sum_{\underline{G}} V(\underline{G}) \exp(i\underline{G} \cdot r),$$

$$V(\underline{G}) = \frac{1}{v_c} \int V(\underline{r}) \exp(-i\underline{G} \cdot \underline{r}) d^3 r.$$

where the integration is carried out over the unit cell.

lattice gauge theory Gauge field theories performed in discrete space-time intervals, i.e., on a lattice, by means of numerical techniques. See also lattice QCD.

lattice QCD Quantum chromodynamics (QCD) is the accepted theory of strong interactions. To facilitate theoretical studies within QCD (which is a highly non-linear theory), numerical calculations are performed in a discrete space-time, namely on a lattice.

lattice vibrations An application of the theory of small oscillations in classical mechanics. The potential energy of a crystal is developed as a quadratic function of the atomic displacements from their equilibrium positions in the lattice (this is often called the harmonic approximation). The kinetic energy is also a quadratic function of the velocities. The periodicity of the crystal requires the atomic displacements to have the wave form

$$\exp\left[i\left(\underline{k} \cdot \underline{n} - \omega t\right)\right]$$

where $\underline{k}$ is a propagation vector of the wave, ω is its frequency, and $\underline{n}$ is an abbreviation for the lattice vector $n_1 \underline{a}_1 + n_2 \underline{a}_2 + n_3 \underline{a}_3$. This reduces the number of equations from $3Ns$ (actually $3Ns - 6$) to $3s$ equations, where N is the number of unit cells in the crystal and s is the number of atoms in a unit cell; s is one for silver and gold, for example, and two for diamond. For a given $\underline{k}$, we obtain $3s$ values of ω, which, when $\underline{k}$ is varied, give $3s$ surfaces or branches. To illustrate, consider a linear chain of atoms of mass m at $x = 0, \pm a, \pm 2a, \ldots$, and atoms of mass M at $x = \pm a, \pm 3/2a$; coupled with springs of spring constants α, we obtain two branches: the lower branch is called an acoustical branch since $\omega = ck$ for small $\underline{k}$ as in sound waves, and the upper branch is called an optical branch by convention. Note that $\underline{k}$ is determined to within $2\pi/a$, or ω as a function of $\underline{k}$ is periodic with the period $2\pi/a$, which is a reciprocal lattice vector for this one-dimensional crystal. The interval $\frac{-\pi}{a} \leq k \leq \frac{\pi}{a}$ is the Brillouin zone for this crystal. In general for a crystal, we obtain three acoustical branches, $3(s-1)$ optical branches, and $3Ns$ harmonic oscillators, which are uncoupled and can be quantized. The specific heat of the crystal is the sum of the specific heats of these oscillators. If we include the potential energy of the crystal cubic terms and atomic displacements, the oscillators will be coupled and lattice waves will scatter each other or break and form other waves. The terms are important in explaining the thermal conductivity of the crystal and the thermal expansion of solids.

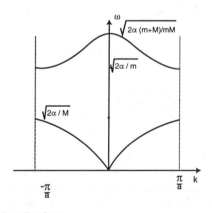

Lattice vibrations.

Laue's condition method In 1912, Max von Laue recognized that a crystal can serve as a three-dimensional diffraction grating for X-rays of wavelengths λ of about 1 Å. Electrons in the atoms of the crystal are excited by the electric field of the incident X-rays and radiate X-rays with the same frequency. The wavelets from different atoms combine (interfere) to form the scattered (diffracted) wave. Constructive interference will result if the phase difference between two wavelets from any two atoms is $2\pi n$, where n is an integer. For atoms A and B separated by a vector r in the crystal and with an incident wave vector k and scattered wave vector $\underline{k}'$ (here, $2\pi/\lambda = \underline{k}' = \underline{k}$, elastic scattering), we see that the phase difference between B and A is $\underline{r} \cdot (\underline{k} - \underline{k}')$, corresponding to a shorter path by $CA + AD$. If we assume, for simplicity, a Bravais lattice, where any vector $\underline{r}$ joining two atoms is given by $n_1 \underline{a}_1 + n_2 \underline{a}_2 + n_3 \underline{a}_3$ where n_1, n_2, and n_3 are integers and $\underline{a}_1, \underline{a}_2$, and $\underline{a}_3$ are three primitive translation vectors, we obtain Laue's conditions:

$(\underline{k} - \underline{k}') \cdot \underline{a}_1 = 2\pi$ (integer)
$(\underline{k} - \underline{k}') \cdot \underline{a}_2 = 2\pi$ (integer)
$(\underline{k} - \underline{k}') \cdot \underline{a}_3 = 2\pi$ (integer)

which are equivalent to the statement $\underline{k} - \underline{k}'$ is a reciprocal lattice vector $\underline{G}$. From the triangle, we see that $\underline{G}$ is perpendicular to the bisector of the angle between $\underline{k}$ and $\underline{k}'$, and $2k \sin \theta = G$.

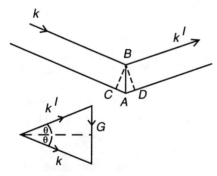

Laue's condition method.

The diffraction appears as reflection from the atomic planes perpendicular to $\underline{G}$ whose spacing $d = 2\pi m/G$, which, when substituted for G gives the Bragg condition $2d \sin \theta = m\lambda$, where m is an integer denoting the order of the reflection. In Laue's method, a well-collimated X-ray beam containing a range of wavelengths (polychromatic) is incident on a single crystal whose orientation has been chosen. A flat film can receive either the reflected or the transmitted beam.

law of corresponding states Hypothesis proposed by Van der Waal that the equation of state expressed in terms of the reduced pressure, temperature, and volume (reduced variables defined as the ratio to the value of the variable at the critical point) becomes the same for all substances. This holds true for Van der Waal's equation of state; real gases do not obey this rule to a high accuracy.

law of mass action In a chemical reaction with ideal gases, the condition of equilibrium can be expressed in terms of the *law of mass action*. Denoting the chemical reaction of the species A_j in terms of the stoichiometric coefficients ν_j as $\sum_j \nu_j A_j = 0$, one can write the equilibrium constant $K(T)$ as

$$k(T) = \prod_j [A_j]^{\nu_j},$$

where $[A_j]$ denotes the concentration of the jth species in the reaction. Note that the stoichiometric coefficients for reactants and products have opposite signs.

law of the wall Variation of velocity in a turbulent boundary layer as given by

$$U/u^* = f(y^+)$$

where $u^* = \sqrt{\tau_o/\rho}$ is the friction velocity and $y^+ = yu^*/\nu$ is the dimensionless distance from the wall. The velocity profiles are divided into two regions, a viscous sublayer near the wall and an outer layer near the free-stream. An overlap layer connects the two. The regions are given by

$U/u^* = y^+$ (viscous sublayer)

$U/u^* = 2.5 \ln y^+ + 5$ (logarithmic layer).

Lawson criterion Attributed to the British physicist J.D. Lawson, this criterion establishes a condition under which a net energy output would be possible in fusion. If n is the ion density and τ is the confinement time (namely, the time during which the ions are maintained at a temperature at least equal to the critical ignition temperature), then the Lawson criterion states that $n\tau > 10^{16} s/cm^3$ for deuterium–deuterium reactions, and $n\tau > 10^{14} s/cm^3$ for deuterium–tritium reactions.

LDV   Laser-Doppler velocimetry. Optical method of measuring flow velocity at a point through use of a crossed laser beam which forms fringes due to interference. Scattered light from particles passing through the laser intersection is measured by a photodetector and processed to determine the velocity.

Le Chatelier's principle   States that the criterion for thermodynamic stability is that the spontaneous processes induced by a deviation from equilibrium must be in a direction to restore the system to equilibrium.

left-handed particle   A particle whose spin is antiparallel to the direction of its momentum. *See* handedness.

Lehmann representation   In the quantum many-particle problem, a standard technique is to use the one-particle Green's function. The space-time Fourier transform of Green's function is useful. The related object is the spectral function defined as follows. Consider a large system of interacting particles. Insert a particle with a fixed momentum in this system. The energy spectrum of the obtained system defines the spectral function. The *Lehman representation* is the expression for the space-time Fourier transformation of the one-particle Green's function in the integral form of the spectral function.

Lennard–Jones potential   The interaction energy between two atoms, such as inert gas atoms, as a function of r, the distance between them, is given by,

$$U(r) = 4\varepsilon \left[(\sigma/r)^{12} - (\sigma/r)^6 \right],$$

where ε and σ are energy and distance parameters. This potential is used in calculating the cohesive energy of inert gas crystals.

lepton   A particle which does not interact via the strong interaction. *Leptons* interact via the weak or electromagnetic interaction. For instance, electrons are *leptons*.

leptonic interactions   Interactions among leptons. *See* lepton.

lepton number   A *lepton number* equal to $+1(-1)$ is assigned to leptons (antileptons), while a *lepton number* equal to zero is assigned to all nonleptons. The *lepton number*, L, is always conserved. That is, reactions or decays that would violate conservation of the *lepton number* have never been observed.

level   In the context of nuclear or atomic physics, it usually denotes an energy level, namely, one of the allowed (quantized) values of the energy a quantum system can have.

level width   The energy of a small quantum system is quantized and is represented as an energy level. In many cases, the system is dynamically coupled with a large degree of freedom. Then the energy of the small system spreads. The distribution function of this energy spread is observed, for example, through an intensity distribution of the emission or absorption of photons. In many cases, the width is defined as the difference between the energies at which the value of the distribution function is one-half its maximum value.

lever rule   In a first order phase transition such as in a liquid–gas system, the ratio of the mole fraction in the coexisting liquid vs. the gas phase, x_l/x_g, for a liquid–gas mixture with total volume is v_T, is inversely related to the ratio of the difference of the volume v_T from the molar volumes of the liquid and gas phases, v_l and v_g, respectively.

Mathematically stated, this gives $x_l/x_g = (v_g - v_T)/(v_T - v_l)$.

Levinson's theorem   In the S-matrix theory of scattering, the angular momentum representation is the most interesting. For the elastic scattering by a potential, the S-matrix is diagonal in this representation. The eigenvalues of S, the S-matrix, are closely related to the phase shifts; $S_l(k) = \exp[2i\delta_l(k)]$, where k is the momentum of the incoming particle, l is the angular momentum of its partial wave, and $\delta_l(k)$ is the phase shift. The Levinson theorem is that

$$\delta_l(0) - \delta_l(\infty) = \text{[number of the bound states with angular momentum } l] \pi \text{ .}$$

levitron Toroidal plasma experimental device that includes a current-carrying coil levitated within the plasma.

lifetime A characteristic time associated with the decay of an unstable system. The law of radioactive decay is

$$N(t) = N(0)e^{-\lambda t}$$

with $N(t)$ symbolizing the number of nuclei present at any time t, $N(0)$ denoting the initial number of nuclei, and λ representing the disintegration constant.

$$\tau = \frac{1}{\lambda}$$

is the *lifetime* or mean life of the sample. *Compare with* half-life.

lift Force perpendicular to the direction of motion generated by pressure differences. The *lift* can be generated by a symmetric body inclined at an angle to the flow, from flow about an asymmetric body, or a combination of both.

lift coefficient Lift non-dimensionalized by dynamic pressure:

$$C_L = \frac{L}{\frac{1}{2}\rho U^2 A} \text{ .}$$

A *lift coefficient* is primarily used to determine the lifting capability of a wing and is plotted vs. the attack angle or drag coefficient (drag polar). *Lift coefficients* for an arbitrary symmetric and cambered wing are shown.

lifting line theory Theory for determining the lift of a wing by assuming the lift is created by a number of discrete line vortices.

lift-to-drag ratio Measure of the efficiency of a airfoil:

$$L/D = \frac{C_L}{C_D} \text{ .}$$

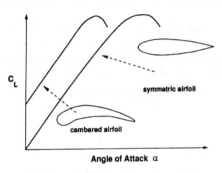

Lift coefficient vs. angle of attack.

The greater the *lift-to-drag ratio* (L/D), the better a wing is at producing lift with minimal drag.

light emitting diode (LED) A p–n junction made from a direct gap semiconductor such as GaAs, where the electron gas (in the n region) and the hole gas (in the p region) are degenerate. When biased in the forward direction (p is connected to the positive terminal and n to the negative terminal), electrons travel to the p side and holes travel to the n side where they recombine with opposite charge carriers emitting radiation. The transition which occurs is that of an electron from the conduction band filling a hole in the valence band. Such a device is a candidate for a laser.

light ion A charged particle obtained from stripping charges from or adding charges to the neutral atom. As opposed to heavy ions, light ions are obtained from lighter atoms. *See* ion.

light quantum *See* photon.

light-water reactor A reactor which uses ordinary water as a moderator, unlike a heavy-water reactor. *Compare with* heavy-water reactor.

limiter Material structure used to define the edge of the plasma and to protect the first wall in a magnetic confinement device. *See also* divertor, plasma divertor.

Lindemann melting formula Assumes that at the melting temperature of a solid, the root-mean-square of the atomic displacement due to vibration is a fraction of the distance between the atoms. For the melting temperature T_m, it gives the formula $T_m = M x_m^2 r_s^2 k\theta^2/(9\hbar^2)$, where M is the mass of the atom, x_m is a fraction 0.2 – 0.25, r_s is the radius of a sphere assigned to an atom in a crystal, k is Boltzmann's constant, and θ is the Debye temperature.

linear accelerator An accelerator which (through electric fields) accelerates particles (typically protons, electrons, or ions) in a straight line, as opposed to a cyclotron or syncrotron, where particle trajectories are bent by magnetic fields into circular shapes.

linear combination of atomic orbitals (LCAO) For example, let $\phi(\underline{r})$ be an s wave function for an atomic level of a single Na atom. For a sodium crystal, we might qualitatively construct from this ϕ a trial Bloch wave function $\Psi_{\underline{k}}(r)$ of an energy band corresponding to this atomic level. Let

$$\Psi_{\underline{k}}(r) = \sum_n \phi\left(\underline{r} - \underline{n}\right) \exp\left(i\underline{k} \cdot \underline{n}\right),$$

where $\underline{k}$ is the wave vector of the Bloch function and $\underline{n}$ is a direct lattice vector, and calculate the energy $E(\underline{k})$ as the expectation value of the single electron Hamiltonian $(p^2/2m)$, the kinetic energy, plus $V(\underline{r})$ the crystal potential. This *LCAO* is known as the tight binding approximation in energy band calculations. See pseudopotential.

linear response theory (1) Most transport problems and other phenomena such as electric and magnetic properties deal with currents produced by forces, or responses to excitations: We assume four things. First, we assume a linear system: if $R(t)$ is a response to excitation $E(t)$, then $c_1 R_1 + c_2 R_2$ is the response to $c_1 E_1 + c_2 E_2$. Second, we assume a stationary medium whose properties are independent of time. If $R(t)$ is the response to $E(t)$, $R(t - t_0)$ is the response to $E(t - t_0)$. If $G(t)$ is the response to $\delta(t)$, then $G(t - t')$ is the response to $\delta(t - t')$. If $E(t) = \exp(-i\omega t)$, then

$$R(t) = \left[\int_{-\infty}^{\infty} G(t) \exp(i\omega t)\, dt\right] \exp(-i\omega t).$$

Third, we assume causality which means that $G(t) = 0$, for $t < 0$, and

$$R(t) = \left[\int_0^{\infty} G(t) \exp(i\omega t)\, dt\right] \exp(-i\omega t)$$
$$= \chi(\omega) E(\omega)$$

where the susceptibility function (χ) is given by

$$\chi_1 + i\chi_2 = \chi(\omega) = \int_0^{\infty} G(t) \exp(i\omega t)\, dt.$$

Finally, we assume that the total response to a finite excitation can be shown to be finite. The above equation shows that $\chi(\omega)$ is an analytic function of ω in the upper half of the ω plane and leads to the dispersion relations,

$$\chi_1(\omega) - \chi_1(\infty) = \frac{1}{\pi} P \int_{-\infty}^{\infty} \frac{\chi_2(\omega')}{\omega'^2 - \omega^2} d\omega'$$

$$\chi_2(\omega) = -\frac{2\omega}{\pi} P \int_{-\infty}^{\infty} \frac{\chi_1(\omega') - \chi_1(\infty)}{\omega'^2 - \omega^2} d\omega'$$

of which the dielectric constant and the index of refraction are examples.

(2) Kubo developed a quantum mechanical *linear response theory* for transport problems without writing a transport equation. The transport coefficients can be obtained from calculating appropriate correlation functions for the system at thermal equilibrium. For example, the electrical conductivity $\sigma_{\mu\nu}(\omega)$ (relating the current density in the μ-direction due to an electric field in the ν-direction) is given by

$$\sigma_{\mu\nu}(\omega) = \int_0^{\infty} e^{-i\omega t} dt \int_0^{\beta} \langle J_\nu(-i\hbar\lambda) J_\mu(t)\rangle\, d\lambda,$$

where β is the reciprocal of the thermal energy kT, and the angular brackets denote an average at thermal equilibrium, namely $< A > = $ trace $(A \exp -\beta H)/Z$, where Z is the trace of the density matrix $\exp(-\beta H)$.

line spectrum A spectrum is obtained by analyzing the intensity of the radiation emitted by a source as a function of its wavelength. A *line spectrum* is observed when a source emits radiation only at specific (discrete) frequencies (or wavelengths).

line tying Boundary conditions for perturbations of magnetically confined plasmas in which the background magnetic field intersects a conducting material wall or a dense gravitationally confined plasma (as in the case of solar prominences). *Line tying* tends to stabilize interchange instabilities in plasmas.

line vortex   *See* vortex line.

Lippmann–Schwinger equation   (1) In quantum mechanical problems of potential scattering or interparticle collisions, we start from a very simple system given by the Hamiltonian H_o for which all eigenvalues and eigenvectors are known. In most cases H_o is the Hamiltonian for all free particles but does not include interactions responsible for collisions. Its eigenvector Φ_n is related to eigenvalue E_n. The real Hamiltonian H is taken to be a sum of H_o and H_I. For large continuous systems where the energy spectrum is continuous, we may safely assume that Φ_n is related to Ψ_n, which has the same energy E_n. Then the *Lippmann–Schwinger equation* gives a formal solution for Ψ_n as

$$\Psi_n^+ = \Phi_n + (E_n - H_o + i\varepsilon)^{-1} \Psi_n^+$$

where Ψ_n^+ represents the state of an incoming wave and ε is a positive infinitesimal. A similar equation holds for Ψ_n^-, the state of outgoing wave, by substituting $-i\varepsilon$ in place of $+i\varepsilon$.

(2) Equation encountered in the context of quantum scattering theory. In operator notation, it reads

$$T = V + VGT$$

where T is the T-matrix (to be solved for) and V is the potential acting between the two scattering particles. G is the Green's function, defined as

$$G = \lim_{\epsilon \to 0} \frac{1}{E - H_0 + i\epsilon}$$

with E representing the energy and H_0 denoting the free-particle Hamiltonian, i.e., the kinetic energy operator.

liquid crystals   Some organic crystals, when heated, go through one or more phases before they melt into the pure liquid phase. These intermediate phases, known as mesophases or mesomorphic phases are called *liquid crystals*. Their structure is less regular than a crystal but more regular than a liquid. Their physical and mechanical properties are intermediate between those of crystals and liquids. There are many types of *liquid crystals*. Nematics have rod-like molecules. They are uniaxial, and the optical axis can be rotated by the walls of a container or an external agent such as an electric field. They can be switched electrically from clear to opaque and are used in image display devices. Smectic *liquid crystals* have many phases. They are soap-like and have a layered structure. Smectic B is almost a crystal, and smectic D is a cubic gel. Hexactic smectic is uniaxial. Cholestics are made from thin layers (one molecule thick). The orientation of the molecules in a layer can change gradually from layer to layer, leading to a helical structure with intriguing optical properties.

liquid drop model   The simplest kind of collective model for the nucleus. Typically, nuclear models can be subdivided in two groups: the independent particle models, and the collective models. The former assume that the nucleons move essentially independently of one another in an average potential. In the collective models, the nucleons are strongly coupled to one another. The nucleons are treated like molecules in a drop of fluid. They interact strongly and have frequent collisions with one another. The resulting motion can be compared to the thermal motion of molecules in a liquid drop.

liquid metals   A fluid of randomly distributed ions with an electron gas glue between them. The thermal and electrical conductivities, though a few times lower than those of the crystals, are still high. The electron screening of the interactions is still as effective as in regular crystals.

L-mode (low mode)   Plasma confinement obtained in tokamak experiments with significant auxiliary heating power (such as neutral beam injection or radio frequency heating) and high recycling or gas puffing of neutrals at the plasma edge.

local gauge transformation   The transformation

$$\psi' = e^{i\epsilon Q}\psi$$

applied to the wave function ψ of a quantum mechanical system, where ϵ is an arbitrary real parameter and Q is an operator associated with the physical observable q, is called a global gauge transformation. Invariance under such a transformation implies conservation of the quantity q. If ϵ is an arbitrary function of space and time coordinates, $\epsilon(\mathbf{r}, t)$, the transformation above becomes a *local gauge transformation*.

locality The property of depending upon the location in space.

localization Local or localized mode or wave, refers to a damped wave such as a localized lattice vibrational wave which is damped away from an atom, which is heavier or lighter than the other atoms, an electron wave around a donor or an acceptor in a semiconductor, or an electron wave localized by disorder (Anderson *localization*).

local thermal equilibrium (LTE) model
Model for computing radiation from dense plasmas in which it is assumed that the population of electrons in bound levels (such as the electrons still attached to impurity ions) follows the Boltzmann distribution.

Londons' equations Hans and Fritz London obtained the following two equations for superconductivity:

$$\Lambda \frac{\partial \underline{J}_s}{\partial t} = \underline{E}$$

$$\underline{A} = -c\Lambda \underline{J},$$

where $\underline{J}_s$ is the supercurrent density, $\underline{E}$ is the electric field, $\Lambda = m/(n_s e^2)$, c is the speed of light, e is the charge, m is the mass of the carrier of supercurrent, n_s is the density of the carriers, and $\underline{A}$ is the vector potential with $\nabla \cdot \underline{A} = 0$. The above equations, together with the Maxwell equations, show that the magnetic fields and currents penetrate a superconductor only to distances of around λ_L, where $\lambda_L^2 = \Lambda c^2/4\pi$.

longitudinal polarization A particle is said to be *longitudinally polarized* when the direction of its spin is parallel to the direction of propagation.

longitudinal wave Wave in a plasma in which the oscillating electric field is partially or totally parallel to the wave number (the direction of wave propagation). Examples include electron plasma oscillations and sound waves.

long wavelength limit This term describes the situation where the wavelength of the electromagnetic radiation is much larger than the nuclear dimensions. This is a valid assumption up to several MeV and therefore applies to most nuclear γ-rays.

Lorentz force Force acting on a charged particle moving through a magnetic field. The Lorentz force is given by $q \, \mathbf{v} \times \mathbf{B}$, where q is the particle charge, $\mathbf{v}$ is the particle velocity, and $\mathbf{B}$ is the magnetic field.

Lorentz invariance The property of being invariant upon a Lorentz transformation between reference frames.

Lorentz ionization The process of ionizing neutral atoms by using the electric field associated with their motion through a background magnetic field.

Lorentz–Lorenz formula The formula $4\pi N\alpha/3 = (n^2-1)/(n^2+2)$, where N is the number of molecules (atoms) per unit of volume, α is the molecular polarizability, and n is the index of refraction, was discovered independently by H.A. Lorentz and L. Lorenz in 1880. A formula which replaces n^2 by ε, the dielectric constant, is known as the Clausius–Mossotti relation. For the field polarizing, the formula uses a molecule of the local field which is $\underline{E}$, the external applied field, plus $4\pi \underline{P}/3$, where $\underline{P}$ is the polarization which is the electric dipole moment per unit of volume.

Lorentz model (Lorentz gas approximation) Kinetic theory model for the collisions of charged particles off cold charged particles with infinite masses.

Lorentz scalar Term used in the context of special relativity. It is the scalar product between two four-dimensional vectors. Namely,

$$A \cdot B = A_\mu B^\mu$$

with $\mu = 1, \ldots, 4$. A *Lorentz scalar* is invariant under Lorentz transformations. *See* Lorentz transformations.

Lorentz transformations Relativistically valid transformations between inertial observers. They reduce to the Galilean transformations in the non-relativistic limit. For instance, if a primed system moves with speed v along the xx' axes, the *Lorentz transformations* between the space-time coordinates of a point are

$$x' = \gamma(x - vt)$$
$$y' = y$$
$$z' = z$$
$$t' = \gamma\left(t - \frac{v}{c^2}x\right)$$

with

$$\gamma = \frac{1}{\sqrt{1 - (v^2/c^2)}}.$$

Lorenz number (L) The ratio $K/(\sigma T)$, where K is the electron thermal conductivity, σ is the electrical conductivity, and T is the absolute temperature. For metals, this number is $L = K/(\sigma T) = (\pi^2/3)(k/e)^2 = 2.7 \times 10^{-13}$ e. s. u, which is an expression of the Wiedemann–Franz law of 1853. For semiconductors and nondegenerate electron gases, where the relaxation time varies as v^p where v is the speed of the electron, $L = 1/2(p + 5)(k/e)^2$. This remarkable result depends on the existence of an isotropic relaxation time.

loss coefficient Dimensionless coefficient of the head or pressure loss in a piping system,

$$K = \frac{h}{U^2/2g} = \frac{\Delta p}{\frac{1}{2}\rho U^2}$$

where h and Δp are the measured head loss and pressure drop, respectively. Values of K are generally determined experimentally for turbulent flow conditions for various pipe types and sizes.

loss cone Region in velocity space of the plasma in a magnetic mirror device in which the charged particles have so much velocity parallel to the magnetic field that they pass through the magnetic mirror.

loss, minor Any loss in a pipe or piping system not due to purely frictional effects of the wall, including pipe entrances and exits, sudden and gradual expansions and contractions, valves, and bends and tees. Values of the loss coefficient K for each loss must be determined experimentally.

low energy electron diffraction (LEED) A slow electron whose kinetic energy is V electron volts has a de Broglie wavelength λ which equals $(12.26/\sqrt{V})$Å. Thus, electrons in the energy range 5–500 eV have wavelengths in the range of 6 to 1/2 Å which is comparable to the distances between the atoms in crystals. However, such electrons, unlike X-rays and slow neutrons, penetrate only a few angstroms in a crystal, and therefore are not suited for obtaining diffraction patterns from crystals. They are, however, highly suited to study crystal surfaces by diffraction methods. Assume that the atoms on the surface have a two-dimensional lattice whose primitive translation vectors are $\underline{a}_1$ and $\underline{a}_2$ (*see* lattice, crystal lattices). If the incident electron wave vector is $\underline{k}$ (usually normal to the surface) and $\underline{k}'$ is the scattered wave vector, then we have only two Laue conditions for constructive interference: $(\underline{k} - \underline{k}') \cdot \underline{a}_1 = 2\pi$ (integer) and $(\underline{k}' - \underline{k}) \cdot \underline{a}_2 = 2\pi$ (integer). This means that $\underline{k} - \underline{k}'$ is an integral combination of the reciprocal lattice vectors $\underline{b}_1$ and $\underline{b}_2$, but has an arbitrary component in the $\underline{b}_3$ direction, and the diffraction pattern consists of lines or rods. Note that $|\underline{k}'| = |\underline{k}|$ as we deal with elastic scattering. *LEED* is now a highly developed technique which reveals many unusual features of surfaces.

Electron diffraction is also carried out using medium energy electrons (500 eV – 5 keV), MEED, and high energy electrons (5 keV – 500 keV), HEED.

lower hybrid frequency Frequency of a longitudinal plasma ion oscillation propagating perpendicular to the background magnetic field. The *lower hybrid frequency* is intermediate to the high frequency of the electron extraordinary wave and the low frequency of the magnetosonic wave. At a sufficiently high plasma density, the lower hybrid frequency is approximately the square root of the ion cyclotron frequency times the electron cyclotron frequency.

lower hybrid resonant heating Plasma heating by lower hybrid waves in which power is absorbed at the lower hybrid resonant frequency for sufficiently high density plasmas or by Landau damping at lower densities.

LS coupling A possible coupling scheme for spins and angular momenta of the individual nucleons in a nucleus, alternative to the j–j scheme. In the LS scheme, the orbital angular momenta of all nucleons are added together to provide the total orbital angular momentum L. The same is done with the intrinsic spins, which yields the total nuclear spin S. L and S are then coupled with each other to give the total angular momentum of the nucleus.

lubrication theory Hydrodynamic theory relating the motion of two solid surfaces separated by a liquid interface, where the relative motion of the solid surfaces generates an excess pressure in the fluid layer. This pressure allows the fluid to support a load force. In hydrostatic lubrication, the excess pressure is maintained with an external pressure source.

luminescence An excitation of a system resulting in light emission which does not include black body radiation. The excitation can be due to photons, cathode rays (electrons), electric field, chemical reactions, heat, or sound waves, for example, and the *luminescence* which results is called photoluminescence, cathodoluminescence, electroluminescence, chemiluminescence, thermoluminescence, and sonoluminescence respectively. *Luminescence* occurs in gases, liquids, and solids. The radiative transitions causing *luminescence* are simple in gases and are given by atomic spectroscopy, but they are more complex in liquids and solids due to the strong interactions between the atoms. Luminescent solids, such as ZnS and CdS, are known as phosphors. They contain impurity activators such as Ag and Cu which act as luminescent centers. They usually absorb ultraviolet light and emit light in the visible range with an efficiency of about 1/2 (one photon emitted for two absorbed).

Fluorescence and phosphorescence are two forms of *luminescence*. After the excitation ceases, fluorescence decays exponentially with a time constant independent of temperature, but phosphorescence (afterglow) persists, and the decay is temperature-dependent.

luminosity Term used within the context of accelerator physics in conjunction with the operational costs of the accelerator. The rate at which a reaction takes place is written as

$$R = l\sigma$$

where l is the *luminosity* and σ is the cross-section. The luminosity is a characteristic of the particular accelerator and its working conditions. It can be determined by calibration using a known cross-section.

Lundquist number A dimensionless plasma parameter equal to the Alfvén speed times a characteristic scale length divided by the plasma resistivity.

Lyddane–Sachs–Teller relation For a cubic polar crystal with two atoms/unit cell, the relation $\omega_L^2/\omega_T^2 = \varepsilon(0)/\varepsilon(\infty)$ is known as the Lyddane–Sachs–Teller relation. Here, ω_L and ω_T are the longitudinal and transverse optical (angular) frequencies, $\varepsilon(0)$ is the static dielectric constant, and $\varepsilon(\infty)$ is the dielectric constant at optical frequencies.

M

Mach angle The angle of a Mach cone, given by

$$\mu = \sin^{-1} \frac{1}{M}.$$

As $M \to 1$, $\mu \to 90°$.

Mach cone For a moving supersonic disturbance, the projection of the sound waves forms a conical volume inside of which the presence of the disturbance is felt by the fluid. Outside the *Mach cone*, the fluid is unaware of the disturbance. The interior of the *Mach cone* is often called the zone of action, while the exterior is known as the zone of silence. This phenomena is related to the Doppler effect.

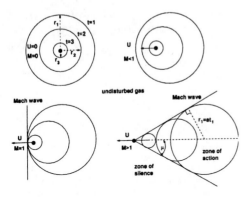

Generation of a Mach cone.

machine A thermodynamic device that converts potential energy into work. Simple examples include a pulley (converts gravitational potential energy into work), an electric motor (converts electrical potential energy into work) and a fuel cell (converts chemical potential energy into work).

Mach line Characteristic lines in supersonic flow along which information propagates and whose orientation is given by the Mach angle.

Mach number The ratio of the local flow velocity U to the speed of sound a

$$M \equiv \frac{U}{a}$$

M is an important dimensionless parameter governing the importance of compressibility effects in the flow.

$$M \sim \frac{\text{inertia}}{\text{compressibility}}.$$

At a low *Mach number*, compressibility forces of the fluid are greater than inertial forces of the flow. Thus, the flow cannot change the fluid's density and the flow can be considered incompressible. As M increases, the inertial forces become large enough to overcome compressibility and can alter the fluid's density. For $M > 0.3$, the flow is considered compressible and the density and temperature may vary with the flow velocity along with pressure.

$M < 0.3$	subsonic incompressible
$0.3 < M < 0.8$	subsonic compressible
$0.8 < M < 1.2$	transonic
$M = 1$	sonic
$M > 1$	supersonic
$M > 5$	hypersonic

Above $M > 5$, ionization becomes important due to high temperatures, and the fluid begins to behave as a plasma in certain regions.

Mach–Zehnder interferometer A special type of two-beam interferometer. An incoming light beam is split into two components that are then recombined with a second beam splitter. When the properties of the *Mach–Zehnder interferometer* are discussed, care must be taken that the phase relationships at the beam splitter, as well as the vacuum radiation incident on the unused input port, are properly taken into account.

macroscopic instability A large scale plasma instability that does not depend on kinetic or microscopic effects.

macrostate A state of existence of the system defined by the values of the principle thermodynamic properties of that system. For an ideal gas, the principle thermodynamic properties are

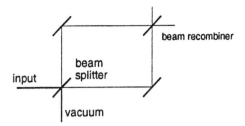

The Mach–Zehnder interferometer.

pressure (P), temperature (T), and volume (V). The macrostate is defined by the probabilities of its constituent microstates.

Madelung constant A constant which is introduced in calculating the electrostatic energy of ionic crystals. Consider a crystal of unit cells containing two ions of opposite charge per cell. Denote the positions of the ions by the vectors $\underline{r}_i$ and their charges by q_i, where $q_i = \pm q$. The electrostatic energy of the crystal U can be written as

$$U = Nq^2 \sum_j {}' \pm 1/r_{ij}$$

where $r_{ij} = |\underline{r}_i - \underline{r}_j|$, the prime excludes $j = i$, i is arbitrary, the plus sign applies if the ions i and j are identical, and the minus sign applies otherwise. By measuring r_{ij} in a suitable unit a, such as the nearest neighbor distance, we define the *Madelung constant* α by,

$$\alpha = \sum_j {}' \pm a/r_{ij}$$

which gives, for the energy,

$$U = Nq^2 \alpha / a \ .$$

magic numbers Nuclei that have certain special numbers of protons or neutrons show an unusually high stability. The numbers for which this stability occurs are called *magic numbers*. This behavior can be accounted for by shell models of the nuclei. The *magic numbers* for the proton number (Z) or the neutron number (N) are $Z, N = 2, 8, 20, 28, 50, 82, 126$.

magnetic axis Magnetic field line surrounded by simply-nested magnetic surfaces. Nearby magnetic field lines wrap around a *magnetic axis*.

magnetic beach Region in which magnetic field strength is decreasing to the extent that the ion cyclotron frequency decreases below the wave frequency, and ion cyclotron wave energy is thermalized by ion cyclotron damping.

magnetic breakdown When a magnetic field B is applied to a crystal containing free electrons (or holes), the electrons follow orbits in $\underline{k}$ space which are obtained by cutting the constant energy surfaces by planes perpendicular to the magnetic field. If an energy surface has more than one sheet, more than one orbit results. For weak fields, an electron would follow only one of these orbits. However, for strong fields, an electron can jump from one orbit to another, resulting in what is known as *magnetic breakdown* or breakthrough. The electron tunnels through regions of forbidden energy states (imaginary k) to an allowed orbit. This is similar to Zener tunneling in a strong electric field E, where the electron tunnels from the valence band to the conduction band in semiconductors. In both cases, the tunneling probability $P \sim \exp(-A/F)$, where A stands for parameters in the problem and F is the field B or E. For example, consider the constant energy contours in the $k_x k_y$ plane ($k_z = 0$):

$$E(k_x k_y) = \frac{\hbar^2}{2m}k_x^2 + \frac{\hbar^2}{2m'}k_y^2 = E_o$$

$$E(k_x k_y) = -\frac{\hbar^2}{2m}k_x^2 + \frac{\hbar^2 k_y^2}{2m'} - V_o = E_o$$

where all the quantities E_o, V_o, m, and m' are positive.

The first equation gives an ellipse for an orbit, and the second equation gives a hyperbola, and breakdown occurs when an electron jumps from one orbit to the other, as depicted by the dotted line.

magnetic buoyancy The tendency for the plasma in regions of strong magnetic field to rise through a gravitationally confined plasma. When this process occurs near the surface of the sun, it leads to the formation of sunspots.

magnetic confinement Confinement of plasma within a magnetic field, which inhibits the flow of charged particles and heat to the surrounding walls of the device.

magnetic dipole A pair of poles, a north pole and a south pole, like in a bar magnet separated by a distance. A loop-carrying electric dipole moment can act as a dipole.

magnetic drift mode Mode of plasma oscillation or instability with wavelengths comparable to Larmor radii and frequencies determined by plasma drifts in the confining magnetic field. These microinstabilities are usually driven by gradients in temperature or density or by non-Maxwellian particle distribution functions.

magnetic energy Product of the magnetic field strength and flux density for points on the demagnetization curve of a permanent magnet which provides a measure of the energy generated in a magnetic circuit. The *magnetic energy* is usually required to be a maximum for the amount of magnetic material used.

magnetic energy density Contribution to the local energy density that is proportional to the magnetic field strength squared.

magnetic field energy The work done in establishing the presence of a magnetic field. If we consider that the magnetic field is a consequence of circulating currents, then the magnetic field energy is the work done in establishing these currents *insitu* from zero. Typically, these currents are of two types. Internal magnetization currents are the microscopic currents (typically in the form of electronic angular momenta) that lead to the magnetization field of the material, $\underline{B}_M$. External currents are macroscopic currents (typically flowing through some magnetization coil) that lead to some external applied field, $\underline{B}_E$.

The total field energy is stored in space with a density given by:

$$U_M = \frac{1}{2}\left(\frac{|\underline{B}|^2}{\mu_0}\right)$$

where:

$$\underline{B} = \underline{B}_M + \underline{B}_E .$$

magnetic gradient drift (grad-B drift) Charged particle drift associated with the component of the magnetic field gradient that is perpendicular to magnetic field lines. This drift velocity is perpendicular to the field line and the magnetic gradient. It is proportional to the perpendicular kinetic energy times the perpendicular gradient of the magnetic field, and it is inversely proportional to the particle charge and magnetic field strength squared.

magnetic island Filament of magnetic field lines forming their own set of nested magnetic surfaces surrounding their own local magnetic axis.

magnetic mirror Region with increased magnetic field strength used to trap charged particles along magnetic field lines.

magnetic moment The parameter used to measure the strength of a magnetic dipole. The earliest known magnetic dipoles were compass needles. Later it was discovered that atoms, neutrons, and electrons can act as magnetic dipoles and exhibit *magnetic moments*. The potential energy E of a magnetic dipole with *magnetic moment* $\vec{m}$ in a magnetic field $\vec{B}$ is

$$E = -\vec{m}\vec{B} .$$

magnetic monopole A hypothetical particle which carries an isolated magnetic charge (an isolated north or an isolated south magnetic pole). Although *magnetic monopoles* have not been seen experimentally, their existence would endow Maxwell's equations with sources with dual symmetry between electric and magnetic quantities. The consequences for quantum mechanics of the existence of magnetic charges were studied by P.A.M. Dirac. The existence of magnetic charges arising from non-Abelian gauge theories was investigated by A.M. Polyakov and G.t'Hooft.

magnetic multipole radiation Radiation due to higher terms in the series expansion of the interaction operator between an atomic system and electromagnetic radiation. The Einstein A^ν coefficient in terms of frequency units for the magnetic dipole radiation is given by

$$A^\nu = \frac{16\pi^3 \mu_0 \nu^3}{3hc^3 g_2} S_{md} ,$$

where ν is the frequency of the transition in Hz, μ_0 is the permeability of the vacuum, h is Planck's constant, c is the speed of light, and S_{md} is the line strength for magnetic dipole transitions. The selection rules for these transitions are given by

$$\Delta J = 0, \pm 1$$
$$\Delta L = 0$$
$$\Delta m = 0, \pm 1 \,.$$

magnetic permeability The ratio of the magnetic field $\underline{B}$ to $\underline{H}$, which is denoted by μ and is a scalar for isotropic media and a tensor otherwise. For ferromagnetic materials, μ must be carefully defined.

magnetic pressure *See* magnetic energy density.

magnetic probe Small insulated coil of wire used to measure local changes in magnetic flux and, consequently, to measure local magnetic field strength and direction.

magnetic pumping Method of heating plasmas by oscillating the magnetic field strength. During this process, there is a net gain of plasma energy because the changing magnetic field causes parallel and perpendicular components of charged particle kinetic energy to change at different rates, while particle collisions transfer energy between these components.

magnetic reconnection Change in magnetic field topology caused by plasma resistive or turbulent dissipation. The formation and growth of magnetic islands, for example, requires *magnetic reconnection*.

magnetic resonance An important tool in physical chemistry and biochemistry. One distinguishes between electron spin resonance (ESR) and nuclear magnetic resonance (NMR). The spin of an elementary particle interacts with a magnetic field. Depending upon how the spin is oriented with respect to the magnetic field, the energy levels will split into two states which correspond to parallel and anti-parallel spin with respect to the magnetic field axis. Electro-magnetic radiation can now cause spin-flips between these energy levels. The resonance frequency depends on the strength of the magnetic field, the magnetic moment, and the nuclear spin of the particle under consideration. The proton has a resonance frequency of 4.2577 kHz/gauss. Depending on the exact environment of a proton within a molecule, this resonance frequency will shift. By mapping out the resonance frequencies of the proton spins, the structure of molecules can be investigated.

magnetic Reynold's number A dimensionless plasma parameter equal to the plasma fluid speed times a characteristic scale length divided by the plasma resistivity. When this parameter is much greater than unity, magnetic field lines move with the plasma. Otherwise, the magnetic field diffuses through the plasma.

magnetic ripple Localized spatial variation in magnetic field strength due to the discrete structure of magnetic coils.

magnetic ripple transport Transport that is enhanced by magnetic ripple in magnetic plasma confinement devices.

magnetic sublevels Or Zeeman sublevels. The substructure of atomic levels with $J > 0$. In the absence of a magnetic or electric external field, these levels are degenerate. In the presence of an external magnetic or electric field, one finds, with the help of perturbation theory, that these levels split in energy into components with different projections m_J of the spin onto the quantization axis, i.e., generally the axis of the external fields. This splitting is due to the interaction of the spin with the external fields

One can distinguish effects for weak and strong magnetic fields. In the case of weak magnetic fields, i.e., when the splitting of the levels due to the magnetic fields is small compared to the fine structure, the magnetic field resembles a small perturbation. For strong fields, however, the internal LS coupling is destroyed and replaced by an independent precession of the angular momenta and electron spins around the field axis. Therefore, the total angular momentum J is not a good quantum number anymore. The magnetic energy is much larger than the fine structure splitting. This regime is called the Paschen–Back effect.

For weak fields, one observes the normal Zeeman effect for singlet systems, i.e., $S = 0$, $J = L$. For non-spin singlet systems, i.e., in the presence of fine structure, the anomalous Zeeman effect is observed. A weak magnetic field B shifts the energy of a Zeeman sublevel by an amount

$$\Delta E(m_J) = g\,\mu_B\,m_J\,B$$

with

$$g = 1 + \frac{J(J+1) + S(S+1) - L(L+1)}{2J(J+1)}$$

where g is the Landé factor and μ_B is the magnetic moment. In the case of an additional hyperfine structure with total angular momentum F due to a nonzero nuclear spin I, one finds

$$\Delta E(m_F) = g_F\,\mu_B\,m_F\,B$$

with

$$g_F = g\,\frac{F(F+1) + J(J+1) - I(I+1)}{2F(F+1)}$$
$$- g_I\,\frac{\mu_K}{\mu_B}\,\frac{F(F+1) + I(I+1) - J(J+1)}{2F(F+1)},$$

where $\mu_K = e\hbar/2m_p$ is the nuclear magneton and g_I is the g-factor for the nucleus for the particular species. For example, for hydrogen, $g_I = +5.58$.

For the Paschen–Back effect, one finds an additional energy of the energy levels dependent on m_l and m_s:

$$\Delta E(m_l, m_s) = (m_l + 2m_s)\mu_b B\;.$$

If an additional hyperfine structure is observed, the additional energy is found to be

$$\Delta E_{HFS} = g_J \mu_b m_J B + a m_I m_J - g_I \mu_k B m_I$$

where

$$a = \frac{g_I \mu_k B}{\sqrt{J(J+1)}}\;.$$

In the latter case, the lines split into groups of lines corresponding to different projections for a particular m_J since the electron spin interacts more strongly with the magnetic field. Within

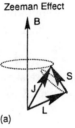

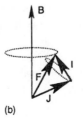

Coupling scheme for the case (a) without and (b) with hyperfine structure in an atom subject to a magnetic field B. For the Paschen–Back effect, the two angular momenta decouple. They then precess around the magnetic field axis with differing frequency.

these groups, constant level separations of $m_J a$ are observed.

Optical dipole transitions between different Zeeman sublevels of electronic states are only possible under the selection rule $\Delta m_J = 0, \pm 1$. One speaks of σ and π transitions, depending upon whether or not the selecting rule 0 or ± 1 applies.

In the case of an external electric field, one distinguishes between the linear and quadratic Stark effects depending on the proportionality of the energy splitting as a function of the electric field axis. For electric fields, the dependence is only on the modulus of m_j.

The linear Stark-effect is observed for degenerate levels, i.e., in hydrogen atoms. For non-degenerate levels, the first order term in the perturbation theory gives no contribution and the quadratic Stark effect is observed. Qualitatively, the quadratic Stark effect is connected with the polarizability of the atom. The electric field induces a dipole moment in the atom, with which the electric field then interacts, leading to a E^2 dependence.

magnetic surface Also called flux surface. Surface that is tangent to magnetic field lines everywhere. *Magnetic surfaces* are toroidally nested in tokamaks and other toroidal plasma confinement devices.

magnetic susceptibility A linear relation between the magnetic moment per unit of volume $\underline{M}$ and the exciting field $\underline{H}$. Note $\underline{B} = \underline{H} + 4\pi \underline{M}$ and $M_i = \sum_j \chi_{ij} H_j$, where χ is the susceptibility tensor (scalar in cubic crystals). For alternating fields, χ has a real and imaginary part $\chi_1(\omega) + i\chi_2(\omega)$ and satisfies the Kramer–Kronig relations.

magnetic tension Force proportional to the curvature of magnetic field lines times the field strength squared.

magnetic trap Trap for neutral particles based on the state-dependent interaction of the magnetic dipole moment with an inhomogeneous magnetic field. The typical trap depths for *magnetic traps* are around 100 mK. They are therefore used to pre-cool atomic samples before evaporative cooling can be used to produce Bose–Einstein condensation. Since the trapping mechanism relies on the internal structure of the atom, not all atoms can be trapped magnetically. The atom or molecule to be trapped must have a magnetic moment.

magnetic well Minimum in the negative line integral of the reciprocal of magnetic field strength. *Magnetic well* configurations tend to stabilize plasma interchange instabilities.

magnetism A property of matter due to electric currents and magnetic dipoles. Some atoms and ions have magnetic moments due to the spin or the orbital motion of their electrons. Some nuclei have magnetic moments, but they are about 2000 times weaker than those of the electrons. At thermal equilibrium, these magnetic moments are randomly oriented and the average magnetic moment per unit of volume (the magnetization) is zero. When an external magnetic field $\underline{B}$ is applied, energy considerations favor alignment of the magnetic moments, and a net magnetic moment in the direction of $\underline{B}$ results. Such systems are called paramagnetic and the susceptibility χ is positive and is approximately 10^{-5}. χ varies with temperature as C/T, Curie's law, where C is a constant. See Langevin–Debye formula.

For crystals where the atoms or the ions have completed (closed) shells, the angular momentum and the magnetic moment are zero. An external magnetic field induces a magnetic moment opposite to the field, and the magnetic susceptibility is negative and approximately 10^{-6}. Examples are alkali halides crystals and solid neon and argon. The susceptibility is independent of temperature and is proportional to the squares of the electron distances from the nucleus. For solids with unfilled d or f shells, the atoms possess magnetic moments and interact with their neighbors by the exchange interaction (due to Heisenberg), which is an electrostatic interaction but of quantum mechanical origin. The form of the interaction is $-J$, $\underline{S}_i \cdot \underline{S}_j$, where J is a constant of 0.01 eV, and $\underline{S}_i$ and $\underline{S}_j$ are the nearest neighbor spins. If J is positive, the magnets are aligned parallel to each other and we have ferromagnets such as Fe, Co, and Ni. If J is negative, the spins align anti-parallel and if their magnitudes are the same, the net magnetic moment is zero and we have anti-ferromagnetism as in MnO. If J is negative and the spins are not equal in magnitude, we have a finite magnetic moment and ferrimagnetism as in Fe_3O_4 (magnetite). This magnetic ordering disappears at high temperatures. See molecular field, Néel temperature.

magnetization ($\underline{M}$) The magnetic moment per unit of volume. The process of getting a sample magnetized. When a ferromagnet (of zero net magnetic moment) is subjected to an increasing $\underline{H}$ field, $\underline{M}$ and $\underline{B}$ rise; as we decrease $\underline{H}$, $\underline{B}$ follows a path which lies above the rising curve. As $\underline{H}$ goes through a cycle of negative and positive values, $\underline{B}$ traverses a hysteresis loop.

magnetization current Plasma current caused by diamagnetic drifts, which are perpendicular to the local plasma pressure gradient and magnetic field.

magneto-acoustic effect Study of sound wave attenuation in metals in a magnetic field. For a transerve sound wave whose directions of propagation and polarization (direction of the oscillation of the ions in the metal) are perpendicular to the magnetic field, the wave attenuation by the electron gas can give some information about the shape of the Fermi surface.

magnetocrystalline anisotropy A change in the orientation of the spin direction (magnetization) in a crystal relative to the crystal axes changes the exchange energy and the electrostatic interactions between atoms. This is reflected in the easy and hard directions of magnetization. For a cubic crystal such as iron, the anisotropy energy U_K can be expressed as:

$$U_K = K_1 \left(\alpha_1^2 \alpha_2^2 + \alpha_2^2 \alpha_3^2 + \alpha_3^2 \alpha_1^2 \right) + K_2 \left(\alpha_1^2 \alpha_2^2 \alpha_3^2 \right)$$

where $(\alpha_1 \alpha_2 \alpha_3)$ are the cosines of the angles the magnetization makes with the axes of the crystal, and K_1 and K_2 are positive constants of about 10^{-5}erg/cm^3. U_K is larger for magnetization along a body diagonal than along a face diagonal and is zero along a cube edge.

magnetohydrodynamics The branch of fluid dynamics that deals with ionized fluids where the fluid is treated as a continuum. The electromagnetic effect produces an additional force.

magneton The combination of factors $e\hbar/2mc$, where e is the electron charge, $\hbar = h/2\pi$, m is the particle mass, and c is the speed of light. The magneton has the dimension of a magnetic moment. Multiplied with the dimensionless g-factor for a particle, one finds the particle's magnetic moment. For electrons one finds $g_e = -2.0023193043737(82)$, for protons $g_p = 5.585694675(57)$, and for neutrons $g_n = -3.82608545(90)$.

magneto-optical effects Changes in optical behavior of materials with external magnetic fields. Examples of this effect are the Faraday effect and Kerr effects. *See* Kerr effect, Landau levels.

magneto-optical trap An optical trap for neutral atoms which is produced by superposing an inhomogeneous magnetic field with six red-detuned light beams in a three-dimensional arrangement, as shown in the picture. A minimal requirement for building a *magneto-optical trap* is that the atom possesses a $J = j \rightarrow J = j+1$ transition. Depending on the electronic level structure of the species under consideration, additional repumping of laser beams might be necessary in order to avoid the trapping of the populations in levels not accessible to the trapping laser.

Slow neutral atoms can be forced to undergo a diffusive motion with the help of two counter-propagating, red-detuned laser beams in a $\sigma^- - \sigma^+$ configuration, i.e., with circular polarized light with opposite chirality. Due to the effect of the absorption of a laser photon, the photon transfers a momentum kick to the atom. Spontaneous decay results in a momentum kick in a random direction within 4π, i.e., has no net-effect. Due to the red-detuning of the laser beams, the atom is more likely to absorb a photon out of the laser beam that is counterpropagating to its motion. Therefore, the momentum transfer tends to slow down the atom as it always opposes the motion of the atom. Due to the Doppler shift of the resonance line, this decelerating force is velocity-dependent. One solves for the net force from two counterpropagating laser beams with frequency ω_L, and a detuning $\Delta = \omega_L - \omega_0$ compared to the resonance frequency ω_0 of an atom moving with speed v is given by

$$F = \frac{2\pi \hbar \Gamma}{\lambda} \frac{I}{I_{\text{sat}}}$$

$$\left(\frac{1}{1 + 4\left(\frac{\Delta - 2\pi v/\lambda}{\Gamma}\right)^2} - \frac{1}{1 + 4\left(\frac{\Delta + 2\pi v/\lambda}{\Gamma}\right)^2} \right),$$

where Γ and λ are the width and wavelength of the resonance transition respectively. The saturation intensity of the transition is defined as $I_{\text{sat}} = \frac{\pi \hbar c}{3\lambda^3} \Gamma$.

Arranging six beams in a three-dimensional configuration leads to a three-dimensional diffusive motion. In order to build a trap, a spatially dependent force is necessary, which drives the atoms to one common point in space. This is accomplished by the superposition of an inhomogeneous magnetic field. Most often this inhomogeneous field is generated by anti-Helmholtz coils (*see* figure), which produce a quadrupole magnetic field of the form

$$B(r) \approx \left.\frac{\partial B_z}{\partial r}\right|_{z=0} \left(z\hat{z} - \frac{1}{2}x\hat{x} - \frac{1}{2}y\hat{y}\right),$$

where $\hat{x}$, $\hat{y}$, and $\hat{z}$ are the unit vectors in the x-, y-, and z-directions.

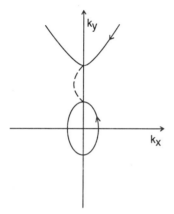

Principle of a magneto-optical trap.

Initially, it was believed that the lowest temperatures achievable in *magneto-optical traps* are defined by the Doppler limit. Experiments showed, however, that lower temperatures are possible due to the internal structure of the atoms. These advanced cooling schemes are called polarization gradient cooling and Sisyphus cooling.

The typical density achievable in a *magneto-optical trap* is approximately 10^{11} atoms/cm^3. The main factor of limitation is the radiation pressure from the spontaneously fluorescing atoms in the trap center. This can be avoided by the use of a dark spot trap. Ultimately, collisions with the background gas are the limiting factor.

Applications of cooling and trapping neutral atoms in *magneto-optical traps* are, among others, the investigation of cold collisions, frequency standards, and photoassociation of cold molecules. Furthermore, the trapping of neutral atoms in *magneto-optical traps* is a prerequisite for the formation of Bose–Einstein condensates.

magnetopause Shock wave boundary between the solar wind and a planetary magnetosphere.

magnetoresistance (magnetoconductivity)
The effects of an external magnetic field $\underline{B}$ on the conductivity (or resistance) of a crystal. The simplest case is an electric current J in the x-direction with an applied field E_x and a magnetic field $\underline{B}$ in the z-direction. The Lorentz force induces an electric field E_y (by charge accumulation on the faces of the sample perpendicular to the y-direction) but no current ($J_y = 0$). Standard calculations for conduction by one band shows that there is no change in the resistivity of the sample. However, if conduction takes place in two bands, the resistivity increases by a quadratic term in the magnetic field, but reaches a limit (saturates) for large fields. This is the transverse *magnetoresistance*. The argument is correct if the electron orbits in the magnetic field are closed, but for open orbits, the *magnetoresistance* increases as B^2. Thus, *magnetoresistance* studies are valuable in determining the shapes of the Fermi surfaces in metals.

magnetosonic wave (magnetoacoustic wave)
Electromagnetic plasma wave propagating perpendicular to the background magnetic field involving compression of the magnetic field as well as plasma pressure.

magnetosphere Plasma confined in the magnetic field of a planet.

magnetostriction The magnetization of ferromagnetic crystals is accompanied by an elastic strain leading to changes in the dimensions of the crystal. In a state of strain, the anisotropy energy is lowered by more than the increase in elastic energy. The effect was discovered by Joule.

magnetotail Comet-shaped wake of the magnetosphere drawn out by the solar wind flowing away from the sun.

magnon Energy quantum for a spin wave which is an excitation in magnetically-ordered materials. It is characterized by a frequency ω and wave vector k. The energy $\hbar\omega \propto k^2$ in ferromagnetic and ferrimagnetic crystals, and leads to a decrease in the magnetization by a term $\propto T^{3/2}$ and an increase in the specific heat by a term $\propto T^{3/2}$. For antiferromagnetic crystals,

$\omega \propto k$, and the specific heat $\propto T^3$. *Magnons* can be detected by inelastic neutron scattering.

Magnus effect A lateral force generated by a rotating body from altered pressure forces. The force is perpendicular to both the rotation axis and the direction of fluid/body motion. Often used to refer to the effect of lift generation on a rotating cylinder. *See* Robins effect.

Majorana fermions A hypothetical spin 1/2 particle described by a Majorana spinor. A Majorana spinor satisfies the condition that the spinor is equal to the charge conjugate of itself — $\psi = \psi^c$. The conditions implies that a *Majorana fermion* would be its own antiparticle. Particles which carry any kind of charge cannot be *Majorana fermions*.

Malmberg–Wharton experiment Experimental verification of Landau damping, published in 1965.

Mandel Q-parameter Measure of the photon statistics of a light source, which can be measured using a Hanbury–Brown–Twiss experiment. It is given by

$$Q = \frac{\langle n^2 \rangle - \langle n \rangle^2}{\langle n \rangle} - 1 = g^{(2)}(0) - 1$$

where $g^{(2)}(\tau) = \frac{\langle n(\tau) n(0) \rangle}{\bar{n}^2}$

$g^{(2)}(\tau)$ is called the second order correlation function. It can be measured in a Hanbury–Brown–Twiss experiment. The photons impinge on a beam splitter and are detected in two photo detectors. The signal of one acts as the start signal for the detection electronics. Essentially, the probability is measured for how long it takes for detection of a photon right after detection of the "start" photon. The beam splitting setup is necessary since photodetectors typically have a dead time between the detection of two photons. It can be shown that detection efficiencies smaller than one for photons do not alter the measured photon statistics. The beam splitter has no effect on the photon statistics either.

Thermal light sources, for which a certain average number of photons $\bar{n}$ is detected during a fixed time interval, follow Bose–Einstein statistics. For thermal light sources, the *Mandel Q-parameter* has a value of one and decays with the characteristic coherence time of the light source to 0. There is a higher probability for detection of photons one right after another, which is referred to as photon-bunching. This behavior is a consequence of the Bose–Einstein statistics.

A laser beam or a coherent light source shows statistics that follow Poisson statistics. The number of photons arriving during a fixed time interval follows a Poisson statistic centered around this average value. In a Hanbury–Brown–Twiss experiment, this will manifest itself in a more regular stream of photons. The *Mandel Q-parameter* has the constant value of 0.

Non-classical light has a *Mandel Q-parameter* of -1 for short times which rises to 0 for long times. An example of a for non-classical light source is the fluorescence of a single atom or ion. After spontaneous emission of a photon, the atom or ion first has to be excited again before another emission process can take place. Consequently, photons tend not to be emitted right after one another. Since the probability of two photons being emitted right after each other is small, we call this anti-bunching.

Mandelstam variables In a two-body reaction where the incoming particles have four momenta, p_1 and p_2, and the outgoing particles have four momenta, p_3 and p_4, the three Lorentz-invariant *Mandelstam variables* are defined by $s = (p_1 + p_2)^2 t = (p_1 - p_3)^2$, and $u = (p_1 - p_4)^2$. These variables satisfy the condition $s + t + u = m_1^2 + m_2^2 + m_3^2 + m_4^2$, where m_1, and m_2 are the rest masses of the incoming particles, and m_3, and m_4 are the rest masses of the outgoing particles. These variables are useful in studying two-body scattering processes.

Manning roughness factor Coefficient quantifying the effect of surface roughness on flow rate in open channel flows.

manometer Any device used to measure a pressure difference between two points in a fluid. The liquid *manometer* uses the variation of pressure to the height of a column of liquid $\Delta p = \rho g h$ to determine a pressure difference.

Marangoni convection Convection in a thin horizontal layer of liquid heated from below driven by the release of surface free energy resulting in hexagonal-shaped circulation cells.

marginal instability The boundary between a stable and unstable state where a slight change in parameters can induce an instability.

marginal stability The borderline between stability and instability.

Markov process A statistical process without any memory, i.e., each instant is independent of the past evolution of the process. An example of such a process is the random walk.

maser Acronym that stands for **m**icrowave **a**mplification by **s**timulated **e**mission of **r**adiation. The principle of operation is very similar to the laser except that instead of light, microwave radiation is emitted. The *maser* was invented a couple of years earlier than the laser by Schawlow and Townes. Since the rate of spontaneous emission scales with the third power of the frequency, it is much easier to produce a population inversion at micro-wave frequencies. On the other hand, due to the long wavelength in *maser* cavities, losses by diffraction are more critical than for the laser. This problem can be solved by building completely closed cavities with only small openings to let the excited atoms enter the cavity.

Maser cavities are also the basis for the single and two atom *masers* under investigation in cavity quantum electrodynamics.

mass operator An operator which is added to the Lagrangian describing a quantum field of a massive particle interacting with another field in order to eliminate certain infinite quantities which appears as a result of the interaction, and to give the sum of the added mass with the mechanical mass the observed value of the particle mass.

mass renormalization An electron is described by a quantum field coupled with another quantum field of electromagnetic waves. Hence the electron creates the electromagnetic field which will, in turn, act on the electron. Since the electron is a point charge and the wavelengths of the electromagnetic field can be endlessly short, the reaction may create infinite mass correction of the electron. The infinity is found to be unphysical, however, so that it can be subtracted. Thus, the artificial infinite mass is added to the mechanical mass of the electron in order to obtain the measured mass of the electron. This mathematical operation is called the *mass renormalization*. This kind of operation is not limited to quantum electrodynamics, but applies also to certain other interacting quantum fields.

mass shell The relativistic condition that a real particle must satisfy where the square of its four-momentum equals the difference between the square of it energy minus the square of its three-momentum, which is also equal to the square of the rest mass of the particle: $p_\mu^2 = E^2 - p^2 c^2 = m^2 c^4$. A particle which satisfies this condition is said to be on *mass shell*. Virtual particles do not necessarily satisfy this condition.

mass tensor Every energy band has critical points where $\nabla E(\underline{k}) = 0$. At these points, such as $\underline{k}_o$, we can expand $E(\underline{k})$ as a quadratic,

$$E(\underline{k}) = E(\underline{k}_o) + \frac{1}{2} \sum_{ij} \frac{\partial^2 E}{\partial k_i \partial k_j} (k_j - k_{oj})[k_i - k_{oi}],$$

which defines the reciprocal *mass tensor*

$$\left(\frac{1}{m}\right)_{ij}, \left(\frac{1}{m}\right)_{ij} = \frac{1}{\hbar^2}\left(\frac{\partial^2 E}{\partial k_i \partial k_j}\right)$$

evaluated at $\underline{k}_o$. It is usually easy to find axes which diagonalize the *mass tensor*.

master equation A non-linear differential equation that describes changes of probabilities for physical processes.

master group A mathematical group which nontrivially combines the Poincaré group and some internal group such as SU(N). There are no-go theorems which forbid such a construction. These theorems can be evaded if one employs supersymmetry.

Matheissen rule The resistivity in the presence of two scattering mechanisms is the sum of the resistivities when each scattering mechanism acts alone. For example, if τ_1 is the relaxation time for phonon scattering and τ_2 is the relaxation time for impurity scattering, then $\rho_1 = \frac{m}{ne^2\tau_1}$ and $\rho_2 = \frac{m}{ne^2\tau_2}$ and the total resistivity $\rho = \frac{m}{ne^2\left(\frac{1}{\tau_1}+\frac{1}{\tau_2}\right)}$, which would be correct if we can define relaxation times which do not depend on directions, and energy.

matrix element In quantum mechanics, the microscopic states are represented as vectors constituting the Hilbert space, and the physical quantities are operators in this space. Any operator, when acting on any vector, will transform it into another vector. The operators are a rather abstract entity. To bring them into numbers (complex numbers, generally speaking), we resort to the matrix representation. Choose a complete orthonormal set of vectors in the Hilbert space. Apply a specified operator to a member of the set. The member is then transformed into another vector. Take a scalar product of the transformed vector with another member of the set. This scalar product is one of the *matrix elements*. By applying all possible processes of this kind, we get the matrix representation of the operator under consideration.

matter wave Equivalent to a light wave consisting of matter. According to quantum mechanics, particles can also exhibit wave characteristics. The wavelength is given by the de-Broglie wavelength. *Matter waves* are the basis of atom optics, in which the physics of *matter waves* analogous to the laws of optics are explored.

maximum and minimum thermometer A device that is designed to measure and record both the minimum and maximum temperatures that occur within a particular time frame. It usually incorporates a bulb containing alcohol attached to a calibrated capillary tube containing mercury. A steel pointer is located at each end of the mercury column. As the temperature rises, the expansion of the mercury causes the upper pointer to rise, remaining in position to record the maximum temperature reached. With falling temperatures, the contraction of the alcohol causes the lower pointer to drop down the column, remaining in position to record the lowest temperature achieved. A magnetic field is used to reset the positions of both pointers.

Maxwell–Bloch equations Set of five coupled partial differential equations that describe the propagation of light and the interaction with a two-level atom. The interaction is described in form of populations $P_1(t)$ and $P_2(t)$ of the atomic levels and a complex valued coherence term $Q(t)$. The atom can also be described in terms of the Bloch variables r_1, r_2, and r_3, which are linear combinations of P_i and Q. We find

$$Q(t) = r_1(t) + r_2(t)$$
$$P_2(t) - P_1(t) = r_3(t) = w(t),$$

where $w(t)$ is also referred to as the population inversion. The underlying picture is that of the Bloch sphere, where the population of the atoms is viewed as a spin vector. Spin-up then corresponds to complete population inversion, i.e., the upper level is populated only, and spin-down corresponds to the atom being in the ground state.

Maxwell distribution (1) Also called the Maxwell–Boltzmann distribution. The specific distribution of velocities for an ensemble of atoms and molecules in thermal equilibrium at a given temperature T (*see* figure). The probability of an atom or molecule having a velocity in the range of v and $v + dv$ is given by

$$f(v) = \frac{1}{N}\frac{dN}{dv} = 4\pi v^2 \left(\frac{M}{2\pi kT}\right)^{3/2} e^{-\frac{Mv^2}{2kT}},$$

where k is the Boltzmann constant and M is the particle mass. The prefactors are derived from the assumption that the distribution has no preferential direction (4π) and the normalization condition which requires that the total number of atoms is equal to N, i.e.,

$$\int f(v) d^3v = 1.$$

Two important characteristics of the distribution are the most likely velocity v_{ml} and the average velocity $\bar{v}$. Both are only functions of

Maxwell equations

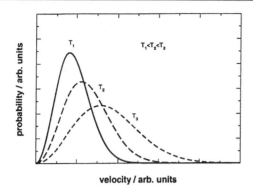

Maxwell–Boltzmann velocity distribution for different temperatures.

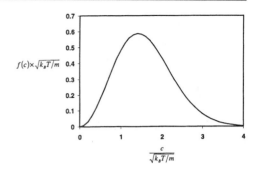

One component of the Maxwell–Boltzmann speed distribution $f_c(c)$ for atoms of mass m at temperature T.

the mass M and the temperature T and scale with $\sqrt{T}$. They are given by

$$v_{ml} = \sqrt{\frac{2kT}{M}}$$

$$\bar{v} = \sqrt{\frac{8kT}{\pi M}}.$$

For growing temperatures, the Maxwell–Boltzmann distribution broadens and the most likely velocity v_{ml} is shifted to higher velocities. The probability of finding this velocity, however, will decrease due to the overall broadening of the distribution.

(2) A description of the distribution of speeds present in a collection of gas molecules. A combination of the statistical arguments proposed by James Clerk Maxwell (1831–1879) and Ludwig Boltzmann (1844–1906). The probability that the speed of any given gas molecule lies in the range of c to $c + dc$ is given by

$$f_c(c)\,dc = \left(\frac{m}{2\pi k_B T}\right)^{3/2} \exp\left(\frac{-mc^2}{2k_B T}\right) 4\pi c^2\,dc$$

for independent gas molecules of mass m at temperature T, where k_B is the Boltzmann constant.

Maxwell equations A set of four classical equations that describe electrostatic and electrodynamic phenomena. They can be written in integral and differential form. The four equations in differential form are:

Gauss' law $\quad \nabla \vec{B} = 0$
Gauss' law $\quad \nabla \vec{D} = \varrho$
Faraday's law $\quad \nabla \times \vec{E} = -\frac{\partial \vec{B}}{\partial t}$
Ampere's law $\quad \nabla \times \vec{H} = \frac{\partial \vec{D}}{\partial t} + \vec{j}$

where $\vec{E}$ is the electric field, $\vec{B}$ is the magnetic field, $\vec{H}$ is the magnetic intensity, $\vec{D}$ is the electric displacement, $\vec{j}$ is the current density, and ϱ is the charge density.

In a vacuum, we have, in addition to the above equations, $\varrho = 0$ and $\vec{j} = 0$. Using these conditions, the wave equations for electromagnetic waves can be derived from the Maxwell equations. For the E and H fields one finds:

$$\Delta E - \mu_0 \epsilon_0 \frac{\partial^2 \vec{E}}{\partial t^2} = 0$$

$$\Delta \vec{H} - \mu_0 \epsilon_0 \frac{\partial^2 \vec{H}}{\partial t^2} = 0.$$

Maxwellian distribution Distribution of particles proportional to the exponential of kinetic energy divided by temperature.

Maxwell relations The four mathematical relationships that relate pressure, volume, temperature, and entropy for a system in thermodynamic equilibrium are:

$$\left(\frac{\partial T}{\partial V}\right)_S = -\left(\frac{\partial P}{\partial S}\right)_V, \quad \left(\frac{\partial S}{\partial V}\right)_T = \left(\frac{\partial P}{\partial T}\right)_V$$

$$\left(\frac{\partial T}{\partial P}\right)_S = \left(\frac{\partial V}{\partial S}\right)_P, \quad \left(\frac{\partial S}{\partial P}\right)_T = -\left(\frac{\partial V}{\partial T}\right)_P.$$

Maxwell's demon A thought experiment proposed by Maxwell to illustrate how the second law of thermodynamics might be defied. Maxwell proposed the existence of an intelligent creature small enough to actually observe the motion of individual atoms. Using this imaginary creature, Maxwell proposed the following experiment: Consider two containers of gas separated by a wall with a door that is operated by the demon. Any gas at thermal equilibrium consists of atoms with a distribution of velocities. If the demon only opens the door when he sees a fast atom approaching from the right (or a slow molecule from the left), then, in principle, he is able to make the left-hand gas hotter than the right hand gas without expending any work. In other words, *Maxwell's demon* allows the entropy of the whole system to decrease. Although, of course, such a demon does not exist, an analogous situation is created when we measure the microscopic details of the same system. If we could know all velocities of all particles, it seems as though we could take advantage of the corresponding entropy reduction. This viewpoint prevailed until Brillouin showed (in the early 1950s) that Maxwell's treatment assumed that the demon generated entropy and that this assumption was indeed false. By calculating the minimum entropy generated by the demon (or any analogous information storage system), Brillouin was able to show that this minimum entropy greatly exceeded the entropy decrease produced by the demon.

Mayer's virial expansion An expression for the equation of state of an imperfect gas. Mayer showed that a general expression for the equation of state of a gas is given by

$$\frac{P}{kT} = n + B_2(T)n^2 + B_3(T)n^3 + \Lambda$$

where P is the pressure of the gas, T is the temperature, n is the number density of gas atoms or molecules, and k is the Boltzmann constant. The virial expansion coefficients, B_i, may also be functions of temperature.

McCleod gauge A device based on Boyle's law for measuring gas pressure. A column of fluid is supported by the compression of a volume of gas at low pressure into a chamber of smaller volume. This gauge typically measures pressure between 10 mbar and 10^{-5} mbar.

M center Two adjacent F centers. *See* color center.

mean free path A collection of particles that undergo random motion in a confined space is characterized by a series of interparticle collisions. The average distance between collisions traveled by such particles is known as the *mean free path*. Examples of physical systems for which the concept of a *mean free path* is important include gas molecules in a container and electrons in a crystalline lattice.

mean free time For particles undergoing random motion, the time between collisions is known as the *mean free time*. This quantity influences a number of thermodynamic processes. For example, in condensed matter physics the *mean free time* plays a role in the conductivity of materials.

mean photon number The average number of photons in one mode of the radiation field. This quantity is dependent on the type of light under consideration. For a thermal light source with temperature T in thermal equilibrium, the probability to detect n photons within a given time interval is given by the Bose–Einstein distribution, since photons are bosons:

$$p(n) = \frac{1}{\bar{n}+1}\left(\frac{\bar{n}}{\bar{n}+1}\right)^n$$

where the *mean photon number* $\bar{n}$ is given by the Boltzmann distribution,

$$\bar{n} = \frac{1}{\exp(h\nu/k_b T) - 1}$$

and the typical fluctuations are given by

$$\sigma^2 = \bar{n} + \bar{n}^2 .$$

The highest probability can always be found at $n = 0$ photons. The photon number fluctuations get larger with increasing temperature T.

For a coherent light source, i.e., the laser, the probability of detecting n photons during a time t is given by the Poisson distribution

$$p(n) = \frac{\bar{n}^n \exp(-\bar{n})}{n!} \qquad n = 0, 1, 2, \cdots$$

where the fluctuations are much smaller

$$\sigma^2 = \bar{n}.$$

measurement theory Quantum mechanics has been successful in understanding nature. It is self-contained, providing self-consistent logics and necessary information of whatever subject. From a practical point of view, there is no problem there. In some extreme cases, however, the logic for relating quantum systems to our cognition is strange at best. From the very start of quantum mechanics there have been troublesome questions about quantum measurements. Our brains, the last device for our cognition, are macroscopic. We have to use a macroscopic device for any measurement of a quantum system, however small it may be. There is a mist between a very small quantum system and a macroscopic device. Furthermore, certain distinguished physicists asserted that we need hidden variables to solve all troublesome problems concerning the measurements of small quantum systems. The assertion is that the current quantum mechanics possesses no practical difficulties but the troubles in its interpretation can only be solved by the hidden variables. All efforts to answer the troublesome questions are classified as the *measurement theory*. The most serious question is that the conventional interpretation of quantum measurements at issue requires non-local events, violating Einstein's hypothesis of prohibiting instantaneous propagation of natural phenomena over the light velocity. A typical example is the Einstein–Podolsky–Rosen experiment, which might have favored the hidden variable theory. The question was greatly clarified by John Bell who found an empirical criterion for the existence of hidden variables by means of mathematical inequalities known as Bell's inequalities. The experiments along this line have not yet rejected the non-locality.

mechanical mass Consider a particle field in the interacting quantum fields. The part of the particle's mass which is supposed to exist in the absence of any interaction is called the *mechanical mass,* or the bare particle mass. The observed mass is different from this. The interaction through the fields gives rise to a mass correction.

Meissner effect The exclusion of magnetic flux in a superconductor. Superconductors are perfect diamagnetic materials.

meniscus Portion of the free surface that extends along a wall due to surface tension effects at the interface of a solid boundary and liquid. The height of the *meniscus* is governed by the balance of pressure and surface tension forces and is given by the relation to the first order

$$h = \sqrt{\frac{2\sigma(1 - \sin\theta)}{\rho g}}$$

where σ is the fluid surface tension, θ is the contact angle, and ρ is the fluid density. The effect of atmospheric density has been neglected, but it can be significant for low liquid densities or high atmospheric pressure.

Mercier stability criterion Plasma conditions needed for stability of localized interchange modes in a toroidal magnetic confinement device.

meson Particle which is a bound state of a quark and an antiquark, held together via the chromodynamic force. The lightest and most common *mesons* are the charged ($\pi^{\pm}$) and neutral (π^o) pions which are composed of up or down quarks/antiquarks. Since *mesons* are formed from quarks/antiquarks they always have integer spin (0, 1, 2, . . .).

metal insulator transition (Mott transition) Can we influence a conductor to become an insulator or an insulator to become a conductor? We know, for example, that intrinsic semiconductors are poor conductors at low temperatures. If a semiconductor is heavily doped with impurities (donors or acceptors), we expect that when the electron wave functions overlap, the semiconductor should become a good conductor. This should happen for impurity densities of $(1/2a_B^*)^3$ where a_B^* is the Bohr radius for the impurity, which is about 30 Å and gives densities of about $10^{18}/\text{cm}^3$.

A similar question concerns H atoms. If compressed in a monoatomic lattice, at what density is a hydrogen crystal a metal? The overlap argument gives $(1/2a_o)^3$, where a_o is the first Bohr radius, which is $\sim 10^{24}/cm^3$. The argument might be refined by including screening, but the orders of magnitude do not change.

Conduction in semimetals and divalent metals takes place in overlapping bands. Are there situations where, by applying pressure to the crystal, one can remove the overlap of the bands or bring it about and thus induce insulator-metal transition?

Another example is disorder: by increasing disorder, a metal can become an insulator, according to Anderson. *See* localization.

metastable state An excited state of an atom or molecule, for which all transitions to lower states are forbidden, which means that they are not coupled via a dipole-allowed transition to a state of lower energy. Lifetimes for *metastable states* are in the millisecond range or longer. Nevertheless, decays of metastable lines can be observed for special conditions. Higher order transitions such as magnetic dipole-allowed transitions or electric quadrupole-allowed transitions are possible. In addition, collisions can quench *metastable states*.

methods of characteristics *See* Riemann invariants.

metric tensor Vector products of the gradients of curvalinear coordinates that characterize the local coordinate geometry.

MHD equations Single fluid model for the interaction between a highly conducting plasma and a magnetic field. The MHD model is used to produce theoretical predictions of plasma equilibrium (force balance) as well as large scale waves and instabilities.

MHD equilibrium Plasma and magnetic field shape determined by force balance using the MHD model.

MHD instability Large scale instability driven by plasma current and pressure gradient as predicted by the MHD model.

Michel parameters Three parameters, usually denoted by $\rho, \delta,$ and ξ, which are used to characterize the differential decay rates for muon decay ($\mu^- \longrightarrow e^- + \overline{\nu}_e + \nu_\mu$). Comparing these experimentally measured parameters with their values predicted from the standard model gives a good test of that theory.

Michelson interferometer A two-beam interferometer. Light is incident on a beam splitter, from where it passes through two different arms. The light is reflected back, and recombines at the original beam splitter. The recombined light beam can be observed with a detector. Depending on the length difference of the two optical paths, one observes constructive or destructive interference of the light. For an ideal light source and mirror surfaces, one observes a constant illumination of the output port. In other cases one observes interference rings or stripes.

If the mirror in one arm is moved, the relative phase between the two contributions to the interference signal changes, and one can observe interference fringes, i.e., a constant change between constructive and destructive interference. Since the principle of conservation of energy must not be violated, the intensity in the other exit of the interferometer (i.e., the path back to the light source) is always 180° out of phase with the other arm. The *Michelson interferometer* was first set up by the American physicist A.A. Michelson in order to measure the diameter of distant stars.

In the history of physics, the *Michelson interferometer* is of utmost importance. Michelson and Morley tried to prove the existence of a medium — the so-called ether — through which electromagnetic waves are propagating using a Michelson interferometer. Their measurements were negative, and only briefly thereafter, Albert Einstein drew the conclusions that ultimately led to the introduction of the theories of special and general relativity.

Currently, gravitational wave detectors based on a Michelson interfometric design are the most promising tools for the detection of gravitational waves predicted by the theory of general relativity.

Before the introduction of the Fabry–Perot interferometer, the *Michelson interferometer* was frequently used as a tool in spectroscopy. It can also be used to measure the temporal coherence of a light source.

It is also of great technical importance, since the flatness of surfaces relative to some reference surface can be tested by observing the structure of the interference fringes. They give a very sensitive contour diagram with a resolution of approximately $\lambda/2$.

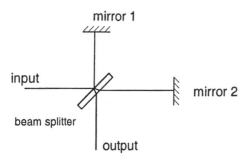

The Michelson interferometer.

microcanonical distribution For an isolated system in equilibrium, all of the states within a small energy band are equally probable and all of the other states have zero probability. An equilibrium distribution of this type is known as a *microcanonical distribution*.

micro-local analysis Let A be an operator in Hilbert space which represents a physical variable. In coordinate representation, its matrix element is $<x'|A|x''>$, where x' and x'' represent the coordinates of a particle or a particle assembly. In the latter case, x stands for a Fourier transformation for a set of coordinates $\{x_1, x_2, \ldots, x_n\}$. In the *micro-local analysis* we take the

$$A(x,p) = <x'|A|x''> \delta\left(x - (x'+x'')/2\right)$$
$$\exp\left[-ip(x'-x'')/\hbar\right] dx' dx''$$

which resembles a classical variable. The *micro-local analysis* is appropriate for semi-classical analysis.

micromaser A maser based on a microwave cavity with an extremely high Q-factor, i.e., photon lifetime, and an extremely small flux of atoms. The parameters are chosen such that typically only one atom at a time interacts with the radiation field.

Atoms are excited to a Rydberg state before entering the cavity. Transition frequencies between two of these Rydberg states are in the microwave region, and due to the long lifetime of Rydberg states, the interaction of a single atom with a single mode of the radiation field can be studied. In order to determine the interaction time of atoms and photons precisely, only atoms with a certain velocity are excited into the Rydberg states. This is facilitated using a Fizeau velocity selection or by making use of the Doppler effect in the excitation process. In order to increase the coupling between cavity modes and atoms, the storage time of the microwave photons within the cavity must be maximized. The cavities are therefore made from niobium, which is kept at cryogenic temperatures and becomes superconducting. This also reduces the background of thermal photons.

Atom–photon interactions are the basis for the Jaynes–Cummings model. Experimentally, many predictions of the Jaynes–Cummings model could be demonstrated with the *micromaser*. Examples are revival, photon trapping states, non-classical light, the maser threshold, power broadening, the Mollow triplet, etc. The first entanglement between atoms was also generated using a *micromaser* setup.

microstate For a given set of constraints (parameters of the thermodynamic system that can be held fixed or varied by some observer), the thermodynamic system still has access to a very large number of microscopic states or *microstates*. For example, for a gas of constant volume, the starting conditions (position and velocity) of the individual molecules could have many different values.

microsystem When evaluating state functions for large systems it is usually advantageous to divide the system conceptually into independent *microsystems,* each with its own set of energy states. For example, when considering the magnetic energy of a paramagnetic salt, the individual ions are considered as an independent *microsystem* in order to obtain the state function of the large system.

microturbulence Fluctuations with wavelengths much smaller than plasma macroscopic dimensions.

Miller indices A plane in direct lattice is specified by three numbers, known as *Miller indices*, in the following way: Choose a set of three convenient axes to describe the crystal with unit lengths a_1, a_2, and a_3 along them. Let the plane intercept these axes at $x_1 a_1$, $x_2 a_2$, and $x_3 a_3$. The *Miller indices* of the plane are the three integers h, k, and l which have no common factor and are inversely proportional to the intercepts of the plane on the axes, namely

$$h : k : l = 1/x_1 : 1/x_2 : 1/x_3 ,$$

Miller indices are enclosed by parentheses (hkl) and negative intercepts are denoted by a minus above the integer such as $(1\bar{3}1)$. Equivalent planes which are obtained by applying symmetry operations to the crystal are denoted by $\{hkl\}$.

The direction of a vector $\underline{r} = n_1 \underline{a}_1 + n_2 \underline{a}_2 + n_3 \underline{a}_3$ in direct lattice is denoted by $[n_1 n_2 n_3]$. A set of equivalent (by symmetry) directions are denoted by $< n_1 n_2 n_3 >$.

The vector $\underline{G} = h\underline{b}_1 + k\underline{b}_2 + l\underline{b}_3$ in reciprocal space is perpendicular to the plane (hkl) in a direct lattice. The spacing of the set of planes (hkl) is inversely proportional to $1/|\underline{G}|$. See Laue's condition method.

In cubic crystals, the crystal axes are three cube edges (forming a right-handed system) with a cube edge as a unit length.

minimal coupling A method of creating an interaction (a coupling) between matter particles and gauge fields which involves replacing the ordinary derivative in the Lagrange density via the covariant derivative. For example, a massless Dirac particle has a Lagrange density of $\overline{\psi}(i\hbar\gamma_\mu \partial^\mu)\psi$. If the ordinary derivative is replaced by the covariant derivative for electromagnetism (i.e., $\partial^\mu \longrightarrow \partial^\mu + ieA^\mu$, where e is the magnitude of the electron's charge and A^μ is the electromagnetic four-vector potential or the gauge field), this introduces a new term into the Lagrange density ($\overline{\psi}$(i.e., $A^\mu)\psi$) which couples the Dirac matter particle with the gauge field.

minimum uncertainty The smallest possible uncertainty in the measurement of two conjugate variables in quantum mechanics. It is given by Heisenberg's uncertainty relation. Generally one finds

$$\Delta A \Delta B \geq \hbar/2 ,$$

where A and B are two conjugate observables, i.e., observables which do not commute: $[A, B] = AB - AB \neq 0$.

Mirnov oscillations Magnetic perturbations detected around the edge of toroidal magnetic confinement devices such as tokamaks.

mirror matter A hypothetical form of matter where every known particle (electron, proton, photon, etc.) has a mirror partner (mirror-electron, mirror-proton, mirror-photon, etc.), which has the same mass, but which interacts via forces which are mirrors of the standard model interactions (e.g., ordinary matter will interact via the electromagnetic interaction, while mirror matter interacts via mirror-electromagnetism). Only the gravitational force operates the same on both matter and mirror-matter. As a result, matter interacts very weakly with mirror-matter.

mirror ratio Ratio of maximum to minimum magnetic field strength along a field line in a magnetically confined plasma.

MIT bag model A phenomenological model for hadrons, where the quarks which constitute the hadron are assumed to be confined within a cavity or bag. Usually, this region in which the quarks are free to move is spherical in shape. The bag model is motivated by an analogy to the Meissner effect in superconductivity: the QCD vacuum outside of the bag is said to expel the color electric field in a manner analogous to the way superconductors expel magnetic fields. Because of this hypothesized expulsion of the color electric field, the quarks remain confined within the bag.

mixed state Or statistical states. States in which pure states are superposed with a probability distribution. The simplest pure state or definite quantum states can be written as a superposition

$$|\Psi(\theta)\rangle = \sum c_n(\theta)|\psi_n\rangle .$$

The wave function for a *mixed state* can then be written as the superposition of different pure states

$$|\Psi_{\text{mix}}\rangle = \int p(\theta)|\Psi(\theta)\rangle \, d\theta ,$$

where $p(\theta)$ is the probability distribution. For *mixed states*, the average values $\langle \hat{O} \rangle$ for an observable $\hat{O}$ are given by

$$\langle \hat{O} \rangle = \int p(\theta)\langle\Psi(\theta)|\hat{O}|\Psi(\theta)\rangle \, d\theta .$$

A completely *mixed state* is represented in the density matrix picture by off-diagonal elements with the value zero.

mixing angle, Weinberg In the $SU(2) \times U(1)$ standard model, an angle, denoted by θ_W, which parameterizes the particular admixtures of the third component of the original $SU(2)$ gauge boson, W_μ^3, and the weak hypercharge gauge boson, B_μ, which make up the electromagnetic field ($A_\mu = \cos\theta_W B_\mu + \sin\theta_W W_\mu^3$) and the field of the Z-boson ($Z_\mu = -\sin\theta_W B_\mu + \cos\theta_W W_\mu^3$).

mixing length In a turbulent flow, the distance traveled by a fluid parcel before losing its momentum. Generally used as a simple analysis in turbulence.

mixing length estimate Estimate for nonlinear saturation of micro-instabilities in which the density perturbation becomes comparable to the background density gradient times the wavelength.

mobility Drift *mobility* is $|q\tau/m|$, where q is the electric charge on the particle, m is its mass (or effective mass), and τ is the average time (relaxation time) between collisions. An isotropic medium is assumed. It is also the ratio of the magnitude of the drift velocity to the magnitude of the electric field (for weak fields).

mobility edge In disordered systems, electron states can be localized or free. A *mobility edge* Ec is an energy value below which a state is localized and above which a state is free (conducting). If the Fermi level lies below E_c, conduction takes place by hopping and the conductivity is low, and if it lies above E_c, we have ordinary conduction. A manipulation of the location of the Fermi level, if possible by external means, can bring about a metal-insulator transition.

mode An eigenstate of the electromagnetic field in a resonator or wave guide. The *mode* is characterized by a wavelength and the spatial distribution of the light. In a resonator, the transverse *modes* are given by the condition that an integer number of half-waves will fit in the resonator (standing wave) or an integer number of wavelengths will fit in the resonator (ring resonator). The lowest order spatial *mode* for a resonator is a Gaussian beam, in which the transverse intensity distribution falls off like a Gaussian function. Parameters characterizing a Gaussian beam are the smallest beam waist ω_0 and the radius of curvature of the wave front. In a wave guide, the lowest order spatial *modes* are Hermite functions.

mode competition In a laser, the mechanism which determines the longitudinal mode characteristics of a laser. When many longitudinal modes are within the gain profile of the laser modes, modes which are populated by spontaneous emission first start to oscillate, receive more of the gain by means of stimulated emission, and grow stronger at the expense of other modes which either never start to oscillate or stop oscillating. In pulsed lasers, this behavior can be used to produce single longitudinal mode output. This technique is referred to as injection seeding. One single longitudinal mode is prepopulated by a weak continuous wave seeding laser. Upon Q-switching the cavity, this one mode will immediately start to oscillate while others would have to build up from the vacuum fluctuations. Due to this *mode competition*, the single mode will be the only one to oscillate. One requirement for this technique to work is that the seed laser is resonant with the pulsed slave laser cavity. Several schemes are reported in the literature to achieve this resonance.

mode degeneracy Refers to the possibility that modes in a resonator or cavity can have the same energy, i.e., resonant frequency. This includes the degeneracy in polarization or the transverse distribution of the intensity. In a spherical resonator, the frequencies of the modes can be found via the condition that the phase of the waves must change by an integer multiple of π for one round trip. This results in the following frequencies for the modes in standing wave cavities with spherical mirrors with radius R_1 and R_2:

$$\nu_j = \frac{c}{2d}\left(n + (n+m+1)\frac{\cos^{-1}\pm\sqrt{g_1 g_2}}{\pi}\right)$$

where $g_{1,2} = 1 - \frac{d}{R_{1,2}}$ and d is the separation of the two mirrors. n are integer numbers characterizing the longitudinal modes with a separation of $c/2d$, and m and l characterize the Gaussian–Hermite transverse eigenmodes of the cavity. The factor $\frac{\cos^{-1}\pm\sqrt{g_1 g_2}}{\pi}$ is called the Guoy phase and takes on the following values for the most important cavity configurations:

$$\frac{\cos^{-1}\pm\sqrt{g_1 g_2}}{\pi} \approx \begin{cases} 0 & \text{planar cavity} \quad g_1 = g_2 = 1 \\ \frac{1}{2} & \text{confocal cavity} \quad g_1 = g_2 = 0 \\ 1 & \text{concentric cavity} \quad g_1 = g_2 = -1 \end{cases}$$

One sees that in the case of the confocal etalon, modes characterized by the integers (n,m,l) are partly degenerate: modes (n, m, l) are degenerate for the same $k = m + l$. Furthermore, those modes for which $m + l$ is an even integer are degenerate with the modes (n+m+l,0,0), while modes (n, m', l'), for which $l' + m' =$ odd integer, fall exactly halfway between the modes with $(n + m' + l' - 1)$ and $(n + m' + l')$ causing a mode spacing of $c/4d$. Due to this *mode degeneracy* an exact mode matching of laser radiation to a confocal Fabry–Perot etalon is not necessary. This has the disadvantage, however, that the free spectral range is reduced to $c/4d$.

mode locking A technique to produce light pulses in the picosecond and femtosecond regime. Phase locking of different longitudinal modes can be regarded as the time expansion of a Fourier series, which, in the time domain, results in light pulses. The frequency distance $\nu_f = c/2L$ between adjacent longitudinal modes phase-locked together, where c is the speed of light and L is the length of the resonator, leads to a pulse train with separation $2L/c$ and where individual pulses have a width of

$$\frac{1}{M\nu_f},$$

where M is the number of modes which are phase-locked.

Experimentally, this phase locking can be achieved by placing an acousto-optic or electro-optic modulator inside the cavity, which is modulated at the free spectral range $c/2L$ of the cavity. Other techniques include placing a saturable absorber inside the cavity. The colliding pulse modulation or CPM is based on the latter method.

mode mismatch The mismatch in spatial profile or frequency of a light beam with respect to the eigenmodes of a resonator or wave guide. Mode matching of laser beams is important in many applications, for instance laser resonators, laser design, build-up cavity for the enhancement of non-linear processes, and coupling to optical fibers.

mode pulling The frequency shift of a laser mode due to a mismatch between the maximum of the gain profile and the longitudinal cavity modes. The sharper the resonator modes, the less severe the *mode pulling*, and the sharper the gain profile, the stronger the frequency pulling. *Mode pulling* also occurs for pulsed injection seeded laser systems in the nanosecond regime. If the slave cavity of the pulsed laser is not perfectly in resonance with the seed laser, a frequency chirp on the pulsed output will be measured that will pull the laser frequency towards the output frequency of the slave cavity.

mode rational surface Magnetic surface in a toroidal magnetic confinement device on which magnetic field lines close on themselves with the same topology as a helical mode of plasma oscillation.

modulation The controlled change of a parameter of the electromagnetic field for the purpose of communication. One distinguishes frequency (FM) and amplitude (AM) *modulation*. In the former, the frequency of a signal is mod-

molasses The arrangement of six laser beams in a three-dimensional arrangement similar to the setup of a magneto-optical trap. However, a magnetic field is not present. The arrangement of the laser beams leads to a velocity-dependent force on the atoms and, consequently, to diffusive motion of the atoms.

mole The amount of substance containing the number of ions, atoms, or molecules, etc. to equal the number of atoms in 12 grams of Carbon 12; SI unit is mol.

molecular beam Generally consists of a directed beam of non-ionized atoms or molecules emerging from a source whose momentum depends solely upon their thermal energy. For a beam of ideal gas atoms at thermal equilibrium, the flux of particles is given by $\frac{1}{4}n\langle c \rangle$, where n is the number density of gas atoms and $\langle c \rangle$ is the mean velocity of the beam assuming a Maxwell–Boltzmann distribution of velocities.

molecular crystals Crystals made from atoms such as Ar, Kr, Ne, and Xe or molecules such as H_2, and N_2, where the atoms or molecules are weakly affected by the formation of the crystal. The binding forces are weak.

molecular dynamics Field which studies the energy flow in molecules after excitation with short light pulses. According to the Born–Oppenheimer approximation which separates the different motions, i.e., rotations, vibrational and electronic, no perturbations with dark states should be allowed. Dark states are defined as background states which are not optically active, formed by rovibrational states in the same or different electronic states. In real molecules, the interaction between bright and dark states does occur. This leads to a flow of energy deposited in molecules into these background states. The possible mechanisms are intramolecular vibrational relaxation (IVR), intersystem crossing (ISC), or internal conversion (IC). One can distinguish three cases: the small, large, and intermediate molecule. For small molecules, the density of states is small and no perturbations are observed; the fluorescence yield, i.e., the ratio of radiative decays to total decays, is one. In the case of the large molecule, which is also called the statistical case, the density of states is so large that the mean separation of states ε is larger than their decay rates Γ_d, such that the states form a quasi-continuum. This leads, after excitation of the Born–Oppenheimer states, to an irreversible energy flow (dissipation) into the background states. The non-radiative decay leads to a reduction in the fluorescence yield and to exponential decays on a much smaller timescale than observed for the small molecule case. Quantum mechanically, this decay can be explained by the dephasing of the different states. A recurrence cannot be observed, due to the irreversible energy flow into the background states.

Finally, in the intermediate case, coupling elements are of the same order of magnitude as the energy separations. The recurrence can be observed as deviations from an exponential decay. In the case of coherent excitations it becomes possible to observe phenomena such as quantum beats and biexponential decays. The intermediate case is particularly interesting since it can be investigated using quantum beat spectroscopy, which is a quasi Doppler-free technique with very high relative frequency resolution.

Typical tools for the investigation of *molecular dynamics* are high resolution, quantum beat and pump-probe femtosecond spectroscopes.

molecular dynamics method In the *molecular dynamics* computational *method,* the distribution of molecular configurations is calculated directly. The molecules are given some initial configuration and each has some definite speed and trajectory. Using an assumed force law (based on the assumed intermolecular potentials), the subsequent trajectory of every molecule is then calculated. A record of the molecular trajectories is kept, and by averaging these over time, it is possible to calculate the equation of state. Using this method it is also possible to calculate non-equilibrium properties such as viscosity or thermal conductivity.

molecular field P. Weiss proposed the idea of a *molecular field* (or a mean field) which acts on the magnetic moments of a ferromagnet in addition to the external field $\underline{B}_a$. The mean field is assumed to be proportional to the magnetization $\underline{M}$. The effective field is thus $\underline{B}_a + \lambda \underline{M}$, where λ is a constant. This hypothesis explains the spontaneous magnetization (when the applied field is zero) at low temperature. It also gives the Curie–Weiss law, $\chi = C/(T - T_c)$, where χ is the paramagnetic susceptibility well above the Curie temperature T_c, C is a constant, and T is the temperature. Heisenberg replaced the mean field by the exchange interaction to align the spins.

molecular heat capacity The heat capacity of a molecule that arises from all of the individual contributions to its internal energy. For a free diatomic molecule, this energy typically comprises contributions from the translational, rotational, and vibrational energies of the molecule. For example, CO at room temperature has a *molecular heat capacity* of 5/2 kB, which arises from a translational energy contribution of 3/2 kB and an unquenched rotational energy contribution of k_B. At room temperature, the vibrational contribution to the *molecular heat capacity* is quenched and only becomes significant for temperatures above 1000 K, whereupon the *molecular heat capacity* approaches 7/2 k_B.

Moller scattering The process in which two initial electrons scatter from one another into a final state of two electrons. This process is written as $e^- + e^- \longrightarrow e^- + e^-$.

Mollow spectrum The three-peaked emission spectrum of a coherently driven two-level atom in the strong field limit. The occurrence of the *Mollow spectrum* can be explained in the dressed state picture. The spectrum consists of three peaks, formed by four contributions. Assuming resonant excitation of the laser, the fluorescence intensity as a function of frequency ω yields:

$$I(\omega) = 2\pi \left(\frac{\Gamma}{\Omega}\right)^2 \delta\omega - \omega_0 + \frac{\Gamma/2}{(\omega - \omega_0)^2 + \Gamma^2}$$
$$+ \frac{3\Gamma/8}{(\omega - \omega_0 - \Omega)^2 + (3\Gamma/2)^2}$$
$$+ \frac{3\Gamma/8}{(\omega - \omega_0 + \Omega)^2 + (3\Gamma/2)^2},$$

where Γ is the line width of the transition at a central frequency ω_0. Ω is the Rabi frequency. The first contribution is due to the elastic scattering of light, which is dominant for a weak excitation ($\Omega \ll \Gamma$), but negligible for a strong excitation. The other three terms are due to incoherent scattering. Two smaller peaks surround a larger central peak located at the atomic resonance frequency. The frequency separation between the central and the outer peaks is given by the Rabi frequency. For resonant excitation, the area ratios of the lines is given by 1:2:1, whereas the peak ratio is given by 1:3:1. For non-resonant excitation one still finds a three-peaked spectrum around the laser frequency. However, the ratio of the main peak to the sideband becomes smaller.

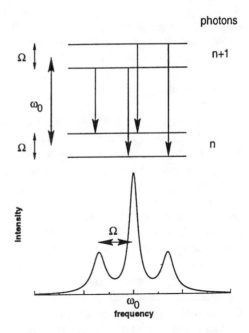

Illustration of the Mollow triplet in the resonance fluorescence of a two-level atom with the help of the dressed atom picture for excitation at the frequency ω_0.

moment equations Fluid equations derived by multiplying a plasma kinetic equation by powers of particle velocity and integrating over all velocities.

momentum equation See Navier–Stokes equations.

momentum integral In a boundary layer, the integral defining a length scale based upon the loss of momentum due to the boundary layer. The momentum thickness is given by

$$\theta = \int_0^\infty \frac{u}{U}(1 - \frac{u}{U})\,dy\,.$$

momentum representation Choose eigenfunctions of momentum as an orthonormal set of vectors in Hilbert space to represent quantum states and quantum variables. Such representation is called the *momentum representation*. The momentum eigenfunctions are simply plane waves.

monochromatic radiation Radiation that contains only the light of one frequency. It can be described by the function $E(t) = E_0 e^{i\omega t}$, where ω is the frequency of the light and E_0 is the field amplitude.

monoclinic lattice A Bravais lattice generated by the primitive translations $\underline{a}_1, \underline{a}_2,$ and $\underline{a}_3$ (whose lengths are a, b, and c respectively). $\underline{a}_3$ is perpendicular to $\underline{a}_1$ and $\underline{a}_2$, but $\underline{a}_1$ is not perpendicular to $\underline{a}_2$ and $a \neq b \neq c$.

monte-carlo method This computational method generates a sequence of configurations of the thermodynamic system (typically a set of atoms or molecules arranged in space) over which equilibrium properties can be averaged. The molecules are started in some initial configuration and are then moved sequentially according to the following rule. If the calculated change in potential energy (ΔE) of the system is negative, then the configurational change of the system is allowed to occur automatically. If the associated potential energy change is positive, however, then the computer is programmed to allow the molecule to move with a probability of $\exp(-\Delta E / k_B T)$. Thus, the system will reach statistical equilibrium when the probability of each configuration is the required Boltzmann probability. The advantage of this technique over a molecular dynamics simulation is that it reaches equilibrium faster, but it cannot be used to calculate non-equilibrium system properties.

Moody chart Plot of the Colebrook pipe friction formula for various surface roughnesses as a function of the Reynolds number for turbulent flow in a pipe.

MOSFET Metal oxide semiconductor field effect transistor.

Mössbauer effect (1) Also called recoil-free gamma-ray resonance absorption. Nuclear process permitting the resonance absorption of gamma rays. It is made possible by fixing atomic nuclei in the lattice of solids so that energy is not lost in recoil during the emission and absorption of radiation. The process, discovered by the German-born physicist Rudolf L. Mössbauer in 1957, constitutes a useful tool for studying diverse scientific phenomena.

In order to understand the basis of the *Mössbauer effect*, it is necessary to understand several fundamental principles. The first of these is the Doppler shift. When a locomotive whistles, the frequency, or pitch, of the sound waves increases as the whistle approaches a listener and decreases as the whistle recedes. The Doppler formula expresses this change, or shift in frequency, of the waves as a linear function of the velocity of the locomotive. Similarly, when the nucleus of an atom radiates electromagnetic energy in the form of a wave packet known as a gamma-ray photon, it is also subject to the Doppler shift. The frequency change, which is perceived as an energy change, depends on how fast the nucleus is moving with respect to the observer.

Finally, it is necessary to understand the principles governing the absorption of gamma rays by nuclei. Nuclei can exist only in certain definite energy states. For a gamma ray to be absorbed, its energy must be exactly equal to the difference between two of these states. Such an absorption is called resonance absorption. A gamma ray that is ejected from a nucleus in a free atom cannot be resonantly absorbed by a similar nucleus in another atom because its energy is less than the resonance energy by an amount equal to the kinetic energy given to the recoiling source nucleus.

(2) The phenomenon where a nucleus within a crystal lattice undergoes a transition between energy states and emits a high energy photon (usually a γ-ray photon) without significantly recoiling. This nearly recoilless emission by the nucleus is possible because the entire lattice takes up the recoil momentum, so that the nucleus that emits the photon only recoils an infinitesimal amount. The photons which occur in the *Mössbauer effect* are extremely sharply peaked in energy and frequency.

MOT *See* magneto-optical trap.

motor A machine designed to convert energy into the mechanical form from some other form. For example, an electrical motor converts electrical energy to mechanical energy, whereas a chemical motor converts stored chemical energy into mechanical energy.

motor generator A device for converting electrical energy at one particular voltage and frequency (or number of phases) to another voltage and frequency (or number of phases). Consists of an electrical motor and generator that are mechanically coupled.

Mott scattering The electromagnetic scattering of electrons from heavy nuclei. The nuclei are treated as point positive charges, and are assumed to be heavy enough that their recoil from the collision with the electron can be ignored. In the limit in which the electron is moving at non-relativistic speeds, *Mott scattering* becomes Rutherford scattering. *See* Rutherford scattering.

Mott scattering formula The formula for the differential scattering cross-section for identical charged particles due to a Coulomb force, and the formula for the scattering cross-section for a relativistic electron by a Coulomb potential field.

MS renormalization The minimal subtraction renormalization scheme is a specific method for dealing with the infinities that occur in higher order radiative corrections to physical processes. In this scheme, one only subtracts the infinite terms that arise in the calculation of the radiative corrections.

Mueller matrix A 4×4 matrix which fully describes the attenuation and polarizing properties of a medium such as polarizers or scattering media. The polarization state of the incoming light is described with the help of the Stokes vector, which is a vector with four components having the following meaning. The first component gives the intensity of the light; the second component is the difference between intensities of the horizontal and vertical polarizations of the beam; the third is the difference of the intensities as measured after polarizers oriented at $\pm 45°$; the last component is the intensity differences with respect to left and right circular polarizations of the beam.

In general, the components of the Stokes vector are normalized with respect to the total intensity, such that the first component has the value of one. For certain polarization states, the following Stokes vectors can be found:

$$\text{unpolarized} \quad \begin{pmatrix} 1 \\ 0 \\ 0 \\ 0 \end{pmatrix}$$

$$\text{horizontally polarized} \quad \begin{pmatrix} 1 \\ 1 \\ 0 \\ 0 \end{pmatrix}$$

$$\text{vertically polarized} \quad \begin{pmatrix} 1 \\ -1 \\ 0 \\ 0 \end{pmatrix}$$

$$\text{polarized at } 45° \quad \begin{pmatrix} 1 \\ 0 \\ 1 \\ 0 \end{pmatrix}$$

$$\text{polarized at } -45° \quad \begin{pmatrix} 1 \\ 0 \\ -1 \\ 0 \end{pmatrix}$$

$$\text{right circular polarized} \quad \begin{pmatrix} 1 \\ 0 \\ 0 \\ 1 \end{pmatrix}$$

left circular polarized $\begin{pmatrix} 1 \\ 0 \\ 0 \\ -1 \end{pmatrix}$

Any polarizer or scattering medium can now be described as a *Mueller matrix* M such that incident light described by a Stokes vector S will be transformed to a state $S' = MS$ after passage through the medium. Examples of such *Mueller matrices* are:

ideal horizontal polarizer $\frac{1}{2}\begin{pmatrix} 1 & 1 & 0 & 0 \\ 1 & 1 & 0 & 0 \\ 0 & 0 & 0 & 0 \\ 0 & 0 & 0 & 0 \end{pmatrix}$

ideal vertical polarizer $\frac{1}{2}\begin{pmatrix} 1 & -1 & 0 & 0 \\ -1 & 1 & 0 & 0 \\ 0 & 0 & 0 & 0 \\ 0 & 0 & 0 & 0 \end{pmatrix}$

ideal polarizer at 45° $\frac{1}{2}\begin{pmatrix} 1 & 0 & 1 & 0 \\ 0 & 0 & 0 & 0 \\ 1 & 0 & 1 & 0 \\ 0 & 0 & 0 & 0 \end{pmatrix}$

ideal $\lambda/4$ plate $\begin{pmatrix} 1 & 0 & 0 & 0 \\ 0 & 1 & 0 & 0 \\ 0 & 0 & 0 & -1 \\ 0 & 0 & 1 & 0 \end{pmatrix}$.

muffin tin potential   The crystal potential which an electron sees is often approximated by nonoverlapping potentials centered at the equilibrium positions of the ions.

multi-photon transition   Transition caused by a multi-photon process, i.e., by absorption or emission of two or more photons. Multi-photon processes can occur through virtual levels. Selection rules for these transitions are different from one-photon transitions. The transition probability for two-photon transitions from state $|i\rangle$ to state $|f\rangle$ can be found, by second order perturbation theory, to be proportional to the intensity I:

$$P_{if} = I^2 \sum_k \frac{\hat{r}|k\rangle\langle k|e\hat{r}|f\rangle|^2}{(\omega_{ki} - \omega_{fi})^2},$$

where the sum extends over all real levels $|k\rangle$. In the case of a near resonance transition, an additional damping term must be included, which prevents the sum from blowing up.

There are three basic types of *multi-photon transitions:* In a case where the initial level is higher in energy than the final level, one speaks of multi-photon emission. In the opposite case, one speaks of multi-photon absorption. In either case, the energy of the photons adds up to result in the energy difference of the atomic final and initial levels.

In the third process type, the Raman process, emission and absorption events are combined. Most common are two-photon Raman processes where one photon at one frequency is annihilated and another one is created.

multiplet   (1) A collection of relatively closely spaced energy levels, their splitting from a single energy level caused by a weak interaction. Examples are spin-orbit *multiplets* in the electronic states in atoms and isospin *multiplets* in nuclear level structures.

(2) A collection of hadrons grouped into some representation of a mathematical group. For example, the SU(3) flavor quark model contains the up, down, and strange quarks in the fundamental representation, $\mathbf{3} = (u, d, s)$. A baryon is formed from various combinations of three of these quarks, which mathematically can be represented via the tensor product $\mathbf{3} \otimes \mathbf{3} \otimes \mathbf{3}$. This tensor product can be decomposed into the direct product as $\mathbf{10} \oplus \mathbf{8} \oplus \mathbf{8} \oplus \mathbf{1}$, where $\mathbf{10}$ is the decuplet representation of SU(3), $\mathbf{8}$ is the octet representation of SU(3), and $\mathbf{1}$ is the singlet representation of SU(3). Each of these representations contains a number of particles equivalent to the number of the representation. *See* octet; nonets.

multipole expansion   The expansion of the interaction Hamiltonian of an atom with light to higher order terms. These higher order terms represent changes of the electromagnetic field across the dimension of the atom. The relevant expansion yields

$$e^{i\vec{k}\vec{r}} = 1 + i\vec{k}\vec{r} + \frac{1}{2}\left(i\vec{k}\vec{r}\right)^2 + \cdots$$

to a higher order than the zeroth order. The latter is called the dipole approximation. The higher order terms correspond to magnetic dipole, electric quadrupole, etc. transitions and have much lower probabilities than electric dipole-allowed transitions.

multipole selection rules   Govern the higher order transitions possible due to higher order terms in the operator governing the interaction of an electromagnetic wave and an atom. Most prominent are the magnetic dipole and electric quadrupole transitions. Their selection rules are given by

$$\text{magnetic dipole} \Delta L = 0$$
$$\Delta J = 0, \pm 1$$
$$\Delta m_J = 0, \pm 1$$
$$\text{electric quadrupole} \Delta L = \pm 2$$
$$\Delta m = 0, \pm 1, \pm 2 \ .$$

muon   An apparently fundamental particle which carries an electric charge equal to that of the electron and has a spin of 1/2, making it a fermion. Except for its mass, which is roughly 200 times that of the electron, the *muon* appears to be similar to the electron. The *muon* decays with a lifetime of approximately 2.2×10^{-6} seconds, and it decays predominantly into an electron, an anti-electron neutrino, and a *muon* neutrino.

muonium   A system consisting of a positively charged antimuon bound to an electron. This system is similar to the hydrogen atom except the proton is replaced by the antimuon. This system is a good test system for the accuracy of QED since the electron and antimuon do not interact via the strong interaction.

muon neutrino   A neutral spin 1/2 particle which, together with the charged muon, forms the second family or second generation of leptonic matter particles. *See* neutrino.

N

Nambu–Goldstone boson A massless boson that arises in theories where a global symmetry is broken. If a theory initially possesses some global symmetry group, G, which has N generators, and if this original symmetry is broken down to a global symmetry subgroup, H, which has M generators, then the final theory will contain $N-M$ massless bosons which are called *Nambu–Goldstone bosons*.

nanostructure Structure of condensed matter with its smallest dimensions approximately 10 to 100 Å, such as fine wires, thin films, and dots. These usually display quantum mechanical effects vividly.

natural line width Linewidth of a transition in the absence of any broadening effects such as Doppler broadening or collisional broadening. The line width of a transition is linked to the lifetime τ of the upper state by

$$\Delta \nu = \frac{1}{2\pi \tau},$$

where the line width is in frequency units of Hz.

natural variables The thermodynamic potentials (internal energy, enthalpy, Helmholtz, and Gibb's free energies), when given in their differential forms, are expressed in terms of their *natural variables*. If the thermodynamic potential is known as a function of its natural variables, it is possible to immediately obtain all of the other significant variables. For example, consider the Helmholtz free energy, F, whose differential form is given by

$$dF = -S\,dT - P\,dV.$$

For the Helmholtz free energy, therefore, the natural variables are T and V, and if F is known as a function of these variables (i.e., $F(T, V)$), then it is immediately possible to calculate the remaining significant variables S and P since $S = -\left(\frac{\partial F}{\partial T}\right)_V$ and $P = -\left(\frac{\partial F}{\partial V}\right)_T$.

Navier–Stokes equations (1) Fluid equations for momentum including viscosity.

(2) Formulation of the conservation of momentum equation applied to a Newtonian fluid. In a general form, this can be written as a second order, non-linear, partial differential equation of the form

$$\rho \frac{D\mathbf{u}}{Dt} = \text{div}\, \overleftrightarrow{\sigma} + \rho \mathbf{g}$$

where σ is the stress tensor. For an incompressible flow in which variations in viscosity can be neglected, this is simplified to

$$\frac{\partial \mathbf{u}}{\partial t} + \mathbf{u} \cdot \nabla \mathbf{u} = -\frac{1}{\rho}\nabla p + \mathbf{g} + \nu \nabla^2 \mathbf{u}.$$

The *Navier–Stokes equations* may be non-dimensionalized as follows. Take the incompressible unsteady momentum equation (neglecting body forces),

$$\frac{\partial}{\partial t}\mathbf{u} + (\mathbf{u} \cdot \nabla)\mathbf{u} = -\frac{1}{\rho}\nabla p + \nu \nabla^2 \mathbf{u}.$$

Each variable should be non-dimensionalized. Use $t^* = t/t_o = tU/L$, $r^* = r/L$, $u^* = u/U$, and $p* = p/\rho U^2$ to obtain the following equation (where the $*$ variables are the non-dimensional ones):

$$\frac{\partial}{\partial t^*}\mathbf{u}^* + \left(\mathbf{u}^* \cdot \nabla^*\right)\mathbf{u}^* = -\nabla^* p^* + \frac{\nu}{UL}\nabla^{*2}\mathbf{u}^*$$

or

$$\frac{\partial}{\partial t^*}\mathbf{u}^* + \left(\mathbf{u}^* \cdot \nabla^*\right)\mathbf{u}^* = -\nabla^* p^* + \frac{1}{\text{Re}}\nabla^{*2}\mathbf{u}^*.$$

Thus, the equation is wholly dimensionless with the dimensionless parameter Re. (This is another way to arrive at the definition of the Reynolds number.) If Re is high, $\text{Re} \gg 1$, the equation is reduced to

$$\frac{\partial}{\partial t^*}\mathbf{u}^* + \left(\mathbf{u}^* \cdot \nabla^*\right)\mathbf{u}^* = -\nabla^* p^*$$

which significantly simplifies the analysis of the equation. This may also be done with a low Re to arrive at the creeping flow equation. With the proper selection of dimensionless relations, any equation can be non-dimensionalized in this manner.

nearly free electrons (NFE) In general, the *Navier–Stokes equations* are intractable and can only be solved exactly for a few specific cases. This requires approximations by simplification or other methods to obtain solutions for most flows of interest.

nearly free electrons (NFE) A method for calculating the electron energies in simple metals for electron wave vectors satisfying the Bragg reflection condition. In the first approximation, electrons are described by plane waves $\exp(i\underline{k} \cdot \underline{r})$, and the effect of the crystal potential is treated by standard (nondegenerate) perturbation theory which couples electron states $\underline{k}$ and $\underline{k}'$ that differ by a reciprocal lattice vector G. Near the Brillouin zone, boundaries $\underline{k}$ and $\underline{k}'$ are nearly equal in magnitude and so are their unperturbed energies ($E = \hbar^2 k^2 / 2m$), and degenerate perturbation theory must be used. Wave functions using only those plane waves which are nearly Bragg reflected are constructed, and energy solutions are found which exhibit energy discontinuities at the Bragg reflection.

For example, assume attractive (negative) symmetric potential wells are located at $x = 0, \pm a, \pm 2a \ldots$ of a one-dimensional crystal. The Fourier coefficients of this potential $V(2\pi n/a)$ are all real and negative. At $k = \pi/a$, we have a Bragg reflection with $k' = -\frac{\pi}{a}$ and $G = -2\pi/a$. The linear combinations $\cos \frac{\pi}{a} x$ and $\sin \frac{\pi}{a} x$ diagonalize the energy operator (Hamiltonian) and we obtain the two energies $E_1 = E_0 - |V(G)|$ and $E_2 = E_0 + |V(G)|$ with $E_0 = (h^2 \pi^2 / a^2 m) + V(0)$ showing an energy gap of $2|V(G)|$ in the free electron spectrum.

Néel temperature The transition temperature T_N below which a paramagnetic state becomes an ordered antiferromagnetic state. For $T > T_N$ the magnetic susceptibility $\chi = C/(T + \theta)$ where C and θ are constants with θ of the order of T_N. *See* ferromagnetism.

negative energy states According to P.A.M. Dirac, the relativistic equation of motion for electrons has negative energy states. The Klein paradox claims that electrons with positive energies could penetrate into negative energy states if these states were empty. To solve this paradox, Dirac concluded that the negative energy states are filled by electrons. The positive and negative energy states are separated by an energy gap of twice the rest mass energy. The gap is about 1,000 Mev. A gamma ray photon with energies larger than this may excite a negative energy electron to a positive energy state. Thus, the gamma ray creates an electron–hole pair from a vacuum. The hole is called the positron. Its existence proved the Dirac theory to be valid.

negative energy wave Wave for which the derivative of the dielectric function with respect to frequency is negative. This typically occurs in an inhomogeneous plasma with non-Maxwellian particle distribution. Under some conditions, the three-wave interaction between negative and positive energy waves results in explosive instability.

negative resistance or conductivity Regions where the derivative of the electric current density $\underline{J}$ with respect to the electric field $\underline{E}$ is negative are called *negative conductivity* or resistivity regions. A plot of J versus E starts as linear for a small value of E (Ohm's law), then rises to an infinite slope and turns back in one S shape or more, giving a *negative resistance* region of type S or type a. Alternatively, J could rise gradually to a maximum, drop to a minimum, then rise again giving *negative resistance* of type N (or type b). N-type *negative resistance* is realized in the tunnel diode of Esaki, which is a heavily doped (degenerate) p–n junction, and in the Gunn effect where carriers, when heated, are transferred from one band to another band of slightly higher energy but lower mobility. These *negative resistance* regions show instabilities which can be exploited to build oscillators or amplifiers. Obtaining *negative resistance* regions for carriers in a single energy band is unlikely unless one can invent unusual scattering mechanisms.

negative temperature Effective *negative temperatures* are said to occur for systems with a finite number of energy levels for which population inversion has been achieved, i.e., there are more particles in the upper energy states than in the lower ones. For a finite energy level system obeying Boltzmann statistics, *negative temperatures* correspond, in an abstract sense, to effective temperatures greater than infinity.

NEL steam tables   Provide data on the properties of water in its liquid and vapor phases for designers using steam, for example, as a working fluid in heat engines.

neoclassical transport   Collisional transport in toroidal magnetic confinement devices that is enhanced by the fact that some of the charged particles are trapped in local magnetic mirrors. The drift orbits of these trapped particles are called banana orbits, which deviate further from magnetic surfaces than the orbits of those particles that are not trapped.

Nernst effect   The potential difference created by a temperature gradient along an electrical conductor or semiconductor placed in a perpendicular magnetic field. The potential difference is generated in the direction that is perpendicular to both the magnetic field and temperature gradient directions. This effect is actually the analog of the Hall effect and was discovered by Walter Nernst in 1886.

Nernst equation   The zero-current cell potential, E, for an electrochemical cell comprised of a cell reaction with a reaction quotient, Q, (in terms of the activities of the participating species, a_J, and corresponding stoicheometries, ν_J)

$$Q = \prod_J a_J^{\nu_J}$$

is given by:

$$E = E° - \frac{RT}{\nu F} \ln Q$$

where $E°$ is the standard cell potential, R is the ideal gas constant, T is the cell temperature, F is Faraday's constant (charge per mole of electrons), and ν is the number of electrons transferred in the cell reaction.

Nernst heat theorem   No change in entropy takes place when a chemical change occurs between two crystalline materials at absolute zero. A more restricted statement of the third law of thermodynamics.

Nernst theorem   An explanation for the origin of electrode potentials, based upon the hypothesis that a dynamic equilibrium is established between ions moving into solution from the electrode and ions depositing onto the electrode from solution.

neutral beam injection   Heating, fueling, current, and momentum driven in magnetically confined plasmas by the injection of high energy neutral atoms which pass freely through the magnetic field before they are captured within the plasma.

neutral-current interaction   The portion of the weak interaction which is mediated by the exchange of the electrically neutral Z^o boson. The effective Lagrangian for these processes can be written as $\Delta \mathcal{L}_Z = \frac{g^2}{m_Z^2} J_Z^\mu J_{Z\mu}$, where g is the $SU(2)$ coupling constant and m_Z is the mass of the Z^o boson. The neutral current is $J_Z^\mu = \sum_f \overline{f} \gamma^\mu (T^3 - \sin \theta_W^2 Q) f$, where f is the spinor associated with a particular flavor of quark or lepton, T^3 is the 3rd $SU(2)$ generator, Q is the $U(1)$ generator associated with electric charge, and θ_W is the Weinberg angle.

neutralino   Hypothetical neutral particle that arises within supersymmetric models. The supersymmetric partners of the electroweak gauge bosons and the Higgs bosons (called gauginos and Higgsinos respectively) can mix, forming neutral, linear combinations called *neutralinos*.

neutrino   Any of three neutral spin 1/2 particles thought to be fundamental matter particles in nature. The *neutrino* comes in three flavors or types: the electron neutrino, the muon neutrino, and the tau neutrino. Each neutrino and its associated charged lepton are grouped together as a family or generation. It is believed that the neutrinos interact only via the weak and gravitational interactions. Recent experiments seem to indicate that the electron neutrino (and possibly also the muon and tau neutrinos) has a small rest mass. Although no exact value for this mass can yet be given, the electron neutrino is very light compared to the electron.

neutron   A spin 1/2 baryon with a rest mass slightly greater than a proton. In free space, the *neutron* is unstable and decays with a mean life of about 890 seconds. Along with the proton, the *neutron* is the fundamental building block of the nuclei of atoms. In the quark model, the *neutron* is a bound state of two down quarks and one up quark.

neutron diffraction   Thermal neutrons with energy of $\sim (1/40)eV$ have a deBroglie wavelength of ~ 2Å which is comparable to interatomic distances in solids. They interact elastically with the nuclei and the magnetic moments in a crystal. Thus, *neutron diffraction* reveals, as in X-rays, the location of the atoms in the crystal as well as the magnetic structure of the crystal. In addition, the thermal neutrons scatter inelastically by absorbing or emitting phonons of comparable energies, which allows accurate measurements of the frequencies of the lattice vibrations and their wave vectors. The same is possible for magnons (spin waves) through the neutron magnetic interaction.

Newtonian fluid   Fluid in which the shear stress varies linearly with the velocity gradient such that

$$\tau \propto \frac{du}{dy}.$$

Air, water, and many common gases and liquids exhibit Newtonian behavior.

nine-j symbols   Used to calculate the coupling of multiple angular momenta. Just as six-j symbols are used to calculate the coupling of three angular momenta, *nine-j symbols* are used to calculate the coupling properties of four angular momenta, j_1, j_2, j_3, and j_4, to form a resultant total angular momentum J.

$$J = j_1 + j_2 + j_3 + j_4.$$

The *nine-j symbols* are defined by

$$\begin{Bmatrix} j_1 & j_2 & J_{12} \\ j_3 & j_4 & J_{34} \\ J_{13} & J_{24} & J \end{Bmatrix} =$$

$$\frac{\langle (j_1 j_2) J_{12}, (j_3 j_4) J_{34}, JM | (j_1 j_3) J_{13}, (j_2 j_4) J_{24}, JM \rangle}{\sqrt{2J_{13}+1}\sqrt{2J_{24}+1}\sqrt{2J_{12}+1}\sqrt{2J_{34}+1}}$$

where J_{12}, J_{23}, J_{13}, and J_{24} are the possible coupling schemes. Even permutations of rows and columns or their transposition leave the value of the *nine-j symbols* unchanged.

When one of the arguments j_1, j_2, j_3, or j_4 is zero, the *nine-j symbols* reduce to six-j symbols.

Noether's theorem   Theorem which connects continuous symmetries of a physical system with conserved quantities. If the Lagrangian of a physical system remains invariant under some continuous transformation, then connected with this symmetry there will be a conserved quantity. For example, if a system is unchanged under spatial translations, the momentum of the system will be conserved.

noise   A general term for describing any signal that affects the efficient working of an electronic device. There are two types of *noise:* white noise and impulse noise. White noise is distributed across a wide frequency level. There are numerous types of white noise. Thermal noise, for example, is caused by random thermal motion of electrons. Impulse noise is a result of sudden electrical impulses.

non-Abelian gauge fields   Fields which arise when the non-Abelian symmetry of a theory is gauged or made into a local symmetry. Similar to the quanta of Abelian gauge fields (such as the photon from the Abelian theory of electromagnetism), the quanta of *non-Abelian gauge fields* are spin 1 objects. In contrast to the quanta of Abelian gauge fields, a non-Abelian gauge field also carries the charge for which it is the mediating particle. For example, photons of the Abelian theory of electromagnetism do not carry electric charge, but gluons of the non-Abelian theory of chromodynamics do carry color charge.

non-Abelian groups   Mathematical groups which are marked by the fact that the generators of the group do not commute with one another. If a *non-Abelian group*, G, has N generators, T_a ($a = 1, \ldots, N$), they will satisfy the following commutation relationship — $[T_a, T_b] = if_{abc}T_c$, where f_{abc} are the antisymmetric structure constants of the group.

non-classical light   Light whose photon-statistics do not behave in a classical way, i.e., whose statistical fluctuations Δn caused by the arrival times of the photons at the detector are smaller than that of a Poisson distribution with the same average number n of photons per time interval:

$$\Delta n < \sqrt{n}.$$

The photon-statistics can be measured in a Hanbury–Brown–Twiss experiment. Since photons are bosons, they behave according to the Bose–Einstein statistics. This means that photons show bunching, i.e., detecting one photon is more likely if another photon has just been detected. For *non-classical light,* this is not the case. There is a greater likelihood of not detecting a second photon right after the detection of a first photon. The fluorescence from a single atom is an example of *non-classical light.* Right after the spontaneous emission of a photon, the atom has to be excited before it can emit another photon.

noncrystalline solids Refers to solids with no regular structure, but which share some of the properties of regular crystals to some degree. Among these are amorphous semiconductors, glasses, which are liquid solids, and amorphous ferromagnets.

non-degenerate States that have slightly different energies.

nonets A grouping of nine mesons based on the property that they can be grouped in the direct product $(3 \otimes \bar{3})$ of the fundamental representation (3) and its conjugate representation $(\bar{3})$ in the flavor $SU(3)$ quark model. The direct product $3 \otimes \bar{3}$ can be decomposed into the direct sum $8 \oplus 1$ that contains an octet (e.g., $\pi^+, \pi^0, \pi^-, \eta, K^+, K^-, K^0, \overline{K}^0$) and a singlet (e.g., η') which, taken together, form a *nonet.*

non-flow energy equation A corollary of the first law of thermodynamics which states that there exists a property of a closed system such that a change in its value is equal to the difference between the heat supplied and the work done during any change of state, and is given by the expression:

$$Q - W = (U_2 - U_1)$$

where Q is the heat supplied, W is the work done, and U is the internal energy of the system.

non-flow process The process undergone by a fluid in a closed system.

non-leptonic decay A decay which does not involve leptonic particles in either the initial or final state. For example, the decay of a kaon into pions — either $K \longrightarrow \pi + \pi$ or $K \longrightarrow \pi + \pi + \pi$.

non-linear effects Deviations from linear response, as illustrated by the electrical conductivity and polarizability. Ohm's law holds well in metals, but in semiconductors we can produce small or significant departures from Ohm's law in the warm or hot electron regimes. The electrical polarization is linear in the electric field for small fields, but quadratic and cubic terms appear when a laser beam excites a crystal. For quadratic terms (and higher), frequency doubling (tripling) and mixing result.

non-linear optics Branch of optics that focuses on effects that have a non-linear dependence on the electric field. The most notable examples are non-linear frequency conversion and four-wave mixing.

non-linear saturation The state that results from the growth of an instability to its maximum or long-term amplitude.

non-Newtonian fluid Any class of fluid in which the shear stress does not vary linearly with the velocity gradient. Suspensions, gels, and polymers often exhibit non-Newtonian behavior.

norm Quantum mechanics was formulated within the geometric framework of Hilbert spaces. A Hilbert space is a vector space. Each vector in this space represents a quantum mechanical state. The vector can be multiplied by any complex number and represents the same quantum state. The scalar product of a vector by itself is called the *norm* of the vector. By definition, the norm is a real and positive number. In the non-relativistic quantum mechanics of electrons, the coordinate representation of a state vector is a scalar function of space coordinates and spin variables. The scalar function is usually called the wave function. The *norm* is given by integrating the square of the modulus of such a wave function over the space coordinates and then summing over the spin variables.

normalization Refers to the fact that every wave function of a system must obey the relationship

$$\int_{-\infty}^{\infty} |\Psi|^2 \, dx = \langle \Psi | \Psi \rangle = 1 \, .$$

The origin of this *normalization* is in the Copenhagen interpretation of quantum mechanics. According to this interpretation, $|\Psi(r)|^2$ can be interpreted as the probability of finding a particle at a location r. The probability must be normalized to one.

normalization condition The sum of all the probabilities of obtaining a measurement is equal to unity. Thus, if u is a variable that can assume any of the M discrete values,

$$u_1, u_2, K, u_M$$

with respective probabilities

$$P(u_1), P(u_2), K, P(u_M) \, .$$

The *normalization condition* is given by:

$$\sum_{i=1}^{M} P(u_i) = 1 \, .$$

normalize To choose a state vector with its norm equal to unity. In other words, to multiply a wave function by a constant so that its norm is equal to unity.

normal order A specific ordering of the annihilation (a) and creation ($a^\dagger$) operators for the quantized radiation field. An operator, which can be written as a superposition of products of the annihilation and creation operators, i.e.,

$$\hat{O} = \sum \sum c_{mn} {a^\dagger}^n a^m$$

is said to be *normal-ordered* when first the annihilation operators are applied and then the creation operators (all annihilation operators are written to the right and creation operators to the left, as in the equation above).

Any operator consisting of a sum of products between the creation and annihilation operators can be cast into *normal order* with the help of the commutation relation

$$\left[a, a^\dagger\right] = aa^\dagger - a^\dagger a = 1 \, .$$

normal ordering A process whereby a product of operators is rearranged so that all annihilation operators ($a, b, c, \ldots$) are to the right of all creation operators ($a^\dagger, b^\dagger, c^\dagger \ldots$). For example, the *normal ordering* of the product $c^\dagger a a^\dagger b a$ would be $N(c^\dagger a a^\dagger b a) = abac^\dagger a^\dagger$.

normal processes Processes for which the conservation of the $\underline{k}$ vector does not require a reciprocal lattice vector. For example, if an electron of wave vector $\underline{k}$ emits a phonon of wave vector q and is scattered to state $\underline{k}'$, the process is normal if $\underline{k} = \underline{k}' + q$. It is an Umklapp process if a reciprocal lattice vector $\underline{G}$ is added to one side of the equation.

normal product Let $a, b, c, \ldots$, denote arbitrary creation or annihilation operators of the Fermi field. The *normal product* is defined by

$$N(abc\ldots) = (-1)^P \alpha \beta \gamma, \ldots ,$$

where $\alpha, \beta, \gamma, \ldots$ denote the same operators as $a, b, c, \ldots$, rearranged in such a way that, reading from left to right, all the creation operators come first, followed by all the annihilation operators. Furthermore, P denotes the number of transpositions of fermion operators that are necessary to achieve the rearrangement.

no-slip condition Requirement that a fluid immediately adjacent to a boundary must move with that boundary as required by viscous or frictional forces. Potential or inviscid flow theory does not require the *no-slip condition*.

nozzle Any duct which accelerates a flow by changing the duct area over a fixed distance. For subsonic flow, the duct area decreases in the direction of the fluid motion. For supersonic flow, the duct area increases in the direction of the fluid motion.

N-P product The product of the electron density n and hole density p in an intrinsic semiconductor of ellipisoidal energy surfaces of masses m_1, m_2, and m_3 for the conduction band, and m'_1, m'_2, and m'_3 for the valence band, is given by

$$np = \frac{1}{2\pi^3 \hbar^6} \left[(m_1 m_2 m_3)(m'_1 m'_2 m'_3) \right]^{1/2} (kT)^3 \exp(-E_g/kT)$$

where kT is the thermal energy, E_g is the energy gap, and $E_g \gg kT$.

nuclear demagnetization Cooling by an adiabatic demagnetization of a nuclear spin system. It is usually a second stage in cooling, following cooling by an adiabatic demagnetization of an electron spin system. Temperatures of approximately 10^{-7} K can be reached by this method.

nuclear energy Ideally, the binding energy of the subatomic particles (nucleons) that make up the atomic nucleus. In general, however, this is the energy released by these particles as they undergo nuclear reactions that mainly involve regrouping, for example, during nuclear fission (separation of nucleons) or nuclear fusion (merging of nucleons).

nuclear force The short-range, strong force which binds protons and neutrons together into nuclei against the electrostatic repulsion of the protons. The *nuclear force* is thought to be a residual force resulting from the more fundamental quantum chromodynamic force that arises from the exchange of gluons. See quantum chromodynamics (QCD).

nuclear magnetic resonance A phenomenon where nuclei, with net spin angular momenta and, therefore, magnetic moments, are placed in a large magnetic field. The interaction between this large, external magnetic field and the magnetic moment of the nuclei will split the spin energy levels. By applying a radiofrequency (RF) pulse of electromagnetic radiation, whose frequency matches the energy level difference of the split spin energy levels, this sample will exhibit resonant absorption of the RF beam energy. This phenomenon has important applications in medical imaging.

nuclear magneton A quantity usually denoted by μ_n and defined as $\mu_n = \frac{e\hbar}{2m_p c} = 5.05 \times 10^{-27}$ joules/tesla. Here, e is the magnitude of the electron's charge, $\hbar$ is Planck's constant divided by 2π, m_p is the mass of the proton, and c is the speed of light. This quantity gives a useful scale for measuring magnetic moments associated with nuclear particles such as the proton or neutron.

nuclear moment The magnetic moment given by

$$\mu_k = \frac{e\hbar}{2m_{\text{proton}}} = 3.15245166(28) * 10^{-14} \text{ MeV/T},$$

where e is the elementary charge and m_{Proton} is the mass of the proton. The *nuclear moments* of protons and neutrons are given by

$$\mu_{\text{proton}} = 5.585694675(57)\mu_k$$
$$\mu_{\text{neutron}} = -3.82608545(90)\mu_k.$$

The fact that μ_{proton} deviates from the nuclear moment and the neutron has a nuclear moment indicates that they possess internal structure.

nuclear quadrupole magnetic resonance Nuclei with spin $I \geq 1$ can have an electric quadrupole moment which interacts with the electric field of the crystal. If the interaction is strong (larger than the Zeeman energy in the magnetic field), one can observe pure electric quadrupole resonance without an external magnetic field as the electrostatic field (which we assume is axial) splits the energy levels by amounts proportional to m^2, where m is the z component of I and transitions can occur between these levels. If, on the other hand, the Zeeman energy is dominant, then the electrostatic field gives the Zeeman levels energy corrections proportional to m^2, and instead of one resonance line we have $2I$ lines.

nuclear spin The spin connected with the nucleus of an atom. A *nuclear spin* different from zero can cause the hyperfine splitting of atomic or molecular levels.

nucleation The mechanism by which a thermodynamic process is initiated. For example, in order for bubbles to form when a liquid boils, *nucleation* of the bubbles must first occur.

nucleation sites Locations on a surface which allow nucleation to occur. In the absence of any *nucleation sites,* it is possible to superheat a liquid beyond its normal boiling point before boiling commences. In practice, dissolved gases or vapor trapped in surface roughness cavities provide the nucleation sites needed for bubbles to form and grow with very little superheat.

nucleon A general term which refers to both protons and neutrons, the constituents of the nucleus. To a certain approximation, the proton and neutron may be regarded as two different states of a single particle — the nucleon. This is analogous to how a spin-up and spin-down electrons are just two different states of the same particle — the electron. The mass difference between the proton and neutron spoils this analogy.

nucleosynthesis The process whereby lighter nuclei are fused into heavier nuclei. This process occurs inside the cores of stars and is believed to have occurred during the early phase of the Big Bang.

number operator The operator whose eigenvalues are the number of photons in the radiation field. Naturally, the eigenstates are the number states. The *number operator* is given by

$$a^\dagger a,$$

where $a^\dagger$ and a are the creation and annihilation operators for a photon, respectively. We find

$$a^\dagger a |n\rangle = n|n\rangle .$$

number state The state of the radiation field with a precisely determined number of photons which is therefore an eigenstate for the number operator $a^\dagger a$:

$$a^\dagger a |n\rangle = n|n\rangle .$$

These states are also called Fock states. Since their photon number is precisely known, the complimentary observable, i.e., the phase, must be completely undetermined. Because of the quantization of the radiation fields, we find the following relationships:

$$a|1\rangle = |0\rangle$$
$$a|n\rangle = \sqrt{n}|n-1\rangle$$
$$a^\dagger|n\rangle = \sqrt{n+1}|n+1\rangle$$

where $a^\dagger$ and a are, respectively, the annihilation and creation operators for the quantized field. The pre-factors can be derived from the normalization condition, i.e.,

$$\langle n|n\rangle = 1 \text{ for } n = 0, 1, 2, \ldots .$$

Accordingly, one finds by multiple application of the creation operator

$$|n\rangle = \frac{\left(a^\dagger\right)^n}{\sqrt{n!}}|0\rangle .$$

The *number states* form a complete basis set of the radiation field, which means that

$$\sum |n\rangle\langle n| = 1 .$$

Nusselt number A non-dimensional parameter used in convective heat loss problems. It is defined by the ratio $Qd/k\Delta T$, where Q is the heat loss rate from a solid body per unit or area, ΔT is the difference in temperature between the solid body and its surroundings, k is the thermal conductivity of the surrounding fluid, and d is the significant linear dimension of the solid.

Nyquist diagram A mathematical technique for determining whether or not there are unstable roots to a dispersion relation.

Nyquist formula Relates the thermal or Johnson noise in a resistor at any frequency with the temperature of the resistor. In general, the fluctuating noise EMF, V, generated in a resistor of resistance, R, over a frequency range df is given by:

$$d\langle V^2\rangle = 4R\left(\frac{hf}{\exp(hf/kT)-1}\right)df .$$

At normal temperatures and frequencies, the relationship reduces to $P = kT\,df$, where k and h are the Boltzmann and Planck constants respectively.

O

O(N) group The N-dimensional orthogonal group. This group is the rotation group in N dimensions. A point in N dimensions is given by N numbers, x_i (where $i=1,\ldots,N$). The *O(N) group* transforms x_i to x'_i ($x_i = O_{ij}x'_j$) in such a way that $x_i \cdot x_i = x'_i \cdot x'_i$ (i.e., the N-dimensional length remains invariant). O(N) has $\frac{1}{2}N(N-1)$ independent generators.

oblique shock A shock wave in which the upstream and downstream flow vectors are not perpendicular to the jump. For supersonic flow approaching a wedge (acute bend) of angle θ, a shock forms at angle β

$$\tan\theta = 2\frac{M_1^2\sin^2(\beta)-1}{M_1^2(\gamma+\cos 2\beta)-2}\cot\beta.$$

The upstream and downstream Mach numbers can be related by the component of the flow vector normal to the shock where

$$M_{1n} = \frac{u_{1n}}{a_1} = M_1 \sin\beta$$
$$M_{2n} = \frac{u_{2n}}{a_2} = M_2 \sin(\beta-\theta).$$

The normal shock relations with M_{1n} and M_{2n} can then be used to obtain the changes in thermodynamic variables (p, T, ρ, s) across the *oblique shock;* the isentropic flow relations with M_1 and M_2 are used to determine the respective stagnation values of the dynamic variables (p_{01}, p_{02}, T_{01}). Note that $T_{02} = T_{01}$.

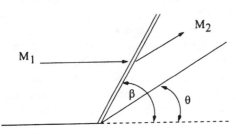

Oblique shock formation.

observable (1) A quantity for which a value can be determined experimentally in a given system, such as position, linear and angular momentum, energy, number of particles, etc. An *observable* is represented by an operator. The average value for the *observable* obtained from a large number of measurements on identically prepared systems is given by the expectation value of the operator, while the eigenvalues of the operator give the possible values obtainable in individual measurements.

(2) A Hermitian operator with a complete orthonormal set of eigenfunctions on the Hilbert space spanned by the set of vectors representing the quantum states under consideration. Any physically observable variable must be represented by one of such operators. If we measure the variable, the result is one of its eigenvalues. If the quantum state is one of its eigenfunctions, the result is the corresponding eigenvalue.

observation Experimental determination of the value of one or more observables in a system. Only compatible observables, represented by commuting operators, can be measured simultaneously for any given system. After *observation* the system finds itself in an eigenstate of the relevant operators with eigenvalues equal to the measured values.

observer entropy The entropy of the observer (or automated system) used during any measurement of a thermodynamic observable. The *observer entropy* explains the discrepancy between the concept of Maxwell's demon and the second law of thermodynamics.

occupation density The product of the Fermi or Bose distribution function, $f(\varepsilon)$, and the density of particle states, $g(\varepsilon)$, is the density of occupation in energy. The term $f(\varepsilon)g(\varepsilon)$ plotted as a function of energy, ε, represents the total number of particles in the system.

occupation number The number of identical particles in a state which can be occupied by more than one particle. Such a state is often called degenerate.

octane number Provides an empirical measure of the ability of a fuel to resist pre-ignition in an internal combustion engine (also known as knocking). It is the volume percentage of isooctane (C_8H_18) in pure heptane (C_7H_16) that matches the knocking behavior of the fuel being tested in a standard single cylinder four-stroke engine.

octet A grouping of eight hadrons which can be fit into a common representation of the flavor group $SU(3)$. For example the pions (π^+, π^0, π^-), the kaons ($\overline{K}^-, \overline{K}^0, K^+, K^0$), and the eta ($\eta^0$) form an $SU(3)$ *octet*. Within the $SU(3)$ flavor quark model, *octets* arise from the group decomposition of the tensor product of two or more fundamental representations of the $SU(3)$ group. In the above example, these *octet* mesons are considered to be composites of a quark (which is in the **3** representation) and an antiquark (which is in the $\overline{\mathbf{3}}$ representation). The tensor product, $\mathbf{3} \otimes \overline{\mathbf{3}}$, can be decomposed into $\mathbf{8} \oplus \mathbf{1}$, where the pions, kaons, and eta belong to the *octet*, **8**, part of this decomposition.

Ohmic heating Heating produced by electrical current passing through a resistive plasma.

Ohm's law The electric current density $\underline{J}$ is proportional to the applied electric field $\underline{E}$ in isotropic conductors and the proportionality constant is the electrical conductivity σ. For anisotropic systems, σ is a tensor and $J_I = \sum_j \sigma_{ij} E_j$, where J_i and E_j are components in the ith and jth directions. It is assumed that $\underline{E}$ is weak and does not cause strong departures from equilibrium.

Okubo–Zweig–Iizuka (OZI) rule An empirical rule which states that in the decay of hadrons, certain decays are suppressed if the quarks of the initial state hadron are not directly connected to the quarks of the final state hadron. If the quarks of the initial hadron are connected to the final quarks only via gluons, the process is said to be OZI-suppressed.

omega minus A negatively charged spin of 3/2 particle. The *omega minus* is unstable, decaying with a mean life of 0.82×10^{-10} seconds. In the quark model it is composed of three strange quarks.

one atom maser The amplification of electromagnetic waves by exciting atoms in a superconducting microcavity. A highly collimated beam of Rydberg atoms passes through a Fizeau velocity selector and is injected in the cavity. There is only one atom at a time interacting with the cavity field in the cavity.

one-particle irreducible diagram A Feynman diagram which cannot be reduced into two more basic diagrams simply by removing one line (i.e., propagator) from the diagram.

The top figure is reducible by cutting the middle wavy line, while the lower figure is one-particle irreducible.

Onsager equations The heat current in a wire depends on both the temperature difference and the electrical potential difference across the wire. This relationship is expressed by the *Onsager equations*:

$$I_S = \frac{I_Q}{T} = L_{11}\frac{\Delta T}{T} + L_{12}\frac{\Delta V}{T}$$
$$I = L_{21}\frac{\Delta T}{T} + L_{22}\frac{\Delta V}{T}$$

where I_S is the entropy flow, I_Q is the heat flow, T is the temperature of the heat reservoir in contact with the wire, ΔT is the small temperature difference across the wire, and ΔV is the potential difference across the wire. The L_{ij} terms are coefficients related to the electrical resistance, thermal conductivity, and thermoelectric properties of the wire. The *Onsager equations* express the linearity between the flows and the generalized forces $\frac{\Delta T}{T}$ and $\frac{\Delta V}{T}$.

Onsager–Lax regression theorem A general statement of the quantum regression theorem is that if for the operator $\hat{O}$, $<\hat{O}(t+\tau)> = \sum_j a_j(\tau) <\hat{O}(t)>$, then a two-time correlation function can be determined from the single-time expectation value. Consequently, the two-time correlation function satisfies the same equation of motion as the single-time expectation value.

Onsager's reciprocal relation (1) In 1931, Onsager proved symmetry relations between conductances relating currents to forces where there are interference effects. For example, if two forces such as an electric field and a thermal gradient produce an electric current and a heat current, we can define the forces and currents in such a way that the currents are related to the forces by a symmetric matrix. If X_1 and X_2 are the forces and J_1 and J_2 are the currents we write,

$$J_1 = L_{11}X_1 + L_{12}X_2,$$
$$J_2 = L_{21}X_1 + L_{22}X_2,$$

where $L_{12} = L_{21}$.

Kohler considered anisotropic cases with magnetic fields and gave symmetry relations between the conductance (kinetic) coefficients.

(2) If the departure from equilibrium is small, the coefficients L_{12} and L_{21} in the Onsager equations are equivalent. The relationship:

$$L_{12} = L_{21}$$

is known as the *Onsager reciprocal relation*.

open channel flow Any flow of liquid in which there is a free surface, such as a river or canal, requiring a balance between gravitational and frictional forces.

open cycle Any thermodynamic cycle that involves a system that is not isolated from its surroundings, for example, a heat engine cycle that does not use the working fluid as the oxidant for the combustion of the fuel. In practice, fresh air must be allowed to enter the heat engine cycle to ensure that combustion continues to occur.

open orbit The orbits of electrons in a magnetic field are not always closed. Such orbits are called open, and the electron path is extended in the crystal. This can happen when the intersection of a plane with the Fermi surface is not a closed curve. The orbit is usually drawn in the repeated zone scheme (periodicity in the reciprocal space of energy surfaces).

open system A region of space defined by a boundary across which matter may flow in addition to work and heat.

operator A mathematical object that acts on a state or wave function, producing another one in the same Hilbert space: $\mathcal{O}|\alpha> = |\beta>$. The new state $|\beta>$ is not necessarily a multiple $(a|\alpha>)$ of the original, except in special cases; in such cases, the state is called an eigenstate of the *operator* and the constant of proportionality a is the eigenvalue. Physical observables are represented by a special class of Hermitian *operators*, which have real eigenvalues. A measurement of the observable represented by an *operator* always returns one of the eigenvalues of the *operator*; after the measurement, the system is found in the corresponding eigenstate. The average value of many measurements on identically prepared systems is called the expectation value of the *operator*. In general, the order in which *operators* are applied on a state is significant. Commuting *operators* are pairs of *operators* for which this order is immaterial. Only observables represented by commuting *operators*, called compatible observables, can have their values simultaneously determined. The impossibility of simultaneously determining the values of two incompatible observables, such as position and momentum, is the origin of the uncertainty principle.

operator algebra A mathematical approach to constructing quantum mechanics of the system with an infinite degree of freedom, such as quantum field theories. Commutation relations for operators are the basis for such an approach.

operator product expansion A mathematical technique for expanding the products of currents in terms of a sum of local operators. In mathematical terms, $A(x)B(y) = \sum_n C_n(x - y)\mathcal{O}_n(x, y)$, where $A(x)B(y)$ is a product of

optical activity Certain crystals and compounds in solution exhibit a property to rotate the plane of polarization of incoming light. The molecules of these materials (crystals and compounds) have asymmetric molecular structures. The molecules can exist in left- and right-handed forms. Particular forms of the compounds are classified as left-handed or right-handed. These materials accordingly rotate the plane of polarization to the left or right (as viewed from the direction facing the light).

optical bistability The optical frequency output of a laser results from a compromise between the atomic and cavity natural frequencies. There are laser systems for which a single input intensity gives rise to two or more stable output intensities. Two features are important for such lasers: the medium in the cavity has an intensity-dependent refractive index and the cavity output depends on an injected signal for its energy and output frequency.

optical Bloch equations Provide an exact description of the state of a few-level atom interacting with a classical, oscillating electric field. The solutions to the first order, linear differential equations for the atomic populations give the time development, as well as, the steady-state solutions of the atomic levels.

optical cavity Resonator-like devices used for generating and/or amplifying light by atomic sources inside the cavity.

optical coherence A process is coherent if there are definite phase relationships between the elements characterizing the electromagnetic radiation. The phenomena of interference in quantum mechanics is another example of coherence, and the phase relationship is in the superposition of states in the wave function. The Young double-slit experiment with classical light is an example of *optical coherence*. Classical or quantum coherence can be characterized by first and second order correlation functions.

optical communication The modulation of a laser field to carry information through an optical transmission line. Amplification of the signal is necessary to counteract losses. The minimization of noise, which results from both amplification and losses, may be accomplished with the use of non-classical light.

optical cooling Atoms moving towards an intense laser beam absorb photons and are slowed down. The slowed atoms emit photons randomly and proceed to absorb further photons. Atoms are slowed down further with each photon absorption and subsequent photon emission.

optical fiber A light-carrying optical cable that prevents light from escaping the fiber through the total internal reflection of light through angles greater than the critical angle.

optical homodyne tomography The density operator $\hat{\rho}$ contains the maximal information about the radiation field and is important to determine how to measure it experimentally. By means of balanced homodyne four-port detections, one can determine the field probability distribution function $p(\mathcal{F}, \varphi_\mathcal{F})$. Using an inverse Radon transformation on the integral connecting the Wigner function to the field distribution function, one can determine the Wigner distribution which, in turn, determines $\hat{\rho}$.

optical Kerr effect A non-linear process that occurs when light that passes through a Kerr medium undergoes an intensity dependent phase shift through a modification of the index of refraction owing to the intensity of the beam. The intensity of light remains unchanged provided that absorption can be neglected.

optical parametric oscillator Converts laser light with frequency ω_l into signal and idler light with frequencies ω_s and ω_i respectively, where $\omega_l \approx \omega_s + \omega_i$. This is a preferred method for generating squeezed light and the amplification is accomplished when a non-linear medium is placed within an optical cavity.

optical parametric process The second-order susceptibility tensor $\chi^{(2)}$ of a non-linear medium plays a fundamental role in the description of the non-linear interaction between optical waves. The susceptibility is connected with the parameters of the atomic system. Parametric amplification is obtained in the interaction of three light waves. A non-linear medium is irradiated by an intense pump mode of frequency ω_1 and the interaction generates two light waves of frequencies ω_2 and ω_3 with $\omega_1 = \omega_2 + \omega_3$. Parametric fluorescence can be obtained by irradiating a thin crystal with a strong pump wave and by properly taking into account a variety of signal and idler modes.

optical properties The interaction of light with wavelengths, ranging from the long wavelengths in infrared light to the short wavelengths in X-rays, reveals a great deal, such as reflection, refraction, transmission, absorption, emission, energy levels, absorption and emission bands, etc, about the nature of the specimens studied. All of this factors can be correlated with the electromagnetic nature of the materials studied. Synchrotron radiation offers good light sources for these studies. Lasers allow precision spectroscopy and observation of non-linear effects.

optical pulse Electromagnetic waves localized in space and time. Pulse propagation in a non-linear medium is governed by non-linear partial differential equations. The broadening of the pulse in optical fibers is an important factor in data transmission. Solitary waves (solitons) can propagate unchanged through an absorber.

optical pumping The action of an external light beam when it raises atoms from the ground state to an excited state by photon absorption. Laser action follows from the population inversion that occurs when the number of atoms in the excited state exceeds that in the ground state.

optical pyrometer An instrument that measures the temperature of a hot object by measuring its color or by comparing its radiant color with a filament of a known temperature.

optical response functions Response functions relate the polarization of a sample to the electric field strength. The polarization is expressed as a power series in the radiation field, $P = P^{(1)} + P^{(2)} + P^{(3)} + \ldots$, and the corresponding terms give rise to the nth order non-linear response functions. Linear optics results from $P^{(1)}$, second order non-linear processes follow from $P^{(2)}$, and $P^{(3)}$ is connected with four-wave mixing.

optical Stark effect Changes brought about on the energy levels of an n-level system due to an external electric field. The treatment is usually to the lowest order in perturbation theory. In the presence of strong fields, the dressed-atom approach is useful since part of the atom–field interaction can be solved exactly.

optical susceptibilities External electric fields produce the polarization of materials. The connection between the field strength and the polarization is, in general, non-linear and is mediated by the susceptibility. The first order polarization, caused by the linear term in the field strength, gives rise to the first order susceptibility. The second order polarization corresponds to the quadratic term in the field strength and gives rise to the second order susceptibility, etc. The polarization of the sample is the microscopic quantity that is measured.

optical theorem States that the imaginary part of the forward scattering amplitude is proportional to the total cross-section. This can be written as $\text{Im}(f(k, \theta = 0)) = \frac{4\pi}{k}\sigma_{\text{total}}$, where $\text{Im}(f(k, \theta = 0))$ is the imaginary part of the forward ($\theta = 0$) amplitude, k is the center of mass momentum, and σ_{total} is the total scattering cross-section.

optoelectronics One of the fastest growing areas of modern technology which is key for much of the communications and information industries. It is based on the quantum phenomena describing the interaction of photons with electrons and employs many of the most advanced device fabrication techniques.

orbital A three-dimensional function describing the spatial distribution of the probability density for a electron in an atom or molecule. The shape of the different *orbitals* varies with both energy and angular momentum quantum numbers.

orbital angular momentum The *orbital angular momentum* operator is related to the position and linear momentum operators $\vec{r}$ and $\vec{p}$ by a relationship identical to their classical counterparts: $\vec{L} = \vec{r} \times \vec{p}$. The eigenvalues of $|\vec{L}|$, the magnitude of the *orbital angular momentum* operator, are always multiples of $\hbar$, Planck's constant h divided by 2π, times $\sqrt{l(l+1)}$, where the integer number l is the *orbital angular momentum* quantum number; the eigenvalues of the x, y, and z components of $\vec{L}$ are integer multiples of $\hbar$. Different components of the *orbital angular momentum* operator, such as L_x and L_y, do not commute with each other, and therefore the values of the corresponding observables cannot be determined simultaneously for a given state due to the uncertainty principle. However, any one component of $\vec{L}$ commutes with $\vec{L}^2$, and, therefore, the *orbital angular momentum* of a state is uniquely determined by the simultaneous measurement of the values of the magnitude of $\vec{L}$ and its projection along any axis.

orbital magnetic moment The magnetic dipole moment associated with the motion of electrons about an origin. This is often simply called the orbital moment. The part of the magnetic moment due to electron spins is excluded from orbital magnetism.

orbital paramagnetism In a free atom or nearly isolated ions in solids, electrons on an open shell may have orbital angular momentum, which produces a magnetic dipole moment. The material containing such atoms or ions shows paramagnetism, which is called *orbital paramagnetism*.

orbital parity Under the inversion transformation of the coordinates, the wave function of electrons in a central force field is either invariant or changes its sign. If it is invariant, the wave function is said to possess even parity. In the case of a sign change, the wave function has odd parity. If the orbital angular momentum quantum number is l, the *orbital parity* is $(-1)^l$.

order disorder The ordered and disordered phases of alloys. For example, in the ordered phase β, brass (CuZn) consists of two interpenetrating cubic lattices with each Zn atom at the center of cube surrounded by eight Cu atoms. In the disordered state, each lattice site is equally likely to be occupied by either Zn or Cu. The phase transition between the ordered and disordered state is of second order and is accompanied by a discontinuity in the specific heat.

order–disorder transformation The transition that occurs with temperature in a solid solution (e.g., a metal alloy) whereby ordered occupancy of lattice sites by the solution components becomes disordered above a certain critical temperature. For example, in a binary alloy, AB, in the ordered state, every atom of component A is surrounded by its nearest neighbors of component B. However, above the transformation temperature there is no such correlation, and the atoms of A and B are distributed randomly across the lattice sites.

ordering process The process whereby a parameter of a system changes from a disordered to an ordered state during a phase transition. The solidification of water into ice at 0°C and the spontaneous magnetization observed in ferromagnetic materials that are cooled below their Curie temperature are both examples of ordering processes.

order parameter The system parameter that defines the ordering process. For example, for the solidification of water into ice at 0°C, the order parameter is density. For the spontaneous magnetization observed in ferromagnetic materials that are cooled below their Curie temperature, the *order parameter* is magnetization.

ordinary wave Electromagnetic plasma wave with a perturbed electric field aligned with the background magnetic field, propagating nearly perpendicular to that magnetic field.

orifice plate Obstruction-type meter used in pipe flow to determine the flow rate by measuring the pressure drop across a sudden contraction and expansion.

Ornstein–Uhlenbeck process It is usual to introduce a fluctuating force $F(t)$ in the Langevin equation that has a zero mean and a δ-function correlation in time, $<F(t)>=0$ and $<F(t)F(t')>=\delta(t-t')$. A random variable $\beta(t)$ with a zero mean is described by the *Ornstein–Uhlenbeck* process if $<\beta(t)\beta(t')>=\Gamma e^{-\Gamma(t-t')}$.

Orr–Summerfeld equation Governing equation for the stability of nearly parallel viscous flows (such as in a boundary layer) written in terms of the complex amplitude of the stream function, ψ,

$$(U-c)\left(\hat{\psi}_{yy}-k^2\hat{\psi}\right)-U_{yy}\hat{\psi} = \frac{1}{ik\mathrm{Re}}\left(\hat{\psi}_{yyyy}-2k^2\hat{\psi}_{yy}+k^4\hat{\psi}\right)$$

where c is the wave speed and k is the wave number.

Orsat apparatus An experimental apparatus for exhaust gas analysis used for investigating the effectiveness of combustion plants. The technique involves measuring the volume change in exhaust gas volume following the successive absorption of CO_2, O_2, and CO to provide a measure of the partial volume of each constituent.

orthogonality In a vector space endowed with an inner product, orthogonality is the property of two vectors that have an inner product of zero for a discrete vector space, or equal to the Dirac δ-function for a continuous space. Eigenstates of a Hermitian operator corresponding to different eigenvalues are orthogonal to each other.

orthogonalized plane waves method (OPW) A method for calculating energy bands which uses as a trial function for the Bloch function $\Psi_{\underline{k}}(\underline{r})$ plane waves from which their overlap with the core states has been subtracted (which is the orthogonalization process). An *OPW* $\chi_{\underline{k}}$ can be written as:

$$\chi_k(\underline{r}) = \exp(i\underline{k}\cdot\underline{r}) - \sum_t b_{t\underline{k}}(\underline{r})\langle b_{tk}(\underline{r})|\exp(i\underline{k}\cdot\underline{r})\rangle$$

where $b_{t\underline{k}}(\underline{r})$ is a core state wave function (usually obtained by the tight binding approximation), and the bracket is the overlap integral between the plane wave and the core state. For the trial wave function, $\psi_{\underline{k}}(\underline{r}) = \sum_{\underline{g}} \chi_{\underline{k}+\underline{g}}(\underline{r})$, where g is a reciprocal lattice vector which insures that $\Psi_{\underline{k}}(\underline{r})$ satisfies Bloch's periodicity condition.

orthonormal basis A complete set of mutually orthogonal vectors, all having unit norm or magnitude. Completeness implies that no other vectors exist that are orthogonal to any in the set, except for multiples of members of the set; however, a different *orthonormal basis,* having an equal number of members, can be constructed by forming linear combinations of the members of the original set. The number of such vectors gives the dimensionality of the space. An arbitrary vector in the space can be expanded as a linear combination of the members of an *orthonormal basis*. The normalized eigenstates of a Hermitian operator corresponding to different eigenvalues form an *orthonormal basis* in the space acted upon by the operator.

orthorhombic lattice Rectangular unit cell with no two sides equal.

ortho states The states of greatest statistical weight in systems where two spins can combine in more than one way. For example, the symmetric, or triplet, spin state of the helium atom, where two electron spins combine to a spin-one state, is called the *ortho state* of helium or orthohelium. There are three such states compared to one antisymmetric state.

oscillations, neutrino A phenomenon where neutrinos can oscillate between the neutrino flavors or types. For example, a beam of initially pure muon neutrinos (ν_μ), could oscillate into electron (ν_e) or tau (ν_τ) neutrinos. This mixing occurs between neutrinos only if they have a rest mass, since in that case the flavor eigenstates are not the same as the mass eigenstates.

oscillator Any system subject to a restoring force depending on the displacement from an equilibrium position, thus resulting in a periodic motion. A special case, the simple harmonic *os-*

cillator, is one of the most commonly occurring systems in physics, either exactly or as an approximation.

oscillatory effects Illustrated by two examples: *oscillatory effects* in a magnetic field and electron density oscillations. The oscillations of the magnetic susceptibility at low temperature are due to the emptying of the Landau levels as the magnetic field is increased. The periodicity of the oscillations are the reciprocal of the magnetic field $1/B$. Whenever $1/B$ changes by $(2\pi e/\hbar Sc)$, where e is the electron charge, S is the area of the orbit in $\underline{k}$ space, and c is the speed of light, the number of occupied Landau levels changes by one. (Actually, the diamagnetic susceptibility of the electron gas consists of a constant term which equals $-1/3$ the Pauli susceptibility plus an oscillatory part.)

The electron density oscillations, known as Friedel's oscillations, result from electron scattering by a surface barrier, an edge dislocation, or an impurity. The amplitude of the oscillation is proportional to the backward scattering amplitude at the Fermi energy and has the form $\cos(2k_f x + \theta)/x^n$, where k_f is the Fermi wave vector, x is the distance from the scatterer, θ is a phase angle, and $n = 3$ for an impurity (3 dimensions), 5/2 for a dislocation (2 dimensions) and 2 for a surface barrier (1 dimension).

Oseen approximation Approximation to Stoke's flow about a sphere such that the inertial advective terms are linearized rather than neglected altogether, thus improving the accuracy of the solution in the far field. The resulting equation is

$$\rho U \frac{\partial \mathbf{u}'}{\partial x} = -\nabla p + \mu \nabla^2 \mathbf{u}'$$

where $\mathbf{u}' = (U + u')\hat{i} + v'\hat{j} + w'\hat{k}$.

Oseen vortex *See* Lamb–Oseen vortex.

osmosis The diffusion of a solvent (usually water) through a semi-permeable membrane from a solution of low ion concentration to one of a high ion concentration. The thermodynamic driving force for *osmosis* acts in the opposite direction to the ion diffusion gradient so as to equalize the concentrations in the two solutions.

osmotic pressure The pressure that, when applied to a solution separated by a semi-permeable membrane from a pure solvent, prevents the diffusion of the solvent through the membrane. For non-dissociating species, the osmotic pressure, (Π) is related to the solution concentration (c), the ideal gas constant (R), and the temperature by the relationship:

$$\Pi = cRT.$$

Ostwald's dilution law The relationship

$$K = \frac{\alpha^2}{(1-\alpha)V}$$

describes the ionization of a weak electrolyte for the situation where two ions are formed. It is derived from a consideration of the law of mass action, where K is the ionization constant, V is the dilution, and α is a parameter that describes the degree of ionization.

Otto cycle The reversible Otto engine is an idealization of the petrol internal combustion engine. Thermodynamically, the *Otto cycle* has a lower efficiency than a Carnot cycle working between the same maximum and minimum temperatures but is a closer approximation to a workable cycle. The *Otto cycle* consists of the four parts shown in the diagram below:

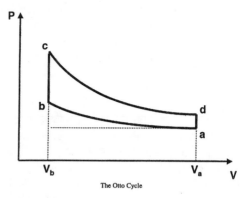

The Otto cycle.

ab Isentropic compression from (V_a, T_1) to (V_b, T_2), where V_a/V_b is known as the compression ratio, r.

bc Heating at constant volume from T_2 to T_3.

cd Isentropic expansion to (V_a, T_1).

da Cooling at constant volume to (V_a, T_1).

outgassing The evolution of gas that occurs when a surface is placed within a vacuum environment. The gas may originate from adsorbed species or from dissolved gas in the bulk of the material that is outgassing. This is a common problem in surface science or other vacuum chambers, which is mediated by periodically baking the chamber to temperatures in excess of 100°C.

overall heat transfer coefficient The net heat conduction of a composite system comprising a series of elements (each with their own thermal conductivity) can be defined in terms of an *overall heat transfer coefficient*, U, such that

$$\frac{1}{U} = \sum_i \frac{\Delta x_i}{k_i}$$

where $\frac{1}{U}$ is the overall resistance to heat flow and is equal to the sum of the individual resistances, $\frac{\Delta x_i}{k_i}$, in series. The *overall heat transfer coefficient* is analogous to the total electrical resistance of a circuit consisting of a number of individual resistors.

overhauser effect In 1953, Overhauser showed that nuclear spins in a metal can be polarized by saturating the spin resonance of the electrons. The electrons interact with the nuclei by the hyperfine interaction $a \underline{I} \cdot \underline{s}$ where a is a constant, I is the nuclear spin (which we assume for simplicity to be $1/2$), and s is the electron spin. Without the contact interaction, in a magnetic field B_o, the ratio of the nuclei with $I = 1/2$ to those with $I = -1/2$ is $\exp(2\delta/kT)$, where 2δ is the Zeeman energy for nuclei. With the contact interaction, this ratio becomes $\exp(2\Delta + \delta)/kT)$, where 2Δ is the Zeeman energy of the electrons; this is the subtle point of the *Overhauser effect*.

overlap integral Consider a particle and its two wave functions which are not orthogonal to each other. They are taken to be normalized. In many cases, two identical particles occupy the two wave functions. Take a scalar product of these two vectors. In other words, we integrate over space the product of one of the wave function with the complex conjugate of the other. The result is the *overlap integral*.

overstability Instability that oscillates as it grows in amplitude.

oxidation In general, any chemical reaction that involves the loss of electrons from a chemical species is known as an *oxidation* reaction. *Oxidation* is most commonly associated with the addition of oxygen to a chemical compound. It is always accompanied by a corresponding reduction process that involves a chemical species gaining electrons.

P

P(α) representation An expansion of the density matrix of the radiation field in terms of the complete set of Glauber coherent states. The representation is diagonal in the coherent states representation and is given by $\rho = \int P(\alpha) |\alpha\rangle\langle\alpha| d^2\alpha$.

packing fraction The ratio of the volume actually occupied by objects in a certain arrangement to the volume of space allotted to the objects. If we place spheres on lattice sites so that each sphere touches its nearest neighbors, the *packing fraction* is maximum and equals 0.74 if the lattice is face-centered cubic or hexagonal close packed.

pair annihilation A process whereby a matter particle and its antimatter counterpart come together and annihilate one another. An example of such a process would be an electron and positron annihilating to form two photons ($e^- + e^+ \longrightarrow \gamma + \gamma$).

pair production A process whereby a matter particle and its antimatter counterpart are created. An example of such a process would be a photon scattering from a nucleus and creating an electron and positron ($\gamma + N \longrightarrow e^- + e^+ + N$).

paradox Frequently used to describe a consequence of quantum physics which is in apparent contradiction with logical deduction based solely on classical arguments. Perhaps the most famous example is the Einstein–Podolsky–Rosen paradox, a thought experiment, subsequently verified empirically, which demonstrates the incompatibility of quantum physics with local causality by showing how a measurement performed on one system can instantaneously affect another measurement performed on a causally disconnected system.

paramagnetism The magnetic property of a material with small susceptibility. In an external magnetic field, a paramagnetic material will preferentially line up its magnetic moments along the direction of the external field. As a result, the sample itself will align parallel to the direction of the field. *Paramagnetism* is due to unpaired electron spins.

parametric amplification The process whereby a non-linear medium, characterized by a second-order susceptibility χ^2, absorbs a pump photon with the simultaneous emission of one signal and one idler photon.

parametric instability Three wave process in which one wave drives an instability in the other waves.

para states The states of smallest statistical weight in systems where two spins can combine in more than one way. For example, the antisymmetric, or singlet, spin state of the helium atom, where two electron spins combine to a spin-zero state, is called the *para state* of helium, or parahelium. There is one *para state*, compared to three symmetric states.

parity A discrete transformation where all spatial coordinates are turned into their negative — (x, y, z)→(-x, -y, -z). A system which is unchanged under a *parity* transformation is said to be *parity*-symmetric. The weak nuclear interaction is the only fundamental interaction which appears not to be symmetric under *parity*.

parity conservation If the wave function of the initial state of a system has even (odd) parity, the final state wave function must have even (odd) parity. This law is called the *parity conservation* rule. It is violated by the weak interaction.

parity selection rules The parity conservation holds true for a total system at any transition of states. In many cases, we are only concerned with a small specified system before and after the transition, and we neglect the radiation or particle emission during the transition. Depending upon the parity of emitted quanta, we get certain selection rules. Rules which specify whether or not a change in parity occurs during a given type of transition of an atom, molecule, or nucleus are called *parity selection rules*. Examples are the Laporte selection rule and the rule that there is no parity change in an allowed β-decay transition of a nucleus.

partial differential For a function, the partial differential is

$$f = f(x, y, z).$$

The *partial differential* of f with respect to x is given by

$$\frac{\partial f}{\partial x} = \lim_{\varepsilon \to 0} \left\{ \frac{f(x+\varepsilon, y, z) - f(x, y, z)}{\varepsilon} \right\}.$$

In other words, the *partial differential* of $f(x, y, z)$ with respect to x is obtained by differentiating $f(x, y, z)$ with respect to x while holding all other parameters constant.

partially ionized plasma A gas in which ions coexist with neutral atoms.

partial pressure The pressure exerted by each component of a gas mixture. Typically given by Dalton's law, which states that the pressure of a gas in a mixture is the same as that exerted by an equivalent isolated volume of the gas at the same temperature.

partial wave A component with definite orbital angular momentum quantum number l in an expansion of a plane wave in terms of spherical waves. This technique, known as *partial wave* expansion, is very useful in the treatment of scattering of an incoming parallel beam of particles, described by a plane wave, from a spherically symmetric potential. This results in a scattering amplitude which is a sum of terms depending only on incident energy, with the angular dependence given by the Legendre polynomial for the appropriate value of l: $f(\vec{k'}, \vec{k}) = \sum_{l=0}^{\infty} (2l+1) f_l(k) P_l(\cos \vartheta)$, where $\vec{k}$ is the momentum vector.

particle A generic term for a body treated as a single entity in a problem. Fundamental entities of nature are usually referred to as elementary *particles* to distinguish them from *particles* that are treated as single units for simplicity in a given problem. Although the term often implies a dimensionless body, it may also be endowed with size, rotational motion, or other properties of extended objects.

particle accelerator A device for accelerating particles such as protons or electrons to high momenta. By colliding these particles with other particles or with fixed targets, one attempts to probe the structure and nature of the particles or their targets. Various types of accelerators are the Van de Graaff accelerator, cyclotron, synchrotron, and linear accelerators.

particle masses The inertial rest mass of a given elementary particle. The mass of the particle determines its inertia or resistance to being accelerated. All the electrically charged matter particles which are believed to be fundamental (the six known quarks and the three known charged leptons) have masses. There is also some evidence that some or all of the three neutral leptons (the three neutrinos) may have non-zero masses. Of the force-carrying or gauge particles, the photon, gluons, and hypothetical graviton are thought to be exactly massless, while the $W^{\pm}$ and Z^0 gauge bosons of the weak interaction have a non-zero mass. In the standard model, all particles obtain their mass through their interaction with the undiscovered massive Higgs bosons, H.

particle–wave duality The concept or idea that objects in nature exhibit both particle properties and wave properties depending on the type of experiment or measurement that is performed. For example, this dual behavior is demonstrated by the photon. In Young's double slit experiment, light behaves like an electromagnetic wave. In the Compton scattering experiment, light behaves like a particle.

partition function The normalization constant of a thermodynamic system whose energy states obey the Boltzmann probability distribution. The *partition function*, Z, is also known as the sum over all states, and is given by the expression

$$Z = \sum_i e^{-E_i/kT}$$

where E_i is the energy of the ith state, k is the Boltzmann constant, and T is the system temperature.

parton Any of the constituents which were thought to make up hadrons, such as protons or neutrons. *Partons* are now thought to be the quarks and gluons which make up hadronic bound states.

pascal Unit of measure of pressure; 1 *pascal* = 1 N/m^2.

Pascal's principle Pressure applied to an enclosed fluid at rest is transmitted undiminished to the entirety of the fluid and the walls of the surrounding container.

passivate To chemically treat a metal surface so as to alter its normal tendency to corrosion. Common *passivates* include surface oxides, phosphates, or chromates that provide enhanced protection from corrosion.

path integral An integration where the integration measure is taken over all possible paths which connect two fixed end points. In general, the integrand will be a functional of the different paths which connect the two fixed end points. The *path integral* provides an alternative quantization method to the canonical creation/annihilation operator method of quantization. For example, the quantum probability for a particle to go from some initial quantum state $\mid q_i \, t_i \, \rangle$ (q_i, and t_i are the fixed initial position and time) to some final quantum state $\mid q_f \, t_f \, \rangle$ (q_f, and t_f are the fixed final position and time) can be written in *path integral* form as $\langle q_f t_f \mid q_i t_i \rangle = N \int \mathcal{D}q \exp\left[\frac{i}{\hbar}\int_{t_i}^{t_f} L(q, \frac{dq}{dt}) dt\right]$, where N is a constant. The integration measure $\mathcal{D}q$ represents an integration over all possible paths which connect the fixed initial and final points. The integrand, $\exp\left[\frac{i}{\hbar}\int_{t_i}^{t_f} L(q, \frac{dq}{dt}) dt\right]$, is a functional of the paths between this end points.

pathline Trajectory of a fluid particle over a period of time.

Pauli anomalous g-factor An additional term which has to be inserted in the Dirac equation to provide for the observed g-value of an electron different from two. The correction is due to the reaction of the electromagnetic field produced by the electron itself.

Pauli exclusion principle The statement that two identical fermions, or particles with half-integer spin, cannot share all their quantum numbers. The formal statement of the principle is that such particles must be in a completely antisymmetric state. The fact that electrons are fermions gives rise to the chemical properties, as well as the stability, of all ordinary matter.

Pauli–Lubanski pseudovector A pseudovector often denoted by W_μ and defined as $W_\mu = -\frac{1}{2} \epsilon_{\mu\nu\alpha\beta} J^{\nu\alpha} P^\beta$, where P^β is the four vector momentum, $J^{\nu\alpha}$ is the angular momentum/boost tensor, and $\epsilon_{\mu\nu\alpha\beta}$ is the totally antisymmetric Levi–Civita symbol in four dimensions. The quantity $W_\mu W^\mu$ is a Casimir invariant of the Poincaré group and is equal to $-ms(s+1)$, where m is the mass of the particle and s is its spin.

Pauli matrices Three 2×2 Hermitian matrices (usually denoted by $\sigma_x, \sigma_y,$ and σ_z) which satisfy the commutation relationships $[\sigma_x, \sigma_y] = 2i\sigma_z$ plus two others obtained by the cyclic permutation of the indices $x, y,$ and z. The *Pauli matrices* are important in studying particles which have half-integer spin.

Pauli spin matrices A set of operators σ_1, σ_2, and σ_3 satisfying the algebraic relations

$$\sigma_1\sigma_2 = i\sigma_3, \quad \sigma_2\sigma_3 = i\sigma_1, \quad \sigma_3\sigma_1 = i\sigma_2$$
$$\sigma_j\sigma_k + \sigma_k\sigma_j = 2\delta_{j,k}.$$

They can be expressed as 2×2 matrices (with two rows and two columns). Such matrices are called the *Pauli spin matrices*. Although the operators applies to fermions with spin 1/2, the eigenvalues of the *Pauli spin matrices* are ± 1.

Pauli susceptibility The electron gas in a metal is a good example of a paramagnetic system. In a magnetic field $\underline{B}$, there is a net magnetic moment of the electrons in the direction of the field. A simple calculation shows that the susceptibility χ_P, named after Pauli, is given by

$\chi_p = \mu_B^2 N(E_f)$, where μ_B is the Bohr magneton and $N(E_f)$ is the density of states at the Fermi energy E_f which is, in the simplest case, $(3n/2E_f)$, where n is the electron density per unit volume.

For an electron gas in a semiconductor obeying Maxwell–Boltzmann statistics, $\chi = n\mu_B^2/(2kT)$, where kT is the thermal energy.

PCAC The partially conserved axial current hypothesis relates the four-divergence of the axial vector current (e.g., $A_\mu^a = \frac{1}{2}\bar{q}\gamma_\mu\gamma_5\lambda^a q$, where q is the quark field and λ^a are the generators of an SU(2) algebra) to the pion field, ϕ^a. The relationship is $\partial^\mu A_\mu^a = f_\pi m_\pi^2 \phi^a$, where m_π is the mass of the pion and f_π is the empirical pion decay constant. If $m_\pi = 0$, then the four-divergence of the axial vector current would be zero and the axial current would be exactly conserved. This relationship is useful in studying pion–nucleon coupling.

PCT theorem A theorem which states that theories having Hermitian, Lorentz-invariant Lagrange densities of local quantum fields will be invariant under the combined operation of parity (P), charge conjugation (C), and time reversal (T).

Peccei–Quinn symmetry A hypothetical non-gauge, Abelian U(1) symmetry which was postulated in order to solve the strong CP problem (i.e., the fact that the strong interaction does not violate CP symmetry despite the existence of instanton effects). The spontaneous breaking of this U(1) symmetry gave rise to a nearly massless Nambu–Goldstone boson called an axion. The axion has not been seen experimentally, which rules out the simple Peccei–Quinn models but not certain extensions.

Peltier coefficient The amount of energy that is liberated or absorbed per unit second when unit current flows through the junction formed by two dissimilar metals.

Peltier effect (**1**) Discovered in 1834 by Jean-Charles A. Peltier. If two metals form a junction and an electric current passes through this junction, heat will be emitted or absorbed at the junction in addition to the Joule heating. The heat current density $Q = \prod J$, where $\prod$ is Peltier's coefficient and J is the electric current density. Since $\nabla \cdot J = 0$, $\nabla \cdot Q$ is not zero since $\prod$ is different for the two metals. The Peltier heat is a reversible heat. In a closed circuit with two junctions, the heat emitted at one junction equals that absorbed at the other junction.

(**2**) The junction of two different metals subjected to an electric current will yield a temperature change across the junction. If the direction of current is reversed, the heating effect switches to a cooling effect. The temperature change is directly proportional to the current.

penetration probability The probability that a particle will pass through a potential barrier through a finite region of space, where the potential energy is larger than the total energy of the particle.

penguin diagram A higher order, radiative correction Feynman diagram whereby a quark of one flavor (e.g., the bottom quark) in the initial state can change into a quark of another flavor (e.g., the strange quark) in the final state. The loop will contain a W boson which is the cause of the flavor change. These diagrams are important in studying CP violation.

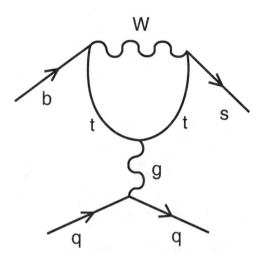

A typical penguin diagram. W is the W gauge boson and t, b, s, and q are the top quark, bottom quark, strange quark, and a generic quark respectively; g is a gluon.

perfect dielectric   A dielectric for which all of the energy required to establish an electric field within the dielectric is reversibly returned when the field is removed. The best real example of a *perfect dielectric* is a vacuum since all other dielectrics irreversibly dissipate energy during the establishment or removal of an electric field within them.

perfect differential   For a function, the *perfect differential* is

$$f = f(x, y, z) .$$

The *perfect differential* of f with respect to x is given by

$$df = \frac{\partial f}{\partial x} dx + \frac{\partial f}{\partial y} dy + \frac{\partial f}{\partial z} dz .$$

perfect gas   In the *perfect* (or ideal) *gas* equation, the individual gas atoms are assumed to behave as non-interacting ideal point particles. Furthermore, any collisions that occur either between gas atoms or between gas atoms and the wall of the container are assumed to occur instantaneously. Given these assumptions, it is possible to write down (from first principles) an equation of state relating the three state variables, pressure (P), temperature (T), and volume (V), in terms of the *perfect gas* constant (R), such that

$$PV = nRT$$

where n is the number of moles of gas present.

periodic boundary conditions   In discussing wave propagation in a crystal of sides $N_1 \underline{a}_1$, $N_2 \underline{a}_2$, and $N_3 \underline{a}_3$, where $\underline{a}_1, \underline{a}_2$, and $\underline{a}_3$ are the primitive translations, it is a standard procedure to assume any function we seek, such as $\Psi(\underline{r})$, is periodic with the periodicity $N_1 \underline{a}_1$, $N_2 \underline{a}_2$, and $N_3 \underline{a}_3$. $\Psi(\underline{r})$ can be an electron wave function or an amplitude of a lattice vibration wave, for example.

periodic table   A table of all chemical elements arranged in ascending order of atomic number and organized in columns by similar chemical properties, originally invented by Mendeleev. The periodicity of chemical behavior is understood in terms of similar electronic structure for the outer, or valence, electrons of elements in the same column.

permeability   Symbol for this quantity is μ. In SI units, absolute *permeability* is defined as the ratio of magnetic flux density (B) to magnetic field strength. Thus, $\mu = B/H$. The *permeability* of free space is given by the constant $(\mu_0) 4\pi \times 10^{-7}$. The relative *permeability* of a material is defined as the ratio of *permeability* (μ_r) to the *permeability* of free space (i.e., $\mu_r = \mu/\mu_0$).

permittivity   According to Coulomb's law, two point charges Q_1 and Q_2, separated in space by a distance r, are subjected to an electrical force (repulsive or attractive depending on the sign of the charges involved) given by $F = Q_1 Q_2 / 4\pi \varepsilon r^2$. The constant ε is called the *permittivity* of the medium. The *permittivity* of free space ε_0 has the value of 8.854×10^{-12} F/m. Relative *permittivity* is a measure of the effect of the electric field on a material compared to free space. It is given by the ratio $\varepsilon/\varepsilon_0$. It is denoted by the symbol ε_r.

permutation operator   An operator which, when applied to a many-particle wave function of identical particles, rearranges their coordinate variables. The permutations form a group.

permutation symmetry   In a many-particle system of identical particles, the permutation operation for particle coordinates keeps the Hamiltonian invariant. This fact is useful for the analysis of non-relativistic systems, where the Hamiltonian is free from the spin variables. Then the energy eigenstates are classified in accordance with the symmetry property of the permutation group. In the case of fermions, the Pauli principle, requiring the change of sign of the many-particle wave function for any interchange of particles, has to be considered in connection with the permutation operation for the spin function. This requirement gives rise to a restriction on the accessibility of the orbital eigenstates. For example, totally symmetric orbital states, as though the lowest energy could be achieved by one of them, are not accessible if the number of fermions is more than three.

perpetual motion It is possible to identify two general types of *perpetual motion* machines, both of which are disallowed by the laws of thermodynamics. In the first case, the continual motion of a machine creates its own energy and in doing so contravenes the first law of thermodynamics. In the second case, the complete conversion of heat into work by a machine contravenes the second law of thermodynamics.

perturbation theory A method for solving problems by first deriving a solution for a simplified problem and using it as a starting point for the exact solution. The difference between the original and the simplified processes is treated as a perturbation of the first solution. The approach usually results in a convergent series by repeated application of the perturbation to subsequent solutions. The series can then be used as an approximation of the exact solution to an arbitrary precision. For the method to result in convergence, it is necessary, but not sufficient, for the perturbation to depend on some naturally small parameter. A typical example for electromagnetic processes is expansion in terms of the fine-structure constant $\alpha = 1/137$, resulting in a series of powers of α which usually converges rapidly.

Pfirsch–Schlüter theory Plasma currents and transport caused by the separation of charges driven by charged particle drifts in toroidal plasma confinement devices, not including the effect of magnetic trapping of particles.

phase Quantum states are generally described by complex numbers, such as wave functions. The complex *phase* of the state is understood to be unobservable and is therefore considered arbitrary, as all measurable quantities should be real; all such quantities are obtained as squares of the absolute values of the relevant complex numbers, wave functions, or matrix elements. However, differences in *phase* between two states can be observable, giving rise to quantum interference effects.

phase conjugation The process whereby the phase of an output wave is the complex conjugate of the phase of the input wave. The spatial part of the wave remains unchanged while the sign of the time t is reversed in the temporal part of the wave. *Phase conjugation* is a time reversal operation.

phase equilibrium At equilibrium, the chemical potential of a constituent in one phase must be equal to the chemical potential of the same constituent in every other phase.

phase fluctuation A quantum mechanical *phase fluctuation* is given by $<(\delta\theta)^2> = \frac{<\Delta a_2^2(\theta_0)>}{<n>}$ with $a_2(\theta_0) = \frac{ae^{-i\theta_0} - a^\dagger e^{i\theta_0}}{2i}$, where a and $a^\dagger$ are boson annihilation and creation operators. Amplitude fluctuation is given by $<(\delta a_\parallel)^2> = <\Delta a_1^2(\theta_0)>$, where $a_1(\theta_0) = \frac{ae^{-i\theta_0} + a^\dagger e^{i\theta_0}}{2}$. Amplitude and *phase fluctuations* are important concepts for squeezed states.

phase matching The condition of momentum conservation in processes where several lasers are involved, giving rise to an increased coupling between the different modes.

phase rule First derived by Gibbs in 1875, the phase rule provides a relationship between the number of degrees of freedom of a thermodynamic system, f, the number of intensive parameters to be varied, I, the number of phases, ϕ, the number of components, c, and the number of independent chemical reactions, r, such that

$$f = I - \phi + c - r.$$

As an illustration, for a mixture of water, hydrogen, and oxygen with pressure and temperature varied, $I = 2$, $c = 3$, and $r = 1$. Thus, for this system, it is possible to have up to four phases in mutual equilibrium. For example, at low temperature and pressure we may have solid water, solid hydrogen, and solid oxygen in equilibrium with a vapor of some appropriate composition.

phase shift Consider a scattering of a particle wave by a spherically symmetric potential around an origin. For a partial wave of the particle, the *phase shift* is the difference between the phase of the scattered wave far from the origin and the corresponding phase of the incoming wave, which is a plane wave.

phase space An abstract space whose coordinates are the degrees of freedom of the system. For a two-dimensional simple harmonic oscillator, the positions (i.e., x and y) and the momenta (i.e., p_x and p_y) of the oscillator when combined would form the coordinates (x, y, p_x, p_y) of the phase space.

phase squeezing Phase and amplitude fluctuations are related by the following uncertainty relation, $<(\delta\theta)^2><(\delta a_\parallel)^2> \geq \frac{1}{16<n>}$. Phase squeezing results when $<(\delta\theta)^2> \leq \frac{1}{4<n>}$. Amplitude squeezing requires $<(\delta a_\parallel)^2> \leq \frac{1}{4}$.

phase state The state defined by

$$|\theta> = \frac{1}{(s+1)^{1/2}} \sum_{n=0}^{s} e^{in\theta} |n>,$$

where $|n>$ is a photon number state, a *phase state* that behaves in some ways as a state of definite phase θ for large s.

phase switching Fast changes in the interaction between the electromagnetic field and atomic systems bring out the importance of the nonlinearity in studies of atomic parameters like relaxation time, line widths, and splittings. Fast *phase switching* can be accomplished by irradiating the sample with an appropriate picosecond light pulse.

phase transition (1) Phase changes are routinely observed and their understanding has been limited to a few models. Weiss molecular field theory has led to partial understanding of ferromagnetism. The Ising model, which is used to model many phenomenona, has proved useful, although it was only solved exactly once, by Onsager in 1944, for a two-dimensional square ferromagnetic lattice in zero magnetic field. The theory of *phase transitions* was recently advanced by the work of Kadanoff and Wilson. Wilson, using ideas from the renormalization work on quantum electrodynamics, developed a theory of critical point singularities which describes the behavior of the physical quantities near the critical points and methods for their calculation.

(2) A process whereby a thermodynamic system changes from one state to another which has different properties, over a negligible range of temperature, pressure, or other such intensive variable. Examples include the melting of ice to form water, the disappearance of ferromagnetism at temperatures above the Curie temperature, and the loss of superconductivity in materials in a magnetic field above the critical field density.

phi An unstable, spin 1 meson which is thought to be predominantly the bound state of a strange and an antistrange quark.

phonon A quantized vibrational mode of excitation in a body, which can be described mathematically as a particle of specific momentum, or frequency, analogous to a photon, the quantum of light.

phosphor Luminescent solids such as ZnS.

phosphorescence The absorption of energy followed by an emission of electromagnetic radiation. *Phosphorescence* is a type of luminescence and is distinguished from fluorescence by the property that emission of radiation persists even after the source of excitation is removed. In *phosphorescence,* excited atoms have relatively long life times (compared to atoms exhibiting fluorescence) before they make transitions to lower energy levels.

photino The hypothetical spin 1/2, supersymmetric partner particle of the photon.

photoconductivity In certain materials, conductivity is increased upon illumination of electromagnetic radiation. This is due to the excitation of electrons from the valence to the conduction band.

photodetector Devices that measure the intensity of a light beam by absorption of a portion of the beam, whose energy is converted into a detectable form. Such intensity measurements are not sensitive to squeezing but detect only nonsqueezed light, e.g., antibunching and

photoelasticity When certain materials (such as cellophane) are subjected to stress, they exhibit diffraction patterns relating to the stress applied. This technique is used in locating strains in glass devices such as telescope lenses.

photoelectric detection of light The emission of electrons by light absorption, the photoelectric effect, is used as a means of counting photons and measuring their intensity by measuring the photoelectrons. Such detectors are absorptive and thus constitute destructive measurements of photons.

photoelectric effect The ejection of electrons from the surface of a conductor through illumination by a source of light of a frequency higher than some threshold value characteristic of the material. The discovery that the energy of the ejected electrons is independent of the intensity of incident light but is a linear function of its frequency was the origin of the understanding, due to A. Einstein, of light as consisting of quanta of energy $E = h\nu$, where ν is the frequency and h is Planck's constant.

photoluminescence Luminescence caused by photons.

photomultiplier tube A device used to enhance photon signals. The photomultiplier tube consists of a tube which is kept under vacuum conditions. At the entrance to the tube, a photocathode converts an incoming photon into an electron via the photoelectric effect. This initial electron then strikes a dynode creating more electrons. Further down the tube is a second dynode which is kept at a higher electric potential than the first dynode, so that the electrons created at the first dynode are attracted to it. When these electrons strike the second dynode, they again create more electrons, which are then attracted to a third dynode at a still higher potential. By having a series of these dynodes at increasing potentials, one has a greatly increased number of electrons for an amplified output signal.

photon The quanta of the electromagnetic field. The idea that the electromagnetic field came in quanta called *photons* was originated by Max Planck in order to explain the blackbody radiation spectrum. The energy (a particle property) of the *photon* is related to its frequency (a wave property) via the relationship $E = hf$, with $h = 6.626 \times 10^{-34}$ joules/second being Planck's constant.

photon antibunching Characterized by the correlation between pairs of photon counts as functions of their time separation τ for laser light. The second order coherence is given by $g^{(2)}(\tau) = <n(\tau)n(0)>/\bar{n}^2$, where $\bar{n}$ is the mean number of photon counts in the short time interval τ. *Photon antibunching* corresponds to $1 > g^2(0) \geq 0$. The latter inequality is violated by any classical light field and thus signifies nonclassical light. *Photon antibunching* indicates that an atom cannot emit two photons in immediate succession.

photon bunching Characterized by the inequality $g^2(0) > 1$ for the second order coherence. This criterion is satisfied by every classical radiation field.

photon correlation interferometry The second order coherence associated with the Hanbury–Brown–Twiss effect, interference of two photons, etc.

photon counting The measurement of the photon statistics by photodetectors with the detection of photoelectrons.

photon distribution function The probability of finding n quanta in the radiation field described by the density matrix $\hat{\rho}(t)$ is the *photon distribution function* $<n|\hat{\rho}(t)|n>$.

photon echo The optical analog of spin echoes, which depends on the presence of a group of atoms that give rise to an inhomogeneous broadening of spectral line. The description of the collection of two-level systems is simplified by using the analogy between a two-level atom and a spin 1/2 in a magnetic field, whose dynamic is governed by the Bloch equation. The time-dependent density matrix of a single atom resembles a magnetic dipole undergoing precession in the magnetic field. The echo is generated by the combination of two coherent laser pulses, a sharp $\pi/2$ pulse followed a time τ later by a sharp π pulse. Prior to the first pulse, all the atoms' spins point in the $-\mathbf{k}$ direction; all atoms are in the ground state. The $\pi/2$ pulse rotates all Bloch vectors to the $\mathbf{j}$ direction. Owing to the distribution in the frequency of transitions in the spectral line, the vectors describing the different atoms spread out in the xy plane, with the more detuned atoms precessing faster than the more resonant ones. The application of the sharp π pulse rotates all vectors π radians about the $\mathbf{i}$ axis. After another time τ, the individual atom vectors precess by the same amount and end up at $-\mathbf{j}$, adding constructively and thus radiating an echo at time 2τ.

photonic bandgaps Frequency bands of zero mode density of states preventing the spontaneous decay, via photon emission, in cavities and dielectric materials.

photonic molecules The eigenstates of the atom and the driving field when considered as a single quantum system. Also referred as dressed states of the atom-driving field Hamiltonian.

photon number basis The Hilbert space spanned by the eigenstates $|n>$, with $n = 0, 1, 2, \cdots$, of the photon number operator $a^\dagger a$, where a ($a^\dagger$) is the annihilation (creation) operator of the radiation field.

photon number density operator The operator $a_\mathbf{p}^\dagger a_\mathbf{p}$, where the momentum vector $\mathbf{p}$ is a continuous variable.

photon number operator The operator $a^\dagger a$ of the radiation field, where $[a, a^\dagger] = 1$.

photon number states The eigenstates $|n>$, with $n = 0, 1, 2, \cdots$, of the radiation field for a single mode. The *photon number states* are complete, $\sum_n |n><n| = 1$.

photon–photon correlations The correlation between photons determined by the second order correlation function $G^{(2)}(\mathbf{r}_1, \mathbf{r}_2; t, t) = <E^{(-)}(\mathbf{r}_1, t) E^{(-)}(\mathbf{r}_2, t) E^{(+)}(\mathbf{r}_2, t) E^{(+)}(\mathbf{r}_1, t)>$, where $E^{(+)}(\mathbf{r}, t)$ contains only photon annihilation operators, and its adjoint $E^{(-)}(\mathbf{r}, t)$ contains only creation operators.

photon–photon interferometry Experiments that determine the joint probability for the detection of two photons at two different location in space as a function of the spatial separation. Such experiments establish the quantum nature of light.

photon statistics Properties of light determined by detectors that either annihilate the photon or not. The former experiments are mainly done with detectors based on the photoelectric effect, whereas the latter are denoted by quantum non-demolition measurements. Quantum non-demolition measurements are necessary for the preparation and study of the quantum statistics of light.

photovoltaic cell When light is incident on it, a voltage appears across its terminals or a current flows in the external circuit. A p–n junction would be an example: without any external bias there is a built-in potential difference V_o, the contact potential (which is the difference between the Fermi levels of the p and n regions when at infinite separation). When light generates electron–hole pairs, the electrons drift to the n region and the holes to the p region, creating an open circuit voltage. If the circuit is closed, current would flow in the external circuit.

photovoltaic effect The production of a potential difference between layers of a material when subjected to an electromagnetic radiation.

piezoelectric effect When certain materials are subjected to a stress, a potential difference is formed across the material. This is widely used in gauges to measure pressure.

piezoelectricity Certain anisotropic crystals (i.e., they do not possess a center of symmetry), exhibit an electric polarization upon experiencing a mechanical load. These materials (e.g., quartz, barium titanate, and Rochelle salt) will also exhibit the corollary effect whereby an applied electric field will produce a mechanical deformation. There are many applications for these materials including microphones, loudspeakers, precision transducers, and actuators.

piezometer   An open tube of liquid connected to a vessel under pressure. The height of the liquid in the tube is the pressure head of the liquid.

pion   A group of three unstable spin 0 particles. The neutral *pion* is denoted by π^o, has a mass of roughly 135.0 MeV, and decays with a mean life of about 8.4×10^{-17} seconds. The neutral *pion* predominantly decays into two photons ($\pi^o \longrightarrow \gamma + \gamma$). The positively and negatively charged *pions* are denoted by π^+ and π^- respectively. Both have a mass of roughly 139.6 MeV and decay with a mean life of about 2.6×10^{-8} seconds. The predominate decay mode for the charged pions is into an antimuon and muon neutrino, or into a muon and muon antineutrino ($\pi^+ \longrightarrow \mu^+ + \nu_\mu; \pi^- \longrightarrow \mu^- + \overline{\nu_\mu}$). In the quark model, the *pions* are believed to be composed of various combinations of the first generation of quarks — the up and down quarks.

pipe flow   Viscous flow in a duct driven by a pressure gradient. For laminar flow in a circular pipe, solution of the steady Navier–Stokes equation in circular coordinates gives the velocity profile as

$$u = \frac{1}{4\mu}\frac{dp}{dx}\left(r^2 - R^2\right)$$

where r is the distance from the center of the pipe and R is the pipe radius. Pipe flow becomes turbulent at a Reynolds number of approximately 2300, where the velocity moves from a parabolic to a flat profile. Determination of the friction factor f is of primary interest in pipe flow. For laminar flow, $f = 64/\text{Re}$. For turbulent flow, the Colebrook pipe friction formula or Moody chart are used. In compressible flow of a gas, frictional effects and addition of heating or cooling can be used to accelerate or decelerate the flow field in the pipe. *See* Fanno line and Rayleigh flow.

Pirani gauge   A gauge used to measure low pressures in the range $1 - 10^{-4}$ mbar ($100 - 0.01$ Pa). The operation of the device involves an electrically heated filament that is exposed to the gas requiring pressure measurement. The dissipation of heat from the filament, and hence its temperature, is determined by the surrounding gas pressure. The electrical resistance of the wire is, in turn, determined by its temperature, and thus the filament resistance is ultimately a strong function of the gas pressure. By building the filament in a Wheatstone bridge arrangement, it is possible to measure the pressure in the gas accurately once the system is calibrated.

pitch angle scattering   Scattering in angle due to collisions between charged particles.

Pi theorem   *See* Buckingham's Pi theorem.

Pitot static tube   Combination of a Pitot tube and static tube that measures the fluid velocity through application of Bernoulli's equation. The difference in the static and stagnation pressures can be written in terms of the velocity that can be solved

$$U = \sqrt{2\Delta p/\rho}$$

for negligible elevation differences.

Pitot tube   Slender tube aligned with the flow that measures the stagnation pressure of the fluid through a hole in the front where the fluid stream is decelerated to zero velocity.

PIV   Optical method for measuring fluid velocity by tracking groups of particles in a fluid through the use of Fourier transforms (Young's fringes). The modern digital equivalent is often referred to as digital particle image velocimetry.

Planck distribution   The formula describing the distribution of frequencies in blackbody radiation, where a blackbody is an idealized body which absorbs all radiation incident upon it and emits radiation proportionally to its total energy content or temperature. It is given by

$$u(\nu) = \frac{8\pi h\nu^3}{c^3}\frac{1}{e^{h\nu/k_B T} - 1},$$

where T is the temperature of the blackbody, k_B is the Boltzmann constant, c is the speed of light, and h is a constant named after M. Planck, who first derived the correct formula under the assumption that light can only be emitted or absorbed in units of energy $h\nu$ proportional to its frequency ν. This subsequently gave rise to the quantum theoryo f light when A. Einstein applied the idea to the photoelectric effect.

Planck length The length $(Gh/2\pi c^3)^1/2$, where G is the gravitational constant, h is Planck's constant, and c is the light velocity. It is about 1.6×10^{-35} m. If matter or energy were restricted in this small dimension of space, quantum fluctuations would dominate the geometry of space-time.

Planck mass The mass $(hc/2\pi G)^1/2$, where h is Planck's constant, c is the light velocity, and G is the gravitational constant. It is about 22 μgram, much heavier than any mass of existing elementary particles, or, in rest mass energy, 1.2×10^{19} GeV.

Planck radiation formula A formula for the intensity distribution or spectrum of radiation emitted from a blackbody by thermal fluctuation of charged particles there. Let $I(\nu)\,d\nu$ be the energy of the emitted electromagnetic waves with frequencies in the interval $(\nu, \nu+d\nu)$. The formula states that

$$I(\nu)\,d\nu = [\exp(h\nu/kT) - 1]^{-1} h\nu^3 \left(8\pi/c^3\right) d\nu.$$

Where h is Planck's constant, k is the Boltzmann constant, and T is the temperature of the blackbody.

Planck scale A mass, length, or time scale formed from Planck's constant divided by 2π ($\hbar$), the speed of light (c), and Newton's constant (G_N). The Planck mass is $\sqrt{\hbar c/G_N} = 2.2 \times 10^{-8}$ kg; the Planck length is $\sqrt{\hbar G_N/c^3} = 1.62 \times 10^{-33}$ cm; the Planck time is $\sqrt{\hbar G_N/c^5} = 5.38 \times 10^{-44}$ seconds. The *Planck scale* is thought to be the scale at which the quantum effects of gravity become important.

Planck's constant A number, denoted by h, equal to 6.626×10^{-27} erg/s, giving the constant of proportionality between a photon's frequency and its energy. It also appears, divided by 2π and written as $\hbar$, as the unit of angular momentum. It sets the scale for all quantum effects; the classical limit is often described as the limit where h is very small compared to all other quantities relevant in a problem, which can then be treated classically.

Planck's radiation law The energy spectrum of a perfect (black body) radiation source is given by the relationship

$$E_\omega\,d\omega = \frac{\eta\omega^3}{\pi^2 c^3 (e^{\eta\omega/kT} - 1)} d\omega$$

where E_ω is the energy density radiated at a temperature T into a narrow angular frequency range from ω to $\omega+d\omega$, c is the velocity of light, η is Planck's constant (h) divided by 2π, and k is Boltzmann's constant.

Planck time The constant $(Gh/c^5)^1/2$ with the dimension of time, where G is the gravitational constant, h is Planck's constant, and c is the velocity of light. Its value is approximately 10^{-43} sec.

plane wave A wave of the form $e^{ik\cdot x}$, where x is the space coordinate and k is the momentum divided by $\hbar$. A *plane wave* has infinitely well-defined momentum and therefore describes a particle with infinite uncertainty in its position, due to the Heisenberg uncertainty principle. It serves as a good approximation for a monochromatic, continuous beam of particles moving in parallel. It can also be of use in describing a localized particle, the wave function of which can be constructed as a superposition of *plane waves* with momenta distributed around a central value.

plasma A mixture of free electrons, ions, or nuclei. The glowing region of ions and electrons in a discharge tube is an example of plasma.

plasma dispersion function $Z(\zeta) = \frac{1}{\pi^{1/2}} \int_{-\infty}^{\infty} \frac{e^{-s^2}}{s-\zeta} ds$, where the contour is deformed down and around the pole at $s = \zeta$ when $\text{Im}(\zeta) < 0$.

plasma physics The branch of physics that deals with ionized particles where the flow is treated using kinetic theory as opposed to continuum theory. The electromagnetic effect produces a force on moving particles.

plasma processing Manufacturing process that uses a low-energy plasma to etch computer chips or to change the surface properties of materials.

plasma propulsion The use of a plasma discharge to propel vehicles, especially in outer space. In most *plasma propulsion* devices, electric or magnetic fields are used to increase the speed of the ejected propellant so that a smaller amount of fuel is used for a given thrust.

plasma spraying Manufacturing process using a plasma to spray a surface coating on materials.

plasma torch The use of a plasma discharge for processing waste materials. Hazardous materials are decomposed and dissociated in plasma torches.

plasmon A quantum of longitudinal plasma oscillation of energy $\hbar\omega_p$ where the plasma frequency is given by $\omega_p^2 = 4\pi n e^2/m$ for long wavelengths. Here, n is the electron density, e is the electron charge, and m is the electron mass. *Plasmons* have been observed in surface waves (with frequencies different from ω_p) and in the passage of fast electrons through thin metallic films, which is usually accompanied by energy losses to the excitation of one or more (bulk) *plasmons*. Reflection of fast electrons from thin films shows energy losses to surface *plasmons*.

plateau regime Neoclassical transport in toroidal magnetic confinement devices under conditions where collisions prevent trapped particles from completing their orbits.

p–n junction Imagine two small rectangular blocks of a semiconductor crystal, identical except for doping by impurities. The p crystal is doped with acceptors and conducts by holes. The n crystal is doped with donors and conducts by electrons. The p crystal has its Fermi level close to the top of the valence band. The n crystal has its Fermi level close to the bottom of the conduction band which is higher than the p crystal by an amount V_o eV which is close to the energy gap E_G. If the two crystals are brought into perfect contact, we have a *p–n junction*. On contact, some electrons diffuse to the p region and some holes diffuse to the n region and are annihilated with opposite charges, leaving behind a layer of ionized donors of positive charge on the n side and a layer of ionized acceptors of negative charge on the p side. These ionized donors and acceptors create an electric potential of magnitude V_o rising from the p to the n region (and whose electric field points from the n to the p side). This region is known as the space charge region or the depletion layer, and can be looked upon as a capacitor. If an external bias voltage δV is applied (with p positive and n negative), the built-in voltage V_o is lowered and charges flow (easy flow, forward flow direction) and the electrical current can be shown to have the form, $I = I_0(\exp(\delta V/kT) - 1)$, where I_o is a constant and kT is the thermal energy. This equation is also valid for negative δV (reverse bias). The *p–n junction* thus can be used as a rectifier, or a photovoltaic cell.

If the doping is so heavy that the electrons and holes form degenerate gases, the *p–n junction* can be used as a tunnel diode, a detector, or a laser.

Poincaré group The group of space-time transformations which includes boosts and rotations (the Lorentz group) in addition to space and time translations. The *Poincaré group* is also called the inhomogeneous Lorentz group.

poise Unit of measure of absolute viscosity, $1\,poise = $ dyne $\cdot$ s/cm^2.

Poiseuille flow Viscous flow between parallel plates due to a pressure gradient. Solution of the Navier–Stokes equations gives the velocity profile as

$$u = \frac{dp}{dx}\frac{h^2}{2\mu}\left((y/h)^2 - 1\right)$$

where y is the distance from the line equidistant between the plates and h is the gap width.

Poisson brackets In classical mechanics, the Poisson bracket of two dynamical variables A and B is defined in terms of derivatives with respect to the generalized coordinates and momenta q and p as

$$\{A, B\} \equiv \sum_{i=1}^{N} \left(\frac{\partial A}{\partial q_i} \frac{\partial B}{\partial p_i} - \frac{\partial B}{\partial q_i} \frac{\partial A}{\partial p_i} \right).$$

They appear in Hamilton's equations of motion, which describe the time evolution of a variable as

$$\frac{dA}{dt} = \frac{\partial A}{\partial t} + \{A, H\},$$

in terms of the Hamiltonian H describing the system. A very similar relationship holds true among operators in the Heisenberg picture of quantum mechanics, with the *Poisson brackets* replaced by commutators.

Poisson distribution The probability distribution $P(m)$ for the random variable m given by $P(m) = \frac{\bar{m}^m e^{-\bar{m}}}{m!}$, where $\bar{m}$ is the mean (average) value of m.

Poisson ratio A measure of the change of shape of a material when it is stretched. It is defined as the fractional change in cross-sectional area ($\Delta A/A$) to the fractional change in length ($\Delta L/L$).

polariton Elementary excitations characterized by a dispersion relation for electromagnetic waves that propagate in a medium with a frequency-dependent dielectric constant. The upper and lower *polariton* branches of the dispersion formula are given by the solutions of the relation $\frac{c^2 k^2}{\omega^2} = \varepsilon_\infty + \frac{\Omega_p^2}{\omega_0^2 - \omega^2}$, where ε_∞ is the background dielectric constant, ω_0 is the position of the sharp absorption line, and Ω_p is a constant associated with the imaginary part of the dielectric function.

polarization The degree to which spins of individual particles in an ensemble point to the same direction. For particles with two possible spin orientations along a given axis, it is defined as $P = \frac{N_+ - N_-}{N_+ + N_-}$, where the indices + and − refer to the direction of spin parallel or antiparallel to the positive direction of the axis. When the two numbers are equal, $P = 0$ and the ensemble is said to be unpolarized, or randomly oriented. $P = \pm 1$ implies 100% polarization along one direction or another. *Polarization* can be best achieved with strong magnetic fields and at low temperatures.

polarization drift Drift of charged particles in a magnetic field caused by an electric field that is changing in time. The *polarization drift* velocity is given by the time rate of change of the electric field divided by the product of the cyclotron frequency and the magnetic field strength.

polaron An electron in an ionic crystal can distort the lattice around it by polarization. If the distortion extends a few atomic distances, the electron and its polarization cloud is known as a large *polaron*. If the distortion extends less than an atomic distance, the self-trapped charge is known as a small *polaron*. Feynman was successful in devising variational methods to obtain the *polaron* energies and effective masses.

poloidal beta Plasma pressure divided by the energy density of the poloidal components of a magnetic field in magnetic confinement devices.

polytropic processes Compressions and expansions that have the form:

$$pv^n = \text{constant}$$

where p and v are average values of the pressure and specific volume respectively, and n is a constant called the index of expansion or compression. When $n = 0$ the expression reduces to $p = 0$, and when $n = \infty$ the expression reduces to $v = 0$.

ponderomotive force Force exerted by electromagnetic waves on a plasma.

population inversion A situation where a metastable excited state has a larger occupation number than the ground state. This is usually achieved by continuous excitation, by radiation of the appropriate frequency, or other similar means to a higher-lying state which then decays rapidly to the metastable state. *Population inversion* is the principle behind the operation of a laser, which relies on stimulated de-excitation of the metastable state.

position operator The position of a particle is represented by coordinates, which are real numbers in classical mechanics. In quantum mechanics they are operators. Observed values of coordinates are eigenvalues of the operators representing the coordinates.

position representation A representation of operators and vectors in Hilbert space by choosing eigenfunctions of coordinates as a basis set of representation.

positive feedback *See* feedback.

positron A spin 1/2 particle carrying a positive electrical charge whose magnitude is equal to that of the electron's charge. The *positron* is the antiparticle of the electron and has the same mass as the electron.

positronium A bound state of an electron and a positron. This system is not stable since the electron and positron annihilate one another.

postulate A statement which is not proven from first principles but is offered as a starting point for the development of a theory. Historically, *postulates* were often transformed into theorems when a different formulation of the theory became available. For example, quantum theory can be formulated using the Schrödinger equation as a *postulate,* but it can also be derived from a different set of *postulates,* at the price of accepting as *postulates* statements that would otherwise be provable theorems. The correctness of a set of *postulates* can only be affirmed on the basis of the self-consistency and experimental verification of the resulting theory.

potential A function describing the distribution of forces in space that a particle will experience, in non-relativistic classical or quantum physics. At every point in space, the particle is subject to a force given by the spatial derivatives of the potential at this point.

potential flow Also called irrotational flow. Flow in which the velocity potential satisfies conditions for irrotationality, $\nabla \times \mathbf{u} = 0$, and the stream function ψ and velocity potential ϕ can be solved exactly by Laplace's equation

$$\nabla^2 \psi = \nabla^2 \phi = 0.$$

Potential flow has many aspects similar to inviscid flow, but is distinctly different since it has an absence of vorticity. *Potential flow* theory is often used in high Reynolds number applications such as aerodynamics.

potential scattering Scattering of a particle due to a potential field of force acting on the particle.

Prandtl–Glauert rule Compressibility correction for pressure distribution on a surface at high subsonic Mach number in terms of the incompressible pressure coefficient, C_{p_o},

$$C_p = \frac{C_{p_o}}{\sqrt{1 - M_\infty^2}}.$$

In the supersonic regime,

$$C_p = \frac{2\theta}{\sqrt{M_\infty^2 - 1}}.$$

Prandtl–Meyer expansion fan Expansion region around an obtuse corner in supersonic flow; the region gradually accelerates across the fan. The Prandtl–Meyer function is used to determine the change in Mach number and thermodynamic properties.

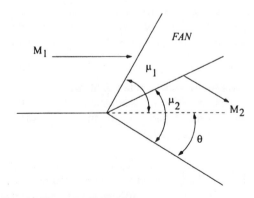

Prandtl–Meyer expansion fan.

Prandtl–Meyer function For isentropic compression and expansion, the *Prandtl–Meyer* *function* $\nu(M)$ is utilized to determine the Mach number change for a change in direction of angle θ

$$\nu(M) = \int \frac{\sqrt{M^2-1}}{1+\frac{\gamma-1}{2}M^2} \frac{dM}{M}$$

$$= \sqrt{\frac{\gamma+1}{\gamma-1}} \tan^{-1} \sqrt{\frac{\gamma-1}{\gamma+1}(M^2-1)}$$

$$- \tan^{-1}\sqrt{M^2-1}$$

where

$$\theta = \nu(M_2) - \nu(M_1)$$

across a Prandtl–Meyer expansion fan.

Prandtl number Ratio of momentum diffusivity to heat diffusivity

$$\mathrm{Pr} \equiv \frac{\nu}{\kappa} = \frac{U_\infty l}{\nu} .$$

Prandtl relation Across a shock wave, the characteristic Mach number can be shown to behave such that

$$M_2' = 1/M_1'$$

and shows that for $M > 1$ on one side of the shock, $M < 1$ on the other side. Observation shows that $M > 1$ upstream of the shock and $M < 1$ downstream of the shock. Though the equation indicates that the reverse is also possible, this would violate thermodynamic restrictions regarding entropy. Thus, flow can only go from supersonic to subsonic across a shock wave.

precession The rotation of the axis of a particle's spin inside a magnetic field. A particle at rest inside a homogeneous magnetic field oriented along the z-axis, $\vec{B} = B\hat{z}$, which is initially in an eigenstate of the x-projection of the spin operator S_x, will find itself after some time in an eigenstate of S_y; eventually it will return to the original state after a period $T = 2\pi/\omega_c$. The cyclotron frequency $\omega_c = |e|B/m_e c$, equal to twice the Larmor frequency, is the frequency at which a charged particle in motion will rotate inside the same magnetic field.

presheath Plasma region that is more than a Debye length away from contact with matter, in which ions are accelerated by an electrostatic potential to a large fraction of the ion sound speed.

pressure Force exerted by a fluid on its surroundings. In an ideal gas, the ideal gas equation relates *pressure* p, density ρ, and temperature T by

$$p = \rho R T$$

where R is the specific gas constant. *Pressure* is a scalar and acts equally in every direction in a fluid; at a boundary with a surface, *pressure* acts normal to the surface.

pressure, absolute Fluid pressure as measured with respect to a zero reference pressure.

pressure, atmospheric Pressure at sea-level due to the weight of the Earth's atmosphere.

$$p_{\mathrm{atm}} = 14.7\,\mathrm{psi} = 101.3\,\mathrm{kPa} .$$

Also referred to as barometric pressure.

pressure broadening Increase in width of a spectral line due to collisions of atoms and molecules. Since the rate of collisions depends on pressure, the change in spectral line width is related to pressure.

pressure, gauge Pressure measured with respect to the atmospheric pressure.

$$p_{\mathrm{absolute}} = p_{\mathrm{gauge}} + p_{\mathrm{atmospheric}} .$$

pressure recovery Increase in pressure in a flow, typically in a diffuser. Diffuser *pressure recovery* is independent of Re but highly dependent upon geometry and inlet conditions. For wide-angle diffusers, guide vanes can be used to reduce the pressure loss and improve the *pressure recovery*. *Pressure recovery* is usually denoted by the *pressure recovery* coefficient C_p

$$C_p \equiv \frac{p_{\mathrm{exit}} - p_{\mathrm{inlet}}}{\frac{1}{2}\rho U_{\mathrm{inlet}}^2}$$

where the rise in static pressure is compared to the inlet dynamic pressure.

pressure, stagnation The pressure in a moving fluid if the flow is brought to rest adiabatically. Measured using a *stagnation pressure* or Pitot probe.

pressure, static The thermodynamic pressure in a moving fluid, measured using a wall tap or static pressure probe. Using Bernoulli's equation, the *static* and stagnation *pressures* can be used to determine the local velocity by

$$U = \sqrt{\frac{2(p_{\text{stagnation}} - p_{\text{static}})}{\rho}}.$$

pressure tensor Components of the plasma pressure in a magnetic field, particularly in plasmas where the perpendicular pressure may be different from the pressure parallel to the magnetic field.

pressure, vacuum Pressure measured with respect to the atmospheric pressure when the absolute pressure is less than the atmospheric pressure.

$$p_{\text{absolute}} = p_{\text{atmospheric}} - p_{\text{vacuum}}.$$

Also referred to as negative gauge pressure.

principle of detailed balance In equilibrium, the power radiated and absorbed by a body must be equal for any particular element of area of the body, for any particular direction of polarization, and for any frequency range.

principle of superposition If $|\psi_1>$ and $|\psi_2>$ are possible states of a system, then by the superposition principle, $a_1|\psi_1> + a_2|\psi_2>$, where a_1 and a_2 are constants, is a possible state of the system.

probability Quantum physics makes predictions for the *probability* for a given process to take place, or for a system to be found in a specific configuration. Determinism in quantum physics refers to the rules for the behavior and time evolution of the *probability* amplitudes, in general complex numbers or functions, which produce *probabilities* when their absolute magnitudes are squared. For example, the Schrödinger equation precisely determines, for a given set of dynamical conditions, the time evolution of the wave function if an initial function is defined; however, the only known consistent physical interpretation of the wave function is a *probability* amplitude.

probability amplitude A complex number, the squared norm of which gives a probability. A wave function gives the *probability amplitude* for a system to be found in a particular space coordinate, or set of coordinates; the square of the norm of the wave function gives the probability, or more precisely the probability density, for the system to be found there: $P(x_i)\,dx = |\psi(x_i)|^2\,dx$.

probability current The probability of finding a particle anywhere in the space is unity, and there is a conservation law of probability. In non-stationary cases, the probability density changes in time. Then the conservation law requires the existence of the *probability current*:

$$(\hbar/2im)[\psi\dagger(x)grad\psi(x) - (grad\psi\dagger(x))\psi(x)],$$

where $\hbar$ is Planck's constant divided by 2π, m is the mass of the particle, $\psi(x)$ is the Schrödinger wave function, with $\psi\dagger(x)$ being its complex conjugate, and *grad* is a coordinate derivative The current is a vector. Its x-component is given by taking the derivative with respect to x in place of *grad*.

probability density *See* probability amplitude.

Proca equation (1) Describes a spin 1 massive particle. If the photon were found to have a rest mass it would be described by the *Proca equation* $(\partial_\mu F^{\mu\nu} + m^2 A^\nu = 0)$ rather than the Maxwell equation $(\partial_\mu F^{\mu\nu} = 0)$. Here $F^{\mu\nu}$ is the field tensor, m is the rest mass, and A^ν is the four-vector gauge field.

(2) Relativistic equation of a quantum field representing a particle of spin 1 and nonzero rest mass. The field requires four-component wave functions. It is similar to the covariant four-vector potential (a set of the vector potential and the scalar potential) of the electromagnetic field. The only difference between the *Proca equations* and the Maxwell equations is the existence of the mass.

projection operator An operator which, acting on a state, isolates the component in a subspace of the space spanned by the whole set of states. If $|i>$ is an eigenstate of some operator with eigenvalue i, then $|\psi_i> = |i><i|\psi> \equiv P_i|\psi>$ produces the component of $|\psi>$ that is a multiple of $|i>$; similarly, the *projection operator* $|i><i| + |j><j|$ projects out the component of a state that is in the subspace spanned by $|i>$ and $|j>$. If $|i>$ is a complete basis, then the sum of all such operators, $I = \sum_i |i><i|$ is the identity operator.

prominences *See* solar prominence.

propagator A function that is the time ordered vacuum expectation value of the product of fields. For example, for a real scalar field $\phi(x)$, the *propagator*, $\Delta(x - x')$ can be written as $-i\Delta(x-x') = \langle 0|T\{\phi(x)\phi(x')\}|0\rangle$, where T is time ordering. The *propagator* represents the physical process of the particle associated with creation of the field out of the vacuum at one space-time point, propagating or moving to some other space-time point, and then being annihilated back into the vacuum. The *propagator* can also be viewed as a Green's function.

proportional counter A device used to measure both the presence and the energy of a charged particle such as those created in radioactive decays or in particle accelerators. The device generally consists of a gas-filled chamber that contains two plates which are maintained at different potentials. When a charged particle passes through the chamber, it ionizes the gas with the positively charged ions moving to one plate and the electrons which have been stripped off going to the other plate. This gives rise to a current which is proportional to the number of atoms of gas which were ionized.

proton A positively charged, spin 1/2 particle which, along with the neutron, is one of the fundamental building blocks of atomic nuclei. The magnitude of the *proton*'s charge is, within experimental limits, equal to the magnitude of the electron's charge, but the *proton* is roughly 2000 times more massive than the electron. The *proton* appears at present to be a stable particle in free space since no experimental evidence of its decay has been seen. However, many theories predict that the proton should be unstable, with a long mean life (current experiments place a model independent-lower bound for the *proton*'s mean life of greater than 1.6×10^{25} years). In the quark model, the *proton* is composed of two up quarks and one down quark.

proton–antiproton collider A machine for protons colliding with antiprotons at high energies in order to probe the structure of matter and its interactions. These machines are generally synchrotrons where the protons and antiprotons are accelerated around a circular path in opposite directions and made to collide with one another at specific interaction stations.

pseudoplastic fluid Non-Newtonian fluid in which the apparent viscosity increases with an increasing rate of deformation. Also referred to as a shear thinning fluid.

pseudopotential The success of the nearly free electron model for s and p electrons in many solids using a weaker potential than the atomic potential to determine energy gaps, suggests that the energy band problem could be approached by solving a Schrödinger equation with the atomic potential replaced by a smoother and weaker potential called the *pseudopotential*, and the resulting smoother wave function called the pseudo-wave function. This becomes apparent when one uses as a trial wave function a set of orthogonalized plane waves (OPW) in the wave equation. By grouping the plane waves into a pseudo-wave function ϕ, and by subtracting the core wave functions part from the potential, one is convinced that the idea is reasonable, although not rigorous. The *pseudopotential* V_{ps} is not unique and this fact can be used to improve the choice of a *pseudopotential*. The method has been successfully used in energy band calculations, in lattice vibrations, in electron phonon interaction, in optical studies such as dielectric constants, and in the conductivity of liquid metals.

V. Heine, M.H. Cohen, and D. Weaire wrote *Solid State Physics,* Vol. 24, Academic Press, New York, 1970, which is devoted to the *pseudopotential*.

pseudorapidity A quantity often denoted by η and defined as $\eta = -\ln(\tan(\frac{\theta}{2}))$, where θ is the angle between the direction of a particle's momentum and the z-axis. This quantity is used in studying collision processes.

psi An unstable, spin 1 meson which is thought to be composed of a charmed and anticharmed quark. The *psi* (Ψ) particle is also called J. The discovery of this meson was evidence for the existence of the charmed quark.

p-state The energy eigenstate with an orbital angular momentum of one. Strictly speaking, this is the angular momentum $\hbar$, Planck's constant, divided by 2π.

pump phase fluctuations Fluctuations in amplitude and phase in a laser pump due to quantum or classical noise.

pump statistics The process of pumping takes atoms from their lower states into their upper excited states. The statistical dependence of the phase and amplitude of the pump laser with time determines the statistical properties of the pump.

pure quantum state The state of maximal information in quantum theory. The density matrix of a pure state is of the form $\rho = |\psi\rangle\langle\psi|$, where $|\psi\rangle$ is any normalized state of the system. In general, $\text{Tr}\rho^2 \leq (\text{Tr}\rho)^2$, where the strict inequality holds for mixed states and the equality holds in the case of pure states.

pure state Any system that can be described completely by a well-defined wave function is said to be in a *pure state*. An ensemble of a large number of particles is in a pure state if all the particles can be described by the same state vector $|a\rangle$, or if the state vectors of all the particles are completely known. Otherwise, the system is in a mixed state, or a mixture (incoherent superposition) of *pure states*. Examples of *pure* and mixed *states* are ensembles of identical particles that are 100% and less than 100% spin-polarized. In general, a state is characterized by a density operator ρ, and the expectation value of an operator A for the state is given by $\langle A \rangle = Tr\,\rho A$, where Tr denotes a sum over diagonal elements. In the case of a *pure state*, the diagonal elements of ρ are simply the probabilities $P_n = |\langle\psi|n\rangle|^2$, where $|n\rangle$ is a complete basis.

pyroelectric crystals Crystals whose electrical polarization is affected by changes in temperature. *Pyroelectric crystals* could be ferroelectric, or piezoelectric. The primitive cell of a pyroelectric crystal has a non-vanishing electric dipole moment.

pyrometer Instrument for measuring high temperatures. *See also* optical pyrometer.

Q

Q(α) representation The representation of the density operator ρ in terms of matrix elements with respect to the pure coherent state $|\alpha>$, i.e., $Q(\alpha, \alpha^*) = \frac{<\alpha|\rho|\alpha>}{\pi}$. The normalization of the density operator to unity, together with the completeness of the coherent states, implies that $\int Q(\alpha, \alpha^*) d^2\alpha = 1$.

Q-factor Ratio of power produced by fusion to the power used to heat the plasma in a fusion reactor.

q, heat The symbol for the quantity of heat that enters a thermodynamic system. Heat is the alternative name for the internal energy of a system whose energy is randomized and free to move from one part of the system to another, or between systems.

Q-machine A magnetic confinement device that produces a quiescent plasma by thermal ionization of cesium or potassium atoms impinging on hot tungsten plates.

Q switch The switching technique used in a laser to change to the resonant mode from the non-resonant mode. An example is a Kerr cell.

quadratic non-linearity Non-linearities in certain quantities that result in the process of solving differential or integral equations by iterative means. The generation of linear, second order, and higher order susceptibilities is one such example.

quadratures The pair of operators defined by $\hat{q} = 2^{-1/2}(\hat{a}^\dagger + \hat{a})$ and $\hat{p} = 2^{-1/2}(\hat{a}^\dagger - \hat{a})$, where $\hat{a}$ and $\hat{a}^\dagger$ are the annihilation and creation operators of the radiation field. The *quadrature* operators are useful to define squeezed states.

quadrupole interaction The effect of interaction between a system of electric or magnetic charges distributed at a certain fashion, e.g., electrons and protons in an atom. *See* electric quadrupole transition.

quantization The restriction to a subset, usually discrete, of the possible values for a variable. Examples are the *quantization* of electric charge in units of the electron charge e, of angular momentum in units of the Planck constant $h/2\pi$, and of energy of the electromagnetic field in units of $h\nu$, where ν is the frequency. Some variables can be quantized in some situations but not others. The momentum of a free particle is not quantized, while restricting the range of motion of the particle between impenetrable walls results in *quantization* of its momentum. In quantum field theories, *quantization* implies that interactions between particles proceed only through the exchange of discrete carriers of the field or force, such as the photon for the electromagnetic interaction.

quantized Hall effect Discovered in 1980; von Klitzing won 1985 Nobel prize. Investigating the conductance properties of two-dimensional electron gases at very low temperature and high magnetic fields, the Hall conductance of such a system plotted as a function of the ratio $r = $ (electron density $*h$)/ (magnetic field $*e$) shows flat plateaus at integer multiples of e^2/h around integer values of the ratio r (h is Planck's constant, e is the electron charge.). The ordinary conductance plotted as a function of r is zero everywhere except where the Hall conductance has a transition from one plateau to another. In other words, there are whole intervals of r where the voltage drop is completely at right angles to the current, with Hall conductance very accurately quantized in terms of the fundamental conductance quantum e^2/h; in between these intervals, the longitudinal conductance has a peak, while the Hall conductance goes from one plateau value to another. In 1982, Tsui, Gossard, and Störmer, working with very pure samples, discovered the so-called fractional quantum Hall effect. Here the conductance is quantized in fractional multiples of e^2/h, like 1/3, 1/5, 2/3, and 2/5, always with an odd denominator. Tsui and Störmer were awarded the 1998 physics Nobel prize for the discovery, sharing it with Laughlin, who developed a theoretical description.

quantum The smallest possible amount of a quantity. The *quantum* of electric charge is the charge of an electron, while the *quantum* of energy for an electromagnetic field is the photon.

quantum (confinement, dots, number, wells, wires, solid, beat) A *quantum well* is a semiconductor structure which confines excitons or carriers in a thin layer (two-dimensional *quantum confinement*). One step further, by considering only a thin wire, we trap particles in (one-dimensional) *quantum confinement*. A carrier is only free to travel along the wire, but for each speed at which it travels, it could have several different ways of being confined. Making a good *quantum wire* with few defects is technologically challenging and is an area of active research. The next step is to increase the *quantum confinement* and make *quantum dots*. The *quantum dot*, although made of many thousands of atoms, has a countable number of allowed states. Each state is separated from the others. This means that it is more like a single atom than like many atoms. Because of this, a *quantum dot* system may be the ideal laser material. Also, it may be an ideal building block for a *quantum computer*. These countable (or discrete) states are so interesting because we can change the properties of these states by varying their environment, e.g., by changing the surrounding electric or magnetic fields. Because the states are separate from each other, it is easy to see them changing.

quantum beats The interference in the decay to a common lower state of two coherently excited atoms by a laser pulse. The radiation power shows a damped oscillation with a beat frequency $v_1 - v_2$, where v_1 and v_2 are the atomic frequency level spacings.

quantum chromodynamics (QCD) (1) A non-Abelian quantum gauge field theory for the strong interaction which governs the structure and interactions of nucleons and nuclei, modeled after the simpler and more thoroughly tested quantum electrodynamics (QED). In *QCD*, the carriers of the strong force are massless, electrically neutral particles called gluons. The strength of the interaction is defined by the strong version of the fine structure constant, called the strong coupling constant α_S; in spite of its name, the coupling is not a constant, depending on the distance, or momentum, scale. The analogous quantity to electric charge is the strong charge or color; there are three different colors and their corresponding anticolors. The name is inspired by the fact that, in addition to the color–anticolor combination, the combination of all three colors also results in a neutral object. In contrast to QED, where the carriers of the field are neutral, gluons themselves carry color. The theory predicts that forces between particles are very weak at short distances, a phenomenon known as asymptotic freedom, increasing roughly linearly with distance at large separations. As a result, only colorless objects are thought to be observable in isolation.

(2) The theory of the strongest of the four known fundamental interactions. This theory postulates that quarks carry a color charge which comes in three varieties: red, blue, and green. The interaction between these color charges is responsible for binding quarks into colorless states such as hadrons, which are composed of three quarks each having a distinct color, or mesons, which are composed of two quarks, one carrying a certain color while the other carries the appropriate anticolor. The quanta of the chromodynamic force are called gluons. The *QCD* interaction is so strong that it is thought to be impossible to liberate individual quarks from a bound state. This hypothesis is called confinement.

quantum coherence The general interference that occurs in quantum mechanics owing to the superposition principle.

quantum cryptography The sending of secret messages between two parties based on the principles of quantum mechanics. The emission of spin 1/2 particles as in the Einstein–Podolsky–Rosen experiment can be used as a signal, and Bell's inequality serves to test for eavesdroppers. Also, the transmission of polarized photons in different polarization states can serve as a means of sending coded messages.

quantum detection Detectors of light based on the photoelectric effect. Two cases can arise. The incident light can excite the electron and extract it from the irradiated surface for detection (external photoelectric effect) or the energy of the light quantum is not sufficient to free the electron from the detector (internal photoelectric effect). In the latter, the detection is via the change of the electrical conductivity of the material.

quantum distribution Representations of the density operator $\hat{\rho}$ in terms of quasi-probability functions that serve as a connection between classical and quantum mechanics. The prototype is the Wigner distribution defined by $W(q, p) = \frac{1}{2\pi} \int_{-\infty}^{+\infty} e^{ipx} < q - \frac{x}{2}|\hat{\rho}|q + \frac{x}{2} > dx$, where $|q>$ is a position eigenstate.

quantum efficiency The efficiency of a quantum detector depends on the characteristic of the detector atoms and the interaction time and is given by the number of photoelectrons created per incident quantum.

quantum electrodynamics (QED) (1) A relativistic Abelian quantum field theory developed in the 1940s as the quantum version of the classical electromagnetic theory. The electromagnetic field is quantized with the forces transmitted between particles through the exchange of field quanta, or photons. With the help of techniques from perturbation theory and renormalization, QED has proven to be one of the most successful theories in physics, achieving remarkable precision in the calculation of fundamental physics quantities such as the anomalous magnetic moment of the electron. It also serves as a model for building theories of more complicated interactions, such as the strong interaction.

(2) The quantum theory of the electromagnetic interaction. In QED, the interaction of electrical charges is taken to occur via quanta of the electromagnetic field called photons. By exchanging real and/or virtual photons, electrical charges interact with one another. The photon is a chargeless spin 1 particle with zero rest mass.

quantum field theory A general term for a classical field theory in which the fields have been quantized either via the canonical method of turning the fields into operators with nontrivial commutation/anticommutation relationships with respect to one another or via the path integral quantization method.

quantum flavordynamics (QFD) A gauge theory of the weak interaction which is an extension or generalization of the original Glashow–Salam–Weinberg theory. QFD takes into account all the known lepton and quark flavors or generations and their mixings with one another.

quantum fluctuation The expected values of bilinear operators in quantum states of the form $< \hat{O}(\xi)\hat{O}(\xi') >$, where the variable ξ can represent a general space-time point. For instance, the derivation of a finite line width of a laser entails the two-time correlation function of the electric field. For squeezed states, the $\hat{O}$ represents the amplitude of the electric field or the quadrature components for the fields.

quantum gravitation A simplified name for the quantum theory of the gravitational field. Also, the study of quantum fields in a curved (Riemann) space representing space-time. Another word for this is quantum gravity.

quantum gravity Any of a set of hypothetical theories which seek to quantize the gravitational interaction or join quantum mechanics with gravity. At present there is no full theory of *quantum gravity*. However, many theories of this sort predict a quanta of the gravitational interaction, generally referred to as the graviton.

quantum interference A fundamental feature of the superposition principle in quantum mechanics where the coherence of the superposition gives rise to destructive or constructive interference. The coherent superposition of atomic states is responsible, for instance, for coherent trapping in three-level atomic systems in the Λ configuration.

quantum jump (1) A quantum system of microscopic size cannot change its states gradually. The change from one stationary state to another is accompanied by emission or absorption of energy quantum or quanta. Hence, the transition is called the *quantum jump*.

(2) A non-Schrödinger-like term in the Liouville equation that represents a discontinu-

quantum Langevin equation Dynamical equation that incorporates system–reservoir interactions quantum mechanically and are of the form

$$\dot{A}(t) = -\frac{\gamma A(t)}{2} + F(t),$$

where $A(t)$ is the annihilation operator in the interaction picture and $F(t)$ represents a quantum noise operator, which is the source of both fluctuations and irreversible dissipation of energy from the system to the reservoir.

quantum Liouville equation The time evolution of the density operator, whether a pure or a mixed state, is governed by the *quantum Liouville equation*

$$\frac{\partial \rho}{\partial t} = -\frac{i}{\hbar}[H, \rho] + \mathcal{L}[\rho],$$

where the Liouvillian $\mathcal{L}[\rho]$ describes the non-Hermitian, irreversible evolution of the system due to couplings to reservoirs.

quantum mechanical operator In quantum mechanics, a physical quantity is represented by a linear Hermitian operator in Hilbert space. The operator is often called the *quantum mechanical operator*. If you measure the physical quantity, the result is one of the eigenvalues of its associated operator. Although a system is in a definite pure state, the observed value of the quantity is not unique but may be any one of the eigenvalues. The state vector representing the pure state can only predict the probability of obtaining any specified eigenvalue at the observation.

quantum mechanics A general term for the foundation of all quantum physics. There are several formulations of *quantum mechanics*, based on different mathematical treatments or even sets of postulates, which are understood to be equivalent. A commonly known set is based on the Schrödinger equation for the wave function describing any system. While *quantum mechanics* is deterministic as far as the behavior of a wave function is concerned, the only known consistent interpretation of the wave function is as a probability amplitude, which has earned *quantum mechanics* a reputation as a probabilistic theory.

quantum noise The random force term $\hat{F}(t)$ that appears in the quantum Langevin equation for the annihilation operator $\hat{a}(t)$ is a form of *quantum noise*. The equation of motion is given by

$$\dot{\hat{a}}(t) = (-i\omega_0 - \kappa)\hat{a}(t) - \hat{F}(t),$$

where κ is a damping constant and the *quantum noise* operator is the source of both fluctuations and dissipation.

quantum nondemolition measurement (QND) A measurement of an observable in quantum mechanics introduces a disturbance in its corresponding conjugate variable. Systems may possess dynamical variables which remain undisturbed after a measurement is performed on the system. Such an observable is called a *QND* observable.

quantum number The eigenvalue for the eigenstate of an operator. A state can be completely determined by its *quantum numbers* with respect to a complete set of commuting, or mutually compatible, operators of which it is an eigenstate. For example, a state of the hydrogen atom can be given in terms of a set of four *quantum numbers* (n, l, m, s) for the eigenvalues of the energy, orbital angular momentum, projection of the orbital angular momentum along some axis, and projection of the spin operator along the same axis.

quantum phase and amplitude fluctuations There are several approaches to addressing the question of how to characterize *quantum mechanical phase and amplitude fluctuations*. One such attempt is by defining amplitude and phase operators. In addition to these abstract definitions, one can also consider measurement-assisted phase observables.

quantum phase operator Operators designed to correspond to the classical phase measurement based on interference experiments.

quantum Rabi flopping The flopping of a two-level atom between the upper and lower levels under the action of an electromagnetic field in the absence of damping. This oscillation between states is in complete analogy to the oscillations of a spin 1/2 magnetic dipole studied by Rabi.

quantum regression theorem States that the expectation value of the two-time correlation function $<A_\mu(t)A_\nu(t')>$ for the system operator $A_\mu(t)$ satisfies the same dynamical equation of motion as the single-time function $<A_\mu(t)>$. The theorem can be derived by using the quantum Langevin equation with the aid of $<F_\mu(t)A_\nu(t')>=0$, where F_μ is the noise operator, or in terms of the reduced density matrix with initial conditions such that $\rho_{sr}(0) = \rho_s(0)\rho_r(0)$ for the system and reservoir-reduced density operators.

quantum state (1) The name used for the purpose of specifying the microscopic state of a physical system in contrast with macroscopic states such as those used in thermodynamics. The *quantum state* is represented by a state vector in Hilbert space.

(2) State of an electron or an atom specified by a unique set of numbers called quantum numbers. Quantum numbers are integer or half-integer numbers that specify the value of a quantized physical quantity such as energy, momentum, angular momentum, etc.

quantum statistics The statistical description of a system of particles whose behavior must be described by quantum mechanics rather than classical mechanics. Typical examples are the Fermi statistics of electrons and the Bose statistics of such bosons as ^{4}He atoms constituting a quantum liquid.

quantum theory of light *See* quantum electrodynamics.

quantum wells Two-dimensional nanostructures where optical properties can be studied in confined geometries.

quantum wires Confinement of electrons, atoms, etc. in restricted spatial dimensions usually in the nanometer range. In *quantum wires*, the movement is confined in two out of three directions in space. In quantum dots, the motion is confined in all three spatial directions.

quantum yield The ratio of the number of quanta produced divided by the quanta lost, e.g., photoluminescence quantum yield = (number of emitted photons)/(number of absorbed photons).

quark A particle which is thought to be the fundamental constituent of hadronic bound state systems. *Quarks* are spin 1/2 particles and they are thought to carry a fractional electric charge of $\pm\frac{2}{3}$ or $\pm\frac{1}{3}$ of an electron's charge. In addition to electric charge, *quarks* are thought to carry color charge. The interaction of *quarks* via their color charge is thought to be so strong that *quarks* are permanently confined in mesons (a *quark* and an antiquark) or hadronic (three *quarks* or three antiquarks). *Quarks* come in three families or generations. In the first generation there are the up *quark* and down *quark*; in the second generation are the charmed and strange *quark*; in the third generation are the top and bottom *quark*. Each generation is similar to the preceding one except for being more massive. It is unknown if there are more families or generations of *quarks* beyond those given above.

quark–gluon plasma A hypothetical state of sub-nuclear matter which is thought to occur at extremely high temperatures and energies. In this state, the quarks and gluons become deconfined and form a gas of many quarks and gluons which are no longer confined to the normal meson or baryon bounds states that arise under normal conditions in QCD. It is thought that this state can be created by colliding heavy nuclei.

quarkonium A bound state of a heavy quark and antiquark, for example, charmonium ($c\bar{c}$) or bottomonium ($b\bar{b}$). The energy spectrum of these systems is often studied using certain phenomenological potentials.

quarter-wave plate The production of circularly or elliptically polarized light from linearly polarized light by introducing a phase difference of $\pi/2$ with the aid of a *quarter-wave plate*. The doubly refracting transparent plates transmit light with different propagation velocities in two perpendicular directions.

quasi-Boltzmann distribution of fluctuations
Any variable, x, of a thermodynamic system that is unconstrained will fluctuate about its mean value. The distribution of these fluctuations may, under certain conditions, reduce to an expression in terms of the free energy, or other such thermodynamic potentials, of the thermodynamic system. For example, the fluctuations in x of an isolated system held at constant temperature are given by the expression

$$f(x) \sim e^{-F(x)/kT}$$

where $f(x)$ is the fluctuation distribution and $F(x)$ is the free energy, both as a function of the system variable, x. Under these conditions, the fluctuation distribution is said to follow a *quasi-Boltzmann distribution*.

quasi-classical distribution   Representations of the density operator for the electromagnetic field in terms of coherent rather than photon number states. Two such distributions are given by the Wigner function $W(\alpha)$ and the Q-function $Q(\alpha)$. The Q-function is defined by $Q(\alpha) = \frac{1}{\pi} <\alpha|\rho|\alpha>$, where $|\alpha>$ is a coherent state. The Wigner function $W(p, q)$ is characterized by the position q and momentum p of the electromagnetic oscillator and is defined by

$$W(p,q) = \frac{1}{2\pi} \int_{-\infty}^{+\infty} dy\, e^{-2iyp/\hbar} <q - y|\rho|q + y>,$$

$W(p, q)$ is quasi-classical owing to the lack of positive definiteness for such distributions.

quasi-continuum   Used to describe quantum mechanical states which do not form a continuous band but are very closely spaced in energy.

quasi-geostrophic flow   Nearly geostrophic flow in which the time-dependent forces are much smaller than the pressure and Coriolis forces in the horizontal plane.

quasi-linear approximation   A weakly non-linear theory of plasma oscillations which uses perturbation theory and the random phase approximation to find the time-evolution of the plasma state.

quasi-neutrality   The condition that the electron density is almost exactly equal to the sum of all the ion charges times their densities at every point in a plasma.

quasi one-dimensional systems   A system that is reasonably confined in one-dimension in order to be considered onedimensional. A typical example would be a polymer chain which is separated from neighboring chains by large sidegroups acting as spacers.

quasi-particle   (1) A conceptual particle-like picture used in the description of a system of many interacting particles. The *quasi-particles* are supposed to have particle-like properties such as mass, energy, and momentum. The Fermi liquid theory of L.D. Landau, which applies to a system of conduction electrons in metals and also to a Fermi liquid of ^{3}He, gives rise to *quasi-particle* pictures similar to those of constituent particles. Landau's theory of liquid ^{4}He postulated *quasi-particles* of phonons and rotons, which carry energy and momentum. Phonons of a lattice vibration could be regarded as *quasi-particles* but they can not carry momentum, though they have wave number vectors.

(2) An excitation (not equivalent to the ground state) that behaves as a particle and is regarded as one. A *quasi-particle* carries properties such as size, shape, energy, and momentum. Examples include the exciton, biexciton, phonon, magnon, polaron, bipolaron, and soliton.

quasi-static process   The interaction of a system A with some other system in a process (involving the performance of work or the exchange of heat or some combination of the two) which is carried out so slowly that A remains arbitrarily close to thermodynamic equilibrium at all stages of the process.

quenching   The rapid cooling of a material in order to produce certain desired properties. For example, steels are typically quenched in a liquid bath to improve their hardness, whereas copper is quenched to make it softer. Other methods include splat quenching where droplets of material are fired at rotating cooled discs to produce extremely high cooling rates.

q-value (magnetic q-value) In a toroidal magnetic confinement device, the ratio of the number of times a magnetic field line winds the long way around the toroid divided by the number of times it winds the short way around, with a limit of an infinite number of times.

R

Rabi oscillation When a two-level atom whose excited and ground states are denoted respectively by a and b, interacts with radiation of frequency ν (which is slightly detuned by δ from the transition frequency $\omega = \omega_a - \omega_b$, i.e., $\delta = \omega - \nu$), quantum mechanics of the problem tells that the atom oscillates back and forth between the ground and the excited state in the absence of atomic damping. This phenomenon, discovered by Rabi in describing spin 1/2 magnetic dipoles in a magnetic field, is known as *Rabi oscillation*. The frequency of the oscillation is given by $\Omega = \sqrt{\delta^2 + R^2}$, where $R = pE_0/\hbar$, p is the dipole matrix element, and E_0 is the amplitude of the electromagnetic field. If the radiation is treated quantum mechanically, the *Rabi oscillation* frequency is given by $\Omega = \sqrt{\delta^2 + 4g^2(n+1)}$, where g is the atom–field coupling constant and n is the number of photons.

radial distribution function The probability, $g(r)$, of finding a second particle at a distance r from the particle of interest. Particularly important for describing the liquid state and amorphous structures.

radial wave equation The Schrödinger equation of a particle in a spherically symmetric potential field of force is best described by polar coordinates. The equation can be separated into ordinary differential equations. The solution is known for the angular variable dependence. The differential equation for the radial part is called the *radial wave equation*.

radial wave function A wave function depending only on radius, or distance from a center. It is most useful in problems with a central, or spherically symmetric, potential, where the Schrödinger equation can be separated into factors depending only on radius or angles; one such case is the hydrogen atom, for which the radial part $R(r)$ obeys an equation of the form

$$\left[\frac{1}{2\mu}\frac{d^2}{dr^2} + \frac{\hbar^2 l(l+1)}{2\mu r^2} + V(r)\right] R(r) = ER(r)$$

and r is the relative displacement of the electron and proton, while μ is the reduced mass of the system.

radiation The transmission of energy from one point to another in space. The *radiation* intensity decreases as the inverse square of the distance between the two points. The term *radiation* is typically applied to electromagnetic and acoustic waves, as well as emitted particles, such as protons, neutrons, etc.

radiation damping In electrodynamics, an electron or a charged particle produces an electromagnetic field which may, in turn, act on the particle. The self interaction is caused by virtual emissions and absorptions of photons. The self interaction cannot disappear even in a vacuum, because of the zero-point fluctuation of the field. This results in damping of the electron motion in the vacuum which is called the *radiation damping*.

radiation pressure De Broglie wave–particle duality of implies that photons carry momentum $\hbar \mathbf{k}$, where $\mathbf{k}$ is the wave vector of the radiation field. When an atom absorbs a photon of momentum $\hbar \mathbf{k}$, it acquires the momentum in the direction of the beam of light. If the atom subsequently emits a photon by spontaneous emission, the photon will be emitted in an arbitrary direction. The atom then obtains a recoil velocity in some arbitrary direction. Thus there is a transfer of momentum from photons to the gas of atoms following spontaneous emission. This transfer of momentum gives rise to *radiation pressure*.

radiation temperature The surface temperature of a celestial body, assuming that it is a perfect blackbody. The radiation temperature is typically obtained by measuring the emission of the star over a narrow portion of the electromagnetic spectrum (e.g., visible) and using Stefan's

radiative broadening law to calculate the equivalent surface temperature of the corresponding blackbody.

radiative broadening An atom in an excited state would decay by spontaneous emission in the absence of photons, described by an exponential decrease in the probability of being found in that state. In other words, the atomic level would be populated for a finite amount of time. The finite lifetime can be represented by γ^{-1}, where γ is the decay rate. The finite lifetime introduces a broadening of the level. Spontaneous emission is usually described by treating the radiation quantum mechanically, and since it can happen in the absence of the field, the process can be viewed as arising from the fluctuations of the photon vacuum. The spontaneous emission decay rate γ, for decay from level two to level one of an atom, is given by $\gamma = e^2 r_{12}^2 \omega^3 / (3\pi \epsilon_0 \hbar c^3)$, where r_{12} is the dipole matrix element between the levels and ω is the transition frequency. γ is also related to the Einstein A coefficient by $\gamma = A/2$.

radiative correction (1) The change produced in the value of some physical quantity, such as the mass, charge, or g-factor of an electron (or a charged particle) as the result of its interaction with the electromagnetic field.

(2) A higher order correction of some process (e.g., *radiative corrections* to Compton scattering) or particle property (e.g., *radiative corrections* to the g-factor of the electron). For example, an electron can radiate a virtual photon, which is then reabsorbed by the electron. In terms of Feynman diagrams, *radiative corrections* are represented by diagrams with closed loops. *Radiative corrections* can affect the behavior and properties of particles.

radiative decay Decay of an excited state which is accompanied by the emission of one or more photons.

radiative lifetime The lifetime of states if their recombination was exclusively radiative. Usually the lifetime of states is determined by the inverse of the sum of the reciprocal lifetimes, both radiative and nonradiative.

radiative transition Consider a microscopic system described by quantum mechanics. A transition from one energy eigenstate to another in which electromagnetic radiation is emitted is called the *radiative transition*.

radioactivity The process whereby heavier nuclei decay into lighter ones. There are three general types of radioactive decay: α-decay (where the heavy nucleus decays by emitting an helium nucleus), β-decay (where the heavy nucleus decays by emitting an electron and neutrinos), and γ-decay (where the heavy nucleus decays by emitting a gamma ray photon).

radius, covalent Half the distance between nuclei of neighboring atoms of the same species bound by covalent bonds.

radius, ionic Half the distance between neighboring ions of the same species.

raising operator An operator that increases the quantum number of a state by one unit. The most common is the *raising operator* for the eigenstates of the quantum harmonic oscillator $a^\dagger$. Harmonic oscillator states have energy eigenvalues $E_n = (n + \frac{1}{2})\hbar\omega$, where ω is the frequency of the oscillator; it is also known as the creation operator as it creates one quantum of energy. The action of the *raising operator* on an eigenstate $|n>$ is $a^\dagger|n> = |n+1>$. In terms of the position and momentum operators, it can be written as

$$a^\dagger = \frac{m\omega}{\sqrt{2\hbar}} \left(x - \frac{ip_x}{m\omega} \right).$$

Its Hermitian conjugate a has the opposite effect and is known as the lowering or annihilation operator.

Raman effect (active transitions) Light interacting with a medium can be scattered inelastically in a process which either increases or decreases the quantum energy of the photons.

Raman instability A three-wave interaction in which electromagnetic waves drive electron plasma oscillations. In laser fusion, this process produces high energy electrons that can preheat the pellet core.

Raman scattering When light interacts with molecules, part of the scattered light may occur with a frequency different from that of the incident light. This phenomenon is known as *Raman scattering*. The origin of this inelastic scattering process lies in the interaction of light with the internal degrees of freedom, such as the vibrational degrees of freedom of the molecule. Suppose that an incident light of frequency ω_i produces a scattered light of frequency ω_s, while at the same time, the molecule absorbs a vibrational quantum (phonon) of frequency ω_v making a transition to a higher vibrational level. The frequencies would be related by $\omega_v + \omega_s = \omega_i$. In this case, the frequency of the scattered light is less than that of the incident light, a phenomenon known as the Stokes shift. Alternately, a molecule can give up a vibrational quanta in the scattering process. In this case the frequencies are related by $\omega_i + \omega_v = \omega_s$, and the scattered frequency is greater than that of the incident light, an effect known as the anti-Stokes shift. *Raman scattering* also exists for rotational and electronic transitions.

Ramsey fringes In a *Ramsey fringes* experiment, an atomic beam is made to traverse two spatially separated electromagnetic fields, such as two laser beams or two microcavities. For instance, if two-level atoms are prepared in the excited state and made to go through two fields, transition from the upper to the lower state can take place in either field. Consequently, the transition probability would demonstrate interference. The technique of *Ramsey fringes* is used in high-resolution spectroscopy.

random phases Consider a quantum system whose state, represented by $|\Psi>$, is written as a superposition of orthonormal states $\{|\varphi_n>\}$, i.e., $|\Psi> = \sum_n a_n |\varphi_n>$. The elements of the density matrix are given by $\rho_{nm} = a_n a_m^*$. The density matrix has off-diagonal elements and the state is said to be in a coherent superposition. The expansion coefficients have phases, i.e., $a_n = |a_n| e^{i\theta_n}$, and if the phases are uncorrelated and random, an average would make the off-diagonal elements of ρ vanish, as would be the case if the system is in thermal equilibrium. The nonzero off-diagonal elements of the density matrix, therefore, imply the existence of correlations in the phases of the members of the ensemble representing the system.

Rankine body Source and sink in potential flow in a uniform stream that generates flow over an oval shaped body.

Rankine cycle A realistic heat engine cycle that more accurately approximates the pressure-volume cycle of a real steam engine than the Carnot cycle. The *Rankine cycle* consists of four stages: First, heat is added at constant pressure p1 through the conversion of water to superheated steam in a boiler. Second, steam expands at constant entropy to a pressure p_2 in the engine cylinder. Third, heat is rejected at constant pressure p_2 in the condenser. Finally, condensed water is compressed at constant entropy to pressure p_1 by a feed pump.

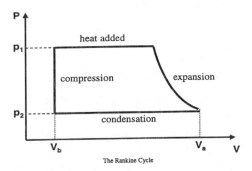

The Rankine cycle.

Rankine efficiency The efficiency of an ideal engine working on the Rankine cycle under given conditions of steam pressure and temperature.

Rankine–Hugoniot relation Jump condition across a shock wave relating the change in internal energy e from the upstream to downstream side

$$e_2 - e_1 = \frac{1}{2}(p_1 + p_2)(v_1 - v_2)$$

where v is the specific volume.

Rankine propeller theory A propeller operating in a uniform flow has a velocity at the propeller disk half of that behind the propeller in the slipstream. Half of the velocity increase is predicted to occur upstream of the propeller and half downstream of the propeller, indicating that the flow is accelerating through the propeller.

Rankine temperature scale An absolute temperature scale based upon the Fahrenheit scale. Absolute zero, 0° R, is equivalent to −459.67° F, while the melting point of ice at −32° F is defined as 491.67° R.

Rankine vortex Vortex model where a rotational core with finite vorticity is separated from a irrotational surrounding flow field. The rotational core can be idealized with a velocity profile

$$u_\theta = \frac{1}{2}\omega_o r_c$$

where r_c is the radius of the core. Matching velocities at $r = r_c$, this makes the irrotational flow outside the core

$$u_\theta = \frac{1}{2}\omega_o \frac{r_c^2}{r}$$

and the vortex circulation

$$\Gamma = \pi \omega_o r_c^2 .$$

This distribution has a region of constant vorticity at $r < r_c$ and a discontinuity at $r = r_c$, beyond which the vorticity is zero. See vortex.

RANS Reynolds Averaged Navier–Stokes. See Reynolds averaging.

Raoult's law The partial vapor pressure of a solvent above a solution is directly proportional to the mole fraction (number of moles of solvent divided by the total number of moles present) of the solvent in solution. If p_0 is the pressure of the pure solvent and X is the solvent mole fraction, then the partial vapor pressure of the solvent, p, is given by:

$$p = p_0 X .$$

Any solution that obeys *Raoult's law* is termed an ideal solution. In general, only dilute solutions obey *Raoult's law*, although a number of liquid mixtures obey it over a range of concentrations. These so-called perfect solutions occur when the intermolecular forces of the pure substance are similar to those between molecules of the mixed liquids.

rapidity A quantity which characterizes a Lorentz boost on some system such as a particle. If a particle is boosted into a Lorentz frame where its energy is E and its momentum in the direction of the boost is p, then the *rapidity* is given by $y = \tanh^{-1}\left(\frac{p}{E}\right)$.

rare-earth elements A group of elements with atomic numbers from 58 to 71, also known as the lanthanides. Their chemical properties are very similar to those of Lanthanum; like it, they have outer $6s^2$ electrons, differing only in the degree of filling of their inner $5d$ and $4f$ shells.

rare earth ions Ions of rare earth elements, viz. lanthanides (elements having atomic numbers 58 to 71) and actinides (elements having atomic numbers 90 to 103).

rarefaction Expansion region in an acoustic wave where the density is lower than the ambient density.

Rarita–Schwinger equation (1) An elementary particle with spin 1/2 is described by the Dirac equation:

$$\left(\gamma_\mu \partial_\mu + \kappa\right)\psi = 0 ,$$

where $\gamma_1, \ldots \gamma_4$ are the Dirac's γ-matrices, obeying the anti-commutation relations $\gamma_\mu \gamma_\nu + \gamma_\nu \gamma_\mu = 2\delta_{\mu\nu}$, κ is the rest mass energy, and ψ is the four-component wave function. A particle with spin 3/2 is described by the *Rarita–Schwinger equation*:

$$\left(\gamma_\mu \partial_\mu + \kappa\right)\psi_\lambda = 0, \quad \gamma_\lambda \psi_\lambda = 0 .$$

Each of the wave functions $\psi_1, \ldots, \psi_4$ have four components (two components represent the positive energy states and the other two represent the negative energy states), and hence the particle is described by 16 component wave functions.

(2) Equation which describes a spin 3/2 particle. The equation can be written as $(i\hbar\gamma_\alpha \partial^\alpha - m_o c)\Psi^\mu(x) = 0$ and the constraint equation

$\gamma_\mu \Psi^\mu = 0$. In these equations, γ_α are Dirac gamma matrices, and $\Psi^\mu(x)$ is a vector-spinor, rather than a plain spinor, $\Psi(x)$, as in the Dirac equation.

Rateau turbine A steam turbine that consists of a number of single-stage impulse turbines arranged in series.

rate constant The speed of a chemical equation in moles of change per cubic meter per second, when the active masses of the reactants are unity. The *rate constant* is given by the concentration products of the reactants raised to the power of the order of the reaction. For example, for the simple reaction

$$A \to B$$

the rate is proportional to the concentration of A, i.e., rate $= k[A]$, where k is the *rate constant*.

rate equation In general, the *rate equation* is complex and is often determined empirically. For example, the general form of the *rate equation* for the reaction $A + B \to$ products is given by rate $= k[A]^x[B]^y$, where k is the rate constant of the reaction, and x and y are partial orders of the reaction.

rational magnetic surface See mode rational surface.

ratio of specific heats The ratio of the specific heat at constant pressure and specific heat at constant volume used in compressible flow calculations

$$\gamma = \frac{C_p}{C_v}.$$

For air, $\gamma = 1.4$.

Rayleigh–Bérnard instability See Bérnard instability.

Rayleigh criteria Relates, for just resolvable images, the lens diameter, the wavelength, and the limit of resolution.

Rayleigh flow Compressible one-dimensional flow in a heated constant-area duct. Assuming the flow is steady and inviscid in behavior, the governing equations simplify to the following:

$$\text{continuity} \quad \rho_1 u_1 = \rho_2 u_2$$

$$\text{momentum} \quad p_1 + \rho_1 u_1^2 = p_2 + \rho_2 u_2^2$$

$$\text{energy} \quad h_1 + \frac{1}{2}u_1^2 + q = h_2 + \frac{1}{2}u_2^2$$

$$\text{total temperature} \quad q = c_p (T_{0_2} - T_{0_1})$$

The behavior varies depending upon whether heat is being added ($q > 0$) or withdrawn ($q < 0$) and whether the flow is subsonic ($M < 1$) or supersonic ($M > 1$). Trends in the parameters are shown in the table below as increasing or decreasing in value along the duct. Note that the variation in temperature T is dependent upon the ratio of specific heats γ.

	$q > 0$ $M < 1$	$q > 0$ $M > 1$	$q < 0$ $M < 1$	$q < 0$ $M > 1$
M	↑	↓	↓	↑
u	↑	↓	↓	↑
p	↓	↑	↑	↓
p_o	↓	↓	↑	↑
T	†	↑	‡	↓
T_o	↑	↑	↓	↓

†: ↑ for $M < \gamma^{-1/2}$, ↓ for $M > \gamma^{-1/2}$;
‡: ↓ for $M < \gamma^{-1/2}$, ↑ for $M > \gamma^{-1/2}$

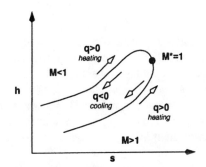

Rayleigh flow Mollier Diagram.

A Mollier diagram shows the variation in entropy and enthalpy for heating and cooling subsonic and supersonic flows. Heating a flow always tends to choke the flow. It is theoretically possible to heat a flow and then cool it to transition from subsonic to supersonic flow and vice-versa.

Rayleigh inflection point criterion To determine flow instability in a viscous parallel flow, a necessary but not sufficient criterion for instability is that the velocity profile $U(y)$ has a point of inflection. See Fjortoft's theorem.

Rayleigh-Jeans law Describes the energy distribution from a perfect blackbody emitter and is given by the expression

$$E_\omega \, d\omega = \frac{8\pi \omega^2 kT}{c^3} \, d\omega$$

where E_ω is the energy density radiated at a temperature T into a narrow angular frequency range from ω to $\omega + d\omega$, c is the velocity of light, and k is Boltzmann's constant. This expression is only valid for the energy distribution at low frequencies. Indeed, attempting to apply this law at high frequencies results in the so-called UV catastrophe, which ultimately led to the development of Planck's quantized radiation law and the birth of quantum mechanics.

Rayleigh number Dimensionless quantity relating buoyancy and thermal diffusivity effects

$$\mathrm{Re} = \frac{g\alpha \Delta T L^3}{\nu \kappa}$$

where α, ν, and κ are the expansion coefficient, kinematic viscosity, and thermal diffusivity respectively.

Rayleigh scattering First described by Lord Rayleigh in 1871, *Rayleigh scattering* is the elastic scattering of light by atmospheric molecules when the wavelength of the light is much larger than the size of the molecules. The wavelength of the scattered light is the same as that of the incident light. The *Rayleigh scattering* cross-section is inversely proportional to the fourth power of the wavelength.

Rayleigh–Schrödinger perturbation expansion Rigorously solving the Schrödinger equation of a system is difficult in almost all cases. In many cases we start from a simplified system described by the Hamiltonian H_0, whose eigenvectors Φ_n and eigenvalues E_n^0 are known, and take account of the rest of the Hamiltonian H_I as a weak action upon the exactly known states. This is perturbation approximation. The *Rayleigh–Schrödinger expansion* is that in the case of the state Ψ_α, its energy E_a, which is supposed to be non-degenerate, is expressed as

$$E_a = E_a^0 + <\Phi_\alpha | H_I | \Phi_\alpha> + \sum_n <\Phi_\alpha | H_I |$$
$$\Phi_n> <\Phi_n | H_I | \Phi_\alpha > / \left(E_a^0 - E_n^0 \right) + \cdots .$$

Rayleigh–Taylor instability Instability of a plane interface between two immiscible fluids of different densities.

ray representation In quantum mechanics, any vector in Hilbert space obtained by multiplying a complex number to a state vector representing a pure state represents the same state. Therefore, we should say that a state is characterized by a ray (rather than a vector) of Hilbert space. It is customary to take a representatives of the ray by normalizing the state to unity. Even so, a phase factor of a magnitude of one is left unspecified. Text books say that a transformation from a set of eigenvectors as a basis for representation to another set for another representation is unitary. That statement is better expressed in operator algebra, where symmetries of our system are clarified in mathematical language. If a symmetry exists it will be described by a unitary or anti-unitary operator, connecting the representations before and after the symmetry operation or transformation. Furthermore, consider groups of symmetry transformations; i.e., a set of symmetry transformations forming a group in the mathematical sense. The set of operators representing the transformations form a representation of the group. This representation is called the *ray representation*.

ray tracing Calculation of the trajectory taken by a wave packet (or, equivalently, by wave energy) through a plasma. Normally this calculation uses the geometrical optical approximation that gradient scale lengths are much longer than the wavelength of the wave.

R-center One of many centers (e.g., F, M, N, etc.) arising out of different types of treatment to which a transparent crystal is subjected to rectify some defects in the form of absorption bands affecting its color. Prolonged exposure with light or X-rays producing bands between F and M bands are responsible for R-centers.

reabsorption Depending on the spectral shape of photon emission and absorption spectra in some media, one observes a strong absorption of emitted photons, i.e., *reabsorption*. This process determines the line width of the electroluminescence of most inorganic light emitting diodes.

real gas *See* perfect gas.

Reaumur temperature scale A temperature scale that defines the boiling point of water as 80° R and the melting point of ice as 0° R.

reciprocal lattice A set of imaginary points constructed in such a way that the direction of a vector from one point to another coincides with the direction of a normal to the real space planes, and the separation of those points (absolute value of the vector) is equal to the reciprocal of the real interplanar distance.

reciprocal relations *See* Onsager's reciprocal relation.

reciprocating engine An engine that uses the pressure of a working fluid to actuate the cycling of a piston located in a cylinder.

recirculating heating system Typically used in industrial ovens or furnaces to maintain the atmosphere of the working chamber under constant recirculation throughout the entire system.

recoil energy The term can be illustrated by the behavior of a system in which one particle is emitted (e.g., hot gas in a jet-engine). The *recoil energy* is determined by the conservation of momentum which governs the velocity of both the gas and the jet. Since the *recoil energy* is equivalent to the kinetic energy of the jet obtained by the emission of the gas, this energy depends on the rifle. If it is held loosely during firing, its recoil, or kick, will be violent. If it is firmly held against the marksman's shoulder, the recoil will be greatly reduced. The difference in the two situations results from the fact that momentum (the product of mass and velocity) is conserved: the momentum of the system that fires a projectile must be opposite and equal to that of the projectile. By supporting the rifle firmly, the marksman includes his body, with its much greater mass, as part of the firing system, and the backward velocity of the system is correspondingly reduced. An atomic nucleus is subject to the same law. When radiation is emitted in the form of a gamma ray, the atom with its nucleus experiences a recoil due to the momentum of the gamma ray. A similar recoil occurs during the absorption of radiation by a nucleus.

recombination The process of adding an electron to an ion. In the process of radiative *recombination*, momentum is carried off by emitting a photon. In the case of three-body *recombination*, momentum is carried off by a third particle.

recombination process The process by which positive and negative ions combine and neutralize each other.

rectification The process of converting an alternative signal into a unidirectional signal.

recycling Processes that result in plasma ions interacting with a surface and returning to the plasma again, usually as a neutral atom.

reduced density matrix For the ground state of an identical particle system described by the wave function $\Psi(x_1, x_2, \ldots, x_n)$, the one-particle *reduced density matrix* is

$$\rho(x'|x'') = \int \Psi(x', x_2, \ldots, x_n) \Psi *(x'', x_2, \ldots, x_n) \, dx_2 \ldots dx_n.$$

The two-particle *reduced density matrix* is

$$\rho(x'_1, x'_2|x''_1, x''_2) = \int \Psi(x'_1, x'_2, x_3, \ldots, x_n) \Psi *(x''_1, x''_2, x_3, \ldots, x_n) \, dx_3 \ldots dx_n$$

reduced density operator Many physical systems consist of two interacting sub-systems. Denoting these by A, and B, the density operator of the total system can be denoted by ρ_{AB}. Quite often, one is only interested in the dynamics of the subsystem A, in which case a reduced density operator ρ_A is formally obtained from the full density operator by averaging over the degrees of freedom of the system B. This can be expressed by $\rho_A = \text{Tr}_B(\rho_{AB})$. For example, consider the interaction of an atom with the modes of the electromagnetic field within a cavity. If the atom is the system A, the many modes of an electromagnetic field could be considered as the other system. While the atom interacts with the field modes, one might be interested in pursuing the dynamics of the atom by considering the density operator ρ_A after formally averaging over the reservoir R of the field modes.

reduced mass A quantity replacing, together with total mass, the individual masses in a two-body system in the process of separation of variables. It is equal to

$$\mu = \frac{m_1 m_2}{m_1 + m_2}.$$

reduced matrix element The part of a spherical tensor matrix element between angular momentum eigenstates that is independent of magnetic quantum numbers. According to the Wigner–Eckart theorem, the matrix element of a spherical tensor operator of rank k with magnetic quantum number q between angular momentum eigenstates of the type $|\alpha, jm>$ has the form

$$<\alpha', j'm'\left|T_q^{(k)}\right|\alpha, jm>$$
$$=<jk; mq|jk; j'm'> \frac{<\alpha' j' \|T^{(k)}\| \alpha j>}{\sqrt{2j+1}}.$$

The double-bar matrix element, which is independent of m, m', and q, is also called the *reduced matrix element*.

reflectance The ratio of the flux reflected by a body to the flux incident on it.

reflection The reversal of direction of part of a wave packet at the boundary between two regions separated by a potential discontinuity. The fraction of the packet reflected is given by the *reflection* coefficient which is equal to one minus the transmission coefficient.

reflection, Bragg The beam reinforced by successive diffraction from several crystal planes obeying the Bragg equation.

reflection coefficient Ratio of reflected to incident voltage for a transmission line. $(Z_0 - Z_R)/(Z_0 + Z_R)$, where Z_0 and Z_R are characteristic and load impedances, respectively.

refractive index When light travels from one medium to another, refraction takes place. The refractive index for the two media (n_{12}) is the ratio of the speed of light in the first medium (c_1) to the speed of light in second medium (c_2). The *refractive index* is thus defined by the equation $n_{12} = c_1/c_2$.

refrigeration cycle Any thermodynamic cycle that takes heat at a low temperature and rejects it at a higher temperature. From the second law of thermodynamics, any refrigeration cycle must receive power from an external energy source.

refrigerator A machine designed to use mechanical or heat energy to produce and maintain a lower temperature.

regenerator A device that acts as a heat exchanger, transferring heat of exit or exhaust gases to the air entering a furnace or the water feeding a boiler. Such a device tends to increase the efficiency of the overall thermodynamic system.

Regge poles A singularity which occurs in the partial wave amplitude for some scattering processes. For some processes, the scattering amplitude, $f(E, \cos\theta)$, where E is the energy and θ is the scattering angle, can be written as a contour integral in the complex angular momentum (J) plane: $f(E, \cos\theta) = \frac{1}{2\pi i} \oint_C dJ \frac{\pi}{\sin \pi J} (2J+1) a(E, J) P_J(-\cos\theta)$, where $a(E, J)$ is the partial wave amplitude. A *Regge pole* is a singularity in $a(E, J)$ for some value of J.

Regge trajectory By plotting the angular momentum (J) and the mass square (m^2) of a given hadron and its rotational excitations, a linear relationship is found to exist of the form $J = \alpha' m^2 + J_0$, where α' is a slope and J_0 is an intercept. This plotted lines form *Regge trajectories*.

regularization A modification of a theory that renders divergent integrals finite. In a quantum field theory, divergent momentum integrals generally arise when radiative corrections are calculated. Some of the more common regularization schemes are Pauli-Villars *regularization*, dimensional *regularization*, and lattice *regularization*.

relative density The density of a material divided by the density of water. It is also known as specific gravity.

relative permeability *See* permeability.

relative permittivity *See* permittivity.

relativistic quantum mechanics A theory that is compatible with both the special theory of relativity and the quantum theory. It is based on the Dirac equation which replaces the Schrödinger equation for spin-1/2 particles with a four-component vector, or *spinor*, as the wave function. Developed in the 1930s, it forms the basis of quantum electrodynamics, the quantum theory of electromagnetism, as well as other modern quantum field theories.

relaxation time The characterisitic time after which a disequilibrium distribution decays toward an equilibrium distribution. The electron *relaxation time* in a metal, for example, describes the time required for a disequilibrium distribution of electron momenta (e.g., in a flowing current) to decay toward equilibrium in the absence of an ongoing driving force and can be interpreted as the mean time between scattering events for a given electron.

relaxation time approximation Approximation to relaxation time, time by which the time-measurable quantity of a physical phenomenon changes exponentially to 1/eth of its original value.

renormalizability Interacting quantum field theories contain technical difficulties, originating from the basic notion of the infinite freedom of field up to endlessly small region of space; up to mathematical points. This is, however, an unphysical difficulty because in an extremely small region certain new field theories or physics would be required. Yet we hope that the known theory can give consistent descriptions and predictions for the phenomena at desired energy range and hence at necessary space dimensions. For some quantum fields, this is shown to be true. In fact, all infinite quantities can be absorbed into a renormalization of physical parameters such as mass and charge. This is the *renormalizability* of the quantum field theories. The quantum electrodynamics is a typical example for providing such renormalizability.

renormalization A rescaling or redefinition of the original bare quantities of the Lagrangian of a theory, such as mass or charge. This rescaling gives a relationship between the original (often infinite) parameters of the theory and the finite real physical quantities.

renormalization group In a particular renormalization scheme R, a renormalized quantity, Γ_R, is related to the unrenormalized quantity, Γ_0, via $\Gamma_R = Z(R)\Gamma_0$, where $Z(R)$ is the renormalization constant associated with R. Under a different scheme R', this relationship becomes $\Gamma_{R'} = Z(R')\Gamma_0$. A relationship can be obtained between Γ_R and $\Gamma_{R'}$, namely $\Gamma_R = Z(R', R) \Gamma_{R'}$. This implies $Z(R', R) = \frac{Z(R')}{Z(R)}$, so that $Z(R', R)$ satisfies the following group multiplication law: $Z(R'', R) = Z(R'', R')Z(R', R)$. The different Zs are the elements of the renormalization group.

representation A choice of a set of quantum numbers, corresponding to a complete set of commuting operators, to describe a state. Typical examples are *representations* in position or momentum space. Since position and momentum operators do not commute, the corresponding quantum numbers cannot be specified simultaneously and a choice of *representation* must be made.

reservoir   A thermal *reservoir* is an idealized large thermodynamic system that can gain or lose heat from the thermodynamic system of interest without affecting its internal energy and hence its temperature. A particle *reservoir* is the analogous case for particle exchange.

residual resistivity   The resistivity of a metal that does not depend on temperature. It is presented even at low temperature and is caused by impurities.

resistance, electrical   The property of a conducting substance determining the magnitude of a current that would flow when a certain potential difference is applied across it.

resistance, minimum   The *minimum resistance* is due to the scattering of conduction electrons showing unexpected features if the scattering center has a magnetic moment given by Kondo theory.

resistance thermometer   A device that uses the dependence of a material's electrical resistance upon temperature as a measure of temperature. For high precision measurements, a platinum wire is typically used, whereas semiconductor materials are the material of choice for high sensitivity (thermistor).

resistive ballooning mode   Pressure-driven mode in which instability is caused or significantly enhanced by electrical resistivity, and the perturbation is concentrated mostly on the outboard edge of a toroidal magnetic confinement device.

resistive drift wave   A magnetic drift mode of plasma oscillation that is unstable because of electrical resistivity.

resistive instability   Any plasma instability that is significantly enhanced or made unstable by electrical resistivity.

resistive interchange mode   Instability driven by plasma pressure gradient together with magnetic reconnection in a magnetic confinement device.

resistivity   The property of a material to oppose the flow of electric current. *Resistivity* (symbol is ρ) depends on temperature. For a wire of length L, cross-sectional area A, and resistivity ρ, the resistance (R) is defined as: $R = \rho L / A$.

resolvent   For the Schrödinger equation $H\Psi = E\Psi$, the resolvent is defined as

$$R(E) = 1/(E - H).$$

resonance   (1) The dramatic increase in a transition probability or cross-section for a process observed when an external applied periodic field matches a characteristic frequency of the system. In particle physics, the term is often used to describe a particle which has a lifetime too short to observe directly, but whose presence can be deduced by an increase in a reaction cross-section when the center-of-mass energy is in the vicinity of the particle's mass.

(2) A particle with a lifetime which is so short that the particle is detected via its *resonance* peak in the cross-section for some process. For example, in the process $\pi^+ + p \to \pi^+ + p$, a *resonance* peak in the cross-section occurs at some particular energy. This *resonance* peak is associated with the Δ^{++} particle which is thought to occur between the initial and final states ($\pi^+ + p \to \Delta^{++} \to \pi^+ + p$).

resonance absorption   The absorption of electromagnetic waves by a quantum mechanical system through its transition from one energy level to another. The frequency of the wave should satisfy the Bohr frequency condition $h\nu = E_2 - E_1$, where E_1 and E_2 are, respectively, the energies of the levels before and after the transition.

resonance fluorescence   The emission of an atom irradiated by a continuous monochromatic electromagnetic radiation. The situation is different from that of spontaneous emission of an atom due to vacuum fluctuations of the electromagnetic field. The *resonance fluorescence* spectrum of a two-level atom weakly excited by a monochromatic field is a single line. The frequency of the fluorescent light is the same as that of the incident light. In reality, the spectral line has a finite width. If the electromagnetic field is strong, then the spectrum of the two-level atom shows a triplet structure — a central peak and two sidebands. The splitting of the line into a triplet can be understood in the dressed atom picture.

resonance level An unstable state of a compound system formed during certain periods in a collision process between two particles. This is found as a peak in a graph of cross-section vs. energy for the scattering of the particles. A typical example is the resonance level of a compound nucleus observed by a peak of the scattering cross-section of a low energy neutron beam by a nucleus.

resonant particles Charged particles with velocities nearly the same as the phase velocity of a longitudinal plasma wave. These particles become trapped within the wave and interact strongly with it.

resonant scattering Scattering of a photon by an atom or nucleus or, in general, a microscopic system, in which the system first absorbs the photon by undergoing a transition from one of its energy states to another state with higher energy, and subsequently reemits the photon by the exactly inverse transition.

resonant surface *See* mode rational surface.

response functions Consider a dielectric system on which an electric field $E(t)$, without any spatial dependence, has been applied. The redistribution of the charges resulting from the applied field causes the medium to be polarized. The polarization of the medium P(t) can be considered as a response to the applied field. The response can be expressed as $P(t) = \int_{-\infty}^{t} K(t', t) E(t')dt'$. $K(t', t)$ is called the *response function* or the after-effect function. In general, the *response function* is a second-rank tensor to allow for the possibility for the two vectors, $E(t)$ and $P(t)$, to be along different directions, as would be the case in a nonisotropic medium. The *response function* is a characteristic property of the physical system. Also, the response has been assumed linear. Quite often, the response is stationary, in which case the *response function* depends on the time difference $(t' - t)$. The concept of the *response function* can be generalized to different physical systems by writing $<g(t)> = \int_{-\infty}^{t} K(t', t) f(t')dt'$. Here, $<g(t)>$ is the ensemble-averaged value of the response. The *response function* is calculated from the quantum mechanics of the atoms or molecules constituting the medium and their interaction with the external field.

Reststrahlen, residual rays Residue left over after successive reflections of infrared rays at quartz or rock salt surfaces. Light also shows such properties at aniline dyes.

reversed field pinch Axisymmetric toroidal magnetic confinement device in which the direction of the toroidal magnetic field reverses near the edge of the plasma.

reversibility The property of a process to proceed in the reverse direction with equal probability. Also known as time-reversal, or motion-reversal, invariance.

reversible process A process whose effects can be reversed such that the system is returned to its original thermodynamic state. An ideal *reversible process* involves a system changing from one thermodynamic state to another through a series of infinitesimally small steps. For example, the transfer of energy by virtue of a temperature difference can only be carried out reversibly if the temperature difference is infinitesimally small and the process is, by definition, infinitely slow.

Reynolds analogy In turbulent stratified flow, the assumption that the eddy coefficients for momentum and heat are nearly equal.

Reynolds averaging Using the Reynolds decomposition and the momentum equation, a time-averaged version of the Navier-Stokes equation is produced suitable for CFD.

Reynolds decomposition In turbulent flow, decomposition of the flow variables into mean and perturbed quantities such that

$$\mathbf{u}(t) = \bar{u} + u'$$
$$p(t) = \bar{p} + p'$$
$$\rho(t) = \bar{\rho} + \rho'$$

where the second and third terms are the steady and unsteady components respectively.

Reynolds experiment Classic experiment in pipe flow demonstrating the difference between laminar and turbulent flows and the importance of the Reynolds number in transition from one state to the other.

Reynolds number The ratio of inertia forces to viscous forces

$$\mathrm{Re} \equiv \frac{\rho U_\infty L}{\mu} = \frac{U_\infty L}{\nu}$$

where U_∞ is a characteristic velocity and l is a characteristic length scale. A critical *Reynolds number* indicates a transition from laminar to turbulent flow. The *Reynolds number* is the most often cited dimensionless group in fluid mechanics. Take a sphere of radius R moving at speed U in a fluid with parameters of density ρ, viscosity μ, pressure p, and temperature T. The Buckingham Pi theorem gives

$$\Pi = R^a U^b \rho^c \mu^d p^e T^f .$$

Using the primary dimensions, we have four equations and six unknowns. This can be simplified by noting that the viscosity and temperature are related ($\mu = \mu(T)$) as are density, pressure, and temperature ($\rho = \rho(p, T)$). Thus, $e = f = 0$. Also, ρ and μ can be combined using kinematic viscosity, $\nu = \mu/\rho$. So,

$$\Pi = R^a U^b \nu^c .$$

The left side is dimensionless while the right now has dimensions $[M]^0[L]^{a+b+2c}[t]^{-b-c}$. Examination shows that $a = b$ and $c = -a$, or $\Pi = (RU/\nu)^a$, where a is left as a variable. Since Π is non-dimensional, choose the simplest solution, $a = 1$. Thus,

$$\Pi = \mathrm{Re} = RU/\nu$$

which is the familiar definition of the *Reynolds number*. This ratio could also be achieved by other means. Typically, one is interested in examining how different effects vary, such as inertial and viscous forces. In equation form,

$$\Pi \sim \frac{\text{inertial forces}}{\text{viscous forces}}$$

where these force are determined by the equations of motion, such that

$$\text{inertial forces} \sim |\mathbf{u} \cdot \nabla \mathbf{u}| \sim U^2/L$$
$$\text{viscous forces} \sim \nu \nabla^2 \mathbf{u} \sim \nu U/L^2$$
$$\Pi \sim \frac{|\mathbf{u} \cdot \nabla \mathbf{u}|}{\nu \nabla^2 \mathbf{u}} \sim \frac{UL}{\nu} = \mathrm{Re} .$$

The *Reynolds number* supplies a relation to compare different physical phenomena by reducing the number of variables. For fluid experiments, instead of varying length scale L, flow velocity U, and viscosity ν, only Re must be varied. Matching geometry, Re may be used such,

$$\mathrm{Re}_\text{model} = \mathrm{Re}_\text{real}$$

where scale effects match prototype effects.

Reynolds number, magnetic *See* magnetic Reynold's number.

Reynolds stress tensor In the Reynolds averaged Navier-Stokes equations, an additional stress term is created with the form $-\rho \overline{u_i u_j}$ whose nine components are

$$-\rho \overline{u_i u_j} = - \begin{bmatrix} \rho \overline{u^2} & \rho \overline{uv} & \rho \overline{uw} \\ \rho \overline{uv} & \rho \overline{v^2} & \rho \overline{vw} \\ \rho \overline{uw} & \rho \overline{vw} & \rho \overline{w^2} \end{bmatrix} .$$

The diagonal components are normal stresses, while the off-diagonal components are shear stresses. In isotropic flow, the off-diagonal components vanish. In observations, the Reynolds stresses are the same size or larger than the viscous stresses.

Reynolds transport theorem For an extensive fluid property Q, the total rate of change of Q is equal to the time rate of change of Q within the control volume plus the net rate of efflux of Q through the control surface

$$\left. \frac{dQ}{dt} \right|_\text{system} = \frac{\partial}{\partial t} \int_V q\rho \, dV + \int_A q\rho \mathbf{u} \cdot d\mathbf{A}$$

where q is the intensive property of Q per unit mass. The transport theorem is fundamental in deriving the fluid dynamic equations of motion for a control volume.

rheopectic fluid Non-Newtonian fluid in which the apparent viscosity increases in time under a constant applied shear stress.

rho meson A family of three unstable spin 1 mesons. There is a neutral *rho meson*, ρ^o, a positively charged *rho meson*, ρ^+, and a negatively charged *rho meson*, ρ^-. The *rho mesons* are thought to be composed of up and down quarks.

rho parameter A parameter in the standard model which is defined as $\rho = \frac{M_W^2}{M_Z^2 \cos^2 \theta_W}$, where M_W is the mass of the W boson, M_Z is the mass of the Z boson, and θ_W is the Weinberg angle. This parameter gives a measure of the relative strengths of the charged and neutral weak currents.

Richardson–Dushman equation The equation describing the thermionic emission from a metallic substance. It gives the number of electrons emitted by the metal in terms of current density (J) as a function of temperature (T)

$$J = AT^2 \exp(-b/T)$$

where A and b are constants depending on the type of material, b being the ratio of work function to Boltzmann's constant.

Richardson number In a stratified flow, the ratio of buoyancy force to inertia force,

$$\text{Ri} \equiv \frac{N^2 l^2}{U^2}$$

where N is the Brunt–Väisälä frequency.

Riemann invariants For finite (non-linear) waves, both expansion and compression and use of the equations of motion and phase-space results in the *Riemann invariants*, J_+ and J_-, where

$$a = \frac{\gamma - 1}{4}(J_+ - J_-) \qquad u = \frac{1}{2}(J_+ + J_-) .$$

Along a C_+ characteristic,

$$J_+ = u + \frac{2a}{\gamma - 1} = \text{constant} .$$

Along a C_- characteristic,

$$J_- = u - \frac{2a}{\gamma - 1} = \text{constant}$$

where a and u are the local speed of sound and flow velocity, respectively.

Righi–Leduc effect The phenomenon of a temperature difference being produced across a metal strip that is placed in a magnetic field acting at right angles to its plane while heat is flowing through it. The type of material used in the strip determines the locations of higher or lower temperatures.

right-hand helicity Property exhibited by a particle whose spin is parallel to its orbital momentum. The eigenvalue of the helicity operator $\sigma_l p_l/|\mathbf{p}|$ is $+1$ in this case.

rigidity modulus *See* elastic modulus.

ring laser cavity A laser cavity consisting of two mirrors set to face each other is referred as a standing wave cavity. In this configuration, the two waves, one traveling in the forward direction and the other in the backward direction, give rise to a standing wave in the cavity. A ring laser utilizes a ring-like cavity with three mirrors. A ring cavity and a standing wave cavity of the same optical path are essentially the same in the sense that one round trip of the ring cavity is the same as a forward and backward path in a standing wave cavity. There are, however, practical advantages in a laser with a ring cavity. When excited, it can oscillate in either of the two distinct counterpropagating directions. A ring laser is an example of a two-mode laser in which the frequencies of the two oppositely-directed waves can be split by rotation of the ring. Also, the ring laser is capable of producing greater single-frequency power output compared with that of a standing wave cavity.

rippling mode A localized MHD instability driven by the gradient of electrical resistivity.

Robins effect Produced when a lateral force on a rotating sphere from altered pressure forces is generated. The force is perpendicular to both the rotation axis and direction of fluid motion. Sometimes referred to in general as the Magnus effect.

rock salt structure Crystalline structure of rock salt, NaCl, sodium chloride, occurring in nature as a mineral.

Rossby number Dimensionless parameter; ratio of the inertial forces to the Coriolis forces in a rotating system

$$\text{Ro} \equiv \frac{U_\infty}{\Omega L}.$$

Commonly used in geophysical applications.

Rossby wave Linear dispersive wave in a rotating system where a vertical displacement of a fluid parcel results in an oscillatory motion in the vertical direction. Rossby waves typically have a wavelength of approximately the size of the planetary radius in the atmosphere, but in the Earth's ocean $\lambda \sim \mathcal{O}$ (100 km). They are also referred to as planetary waves.

rotameter Flow rate meter which utilizes a float in a vertical variable area tube to measure the volumetric flow rate. Also referred to as a variable-area meter.

rotating crystal method The method of analyzing the structure of a crystal with X-rays. The crystal is rotated around one of its axes and the X-ray beam is allowed to fall on it perpendicular to the axis, the reflected radiation being recorded as spots on some photographic device.

rotating wave approximation The interaction Hamiltonian of an atom, specifically a two-level atom, with a single-mode quantized electromagnetic field in the dipole approximation, can be written as

$$V = \hbar \left(g\sigma^+ - g^*\sigma^-\right)\left(a - a^\dagger\right),$$

where g, and g^* are dipole matrix elements, and $\sigma^\pm$ are the atomic transition operators. a, and $a^\dagger$ are, respectively, the photon annihilation and creation operators. The two terms $\hbar g \sigma^+ a^\dagger$ and $\hbar g \sigma^- a$ do not conserve energy. For example, the first term represents an atom that makes a transition from the ground state to the excited state by emitting a photon, a process that would violate conservation of energy. In the interaction picture, the time-dependence of the energy non-conserving terms, respectively, are $e^{\pm i(\omega_0+\omega)t}$, where ω is the frequency of the field and ω_0 is the atomic transition frequency. The energy conserving terms, on the other hand, behave as $e^{\pm i(\omega_0-\omega)t}$. The neglect of the energy non-conserving terms in the Hamiltonian is called the *rotating wave approximation*.

rotational invariance See rotation group.

rotational transform Reciprocal of the magnetic q-value ($1/q$).

rotation group A group formed by rotations of the coordinate axes of space. The quantum mechanics of a spherically symmetric system must be invariant under the rotations. Hence the energy eigenstates should provide an irreducible representation of the rotation group. In fact, the momentum eigenfunctions with a definite eigenvalue of [total orbital angular momentum]^2 span a basis set for the irreducible representation. Generators for the infinitesimal rotations give rise to the angular momentum operators.

Rowland circle A circular shaped magnetic material (e.g., ferromagnet), where the magnetic flux is entirely contained within the material of the ring so that no demagnetization field is present.

R-ratio The ratio of the cross-section for an electron–positron collision to yield hadrons (i.e., $\sigma[e^-e^+ \longrightarrow \text{hadrons}]$) to the cross-section for an electron–positron collision to yield a muon and antimuon (i.e., $\sigma[e^-e^+ \longrightarrow \mu^-\mu^+]$). This ratio, $R = \frac{\sigma(e^-e^+ \longrightarrow \text{hadrons})}{\sigma(e^-e^+ \longrightarrow \mu^-\mu^+)}$, is theoretically proportional to the number of quark flavors that are energetically accessible in the collision times the sum of the squares of the charges of these quarks.

runaway electrons The fast electrons in the tail of the electron distribution function that accelerate to high energy in an electric field because the Coulomb collisional frequency decreases with velocity.

Russel–Saunders coupling   The coupling, in the form of interaction, between the resultant orbital angular momentum of the particles in the atom and the resultant internal or spin angular momentum of the particles.

Rutherford atom   An early model of the atom inspired by the planetary system. It was motivated by experimental evidence from scattering experiments that essentially all the mass of an atom is concentrated in a miniscule positively charged region. It assumed that the negatively charged electrons circulated this positive nucleus in a fashion similar to planets around the sun. The difficulties in explaining the absence of radiation from these electron orbits was one of the main motivations for the development of the quantum theory.

Rutherford scattering   The electromagnetic scattering of an charged particle, which is assumed to be point-like and moving at non-relativistic speeds, from a positively charged nucleus. The nucleus is assumed to be point-like and massive enough that its recoil can be ignored. Ernst Rutherford used this process (scattering positively charged Helium nuclei from gold nuclei) to determine the structure of the atom.

R_ξ-gauge   A general gauge condition which is parameterized via ξ. In terms of the Lagrangian, this gauge fixing condition can be imposed by adding a term to the Lagrangian such as $\mathcal{L}_{GF} = -\frac{1}{2\xi}(\partial_\mu A^\mu + \xi M \phi)^2$ where A^μ is a four-vector gauge potential, M is a mass term, and ϕ is a scalar field. The advantage of fixing the gauge in this general way is that it allows one to study the renormalizability of a theory.

Rydberg atom   A hydrogen-like atom with an electron in a very highly excited state and therefore producing only an average field from the nucleus and all other electrons together.

Rydberg constant   A combination of fundamental constants appearing in the formulas for the energy spectrum of the hydrogen and other atoms. It is equal to $\mathfrak{R} = me^4/2\hbar = 2.18 \times 10^{-11}$ erg $= 13.6$ eV, where m is either the reduced mass of the atom or the mass of the nucleus; in the latter case, the *Rydberg constant* is sometimes written as $\mathfrak{R}_\infty$, since the two definitions coincide exactly for an infinitely heavy nucleus. For example, the energy spectrum of the hydrogen atom is simply $E_n = -\mathfrak{R}/n^2$.

Rydberg states   With the aid of frequency-tunable lasers it is possible to excite atoms into states of high principle quantum number n of the valence electron; n can be very high, of approximately 50–60. Such atoms behave like giant hydrogen atoms. The energy levels can be described by the Rydberg formula, and hence the states are called *Rydberg states*. The energy difference between nearby levels is of the order of R/n^3. Rydberg atoms have rather high values of the electric dipole matrix elements in view of the large atomic size, of the order of qa_0n^2, where q is the charge and a_0 is the Bohr radius. The largeness of the dipole matrix elements coupled with the fact that the emissions are in the millimeter range makes Rydberg atoms ideal for maser experiments in high-Q cavities.

S

Sachs form factor A nucleus form factor. Namely, in the process of electron scattering on nuclei at energies of GeV (and beyond), de Broglie's wavelength electron becomes smaller than the size of a nucleus. In such a case, instead of nuclear form factors, form factors of nucleons are used to describe scattering. Nucleons are particles with spin 1/2, and electrical and magnetic scattering contribute to the cross-section. For this problem, the form factors called *Sachs form factors* are more convenient than standard longitudinal and transversal form factors. *Sachs form factors*, at zero momentum, transfer from the electron to the nucleon ($q = 0$):

$$G_E(0) = \begin{cases} 1 & \text{for a proton} \\ 0 & \text{for a neutron} \end{cases}$$

$$G_M(0) = \begin{cases} \mu_p & \text{for a proton} \\ \mu_n & \text{for a neutron} \end{cases}$$

safety factor The plasma *safety factor*, q, is important in toroidal magnetic confinement geometries, where it denotes the number of times a magnetic field line goes around a torus the long way (toroidally) for each time around the short way (poloidally). In a tokamak, for example, the *safety factor* profile depends on the plasma current profile, and q typically ranges from near unity in the center of the plasma to 2–8 at the edge. The *safety factor* is so named because larger values are associated with higher ratios of toroidal field to plasma current (poloidal field) and, consequently, less risk of current-driven plasma instabilities. The safety factor is the inverse of the rotational transform, ι (iota), and can be expressed mathematically as $q \equiv rB_t/RB_p$, where r is the local minor radius, R is the major radius, and B_t and B_p are the toroidal and poloidal magnetic fields. In stellarator physics, one typically works in terms of the rotational transform instead.

SAGE A joint Russian–American experiment for flux measurement of low energy solar neutrinos using metallic gallium as the detection media. This experiment is based on the reaction

$$\nu + {}^{71}Ga \rightarrow e - + {}^{71}Ge,$$

threshold energy for this reaction is 0.233 MeV. This detector initially runs with 30 T (final 60 T) of metallic gallium. In the initial run (for six months), no event that could be assigned as a solar neutrino (above background level) was detected. In the run with full gallium load (60 T), researchers found a reaction rate below the value predicted by standard solar neutrino models. These results (with the results of GALLEX) could be explained with two models. The first model assumes that neutrinos have sufficiently large dipole magnetic moments that interact with the sun's magnetic field and change its state from left-handed to right-handed. Because only left-handed neutrinos reacts with ^{37}Cl, these newly formed right-handed neutrinos are undetected. This effect has to be correlated with the cyclic variation of sunspots (followed by a change in the sun's magnetic field). A problem is that in this explanation, the magnetic dipole moment of a neutrino has to be 10^8 times larger than the value predicted by the standard model ($10^{-19}\mu_B$).

A second, more plausible explanation is called the Mikheyev–Smirnov–Wolfenstein (MSW) model. This model assumes that electron neutrinos on the way from the sun to earth interact with electrons and convert into muon neutrinos.

Sagnac effect A ring cavity that is rotating will have a phase shift every round trip as the mirrors are constantly approaching or receding from the light beam. As such, the beam suffers a frequency shift $\Delta \nu = \nu \Delta L/L = 4Ac\Omega/\nu L$. Here, L is the cavity path length, A is the area enclosed by the beam, and Ω is the frequency of rotation. One can measure the frequency shift, and hence the rotation rate, for gyroscopic applications.

Saha–Boltzmann distribution Described by the Saha equation, the distribution of ion species for a plasma in local thermodynamic equilibrium, which applies in the (relatively rare) case where the radiation field is in local equilibrium with the ions and electrons.

Saha equation   *See* Saha–Boltzmann distribution.

Sah–Noyce–Schockley current   The current in a bipolar junction transistor arising from the generation of electrons and holes in the depletion region that exists at the emitter base interface. This current adds to the collector current and can be an appreciable fraction of the total collector current at low current levels.

Saint Elmo's Fire   A type of corona discharge originally named by sailors viewing the plasma glow from the pointed mast of a ship. This plasma glow arises when a high voltage is applied to a pointed (convex) object, and the concentration of the electric field at the point leads to ionization and the formation of a corona discharge.

Salam, Abdul   Won the 1979 Nobel Prize in physics for his work on the unified electro-weak theory (*see* Glashow, Sheldon L. and Steven Weinberg, who shared the same prize). Salam, together with Jogesh C. Pati (University of Maryland), made the first model of quark and lepton substructure (1974).

sampling calorimeters   Specific devices for calorimetric measurements in high-energy physics. At very high energies, magnetic measurements become expensive because they require very strong magnetic fields or very long detection arms to measure small trajectory changes. Magnetic detection cannot be used for measurement of energies of neutral particles (neutrons or photons). Calorimetric measurement measures the total energy that was realized in some detection medium. A calorimeter absorbs the full kinetic energy of a particle and produces a signal that is proportional to the absorbed energy. The system of deposition of energy depends on the kind of detected particles. High energy photons deposit energy when they transform into electron–positron pairs. Produced electrons and positrons deposit their energy by ionizing atoms. When they are very energetic, they lose most of their energy through bremsstrahlung. These bremsstrahlung photons can again be converted into electron–positron pairs. Hadrons lose most of their energy through successive nuclear collisions. In most materials with $Z > 10$, the mean free path for nuclear collision is greater than the free path for electromagnetic interactions; because of that, calorimeters for measurement of deposited electromagnetic energy are thinner than calorimeters for measurement of energy deposited by hadrons. Interaction probabilities for neutrons are small, and they can escape undetected. This reduces accuracy in measurement. Calorimeters can function as ionizing chambers (liquid-argon calorimeters) through production of scintillation light or scintillation-sensitive detectors (NaI), or they can relay on the production of Cernikov light (lead glasses). They can be constructed as homogeneous media or *sampling calorimeters*. *Sampling calorimeters* mainly use absorber material that is interspersed with active sampling devices to detect realized energy. This kind of detector is easier and cheaper to build, but has worse resolution than homogeneous detectors.

Sasaki–Shibuya effect   In semiconductors such as silicon or germanium, the lowest conduction band valley does not occur at the center of the reduced Brillouin zone, but rather at an edge. In silicon, the lowest valley is at the so-called x-point, which is the zone edge along the crystallographic direction. Hence the lowest valley is six-fold degenerate in energy since there are six equivalent $< 100 >$ directions.

If a silicon sample is subjected to an external electric field that is not directed along any of the $< 100 >$ directions, then two of the directions will bear a smaller angle with the direction of the electric field than the other four. The effective mass of electrons in these valleys along the direction of the electric field (remember that effective mass is a tensor) will be larger since it will have a larger component of the longitudinal mass as opposed to the transverse mass. Thus, electrons in these valleys remain colder than the electrons in the other four valleys which have a smaller effective mass component along the driving electric field and hence gain more energy from the electric field. Thus, there is a possibility that electrons will transfer from the four hotter valleys to the two colder valleys.

It is even possible that the two colder valleys will contain more electrons than the four hotter ones, even though they constitute a minority of the valleys. If this happens, the average velocity of the carriers (drift velocity) may exhibit a non-monotonic dependence on the electric field (this may happen at well below room temperature) thereby causing a negative differential mobility much like in the case of the Ridley–Watkins–Gunn–Hilsum effect.

saturable absorption Most materials have an absorption coefficient that is dependent on the intensity of the incident light in some non-linear fashion. A common form for the intensity dependence is $\alpha = \alpha_0/(1 + I/I_{sat})$. Here, α is the absorption coefficient, α_0 is the absorption coefficient for small intensities, I is the intensity of the incident beam, and I_{sat} is the saturation intensity, at which the absorption is half the value for vanishingly small intensities.

saturation current The value of the current which cannot increase any further even when the outside signal increases, e.g., in a transistor. The drain current will not increase when the applied voltage is increased.

saturation current, electron or ion When a positive electrical potential is applied to a surface in contact with a plasma (the electrode), the surface attracts electrons in the plasma. Above a certain voltage, the electron current is observed to saturate; this is the *electron saturation current*. Similarly, when a negative potential is applied, the surface attracts ions, and the limiting current is the *ion saturation current*. The exact values of the *saturation currents* depend upon many factors, including the surface geometry and sheath effects, the plasma density, magnetic fields (if any), and the plasma composition, but the basic mechanism for the saturation is that the Debye shielding of the electrode by the surrounding plasma prevents distant ions and electrons from being affected by the electric field of the electrode, so that only ions or electrons drifting into the Debye sheath can be collected by the electrode.

saturation intensity The intensity at which a saturable absorber has half the small intensity absorption coefficient. For a two level atom, the saturation intensity is given by $I_{sat} = (c\hbar^2/8\pi |\mu_{eg}|^2 T_1 T_2)$. Here, c is the speed of light in a vacuum, $\hbar$ is Planck's constant, μ_{eg} is the transition matrix element, and T_1 and T_2 are the population and dipole decay rates respectively.

saturation magnetization The maximum magnetization resulting from the alignment of all the magnetic moments in the substance.

saturation spectroscopy A type of spectroscopy where a strong pump beam (frequency ν) and a weaker probe beam (frequency $\nu + \delta$) are incident on a sample, and the transmission at $\nu + \delta$ is measured. Sub-Doppler precision is possible.

sawtooth When a tokamak runs with enough current to achieve a safety factor of $q < 1$ on the magnetic axis, the plasma parameters (n, T, B) are often observed to oscillate with a *sawtooth* waveform, with long steady increases followed by sudden short decreases, known as *sawtooth* crashes. Similar phenomena are seen in some other toroidal magnetic confinement systems. The oscillation is localized to a region roughly within the $q = 1$ magnetic flux surface, and arises from internal magnetohydrodynamic effects. Plasma confinement is degraded within the *sawtooth* region. Empirically, it is found that the interval between *sawteeth* increases when a sufficient number of superthermal ions are present, but in that case, the subsequent *sawtooth* amplitude is correspondingly increased.

Saybolt viscometer Device used to measure viscosity by measuring the length of time it takes for a fluid to drain out of a container through a given orifice; greater viscosity results in a longer time to drain. The Saybolt Seconds Universal (SSU) scale is the most common unit using this method.

scalar potential In electrostatics, with only static charge distributions or steady currents, Maxwell's equations yield $\vec{\nabla} \times \vec{E} = -\partial \vec{B}/\partial t$. As the curl of $\vec{E}$ vanishes in this case, the electric field can be written as the gradient of a scalar function. The usual choice is to define the *scalar* *potential* ϕ via $\vec{E} = -\vec{\nabla}\phi$. The *scalar potential* is not uniquely defined by this relation, as any ϕ' related to ϕ by a gauge transformation will produce the same electric field.

scanning electron microscopy (SEM) An optical microscope cannot usually resolve features smaller than a wavelength of light. The one exception to this is the case when the sample to be inspected is placed very close (closer than a wavelength) to the microscope. This situation (which is called near field optical microscopy) allows the resolution of features smaller than the wavelength.

Electron microscopy benefits from the much smaller wavelength of electrons (deBroglie's wavelength) compared to that of visible light. The *scanning electron microscope* generates an electron beam and then collimates it to a diameter of only 200–300 Å by passing the beam through a collimator consisting of several electron lenses for focusing. The beam can be rastered over the surface of the sample by magnetic coils or electrostatic plates. When the beam strikes a sample, there is a possibility of extracting several different kinds of signals. Some electrons are reflected at the surface without significant energy loss and can be collected by a surface barrier diode. Low energy secondary electrons that are knocked off by the primary beam can be collected by a wire mesh biased to a few hundred volts. They are then accelerated by several thousand volts before striking a scintillator crystal. The intensity of light emitted as they strike the crystal is proportional to the number of secondary electrons emitted and this intensity can be measured by a photomultiplier tube. Finally, the currents and voltages generated on the sample surface owing to the incident electron beam can be measured.

The selected signal, which may be a composite of two or more of the signals just described, is used for display. Typical display units are cathode ray tubes (CRT). For two-dimensional coverage of a surface, one beam across the face of the CRT will be synchronized with one sweep across the sample surface. For two-dimensional coverage, TV rastering of the beam is used. Magnification is determined by the ratio of beam movement on the surface of the sample to the spot movement across the face of the CRT.

Contrast is achieved because the yield of the secondary electrons depends on the angle of incidence. This allows resolving an angle change of 1° which then provides a depth contrast.

A variation of conventional *SEM* is the field-emission *SEM* where much lower voltages are used. As a result, resulting samples do not charge up, which they do if a large voltage is used. Thus, field emission *SEMs* are more suitable for resistive samples and typically give better resolution.

Other than microscopy, a major application of *SEM* is in fine line electron-beam lithography. The electron beam exposes a resist film (typically PMMA) which consists of long chains of organic molecules. The beam breaks up the chains where it hits and makes those regions dissolvable in a suitable chemical (exposing the resist). Thus, one can delineate nanoscale patterns on a resist film and subsequently develop them to create patterns. What limits the resolution is the emission of secondary electrons which also expose the resist film. Field emission *SEMs* use lower energy and hence cause less secondary electron emission, thereby improving the resolution.

scanning tunneling microscope (STM) A device in which a sharp conductive tip is moved across a conductive surface close enough to permit a substantial tunneling current (typically a nanometer or less). In a common mode of operation, the voltage is kept constant and the current is monitored and kept constant by controlling the height of the tip above the surface; the result, under favorable conditions, is an atomic-resolution map of the surface reflecting a combination of topography and electronic properties. The *STM* has been used to manipulate atoms and molecules on surfaces.

scanning tunneling microscopy (STM) A microscopy technique that allows literal atomic resolution. A metal tip (which ideally has a single or few atoms at the end of the tip) is mechanically scanned over a conducting surface. Current is passed between the tip and the surface at a constant voltage. The current is a tunneling current which tunnels through the air (or partial vacuum) gap between the tip and the surface. The magnitude of this current depends exponentially on the width of the gap which is the tunneling barrier. Thus, the current is very sensitive to the distance between the tip and the surface and hence one can map out the crests and troughs on the surface (surface features).

In the above mode, the tip is scanned horizontally and has no vertical motion. In another mode, the tip is allowed to move vertically to keep the current always constant. A feedback loop is used to achieve this. Thus, the tip follows the surface contour and its vertical motion maps out the surface features.

scattering amplitude A function $f(\mathbf{n}, \mathbf{n}')$, generally of the energy and the incoming and outgoing directions $\mathbf{n}$ and $\mathbf{n}'$ respectively, of a colliding projectile, which multiplies the outgoing spherical wave of the asymptotic wave function $\psi \propto e^{ikr\mathbf{n}\cdot\mathbf{n}'} + f(\mathbf{n}, \mathbf{n}') e^{ikr}/r$. Its squared modulus is proportional to the differential scattering cross-section.

scattering angle The angle between the initial and final directions of motion of a scattered particle.

scattering coefficient A measure of the efficiency of a scattering process. The *scattering coefficient* is defined as $R = I_s L^2 / I_0 V$, where I_s is the scattered intensity, I_0 is the incident intensity, L is the distance to the observation point, and V is the volume of the interaction region.

scattering cross-section The sum of the cross-sections for elastic and inelastic scattering.

scattering length A parameter used in analyzing quantum scattering at low energies; as the energy of the bombarding particle becomes very small, the scattering cross-section approaches that of an impenetrable sphere whose radius equals this length.

scattering matrix A matrix operator $\hat{S}$ which expresses the initial state in a scattering experiment in terms of the possible final states. Also known as collision matrix or *S*-matrix. The operator $\hat{S}$ has to satisfy certain invariance properties and other symmetries, e.g., unitarity condition $\hat{S}^\dagger \hat{S} = \hat{S}\hat{S}^\dagger = 1$.

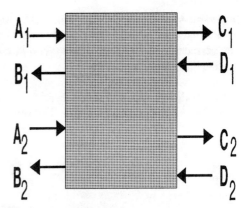

Depiction of a scattering matrix.

scattering operator An operator $\hat{S}$ which acts in the vector space of solutions of a wave equation, transforming solutions representing incoming waves into solutions representing outgoing waves: $\Psi_{\text{final}} = \hat{S}\Psi_{\text{initial}}$.

Schawlow–Townes line width For a four-level laser well above threshold, Schawlow and Townes showed that the lower limit for the laser line width is given by $\Delta\omega = \kappa/2\bar{n}$. Here, κ is the decay rate of the electric field in the cavity and $\bar{n}$ is the mean photon number.

Schmidt values The magnetic dipole moment of a nucleon is given by Schmidt's values. For even–even nuclei, the magnetic moment of nuclei is zero and nuclear spin is also zero. With odd number of nucleons the magnetic dipole moment arises from the unpaired nucleon (a proton, or neutron). For a case of an unpaired neutron, there is only spin contribution; for an unpaired proton there are both orbital and spin contributions. The magnetic dipole moment of nucleon is:

$$\mu = \begin{bmatrix} \mu_n \cdot \left[g_l \left(j - \frac{1}{2} \right) + \frac{1}{2} g_s \right] & j = l + \frac{1}{2} \\ \mu_n \cdot \left[g_l \left(j + \frac{3}{2} \right) - \frac{1}{2} g_s \right] \cdot \frac{j}{j+1} & j = l - \frac{1}{2} \end{bmatrix},$$

where μ_n is a nuclear magneton, $g_l = 1$ and $g_s = 5.586$ for a proton are orbital and spin contributions, $g_l = 0$ and $g_s = -3.826$ for a neutron, j is total angular momentum, and l is the orbital angular momentum.

Schottky barrier A potential barrier at the interface between a metal and a semiconductor that must be transcended by electrons in the metal to be injected into the semiconductor.

Schottky barrier diode A *p-n* junction diode used as a rectifier where the forward bias does not cause any storage of charge, while a reverse bias turns it off quickly.

Schottky defect A point vacancy in a crystal caused by a single missing atom in the lattice.

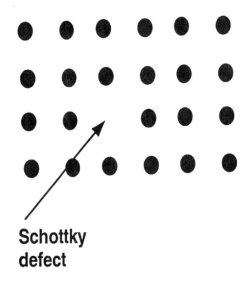

Schottky defect

A missing atom in a lattice of atoms is a Schottky defect.

Schottky Noise An effect used in nondestructive diagnostics of beam parameters in circular accelerators. In circular accelerators, motion of charged particles establish electrical current. Electrical charge of particles gives an increase to statistical variations of current. Now it is a standard method of beam diagnostics.

Schrödinger cat Generally taken to be a macroscopic system in a quantum superposition of states preserving the coherence between two or more discernable outcomes to a measurement. The name comes from Schrödinger's famous thought experiment, where a cat is in a box with a vial of poison which is triggered to open by spontaneous emission of some unstable state.

Schrödinger equation (**1**) A linear differential equation — second order in space and first order in time — that describes the temporal and spatial evolution of the wave function of a quantum particle.

$$i\hbar \frac{\partial \psi}{\partial t} = \left[-\frac{\hbar^2}{2m} \nabla^2 + V(\vec{r}, t) \right] \psi .$$

The left side gives the total energy of the particle and the right side consists of two terms: the first is the kinetic energy and the second is the potential energy. The *Schrödinger equation* is thus nothing but a statement of the conservation of energy. The term within the square brackets on the right side can be viewed as an operator operating on the operand ψ. This operator is called the Hamiltonian.

The solution of the *Schrödinger equation* is the space- and time-dependent wave function $\psi \equiv \psi(\vec{r}, t)$, which is generally a complex scalar quantity. The physical implication of this wave function is that its squared magnitude $|\psi(\vec{r}, t)|^2$ is the probability of finding the quantum particle at a position $\vec{r}$ at an instant of time t. More importantly, in quantum mechanics any physical observable is represented by a mathematical (Hermitean) operator, and the so-called expected value of the operator is what an observer will expect to find if he or she carried out a physical measurement of that observable. The expected value is the integral $\int_{\text{all space}} \psi^* \hat{O} \psi \, d^3\vec{r}$, where the volume integral is carried out over all space, $\hat{O}$ is the operator corresponding to the physical observable in question, and ψ^* is the complex conjugate of ψ.

(**2**) A partial differential equation for the Schrödinger wave function ψ of a matter field representing a system of one or more nonrelativistic particles, $-i\hbar(\partial \psi/\partial t) = H\psi$, where H is the Hamiltonian or energy operator which depends on the dynamics of the system, and $\hbar$ is Planck's constant.

Schrödinger picture A mode of describing dynamical states of a quantum-mechanical system by state vectors which evolve in time and physical observables which are represented by stationary operators. Alternative but equivalent descriptions in use are the Heisenberg picture and the interaction picture.

Schrödinger representation Often used for the Schrödinger picture.

Schrödinger's wave mechanics The version of nonrelativistic quantum mechanics in which a system is characterized by a wave function which is a function of the coordinates of the particles of the system and time, and obeys a differential equation, the Schrödinger equation. Physical observables are represented by differential operators which act on the wave function, and expectation values of measurements are equal to integrals involving the corresponding operator and the wave function.

Schrödinger variational principle For any normalized wave function Ψ, the expectation value of the Hamiltonian

$$< \Psi | H | \Psi >$$

cannot be smaller than the true ground state energy of the system described by H.

Schrödinger wave function A function of the coordinates of the particles of a system and of time which is a solution of the Schrödinger equation and which determines the average result of every conceivable experiment on the system. Also known as probability amplitude, psi function, and wave function.

Schwarz inequality In the form typically used in quantum optics, it states $|V_1|^2 |V_2|^2 \geq |\langle V_1 | V_2 \rangle|^2$, where $|V_1|^2 = |\langle V_1 \rangle|^2$. The brackets can represent a classical or quantum average.

Schwarz, John John Schwarz of the California Institute of Technology, together with Michael Green and Pierre M. Ramond, is an architect of the modern theory of strings.

Schwinger, Julian He developed the gauge theory of electromagnetic forces (quantum electrodynamics, QED). *Schwinger*, Richard P. Feyman, and Sin-Itiro Tomonaga first tried to unify weak and electromagnetic interaction. Schwinger introduced the Z neutral boson, a complement to charged W bosons.

Schwinger's action principle For any quantum mechanical system there exists an action integral operator constructed from the position operators and their time derivatives in exactly the same manner as the corresponding classical action integral W, an integral of the Lagrangian over time from t' to t''. In performing an arbitrary general operator variation, the ensuing change in the action operator δW is the change between the values at t'' and t' of the generator of a corresponding unitary transformation, causing the change in the quantum system. Its classical analog is the generator of a classical canonical transformation.

scientific breakeven One of the major performance measurements in fusion energy research. In steady-state magnetic confinement fusion, *scientific breakeven* means that the fusion power produced in a plasma matches the external heating power applied to the plasma to sustain it, i.e., $P_{\text{fusion}}/P_{\text{heating}} \equiv Q \geq 1$. This concept can be extended to inherently pulsed fusion approaches, such as inertial confinement fusion, in which case *scientific breakeven* can be said to occur when the fusion energy produced in the plasma matches the heating energy that was applied to the plasma. The heating power and energy are only what is actually applied to the plasma; conversion losses are typically neglected. Several other types of breakeven are commonly used. *See* breakeven.

scintillation Emission of light by bombarding a solid with radiation. High energy particles are usually detected by this process in scattering experiments.

scintillation detectors These devices detect charged particles. Scintiallators are substances that produce light after the passage of charged particles. Two types of scintiallators are primarily used: organic (or plastic scintillators, e.g., anthracene, naphthalene) and inorganic (or crystalline scintillators, e.g., NaI, CsI). Activators that can be excited by electron–hole pairs produced by charged particles usually dope crystalline scintillators. These dopants can be de-excited by the photon emission. Organic scintillators have very quick decay times ($\sim 10^{-8}$ s). Inorganic crystal scintillators decay slower ($\sim 10^{-6}$ s). Plastic scintillators are more suitable for a high-flux environment.

scrape-off layer (SOL) The outer layer of a magnetically confined plasma, where the field lines come in contact with a material surface (such as a divertor or limiter). Parallel transport of the edge plasma along the field lines to the limiting surface scrapes off the plasma's outer layer (typically about 2 cm), thereby defining the plasma's outer limit.

screening   Effective reduction of electrical charge and hence the electric field around the nucleus of the atom due to the effect of electrons surrounding it.

screening constant   A correction to be applied to the nuclear charge of an element because of partial screening by inner electrons when orbitals of outer electrons are determined.

screw axis   An axis of symmetry in the crystal lattice structure whereby the lattice does not change even though the structure is rotated around the axis and also subjected to a translational motion along the axis.

screw dislocations   A dislocation is a crystallographic defect whereby a number of atoms are displaced (or dislocated) from their normal positions. A *screw dislocation* is one in which the displacement has come about as if one had twisted one region of the crystal with respect to another.

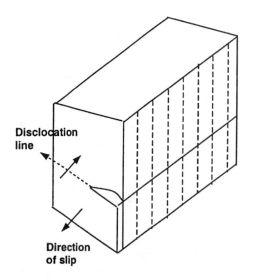

Visualization of a screw dislocation.

screw pinch   A variant on the theta pinch, in which axial currents (as in a z pinch, but less intense) produce a poloidal (azimuthal) magnetic field (in addition to the usual longitudinal field), thus making a corkscrew-type field configuration.

seaborgium   A trans-uranic element ($Z = 106$). It has relativistic deviation in chemistry properties.

secondary electron emission   The ejection of an electron from a solid or liquid by the impact of an incident (typically energetic) particle, such as an electron or ion. The secondary yield is the ratio of ejected electrons to incident particles of a given type. The details of *secondary electron emission* depend upon many factors, including the incident particle species, energy, angle of incidence, and various material properties of the solid or liquid target. *Secondary electron emission* is essential to the operation of electron multipliers and photomultipliers. It is also of great importance in situations where a plasma or particle beam is in contact with the solid or liquid. *Secondary electron emission* is also applied in surface science, materials science, and condensed matter physics for characterizing the target solid. A related process is sputtering, in which ions, atoms, or molecules are ejected from the solid or liquid.

secondary electrons   Electrons emitted from a substance when it is bombarded by other electrons or other particles of light (photons).

second-harmonic generation   A laser beam incident on a material (typically a crystal) that has a second order susceptibility can produce a beam with twice the frequency. This occurs via absorption of two photons of frequency ω and the emission of one photon of frequency 2ω. It can only occur in media that do not posses inversion symmetry.

second order susceptibility   The susceptibility defined by $\vec{P} = \epsilon_0 \chi \vec{E}$ often has a dependence on the applied field. It is often useful to use a Taylor series expansion of the susceptibility in powers of the applied field. For an isotropic homogeneous material, this yields

$\chi = \chi^{(1)} + \chi^{(2)}E + \chi^{(3)}E^2$. The factor $\chi^{(2)}$ is referred to as the *second order susceptibility*, as it results in a term in the polarization second order in the applied field. This factor is only nonzero for materials with no inversion symmetry. For a material that is not isotropic, the *second order susceptibility* is a tensor.

second quantization Ordinary Schrödinger equation of one particle or more particles are described within a Hilbert space of a single particle or a fixed particle numbers. The single electron Schrödinger equation written by the position representation can be interpreted as the equation for the classical field of electrons: we need to quantize the field. Then the field variable or, in short, the wave function is regarded as a set of an infinite number of operators on which commutation rules are imposed. This produces a formalism in which particles may be created and annihilated. We have to extend the Hilbert space of fixed particle numbers to that of arbitrary number particles.

Seebeck effect The existence of a temperature gradient in a solid causes a current flow as carriers migrate along or against the gradient to minimize their energy. This effect is known as the *Seebeck effect*. The thermal gradient is thus equivalent to an electric field that causes a drift current. Using this analogy, one can define an electric field caused by a thermal gradient (called a thermoelectric field). This electric field is related to the thermal gradient according to

$$\mathcal{E} = Q\nabla T$$

where $\mathcal{E}$ is the electric field, ∇T is the thermal gradient, and Q is the thermopower.

seiche Standing wave in a lake. For a lake of length L and depth H, allowed wavelengths are given by

$$\lambda = \frac{2L}{2n+1}$$

where $n = 0, 1, 2, \ldots$.

selection rules (1) Not all possible transitions between energy levels are allowed with a given interaction. *Selection rules* describe which transitions are allowed, typically described in terms of possible changes in various quantum numbers. Others are forbidden by that interaction, but perhaps not by others. For a hydrogen atom in the electric dipole approximation, the *selection rules* are $\Delta l = \pm 1$, where l is related to eigenstates of the square of the angular momentum operator via $\hat{L}^2 \psi_l = l(l+1)\hbar^2 \psi_l$. The rules result from the vanishing of the transition matrix element for forbidden transitions.

(2) Symmetry rules expressing possible differences of quantum numbers between an initial and a final state when a transition occurs with appreciable probability; transitions that do not follow the *selection rules* have a considerably lower probability and are called forbidden.

selection rules for Fermi-type β^- decay Allowed Fermi β^- decay changes a neutron into a proton (or vice versa in β^+ decay). There is no change in space or spin part of the wave function.

$\Delta J = 0$ no change of parity (J total moment);

I (isospin), $I_f = I_i \neq 0$, (initial and final isospin zero states are forbidden);

$I_{zf} = I_{zi} \mu 1 \Delta I_z = 1$ (third component of isospin);

$\Delta \pi = 0$ (there is no parity change)

In this kind of transition, leptons do not take any orbital or spin moment.

Allowed Gamow–Teller transitions:

$\Delta J = 0, 1$ but $J_i = 0; J_f = 0$ are forbidden.
$\Delta T = 0, 1$ but $T_i = 0; T_f = 0$ are forbidden.
$I_{zf} = I_{zi} \mu 1 \Delta I_z = 1$.
$\Delta \pi = 0$ (no change of parity).

s-electron An atomic electron whose wave function has an orbital angular momentum quantum number $\ell = 0$ in an independent particle theory.

self-assembly Any physical or chemical process that results in the spontaneous formation (assembly) of regimented structures on a surface. In *self-assembly*, the thermodynamic evolution of a system driving it towards its minimum energy configuration, automatically results in the formation of well-defined structures (usually well-ordered in space) on a surface without outside intervention. The figure shows the atomic force micrograph of a self-assembled pattern on the surface of aluminum foil. This well-ordered pattern consists of a hexagonal close-packed array of 50 nm pores surrounded by alumina. It was produced by anodizing aluminum foil in oxalic acid with a DC current density of 40 mA/cm^2. This pattern was formed by a non-linear field-assisted oxidation process.

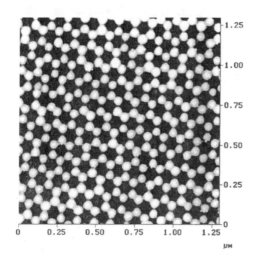

A raw atomic force micrograph of a self-assembled array of pores in an alumina film produced by the anodization of aluminum in an acid.

self-charge A contribution to a particle's electric charge arising from the vacuum polarization in the neighborhood of the bare charge.

self-coherence function The cross-correlation function $\Gamma(\vec{r}_1, \vec{r}_2; t_1, t_2) = \langle V^*(\vec{r}_1, t_1) V(\vec{r}_2, t_2) \rangle$ reduces to the *self-coherence function* for $\vec{r}_1 = \vec{r}_2$. It contains information about the temporal coherence of $V(\vec{r}, t)$, essentially a measure of how well we can predict the value of the field at t_1 if we know its value at t_2. Common choices for V are the electric field amplitude and the intensity of a light field.

self-consistent field See Hartree, Hartree–Fock method.

self-energy The *self-energy* of a charged particle (charge q) results from its interaction with the field it produces. It is expressed in terms of the divergent integral $E_{\text{self}} = (q^2/4\pi^2\epsilon_0) \int_0^{k_c} dk = (q^2 k_c / 4\pi^2 \epsilon_0)$, where k_c is a cutoff wave number that is infinite in principle.

self-focusing A beam of light with a nonuniform transverse intensity distribution may spontaneously focus at a point inside a medium with an intensity-dependent index of refraction, $n = n_0 + n_2 I$. To achieve *self-focusing*, n_2 must be positive. The *self-focusing* increases the intensity of the beam inside the material and can lead to damage of the material, particularly if it is a crystal.

self-induced transparency When a pulse of a particular shape and duration interacts with a non-linear optical material, it may form an optical soliton, which would propagate in a shape preserving fashion. For a gas of two-level atoms, this can be accomplished by a 2π pulse with a hyperbolic secant envelope.

self-similarity Flow whose state depends upon local flow quantities such that the flow may be non-dimensionalized across spatial or temporal variations. Self-similar solutions occur in flows such as boundary layers and jets.

Sellmeier's equation An equation for anomalous dispersion of light passing through a medium and being absorbed at frequencies corresponding to the natural frequencies of vibration of particles in the medium. The equation is given by

$$n^2 = 1 + A_k l^2 / (l^2 - l_k^2) + \cdots + \cdots.$$

Here n is the refractive index of the medium, l is the wavelength of the light passing through the medium where the kth particle vibrates at the natural frequency corresponding to the wavelength of l_k, and A_k is constant.

semiclassical theory Type of theory that deals with the interaction of atoms with light, treating the electromagnetic field as a classical variable (c-number) and the atom quantum mechanically.

semiconductor (1) A solid with a filled valence band, an empty conduction band, and a small energy gap between the two bands. Here, small means approximately one electron volt (1 eV). In contrast, for a conductor, the conduction band is partially populated with electrons, and an insulator has a band gap significantly larger than 1 eV.

(2) Materials are classified into four classes according to their electrical conductivity. The first are conductors, which have the largest conductivity (e.g., gold, copper, etc., these are mostly metals). In conductors, the conduction band and valence bands overlap in energy. The second are semi-metals (e.g., HgTe) which have slightly less conductivity than metals (here the conduction band and valence band do not overlap in energy, but the energy difference between the bottom of the conduction band and top of the valence band (the so-called "bandgap") is zero or close to it. The third are semiconductors, which have less conductivity than semi-metals and the bandgap is relatively large (examples are silicon, germanium, and GaAs). The last are insulators which conduct very little. They have very large bandgaps. An example is NaCl.

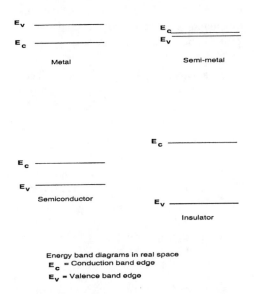

Energy band diagrams in real space
E_c = Conduction band edge
E_v = Valence band edge

The energy band diagram of metals, semi-metals, semiconductors, and insulators.

semiconductor detectors Use the formation of electron-hole pairs in semiconductors (germanium or silicon) to detect ionizing particles. The energy of formation of a pair is only about 3eV, which means that they can provide large signals for very small deposit energy in the detection medium. These devices were first used in high-resolution energy measurements and measurements of stopping power of nuclear fragments. Now they are used for the precise measurement of the position of charged particles. Very thin wafers of semiconductors are used for detection (200 − 300μ m thick). These detectors are quite linear. Two silicon detectors positioned in series can measure the kinetic energy and velocity of any low-energy particle and its rest mass.

semileptonic processes Decays with hadrons and leptons involved. Two types of these processes exist. In the first type there is no change in strangeness of hadrons, in the second type there is change in strangeness of hadrons.

In the first type, strangeness $|\Delta S| = 0$ (strangeness preserving decay), Isospin $\Delta I = 1$, and Z projection of isospin $|\Delta I_z| = 1$. For example, $n \to p + e^- + \bar{v}_e$ ($S_n = 0; I_{z,n} = -1/2 : S_p = 0; I_{z,p} = 1/2$).

In the second type, the strangeness non-conserving decay,

$$|\Delta S| = 1; |\Delta I_3| = 1/2; \Delta I = 1/2 \text{ or } 3/2.$$

For example,

$$K^+ \to \pi^0 + \mu^+ + \nu_\mu \left(S_K^+ = 1; I_{z,K}^+ = -1/2 \right.$$
$$\left. : S_{\pi^0} = 0; I_{3,\pi^0} = 0 \right).$$

semi-metal Elements in the Periodic Table that can be classified as poor conductors, i.e., in-between conductors and non-conductors. Examples are arsenic, antimony, bismuth, etc. *See* semiconductor.

separation In viscous flows under certain conditions, the flow in the boundary layer may not have sufficient momentum to overcome a large pressure gradient, particularly if the gradient is adverse. The boundary layer approximation results in the momentum equation at the wall taking the form

$$\frac{1}{\rho}\frac{dp}{dx} = \nu \frac{\partial}{\partial y}\left(\frac{\partial u}{\partial y}\right).$$

As dp/dx changes sign from negative to positive, the flow decelerates and eventually results in a region of reverse flow. This causes a *separation* of the flow from the surface and the creation of a separation bubble

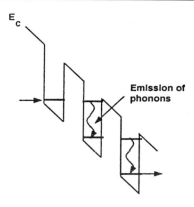

The process of sequential resonant tunneling through a superlattice under the influence of an electric field. The conduction band profile of the superlattice is shown along with the quantized sub-band states' energy levels (in heavy dark lines).

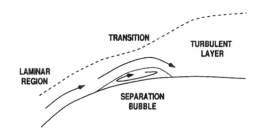

Separated flow in a transition region.

separatrix In a tokamak with a divertor (and in some other plasma configurations), the last closed flux surface is formed not by inserting an object (limiter) but by manipulating the magnetic field, so that some field lines are split off into the divertor rather than simply traveling around the central plasma. The boundary between the two types of field lines is called the *separatrix,* and it defines the last closed flux surface in these configurations.

sequential resonant tunneling In a structure with alternating ultrathin layers of materials (called a superlattice), an electron can tunnel from one layer to the next by emitting or absorbing a phonon, then tunnel to the next layer by doing the same, and so on. The phonon energy must equal the energy difference between the quantized electronic energy states in successive layers. This type of tunneling is called incoherent tunneling because the electron's wave function loses global coherence because of its interaction with the phonon.

The current voltage characteristic of a structure that exhibits *sequential resonant tunneling* has a non-monotonicity and hence exhibits negative differential resistance. This has been utilized to make very high frequency oscillators and rectifiers.

Serpukhov Institute for Nuclear Physics Located 60 miles south of Moscow. It has a 76 GeV proton synchrotron that was the most powerful accelerator in the world for several years. The *Serpukhov Institute* collaborated on the UNK project (accelerated protons up to 400 GeV within one booster synchrotron and then injected in the next synchrotron with energies up to 3 TeV — 3 TeV ring with superconductors magnets. Magnets have been developed in collaboration with Saclay Paris.

Sezawa wave A type of surface acoustic wave with a specific dispersion relation (frequency vs. wave vector relation).

shadow matter Unseen matter in the universe (*see* supersymmetric theories). This matter is visible only through gravitational interaction in the modern theory of superstrings.

shadow scattering Quantum scattering that results from the interference of the incident wave and scattered waves.

shallow water theory *See* surface gravity waves.

shape vibrations of nuclei Vibrational model of nuclei which describes shape vibrations of nuclei. This type of vibration considers oscillations in the shape of the nucleus without changing its density. It is similar to vibrations of a suspended drop of water that was gently disturbed.

Departures from spherical form are described by shape parameters $\alpha_{\lambda\mu}(t)$.

The shape parameters are defined in the following way:

$$R(\theta, \varphi, t) R_0 \cdot \left[1 + \sum_{\lambda,\mu} \alpha_{\lambda,\mu}(t) \cdot Y_{\lambda\mu}(\theta, \varphi) \right],$$

where $R(\theta, \varphi, t)$ is the distance between the surface of the nucleus and its center at the angles (θ, φ) at the time t, and R_0 is the equilibrium radius.

Because of properties of spherical harmonics ($Y^*_{\lambda\mu}(\theta, \varphi) = (-1)^\mu \cdot Y_{\lambda,-\mu}(\theta, \varphi)$), and in order to keep the distance $R(\theta, \varphi, t)$ real, the requirement for shape parameters $\alpha_{\lambda\mu}(t)$ is

$$\alpha_{\lambda\mu}(t) = (-1)^\mu \cdot \alpha_{\lambda,-\mu}(t).$$

For each λ value there are $2\lambda+1$ values of μ ($\mu = -\lambda, -\lambda+1, \ldots, \lambda$).

For $\lambda = 1$, vibrations are called monopole and dipole oscillations (the size of the nucleus is changed, but the shape is not changed for the monopole oscillations, and for the dipole oscillations the nucleus as a whole is moved), $\lambda = 2$ describes quadrupole oscillations of the nucleus (the nucleus changes its shape from spherical → prolate → spherical → oblate → spherical. The value $\lambda = 3$ describes more complex *shape vibrations* which are named as octupole vibrations.

Shapiro steps When a DC voltage is applied across a Josephson junction (which is a thin insulator sandwiched between two superconductors), the resulting DC current will be essentially zero (except for a small leakage current caused by few normal carriers). But when a small AC voltage is superimposed on the DC voltage, the DC component of the current becomes large if the frequency of the AC signal ω satisfies the condition

$$\omega = \frac{2e}{n\hbar} V_0$$

where V_0 is the amplitude of the DC voltage and n is an integer.

The values of the DC voltage V_0 that satisfy the above equation are called *Shapiro* steps after S. Shapiro who first predicted this effect.

shear A dimensionless quantity measured by the ratio of the transverse displacement to the thickness over which it occurs. A *shear* deformation is one that displaces successive layers of a material transversely with respect to one another, like a crooked stack of cards.

sheared fields As used in plasma physics, this refers to magnetic fields having a rotational transform (or, alternatively, a safety factor) that changes with radius. For example, in the stellarator concept, *sheared fields* consist of magnetic field lines that increase in pitch with distance from the magnetic axis.

shear rate Rate of fluid deformation given by the velocity gradient du/dy. Also called strain rate and deformation rate.

shear strain rate The rate at which a fluid element is deformed in addition to rotation and translation. The *shear strain rate* tensor is given by

$$e_{ij} \equiv \frac{1}{2} \left(\frac{\partial u_i}{\partial x_j} + \frac{\partial u_j}{\partial x_i} \right).$$

The tensor is symmetric.

shear stress See stress and stress tensor.

sheath See Debye sheath.

shell model A model of the atomic nucleus in which the nucleons fill a preassigned set of single particle energy levels which exhibit a shell structure, i.e., gaps between groups of energy levels. Shells are characterized by quantum numbers and result from the Pauli principle.

shell model (structures) A model based on the analogous orbital electron structure of atoms for heavier nuclei. Each nucleus is an average field of interactions of that nucleon to other nuclei. This average field predicts formation of shells in which several nuclei can reside. Basically, nucleons move in some average nuclear potential. The coulomb potential is binding for atom, the exact form of nuclear potential is unknown, but the central form satisfies initial consideration.

Experimental evidence shows the following: Atomic shell structure explains chemical periodicity of elements. After 1932, experimental data revealed that there is a series of magic numbers for protons and neutrons that gives stability to nuclei with such numbers Z and N. $Z = 2, 8, 20, 28, (40)50, 82$, and 126 are stable. These numbers are called magic numbers of nuclei.

The spectrum of energies of nuclei forms shells with big energy gaps between them. The *shell model* can be calculated on a spherical or deformed basis, but mathematical convince makes viable spherical approach. In a spherical model, each particle (nucleon) has an intrinsic spin **s** and occupies a state with a finite angular moment **l**. For many nucleon systems, nucleons are bonded in states with total angular moment **J** and total isospin **I**. There are two ways to compute angular moment coupling. One method is LS coupling and the other is $j-j$ coupling.

In an LS scheme, first the total orbital momentum for all nucleons (total L) is calculated, followed by the isospin for all nucleons (S). Finally, the total momentum (**J**) is computed as a vector sum of L and S:

$$\mathbf{J} = \mathbf{L} + \mathbf{S}.$$

Alternately the $j-j$ model computes orbital and intrinsic moments coupled for each nucleon and later sums over all total nucleon moments. In a deformed base the above procedure can be followed:

First, nucleons are divided in two groups: core and valence nucleons. The single particle states are separated into three categories: core states, active states, and empty states.

The low lying states make an inert core. The Hamiltonian can be separated into two parts: the constant energy term made from single particle energies and the interaction between them and the binding energy of active nucleons in the core. This second part is made from the kinetic energy of nucleons and their average interaction energy with other nucleons, including nucleons in the inert core.

Magic numbers are configurations that correspond to stable configurations of nuclei. These numbers are:

$$N = 2, 8, 20, 28, 50, 82, 126$$

$$Z = 2, 8, 20, 28, 80, 82$$

Nuclei that have both magic numbers are called double magic, e.g., $^4He^2$, $^{16}O^8$, and $^{208}Pb^{82}$.

shell models A simple view of atoms in the solid state represents them by neutral point masses interacting via springs. Cochran proposed atoms in a solid be considered to consist of a rigid ion core of finite extension (core shell, cs) surrounded by a shell of valence electrons (valence shell, vs) that can move relative to the core. Interactions between the atoms are therefore represented by three shell–shell interactions: cs–cs, cs–vs, and vs–vs.

Shockley–Read–Hall recombination Electrons and holes in a semiconductor recombine, thereby annihilating each other. They do so radiatively (emitting a photon) or non-radiatively (typically emitting one or more phonons). *Shockley–Read–Hall* is a mechanism for non-radiative recombination. The recombination rate (which is the temporal rate of change of electron or hole concentration) is given by

$$R = \frac{np - n_i^2}{\tau_p(n + n_i) + \tau_n(p + n_i)}$$

where n and p are the electron and hole concentrations respectively, and n_i is the intrinsic carrier concentration in the semiconductor which depends on the semiconductor and the temperature. The quantities τ_p and τ_n are the lifetimes of holes and electrons respectively. They depend on the density of recombination centers (traps facilitate recombination), their capture rates, and the temperature.

shock tube (1) Device used to study unsteady shock and expansion wave motion. A cavity is separated with a diaphragm into a high pressure section and a low pressure section. Upon rupture, a shock wave forms and moves from the high pressure region to the low pressure region, and an expansion wave moves from the low pressure region to the high pressure region. The interface between the two gases moves in the same direction as the shock wave albeit with a lower velocity. A space-time (phase-space) diagram is used to examine the motion of the various structures.

(2) A gas-filled tube used in plasma physics to quickly ionize a gas. A capacitor bank charged

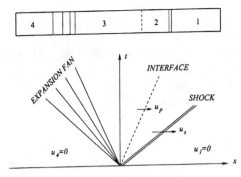

Shock tube with phase-space diagram.

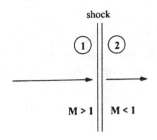

Shock wave.

to a high voltage is discharged into the gas at one tube end to ionize and heat the gas, producing a shock wave that may be studied as it travels down the tube.

shock wave (1) A buildup of infinitesimal waves in a gas can create a wave with a finite amplitude, that is, a wave where the changes in thermodynamic quantities are no longer small and are, in fact, possibly very large. Analogous to a hydraulic jump, this jump is called a shock wave. Shocks are generally assumed to be spatial discontinuities in the fluid properties. This makes it simpler from a mathematical perspective, but physically, shocks have a definite physical structure where thermodynamic variables change their values over some spatial dimension. This distance, however, is extremely small. In general, shocks are curved. However, there will be many cases where the *shock waves* in a flow are either entirely straight (such as in a shock tube) or can be assumed straight in certain sections (such as ahead of a blunt body). In these cases, the shock is normal if the incoming flow is at a right angle to the shock and oblique for all other cases. The figure idealizes a *shock wave* as a discontinuity. The variations from the upstream side of the shock to the downstream side are often called the jump conditions.

(2) A wave produced in any medium (plasma, gas, liquid, or solid) as a result of a sudden violent disturbance. To produce a shock wave in a given region, the disturbance must take place in a shorter time than the time required for sound waves to traverse the region. The physics of shocks is a fundamental topic in modern science; two important cases are astrophysics (supernovae) and hydrodynamics (supersonic flight).

short range order Refers to the probability of occurrence of some orderly arrangements in certain types of atoms as neighbors and is given by the following:

$$s = (b - b_{\text{random}})/(b_{\text{maximum}} - b_{\text{random}})$$

where b is the fraction of bonds between closest neighbors of unlike atoms, b_{random} is the value of b when the arrangement is random and b_{maximum} is the maximum value that b may assume.

shot noise A laser beam of constant mean intensity incident on a detector creates a photocurrent, whose mean is proportional to the beam's intensity. There are fluctuations in the photocurrent as there are quantum fluctuations in the laser beam. For a laser well above threshold producing a coherent state, these beam intensity fluctuations are Poissonian. The resulting photocurrent noise is referred to as *shot noise*. Light fields that are squeezed exhibit sub-*shot noise* for one quadrature, typically over some range of frequencies.

Shubnikov–DeHaas effect The electrical conductance of a material placed in a magnetic field oscillates periodically as a function of the inverse magnetic flux density. This is the *Shubnikov–DeHaas effect,* and the corresponding oscillations are called Shubnikov–DeHaas oscillations. The period of the oscillation $\Delta(1/B)$ is related to an extremal cross-sectional area of the Fermi surface in a plane normal to the magnetic field A according to

$$\Delta\left(\frac{1}{B}\right) = \frac{2\pi^2 e}{h}\frac{1}{A}.$$

If a magnetic field is applied perpendicular to a two-dimensional electron gas, then remembering that the Fermi surface area is $2\pi^2/n_s$ where n_s is the two-dimensional carrier density, one obtains:

$$\Delta\left(\frac{1}{B}\right) = \frac{2\pi^2 e}{h}\frac{1}{n_s}.$$

Thus, Shubnikov–DeHaas oscillations are routinely used to measure carrier concentrations in two-dimensional electron and hole gases. In systems that contain two parallel layers of two-dimensional electron gases, the oscillations will show a beating effect if the concentrations in the two layers are somewhat different. The beating frequency depends on the difference of the carrier concentrations. Beating may also occur if the spin degeneracy is lifted by the magnetic field or some other effect.

Σ baryon There are three *sigma* (triplet) *baryons* (Σ^+ plus sigma baryon (uus), Σ^- minus sigma baryon (dds), and Σ^0 neutral (uds), according $SU(3)$ (flavor) symmetry). Wave functions are:

$$|\Sigma^+> = \frac{1}{\sqrt{3}} \cdot \{|suu> + |usu> + |uus>\},$$

$$|\Sigma^-> = \frac{1}{\sqrt{3}} \cdot \{|dds> + |dsd> + |sdd>\},$$

$$|\Sigma^0> = \frac{1}{\sqrt{6}} \cdot \{|dus> + |uds> + |dsu>$$
$$+ |usd> + ||sdu> + |sud>\}.$$

signal-to-noise ratio The ratio of the useful signal amplitude to the noise amplitude in electrical circuits, the noise is not used anywhere in the circuit.

silsbee effect The process of destroying or quenching the superconductivity of a current carried by a wire or a film at a critical value.

similarity See dynamic similarity and self-similarity.

similarity transformation The relationship between two matrices such that one matrix becomes the transform of the second.

simplex A system of communication that operates uni-directional at one time.

sine operator There is no phase operator in quantum mechanics. In a complex representation, the classical field $E = E_0 e^{i\theta}$ is quantized such that E_0 and $e^{i\theta}$ are separate operators. The imaginary part of the operator $e^{i\theta}$ is $\sin(\theta)$. There is no operator for θ itself.

single electronics A recently popular field of electronics where the granularity of charge (i.e., electric charge comes in quanta of the single electron's charge of $1.61 \times 10-19$ Coulombs) is exploited to make functional signal processing, memory, or logic devices.

Single electronic devices operate on the basis of a phenomena known as a Coulomb blockade which is a consequence of, among other things, the granularity of charge. When a single electron is added to a nanostructure, the change in the electrostatic energy is

$$\Delta E = \frac{(Q-e)^2}{2C} - \frac{Q^2}{2C} = -\frac{Q-e/2}{C}$$

where e is the magnitude of the charge of the electron ($1.61 \times 10-19$ Coulombs), C is the capacitance of the nanostructure, and Q is the initial charge on the nanostructure. Since this event is permitted only if the change in energy ΔE is negative (the system lowers its energy), Q must be positive. Furthermore, since $Q = e|V|$ (V is the potential applied over the capacitor), it follows that tunneling is not permitted (or current cannot flow) if

$$-e/2C \leq V \leq e/2C.$$

The existence of this range of voltage at which current is blocked by Coulomb repulsion is known as the Coulomb blockade.

The Coulomb blockade can be manifested only if the thermal energy kT is much less the electrostatic potential barrier $e^2/2C$. Otherwise, electrons can be thermally emitted over the barrier and the blockade may be removed. In nanostructures, C may be 10^{-18} farads and hence the electrostatic potential barrier is ~ 100 meV, which is four times the room-temperature thermal energy kT. Thus, the Coulomb blockade can be appreciable and discernible at reasonable temperatures.

The phenomenon of the Coulomb blockade is often encountered in electron tunneling across a nanojunction (a junction of two materials with nanometer scale dimensions) with small capacitance. The tunnel resistance must exceed the quantum of resistance h/e^2 so that single electron tunneling events may be viewed as discrete events in time.

single electron transistor Consists of a small nanostructure (called a quantum dot, which is a solid island of nanometer scale dimension) interposed between two contacts called source and drain. When the charge on the quantum dot is nq (n is an integer and q is the electron charge), current cannot flow through the quantum dot because of a Coulomb blockade. However, if the charge is changed to $(n+0.5)q$ by a third terminal attached to the quantum dot, then the Coulomb blockade is removed and current can flow. Since the current between two terminals (source and drain) is being controlled by a third terminal (called gate in common device parlance), transistor action is realized. If it is bothersome to understand why the charge on the quantum dot can ever be a fraction of the single electron charge, one should realize that this charge is transferred charge corresponding to a shift of the electrons from their equilibrium positions. This shift need not be quantized.

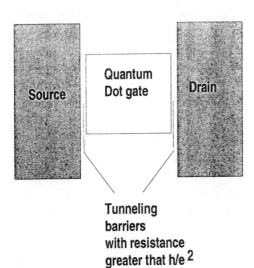

Schematic of a single electron transistor.

single electron turnstile A single electron device consisting of two double nanojunctions connected by a common nanometer sized island. The island is driven by a gate voltage. When an AC potential of appropriate amplitude is applied to this circuit, a DC current results which obeys the relation

$$I = ef$$

where e is the single electron charge and f is the frequency of the applied AC signal. This device, and others like it, have been proposed to develop a current standard with metrological accuracy.

single-mode field A single-mode field is an electromagnetic field with excitation of only one transverse and one longitudinal mode.

singlet An energy level with no other nearby levels. Nearby is a relative term, and the operational definition is that the energy difference between the *singlet* and other nearby states is comparable to the excitation energy. *See also* doublet; triplet states.

singlet state An electronic state of a molecule in which all spins are paired.

singlet-triplet splitting The process of separation of the singlet state and triplet state in the electronic configuration of atom or molecule.

Sisyphus cooling A method of laser cooling of atoms. It utilizes position-dependent light shifts caused by polarization gradients of the cooling field. It takes its name from the Greek myth, as atoms climb potential hills, tend to spontaneously emit and lose energy, and then climb the hills again.

six-j symbols A set of coefficients affecting the transformation between different ways of coupling eigenfunctions of three angular momenta. *Six-j symbols* are closely related to the Racah coefficients but exhibit greater symmetry.

skin depth The depth at which the current density drops by 1 Neper smaller than the surface value, due to the interaction with electromagnetic waves at the surface of the conductor.

skin friction Shear stress at the wall which may be expressed as

$$\tau_w = \mu \frac{\partial u}{\partial y}\bigg|_{y=0}$$

where the velocity gradient is taken at the wall.

skin friction coefficient Dimensionless representation of the skin friction

$$C_f = \frac{\tau_w}{\frac{1}{2}\rho U_\infty^2}.$$

For a Blasius boundary layer solution (laminar flat plate), the skin friction is

$$C_f = \frac{0.664}{\sqrt{\text{Re}_x}}.$$

For a turbulent flate plate boundary layer,

$$c_f = \frac{0.0576}{\text{Re}_x^{-1/5}}.$$

Also referred to as the wall shear stress coefficient.

Slater determinant A wave function for n fermions in the form of a single $n \times n$ determinant, the elements of which are n-different one-particle wave functions (also called orbitals) depending successively on the coordinates of each of the particles in the system. The matrix form incorporates the exchange symmetry of fermions automatically.

Slater–Koster interaction potential Using a Green's function model, one can express the binding energy of an electron to an impurity (e.g., N in GaP). In this case, one needs to express the impurity potential V. If one chooses to express V as a delta function in space via the matrix elements of Wannier functions, the potential is called the *Slater–Koster interaction potential*.

slip A deformation in a crystal lattice whereby one crystallographic plane slides over another, causing a break in the periodic arrangement of atoms (see the figure accompanying the definition of screw dislocation).

slowly varying envelope approximation For a time-varying electromagnetic field that is not purely monochromatic but has a well defined carrier frequency, we may write $E(x, t) = A(x, t)\cos(kx - \omega t + \phi)$, where ω is the carrier frequency and k is the center wave number. $A(x, t)$ is referred to as the envelope function, and in the slowly varying envelope approximation, we assume that the envelope does not change much over one optical period, $dA(x, t)/dt \ll \omega A(x, t)$. A similar approximation can be made in the spatial domain, $dA(x, t)/dx \ll kA(x, t)$

slow neutron capture This capture reaction captures thermal neutrons (with few eV energy). This kind of reaction is responsible for most matter in our world (*see* supernova). An example of this reaction is $^{16}O(n, \gamma)^{17}O$. At higher temperatures, capture of protons and alpha particles is possible.

Elements beyond A $\sim$ 80 up to uranium are mostly produced by slow and rapid neutron capture. Knowledge of these kinds of reactions is very important for synthesis of new elements. The capture of neutrons in uranium can raise the energy of nuclei to start the fission process.

sluice gate Gate in open channel flow in which the fluid flows beneath the gate rather than over it. Used to control the flow rate.

small signal gain For a laser with weak excitation, the output power is linearly proportional to the pump rate. The ratio of output power to input power in that operating regime is referred to as small signal gain.

S-matrix The matrix that maps the wave function at a long time in the past to the wave function in the distant future. Also referred to as the scattering, or S-operator, it is defined as $|\psi(t = \infty)\rangle = \hat{S}|\psi(t = -\infty)\rangle$. It is typically calculated in a power expansion in a coupling constant, such as the fine structure constant for quantum electrodynamics.

S-matrix theory A theory of collision phenomena as well as of elementary particles based on symmetries and properties of the scattering matrix such as unitarity and analyticity.

Snell's law When light in one medium encounters an interface with another medium, the light ray in the other medium traveling in a different direction can be determined from *Snell's Law*, $n_i \sin\theta_i = n_0 \sin\theta_0$. Here, the angles are measured with respect to the normal to the interface, n_i is the index of refraction of the initial medium, and n_0 is the index of refraction of the medium on the other side of the interface. For a given initial angle, there may be no possible ray that enters the other medium. This condition is known as total internal reflection, and it occurs when $n_i/n_0 < \tan \cdot \theta$.

SO(10) symmetry (E_6) A symmetry present in grand unified theory (gravity not included).

SO(3) group A group of symmetry of spatial rotations. This group is represent by a set of 3×3 real orthogonal matrices with a determinant equal to one.

SO(32) Group symmetry (32 internal dimensional generalization of space-time symmetry). In chiral theory SO(32) describes Yang-Mills forces.
These forces can be described with $E_6 X E_8$ symmetry groups product two continuous groups discovered by French mathematician Elie Cartan.

sodium chloride structure *See* rock salt structure.

soft X-ray X-rays of longer wavelengths, the term "soft" being used to denote the relatively low penetrating power.

solar (stellar) energy In the sun, $4 10^{12}$ g/s mass is converted in energy. There are two main type of reactions inside the sun. First is the carbon cycle (proposed by Bethe in 1938):

$$p + {}^{12}C \to {}^{13}N$$
$$ {}^{13}N \to {}^{13}C + e^+ + \nu$$
$$p + {}^{13}C \to {}^{14}N + \gamma$$
$$p + {}^{14}N \to {}^{15}O + \gamma$$
$${}^{15}O \to {}^{15}N + e^+ + \nu$$
$$p + {}^{15}N \to {}^{12}C + \alpha + \gamma .$$

In this process, carbon is a catalyst (number of C stays the same).

The total balance of this process is

$$4p \to \alpha + 2e^+ + 2\nu + 26.7\, MeV .$$

The second type of reaction is the proton–proton cycle:

$$p + p \to d + e^+ + \nu$$
$$p + d \to {}^3He + \gamma$$
$${}^3He + {}^3He \to \alpha + 2p .$$

The effect of this process is the same as in the carbon cycle.

$$4p \to \alpha + 2e^+ + 2\nu + 26.7\, MeV .$$

The prevailing reaction depends on the plasma temperature. The proton–proton cycle dominates below $1.8\, 10^7$ K. The proton–proton cycle produces 96% of the energy in the center of the Sun (temperature $T = 1.5 10^7$ K). Each proton in the reaction contributes 6.7 MeV, which is eight times greater than the contribution of one nucleus in ^{235}U fission.

solar cell A solar cell is a semiconductor p–n junction diode. When a photon with energy $h\nu$ larger than the bandgap of the semiconductor is absorbed from the sun's rays, an electron–hole pair is created. The electron–hole pairs created in the depletion region of the diode travel in opposite directions due to the electric field that exists in the depletion region. This traveling electron–hole pair contributes to current. Thus, the *solar cell* converts solar energy to electrical energy.
Solar cells are among the best and cleanest (environmentally friendly) energy converters. They are also inexpensive. The cheapest cells made out of amorphous silicon exhibit about 4% conversion efficiency.

solar corona The *solar corona* is a very hot, relatively low density plasma forming the outer layer of the sun's atmosphere. Coronal temperatures are typically about one million K, and have densities of approximately 10^8–10^{10} particles per cubic centimeter. The corona is much hotter than the underlying chromosphere and photosphere layers. The mechanism for coronal heating is still poorly understood but appears to be magnetic reconnection. Plasma blowing out from the corona forms the solar wind. *See also* corona.

solar filament A solar surface structure visible in H_α light as a dark (absorption) filamentary feature. The same structures are referred to as solar prominences when viewed side-on and seen extending off the limb.

solar flare A rapid brightening in localized regions on the sun's photosphere that is usually observed in the ultraviolet and X-ray ranges of the spectrum and is often accompanied by gamma ray and radio bursts. *Solar flares* can form in a few minutes and last from tens of minutes to several hours in long-duration events. Flares also produce fast particles in the solar wind, which arrive at the earth over the days following the flare. The energy dumped into the earth's magnetosphere and ionosphere from flares is a major cause of space weather.

solar neutrinos (physics) Neutrinos produced in nuclear reactions in the sun are detected on the earth through neutrino capture reactions. An example of that reaction is the capture of a neutrino by chloral nuclei:

$$\nu + {}^{37}Cl \rightarrow {}^{37}Ar + e^- \quad Q = -0.814\, MeV.$$

This *Ar* isotope is unstable and beta decays into ^{37}Cl with a half-life of 35 days. We observe half as many neutrinos from the sun as are predicted from a nuclear fusion mechanism. There are several possibilities: the nuclear reaction rates may be wrong; the temperature of the center of the sun predicted by the standard solar model may be too high; something may happen to neutrinos on the way from the center of the sun to the detectors; or electron–muon neutrino oscillations may occur if the neutrino has a rest mass different than zero.

The kamiokande II detector shows that neutrinos cannot decay during flight from the sun.

solar prominence A large structure visible off the solar limb, extending into the chromosphere or the corona, with a typical density much higher (and temperatures much colder) than the ambient corona. When seen against the solar disk, these prominences manifest as dark absorption features referred to as solar filaments.

solar wind A predominantly hydrogen plasma with embedded magnetic fields which blows out of the solar corona above escape velocity and fills the heliosphere. The *solar wind* velocities are approximately 100–1000 km/s. The *solar wind's* density is typically around 10 particles per cubic centimeter, and its temperature is about 100,000 K as it crosses the earth's orbit. The *solar wind* causes comet tails to point mainly away from the sun. Storms in the *solar wind* are caused by solar flares.

sol-gel process A chemical process for synthesizing a material with definite chemical composition. The constituent elements of the material are first mixed in a solution and then a gelling compound is added. Residues are evaporated to leave behind the desired material.

solid solubility The dissolution of one solid into another is the process of solid dissolution. *Solid solubility* refers to the solubility (the possibility of dissolving) of one solid into another. Diffusion of impurities into a semiconductor (employed as the most common method of doping an *n*- or *p*-type semiconductor) is a process of solid dissolution. *Solid solubility* is limited by the *solid solubility* limit, which is the maximum concentration in which one solid can be dissolved in another.

soliton (1) Stable, shape-preserving, and localized solutions of non-linear classical field equations, where the non-linearity opposes the natural tendency of the solution to disperse. Solitons were first discovered in water waves, and there are several hydrodynamic examples, including tidal waves. *Solitons* also occur in plasmas. One example is the ion-acoustic *soliton,* which is like a plasma sound wave; another is the Langmuir soliton, describing a type of large amplitude (non-linear) electron oscillation. *Solitons* are of interest for optical fiber communications, where the use of optical envelope *solitons* as information carriers in fiber optic networks has been proposed, since the natural non-linearity of the optical fiber may balance the dispersion and enable the soliton to maintain its shape over large distances.

(2) A wave packet that maintains its shape as it propagates. Typically, a wave packet spreads

as its various frequency components have different velocities $v = c/n(\lambda)$ due to dispersion in a medium. A compensating mechanism, such as an index of refraction that also depends on the intensity of a particular frequency component, allows one to tailor a pulse shape that will not spread during propagation.

(3) A quantum of a solitary wave. Such a wave propagates without any change in the shape of the pulse. In contrast, the pulse shape of an ordinary wave distorts as the wave propagates in a dispersive medium because different frequency components have different velocities. Typically, a dispersive medium has the effect of a low-pass filter which tends to smooth out the shape of a pulse and makes it spread out in time. However, if the medium has a non-linearity that generates higher harmonics, the lost high frequency components are compensated for by the harmonics. If the two effects exactly cancel each other, then a *soliton* can form which travels without any distortion of pulse shapes.

Certain non-linear differential equations have *soliton* solutions. In other words, waves whose evolutions in time and space are governed by such an equation can produce solitons. Examples of non-linear differential equations that have *soliton* solutions are the sine Gordon equation and the Korteweg–DeVries equation.

Sommerfeld doublet formula Equation to account for the frequency splitting of doublets: $\alpha^2 R (Z - \sigma)^4 / n^3 (\ell + 1)$, with the quantities α, R, Z, σ, n, and ℓ indicating, respectively, the fine structure constant, the Rydberg constant, the atomic number, a screening constant, the principal quantum number, and the orbital angular momentum quantum number.

Sommerfeld number The probability for an α particle to tunnel from a nuclei through a Coulomb barrier at low energies is given by transmission coefficient (α decay).

$$T = e^{-2\pi\eta} = \exp\left\{-2\pi \frac{zZ\alpha c}{v}\right\}.$$

The parameter η is called the *Sommerfeld number*.

sonic boom Sound wave created by the confluence of waves across a shock.

sound speed The speed of sound in a general fluid medium is given by the fluid's bulk modulus E (inverse compressibility) and the fluid density

$$a = \sqrt{\frac{E}{\rho}}.$$

In a perfect gas, this reduces to

$$a = \sqrt{\gamma RT}$$

using the isentropic relation

$$\frac{p}{\rho^\gamma} = \text{constant}$$

and the ideal equation of state

$$p = \rho RT$$

where γ, R, and T are the ratio of specific heats, specific gas constant, and temperature of the gas respectively. *See* sound wave.

sound wave Infinitesimal elastic pressure wave whose propagation speed moves at the speed of sound. In a compressible fluid, the square of the speed of sound is given by the rate of change of pressure with respect to density

$$a^2 = \frac{dp}{d\rho}.$$

A *sound wave* can be either compressive or expansive. Also referred to as an acoustic wave. *See* sound speed.

space charge In a plasma, a net charge which is distributed through some volume. Most plasma are electrically neutral or at least quasi-neutral, because any charge usually creates electric fields which rapidly move surplus charge out of the plasma. However, in some applications, one wishes to apply external electric fields to the plasma, and a net *space charge* can be produced as a result. The resulting *space charge* must often be accounted for in the physics of these sorts of devices.

space charge layers Layer of electrical charges that distribute in an electronic device or over a material.

space group A group of symmetry elements developed by a set of operations, e.g., reflection

space potential Also known as the plasma potential, this refers to the electric potential within a plasma in the absence of any probes. The *space potential* is typically more or less uniform outside of plasma sheath regions. The *space potential* differs from the floating potential, which is the potential measured at a probe placed inside the plasma. This is because the faster electron speeds in a plasma cause a net electron current to deposit onto a floating probe until the floating probe becomes sufficiently negatively charged to repel electrons and attract ions. The result is that the floating potential is less than the actual *space potential*.

space quantization The quantization of the component of an angular momentum vector of a system in some specified direction.

space reflection symmetry *See* parity.

space weather The state of the geoplasma space (the ionosphere and the magnetosphere plasmas) surrounding the earth's neutral atmosphere. *Space weather* conditions are determined by the solar wind and can show disturbances (e.g., geomagnetic substorms and storms). Under disturbed *space weather* conditions, satellite-based and ground-based electronic systems such as communications networks and electric power grids can be disrupted.

spatial coherence The degree of *spatial coherence* for a light field is determined by the ability to predict the amplitude and phase of the electric field at a point $\vec{r}_1$ if one knows the electric field at $\vec{r}_2$. The appearance of interference fringes behind a double slit apparatus illuminated by a field is one manifestation of *spatial coherence*.

spatial frequency Also known as the wave number, it is $2\pi/\lambda$, where λ is the wavelength.

spatial translation We assume that space is homogeneous. Then closed physical systems must have translational invariance. Translations of space coordinates form a continuous Abelian group. A direct consequence of this invariance is the momentum conservation.

specific gas constant (R) Equal to the universal gas constant $\mathcal{R}$ divided by the molecular weight of the fluid.

$$R = \frac{\mathcal{R}}{MW}$$

where $\mathcal{R} = 8.314 \, J/mol/K$.

specific gravity Dimensionless ratio of a fluid's density to a reference density. For liquids, water at STP is used, such that

$$specific\ gravity = \frac{\rho_{liquid}}{\rho_{water}}.$$

For gases, air at STP is typically used,

$$specific\ gravity = \frac{\rho_{gas}}{\rho_{air}}.$$

specific volume The volume occupied by a unit mass of fluid; inverse of density.

$$v = \frac{1}{\rho}.$$

specific weight Weight of a fluid per unit volume:

$$specific\ weight = g\rho.$$

speckle When coherent (usually laser) light is scattered from a rough surface, a random intensity pattern is created due to constructive and destructive interference. This tends to make the surface look granular.

spectral cross density The Fourier transform of the mutual coherence function, $W(\vec{r}_1, \vec{r}_2, \omega) \equiv \int_{-\infty}^{\infty} \Gamma(\vec{r}_1, \vec{r}_2, \tau) \exp(-i\omega\tau)$, where $\Gamma(\vec{r}_1, \vec{r}_2, \tau)$ is the mutual coherence function.

spectral degree of coherence Defined in terms of the cross-spectral density function, $W(\vec{r}_1, \vec{r}_2, \omega)$. The *spectral degree* *of coherence* is given by $\mu(\vec{r}_1, \vec{r}_2, \omega) \equiv \frac{W(\vec{r}_1, \vec{r}_2, \omega)}{[W(\vec{r}_1, \vec{r}_1, \omega)]^{1/2}[W(\vec{r}_2, \vec{r}_2, \omega)]^{1/2}}$.

spectral density The spectral cross density $W(\vec{r}_1, \vec{r}_2, \omega)$ with $\vec{r}_1 = \vec{r}_2$, i.e., $S(\vec{r}, \omega) \equiv W(\vec{r}, \vec{r}, \omega)$. It is also referred to as the power spectrum of the light field.

spectral response of a solar cell The number of carriers (electrons and holes) collected in a solar cell per unit incident photon of a given wavelength.

spectroscopy The use of frequency dispersing elements to measure the spectrum of some physical quantity of interest, typically the intensity spectrum of a light source.

spectrum A display of the intensity of light, field strength, photon number, or other observable as a function of frequency, wavelength, or mass. Mathematically, it is the allowed eigenvalues λ in the equation $O\psi = \lambda\psi$, where O is some linear operator and ψ is an eigenstate or eigenvector.

speed of sound *See* sound speed.

spherical Bessel functions $j_l(x)$ Solutions of the radial Schrödinger equation in spherical coordinates. These functions are related to ordinary Bessel functions $J_n(x)$

$$j_l(x) = \sqrt{\frac{\pi}{2x}} \cdot J_{l+\frac{1}{2}}(x).$$

spherical harmonics Eigenstates of the Schrödinger equation for the angular momentum operator L^2 and its z projection L_z in a central square potential:

$$L^2 \cdot Y_{l,m}(\theta, \varphi) = \eta^2 \cdot l \cdot (l+1) \cdot Y_{l,m}(\theta, \varphi),$$
$$L_z \cdot Y_{l,m}(\vartheta, \varphi) = \eta \cdot m \cdot Y_{l,m}(\vartheta, \varphi),$$

where

$$Y_{l,m}(\vartheta, \varphi) = \sqrt{\frac{(2l+1) \cdot (l-m)}{4\pi(l+m)!}} P_{l,m}(\cos\theta) \cdot e^{im\varphi}.$$

$P_{l,m}(\cos\varphi)$ are well known Legendre polynomials.

spherical tokamak A magnetic confinement plasma device based on the tokamak design in which the center of the torus is narrowed down as much as possible, thereby bringing the minor radius as close as possible to the major radius. Also known as low aspect ratio tokamaks, *spherical tokamaks* appear to have favorable magnetohydrodynamic stability properties relative to conventional tokamaks and are an active area of current research.

spherical wave A wave whose equal phase surfaces are spherical. Typically written in the form $E = E_0 e^{i\omega t}/r$.

spheromak A compact toroidal magnetic confinement plasma with comparable toroidal and poloidal magnetic field strengths. The spheromak's plasma is roughly spherical and is usually surrounded by a close-fitting conducting shell or cage. Unlike the tokamak, stellarator, and spherical tokamak configurations, in the *spheromak* there are no toroidal field coils linking the plasma through the central plasma axis. Both the poloidal and toroidal magnetic fields are mainly generated by internal plasma currents, with some external force supplied by poloidal field coils outside the separatrix. The resulting configuration is approximately a force-free magnetic field.

spillway Flow rate measurement device similar to a weir with a gradual downstream slope.

spin Intrinsic angular momentum of an elementary particle or nucleus, which is independent of the motion of the center of mass of the particle.

spin–flip scattering Scattering of a particle with intrinsic spin in which the direction of the spin is reversed due to spin-dependent forces.

spin matrix In quantum mechanics, the phenomenology of electron spin is described in terms of a spin vector

$$\vec{\sigma} = \sigma_x \hat{x} + \sigma_y \hat{y} + \sigma_z \hat{z}$$

where the x-, y-, and z-components of the spin vector are 2×2 matrices given by

$$\sigma_x = \begin{pmatrix} 0 & 1 \\ 1 & 0 \end{pmatrix}$$
$$\sigma_y = \begin{pmatrix} 0 & -i \\ i & 0 \end{pmatrix}$$
$$\sigma_z = \begin{pmatrix} 1 & 0 \\ 0 & -1 \end{pmatrix}.$$

The matrices σ_x, σ_y, and σ_z are called (Pauli) *spin matrices*.

spinor A *spinor* of rank n is an object with 2^n components which transform as products of components of n *spinors* of rank one. The latter are vectors with two complex components which, upon three-dimensional coordinate rotation, transform under unitary, unimodular transformations. *Spinors* are suited to represent the spin state of a particle with half-integer spin.

spin–orbit coupling The interaction between spin and orbital angular momentum of a particle which moves in a confining potential. It is expressed by a term in the Hamiltonian which is proportional to the product $\hat{S} \cdot \hat{L}$ of the corresponding operators.

spin–orbit interaction Critical force to obtain magical numbers in the mean field method. *See* shell model.

spin–orbit multiplet A group of states of an atomic or nuclear system with energies that differ only because of the directional dependence of the spin–orbit coupling term in the Hamiltonian. All members of the multiplet have the same total spin angular momentum quantum number S and total orbital angular momentum quantum number L, but their total angular momentum quantum number J differs. The vector operator $\hat{J}$ is the vector sum of $\hat{L}$ and $\hat{S}$ with only discrete values due to spatial quantization.

spin polarized beams and targets Refers to preferential orientation along some chosen direction in space of the intrinsic spins of the beam or target particles (now up to 90% of particles in beams or target can be polarized).

spin quantum number The largest value of a system's spin observed in a particular quantum state (in units of $\hbar$). It is either an integer or a half-integer.

spin space The two-dimensional complex vector space representing the various spin states of a particle with spin 1/2. The unitary unimodular transformations in *spin space* form a two-dimensional double-valued representation of the three-dimensional rotation group.

spin–spin interaction An energy term proportional to $\hat{S}_1 \cdot \hat{S}_2$, i.e., the dot product of the spin angular momentum operators of two particles.

spin state Quantum state of a system in which its spin and one component of it along a specified direction — usually, but not necessarily, the z-direction — have definite values.

spin-statistics theorem A result of assuming causality, along with the laws of quantum mechanics and special relativity. It states that an ensemble of particles of half-integer spin (fermions) satisfy the Fermi–Dirac distribution function (and hence the Pauli exclusion principle), and that an ensemble of particles of integer spin (bosons) satisfies the Bose–Einstein distribution function.

spintronics The recently popular field where the spin degrees of freedom of an electron or hole in a semiconductor material are utilized to store and process data and realize electronic functionality as opposed to the more conventional charge degrees of freedom.

spin wave Waves of departures in magnetic moment orientations traveling through electron spin couplings.

split gate electrode A technique for fabricating a quasi one-dimensional structure by electrostatic confinement. A metal pattern is defined on the surface of a quantum well heterostructure which contains a buried two-dimensional layer of electrons. When a negative potential is applied to the metal pattern, electrons underneath the metal are pushed away by the Coulomb repulsion, leaving behind a narrow quasi one-dimensional layer of electrons just underneath the region where there is a physical split in the metal pattern.

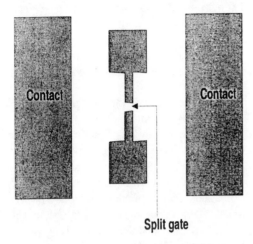

Top view of a structure consisting of a split gate.

spontaneous emission An atom in a quantum state other than the ground state will eventually make a transition to a lower energy state. When this transition results in the emission of a photon, with no external field present, it is called *spontaneous emission*. The emitted photon is random in direction and the time of emission is unknown as well, leading to phase uncertainty. For N_0 atoms initially in an excited state, the number remaining in the excited state at a time t is $N(t) = N_0 e^{-t/\tau}$, where τ is the *spontaneous emission* lifetime, the inverse of the *spontaneous emission* rate. *Spontaneous emission* is the result of radiation reactions and vacuum fluctuations.

spontaneous emission lifetime The inverse of the spontaneous emission rate.

spontaneous emission rate If one has N_0 atoms in the excited state of an atom, the population of the excited state can decay via spontaneous emission to a lower energy state at a rate defined by $\dot{N} = -AN$, where A is the *spontaneous emission rate*. For an atom in free space, this rate is given by $A = (16\pi^3 \nu^3 / 3hc^3)|\mu_{eg}|^2$. Here, ν is the energy of the emitted photon divided by $\hbar$, c is the speed of light, h is Planck's constant, and $|\mu_{eq}|$ is the magnitude of the transition matrix element. This rate can be modified by placing the atom inside an optical cavity or dielectric material.

spontaneous magnetization The phenomenon of maximum magnetization in ferromagnetic materials even though no magnetizing force is applied.

spot size For a Gaussian beam, that is, one whose transverse intensity has a Gaussian distribution $I \propto e^{-2(x^2+y^2)/w^2(z)}$, $w(z)$ is the *spot size*, which is the radius at which I drops to $1/e^2$ of its maximal value.

sputtering The ejection of one or more ions, atoms, or molecule from a solid or liquid by the impact of an ion or atom. The efficiency of this process increases with the mass of the impacting particle. A related process is secondary electron emission, where the ejected particle is an electron.

squeezed state A state which has fluctuations below the standard quantum limit along some direction in phase space. Along the conjugate direction, the fluctuations must be larger than the standard quantum limit to preserve the uncertainty principle. Examples include quadrature *squeezed states* (or two-photon coherent states), amplitude *squeezed states* (also known as photon antibunched states), and phase *squeezed states*.

squeezed vacuum A particular squeezed state, a quadrature squeezed state with a zero average field, but a nonzero photon number.

squeezing spectrum This is the result of a frequency decomposition of the output of a balanced homodyne detector, which is fed by the output of a source with field decay rate κ, and is given by $S_\theta(\omega) = 16\kappa \int_0^\infty d\tau \cos(\omega\tau) \langle : \Delta A_\theta(0) \Delta A_\theta(\tau) : \rangle$. Here, θ is the phase of the local oscillator, the semi-colons denote normal ordering, and the quadrature $A_\theta \equiv (1/2)(ae^{-i\theta} + a^\dagger e^{i\theta})$. In this expression, a and $a^\dagger$ are the annihilation and creation operators for the field mode of interest.

Squire's theorem In viscous flow, for each unstable three-dimensional disturbance there exists a more unstable two-dimensional disturbance. This is typically exhibited by the more rapid growth of the two-dimensional instability than the three-dimensional instability.

stabilized pinch A class of toroidal magnetic confinement plasmas which stabilize the toroidal pinch configuration by adding a toroidal magnetic field and close-fitting conducting shell to stabilize magnetohydrodynamic instabilities. The tokamak and reversed-field pinch can be seen as evolved examples of *stabilized pinches* which no longer rely on the pinch effect for plasma confinement.

stacking faults The stacking mistake in sequencing of atomic planes of hexagonal close-packed device or of face-centered device, by which one device may result in the other.

stagnation point Point at which fluid comes to rest in a flow field. *Stagnation points* can exist anywhere in the flow, but commonly form on surfaces.

stagnation pressure *See* pressure, stagnation.

stall Separation on an airfoil at high angles of attack causing a decrease in the lift and increase in the drag. *Stall* for most airfoils occurs in the range of $\alpha = 10° - -18°$ but may vary depending upon Re, M, airfoil profile, and other parameters such as surface roughness and free-stream turbulence intensity.

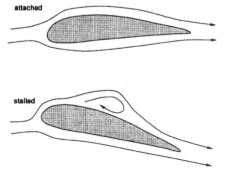

Attached and stalled flow over a wing.

standard quantum limit Defined in terms of the fluctuations of the ground state of the harmonic oscillator. In that state, fluctuations are phase insensitive, the same for any quadrature. A measuring device which uses laser light and is coupled to vacuum modes of the electromagnetic field has a lower limit of sensitivity. That sensitivity can be enhanced by shining a squeezed vacuum on the ports that are normally coupled to ordinary vacuum modes.

standard theory and standard model of particle physics According to this theory, all matter is made up of quarks and leptons, which interact by the exchange of gauge particles. There are four basic interactions: electromagnetic, weak, gravitational, and strong interactions. In electromagnetic interaction, an electron (lepton) interacts with a proton by a photon, which is a gauge particle. Beta decay caused by weak interaction is mediated by a gauge vector particle, a weak vector boson. Hadrons (e.g., protons and neutrons) are made up of three fractionally charged quarks. The interaction of quarks is called color exchange and is described by eight kinds of gauge particles called gluons. Graviton is a particle that mediates gravitational interaction. This model is mainly based on data from CERN, the Fermi lab, and SLAC.

standing wave Nonpropagating surface gravity wave generated by the superposition of two opposite moving waves of identical wave number k and amplitude a. The displacement y of the free surface is given by

$$y = 2a \cos kx \cos \omega t$$

ω is the frequency at which the wave oscillates vertically.

Stanford linear accelerator center (SLAC) This two mile long accelerator accelerates electrons up to 50 GeV. A series of metal tubes (drift tubes) are in a vacuum vessel and connected to alternate terminals of a radio frequency oscillator. Linear accelerators have an advantage in comparison to synchrotrons because energy losses in a form of synchrotron radiation are not present, but they require more radio-frequency cavities and radio oscillators. *SLAC* was completed in 1961 at cost of $115.

Studies in the late 1960s supported Gell-Mann's quark hypothesis (Jerome I. Friedman and Henry W. Kendall from Massachusetts Institute of Technology and Richard E. Taylor of SLAC at SLAC. Bombing with high-energy electrons fixed a proton target. Analysis of the products of decay showed that the proton has constituents with quark properties.) Psi (J at Brookhaven Lab.)) Meson (Burton Richter Jr.) discovered together with people at Brookhaven National Laboratory (Samuel C.C. Ting at al.) Excited states psi' a n psi" are seen only at SLAC. Two Charmonium energy states were discovered at SLAC soon after first state was found at (psi' about 3.7 Gev, and psi" with mass about 4.1 Gev) SPEAR electron-positron storage ring at SLAC to conduct high-energy annihilations experiments. Experiments with charmonium are mostly done at SLAC. SPEAR has two interactions regions MARK II detector and Crystal Ball detector, used to detect electronmuons events. Crystal Ball detector is in 1982 moved to DESY and installed in DORIS e^+e^- storage ring.

Stanford linear collider An electron-positron accelerator which can be used for detection Higgs bosons below 50 GeV.

The collider design has an advantage in comparison to storage rings because beams can be made smaller; in such a way, the probability of interaction is higher (it can produce toponium-t quark in decay of Z^0 bosons up to 100 GeV).

Stanton number Dimensionless number relating the heat transfer

$$St \equiv \frac{\dot{h}}{\rho U_\infty C_p}$$

where C_p is the specific heat at constant pressure and $\dot{h}$ is the heat transfer coefficient.

Stark effect (1) The change in the energy of a material system upon the application of an external electric field. This effect is exploited in semiconductor quantum wells to realize ultrafast optical switches and is an example of wave function engineering. When an electric field is applied perpendicular to the interfaces of a quantum well (which is a narrow bandgap semiconductor sandwiched between two wide bandgap semiconductors), the potential profiles

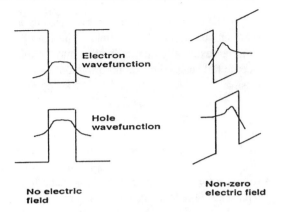

No electric field / Non-zero electric field

The electron and hole wavefunctions and the energy band profile of a quantum well with and without an applied electric field

The quantum-confined Stark effect. When an electric field is applied transverse to the heterointerfaces of a quantum well, the conduction band profile tilts. The electron and hole wave functions are skewed in opposite directions which reduces the overlap between them.

in the conduction and valence bands tilt to accommodate the electric field. In other words, the potential energies of both electrons and holes change, which is the *Stark effect*. The altered potential landscape causes the wave functions of electrons and holes to be skewed since both electrons (negatively charged) and holes (positively charged) will tend to minimize their energies by moving against and along the electric fields respectively. This wave function skewing alters the so-called overlap between the electron and hole wave functions. The overlap is the integral $\int_0^a \psi^*_{\text{electron}}(x)\psi_{\text{hole}}(x)\,dx$, where a is the width of the quantum well and the ψs are the wave functions. The intensity of light emanating from the quantum well (photoluminescence) caused by the radiative recombination of electrons and holes is proportional to the square of the overlap, and the frequency of the light depends on the effective bandgap. Since both these quantities change when the electric field is applied, both the intensity and frequency of the photoluminescence change and can be modulated by the electric field. Thus, both amplitude and frequency modulation of the electromagnetic signal (light coming out of the quantum well) can be achieved via the externally applied electric field.

(2) The change of spectral lines caused by a static or quasistatic electric field. The field is either an externally applied one or may be the electric field caused by neighboring ions as in a plasma.

state preparation   The experimental process of arranging a quantum system to be in some well-defined state at a particular time.

state vector   A ray in a Hilbert space that represents a quantum state of a system.

state vs. level   A physical system is said to be in a particular state when its physical properties fall within some particular range; the boundaries of the range defining a state depend on the problem under consideration. In a classical world, each point in phase space could be said to correspond to a distinct state. In the real world, time-invariant systems in quantum mechanics have a set of discrete states, particular superpositions of which constitute complete descriptions of the system. In practice, broader boundaries are usually drawn. A molecule is often said to be in a particular excited electronic state, regardless of its state of mechanical vibration. In nanomechanical systems, the PES often corresponds to a set of distinct potential wells, and all points in configuration space within a particular well can be regarded as one state. Definitions of state in the thermodynamics of bulk matter are analogous, but extremely coarse by these standards.

static tube   Slender tube aligned with the flow direction with circumferential holes parallel to the fluid motion such that the static pressure is measured.

stationarity   For a stochastic process, the average value of a variable will fluctuate in time, but the statistics of the fluctuations can become time-independent. For example, $\langle V(t)V(t+\tau)\rangle$ can become independent of the time t and depend only on the delay time τ. This property is known as *stationarity*.

stationary state   A state in which $|\psi(x)|^2$ is independent of time. These are eigenstates of a Hamiltonian operator with no explicit time dependence, and satisfy the time-independent Schrödinger equation $\hat{H}\psi(x) = E\psi(x)$. In addition, they are states of definite energy.

steady flow   Flow in which the flow variables (velocity, pressure, etc.) are not a function of time such that $\mathbf{u} \neq u(t)$. A particle on a streamline in *steady flow* will remain on that streamline. In steady flow, pathlines, streaklines, and streamlines are coincident.

Stefan–Boltzmann law   (1) Law that states that the energy density of the radiation from a blackbody is proportional to the fourth power of the absolute temperature of the blackbody.

(2) For a perfect blackbody radiator in thermal equilibrium at temperature T, the *Stefan-Boltzmann law* states that the total emitted intensity is proportional to the fourth power of the temperature, $I_{tot} = \int_0^\infty \frac{c}{4}\rho(\omega)d\omega = \sigma T^4$. Here $\rho(\omega)$ is the Planck spectral energy density. The constant $\sigma = 5.67 \times 10^{-8} W/m^2 K^4$

stellarator   A class of toroidal devices for magnetic confinement of plasmas. As originally invented by Lyman Spitzer (1914–1997), the *stellarator* used either a racetrack-shaped or figure-8 tube. Field coils around the tube provided a magnetic field structure with both an axial (toroidal) field and a rotational transform (poloidal field) to provide stable particle orbits. More recent *stellarators* have a more purely toroidal geometry but retain the notion that the stabilizing poloidal field is supplied by external field coils, in contrast to the tokamak, where a plasma current produces the stabilizing field. The basic idea behind both concepts is that there must be a helical twist in the magnetic field in order to average out particle drift motions that would otherwise take the plasma to the walls of the vacuum vessel. Because of the twist in the external coils, the *stellarator* (unlike the tokamak) is not axisymmetric, that is, not symmetric about the major axis of the torus. A number of different *stellarator* designs and coil configurations are possible. The *stellarator* is at present widely considered the most serious alternative to the tokamak for magnetic confinement fusion. Since the concept is inherently steady-state, it would not have the tokamak's problems with thermal and mechanical cycling or current drive. However, to date, *stellarators* have had poorer energy and particle confinement than tokamaks, due in part to their more complex field geometry and correspondingly complex range of particle orbits. Other toroidal confinement schemes similar to the *stellarator* include the reversed-field pinch (RFP) and the bumpy torus.

stellar wind   The plasma (typically comprised mostly of protons and electrons) flowing outwardly from a star, with or without magnetic fields. The *stellar wind* for our sun is known as the solar wind.

Stern–Gerlach effect   The splitting of a beam of atoms with magnetic moments when they pass through a strong, inhomogeneous magnetic field into several beams.

stiffness constant   Constant coefficients involved in equations that relate stress components as functions of strain components in elasticity.

stimuated emission rate   The rate at which stimulated emission occurs. Typically given by $R_{stim.\ em.} = BU(\omega)$, where B is the Einstein B coefficient and $U(\omega)$ is the electromagnetic energy density at the resonant frequency of the transition.

stimulated Brillouin scattering   Brillouin scattering that is enhanced by an external field. This can occur when a laser beam of frequency ω_L is incident on a medium with an acoustic wave of frequency ω_A inside. The acoustic wave sets up a refractive index variation, leading to a reflected wave that is Stokes downshifted to a frequency of $\omega_L - \omega_A$.

stimulated emission   An atom in an excited stated can be induced to make a transition to a lower state by the presence of electromagnetic radiation (photons). The emitted light is in phase with the incident field and in the same direction, as opposed to the random nature of spontaneous emission.

stimulated Raman scattering   In this process, a photon of frequency ω incident on a medium is annihilated, and a photon at the Stokes frequency $\omega_S = \omega - \omega_\nu$ is created, where ω_ν is typically the frequency difference between two vibrational states of the medium.

stochastic cooling   Very important in building proton–antiproton storage rings. Specifically, antiprotons are produced in the collision of protons and ordinary matter, but these antiprotons have wide interval speeds and directions. Before usage in colliding beams, antiprotons have to be cooled. One method is *stochastic cooling*, invented by Simon van der Meer of CERN. This method of cooling antiprotons includes a small ring with a large aperture for the storage of antiprotons, system detectors, and orbit-correcting
magnets. Detectors detect the average position of the particles if the center of charge strays from the axis of the vacuum chamber; the correction is computed and dispatched to magnets. Some particles could be deflected even more from a proper trajectory, but the majority of the particles are moved in the proper direction.

stochastic differential equation   In many cases, one takes the effects of a system's environment into account by adding a dissipative term to a differential equation. The fluctuation–dissipation theorem requires that a noise term of zero mean and nonzero root mean square fluctuations be added as well. An example is Brownian motion, where the motion of a particle interacting with a background reservoir is described by $d^2r(t)/dt^2 = F_{ext} - \gamma dr/dt + \Gamma(t)$, where $\langle \Gamma(t) \rangle = 0.0$ and $\langle \Gamma^*(t)\Gamma(t+\tau) \rangle$ is proportional to γ.

stochastic electrodynamics   Theory of electrodynamics that tries to replace quantum fluctuations with stochastic processes. It does not agree fully with the predictions of quantum electrodynamics, which have been well confirmed by experiment.

stoichiometric alloys   Alloys that contain component elements in exact ratio as required by their chemical composition.

stoke   Unit of measure of kinematic viscosity, $1\ stoke = cm^2/s$.

Stokes bulk viscosity assumption Assumption that the viscous parameters in the constitutive relations for Newtonian fluid are related such that

$$\lambda + \frac{2}{3}\mu = 0$$

which is accurate in most cases.

Stokes component If photons (quanta of light) impinging on a solid are scattered along with the emission or absorption of a phonon (quanta of ion vibration), then the process is called either Brillouin scattering (if the phonon involved is an acoustic phonon) or Raman scattering (if the phonon involved is an optical phonon). If absorption of a phonon takes place, then the scattered light has a higher frequency than the incident light (blue-shifted) and the process is referred to as the anti-Stokes process (the component of increased frequency in the scattered radiation is called the anti-Stokes component). If emission of a phonon takes place, then the scattered light has a lower frequency (red-shifted) and this process is called the Stokes process.

Stokes drift Advection of fluid parcels in the direction of wave propagation in surface gravity waves. The phenomonon is due to the higher velocity of the periodic motion near the top of the circular orbit causing a nonzero net velocity. The average lateral velocity is given by

$$\bar{u} = a^2 \omega k \frac{\cosh 2k(z_o + H)}{2 \sinh^2 kH}$$

where a, ω, and k are the wave's amplitude, frequency, and wave number respectively. z_o is the distance from the surface and H is the fluid depth. This drift results in an overall mass transport of fluid due to the wave motion.

Stokes flow Steady creeping flow in which Re $\to$ 0, reducing the momentum equation to

$$\nabla p = \mu \nabla^2 \mathbf{u} .$$

Viscous forces are exactly balanced by pressure forces. This characterization describes behavior of an essentially massless fluid. The solution for creeping flow around a sphere is often referred to as *Stokes flow*. In this case, the radial and tangential velocity components can be shown to be

$$u_r = U_\infty \cos\theta \left(1 - \frac{3R}{2r} + \frac{R^3}{2r^3}\right)$$

and

$$u_\theta = -U_\infty \sin\theta \left(1 - \frac{3R}{4r} - \frac{R^3}{4r^3}\right)$$

where R is the sphere radius. The pressure field can be solved exactly to show

$$p = -3R\mu U_\infty \cos\theta / \left(2r^2\right)$$

while the drag force on the sphere is given by

$$D = 6\pi \mu R U_\infty$$

which is also referred to as Stoke's law of resistance.

Stokes shift Shift in a spectral line toward larger wavelengths via absorption of a photon and emission of a second one with lower energy. Typically occurs via a Raman process. *See also* anti-Stokes shift.

Stokes theorem The circulation about a closed loop is equivalent to the flux of vorticity, or vorticity at a point is equal to the circulation per unit area such that

$$\Gamma = \int_A \omega \cdot d\mathbf{A} .$$

Stoner–Wohlfarth model A theoretical model to explain the magnetic properties of small single domain particles with uniaxial symmetry. This model predicts the nature of the hysterisis curves (magnetization vs. magnetic field) when the magnetic field is directed along or perpendicular to the easy axis of magnetization.

stop band The range of wavelength or frequency that is attenuated very heavily so as to almost stop, while the wavelengths or frequencies outside this range are allowed to pass freely through. This is true in case of optical or electrical devices.

stopping power Value defined to characterize the stopping (due to losing energy through ionization) of charged particles in some media $S(T)$. This is defined as the amount of kinetic energy that particle lost per unit of path in some medium:

$$S(T) = -\frac{dT}{dx} = n_{\text{ion}} \cdot \bar{I},$$

where T is the kinetic energy of particle, n_{ion} is the number of electron pairs per unit of path, and I is the average energy of ionization of an atom in the medium.

storage rings One way of building head-on collisions. Beams of particles circulate continuously (similar to synchrotrons). Two storage rings can be tangent to each another and build collision in the place of contact. When particles and their antiparticles are used for collision, one ring can be used (e.g., electrons–positrons). Electrons travel in one direction in the ring while positrons travel in the opposite direction. Collisions are diametrically opposed at two points. (SPEAR, SLAC, and University of California's Lawrence Berkeley Laboratory)

strain rate *See* shear rate.

strangeness changing neutral currents Weak interactions in which the total charge of hadrons stays the same, but the strangeness is changed. Typically s quark goes in d quark with emission of two leptons. (Decay of neutral K-meson into two opposite charged muons. This kind of process is very rare, in comparison with the prediction of unified electroweak theory (three quark flavors; the prediction is one million times greater than the experimental result). Addition of a fourth quark flavor with the same electrical charge as the u quark explains this discrepancy.

strangeness with charm and beauty is the quantum number of the quark-strangeness of a quark Quantum number of the s quark. This quark is part of particles with a strangeness different than zero (kaons and lambda baryons). In strong reactions, this quantum number is conserved.

Stranski–Krastanow growth Epitaxial growth of a solid material on a solid substrate can occur in three distinct modes. If the growth proceeds layer by layer, then the mode is called a van der Merwe mode. This happens if the substrate and the thin film grown atop it are more or less lattice-matched so that the strain in the film is small. If the lattice constants of the film and substrate are significantly different and the film has a higher surface energy than the substrate, three-dimensional islands of the film material nucleate on the substrate. This is called the Volmer–Weber mode. The Stranski–Krastanow mode is a combination of the two previous modes where the growth of a several-monolayer thin film (called the wetting layer) is followed by the nucleation of clusters and then island formation. Which of the three modes predominates depends on the lattice mismatch and differences in surface energy between the film material and the substrate.

Quantum dots (three-dimensional nanostructures) of InAs are routinely grown on GaAs substrates by the *Stranski–Krastanow growth* method. These quantum dots have been shown to possess excellent optical properties for applications in lasers, photodetectors, etc.

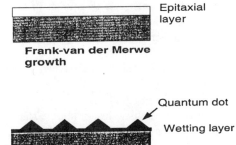

Three types of film growth mode.

stratification   Flow field in which the fluid density is a function of the vertical distance $\rho = \rho(z)$. In analysis of stratified flow, the density is treated as constant using the Boussinesq approximation, except in the vertical momentum equation. The primary dimensionless parameter of interest is the Brunt–Väisälä frequency. Internal waves similar to surface gravity waves may form in regions of large *stratification* gradients.

streakline   Line connecting loci of all fluid parcels that have passed through a fixed point in space at some previous time. Injection of dye at a single point in a flow reveals the *streakline*.

stream function   Scalar quantity Ψ in a flow field used to determine the values of streamlines defined by

$$u = \frac{\partial \Psi}{\partial y} \quad v = -\frac{\partial \Psi}{\partial x}.$$

streamline   A line formed by points tangent to the velocity vectors in a flow field at a given instant. Since the *streamlines* are tangent to the local velocity vector, there is no flow across a *streamline*. Each *streamline* is a line with a constant value of the stream function.

streamtube   A combination of multiple streamlines forming a surface.

strength of crystals   The amount of stress (e.g., tensile, compression, etc.) that crystals can withstand before they break down.

stress   Normal or shear surface force given by the stress tensor.

stress tensor   Tensor describing the state of stress at a point in a fluid. The three-dimensional *stress tensor* consisting of nine components can be written as

$$\tau_{ij} = \begin{bmatrix} \tau_{11} & \tau_{12} & \tau_{13} \\ \tau_{21} & \tau_{22} & \tau_{23} \\ \tau_{31} & \tau_{32} & \tau_{33} \end{bmatrix}$$

where the subscripts i and j denote the direction of the normal component and direction in which the stress force is acting. The diagonal components are normal stresses and the off-diagonal components are shear stresses.

string theories, superstring theories   The structure in *string theories* is 10^{-35} m long, or 10^{20} times smaller than proton diameter. For example, in *string theory* one hadron is understood as a string. This string can rotate and vibrate by laws of relativistic dynamics. Endpoints are moved with light speed. Quantization of the vibration of the string gives different hadrons. This model can describe the confinement of quarks (Gabriele Veneziano, CERN, Yoichiro Nambu, University of Chicago). Tension of the string can explain the binding of quarks (quarks are located on ends of strings). A string by itself is a mathematical one-dimensional object. Gravity in this theory is described as nine-dimensional spatial plus a time dimension of space. The vibrational mode of string corresponds to a particle. The vibrational frequency of the mode corresponds to the mass of the particle. Familiar particle are different modes of single string. *Superstring theory* combines *string theory* and supersymmetry. This allows the possibility of unifying all four forces. There are open and closed strings. Open strings have endpoints. Conservation of charge is connected to Yang-Mills forces that are tied to the endpoints of the string. With vibrational states of an open string are massless spin one gauge particles (but not gravitons). Two strings can interact making a third string, which can split into final strings.

stripping reaction   Type of nuclear reaction when two nucleons interact that allows the possibility of transferring one or more nucleons between them. For example, a deuteron is incident on an ^{16}O target neutron from deuteron can be attached to oxygen in the target. The scattered particle is a proton and the nucleus is ^{17}O. This reaction $^{16}O(d, p)^{17}O$ is an example of a *stripping reaction* (the neutron is stripped from the deuteron). The inverse reaction is called the pick-up reaction.

strong interaction   Strong focusing principle in the construction of synchrotrons which includes dipole and quadruple magnetic fields. This technique made possible the construction of very large machines.

strong interaction forces   Bind nucleons together. Can be described in terms of the ex-

change of mesons. A better understanding is given by QCD.

strong localization The phenomenon whereby time reversed pair trajectories reinforce each other by constructive interference to the extent that virtually all trajectories are reflected, leading to very large resistance. Typically, the resistance increases exponentially with the length of the resistor as opposed to linearly:

$$R \sim \exp[L/L_0]$$

where R is the resistance, L is the length, and L_0 is called the localization length. *Strong localization* is usually observed in quasi one-dimensional conductors. *See* weak localization.

strongly coupled plasma A collection of charged particles whose inter-particle Coulomb potential energy exceeds the particle thermal energy. Unlike the more common weakly coupled plasma, which is gas-like, a *strongly coupled plasma* behaves like a liquid or crystal, and is sometimes termed a Coulomb lattice or Wigner lattice.

Strouhal number Dimensionless frequency

$$St \equiv \frac{fL}{U_\infty}$$

important in flows where periodic motion is involved. *See* Kármán vortex street.

structure factor The amplitude of the scattered wave in a particular direction, in a crystal, when the reflection takes place obeying Bragg law, the incident wave (X-rays or electrons) being of unit magnitude and the scattered amplitude being measured at unit distance with its phase known.

SU(2) The symmetry underlying spin and isospin is the symmetry of a non-Abelian group *SU(2)*. This is a special unitary group in two dimensions. Pauli matrices represent generators of this group in two dimensions.

SU(3) symmetry Prediction of the group theory stating that particles with strong interactions can be grouped into 1, 8, 10, and 27 such that those in each group can be considered as belonging to different states of the same particle.

SU(5) Simplest group that can be used in grand unification theories. This is the five-dimensional analog of isospin.

SU3 This group symmetry describes the internal three-dimensional space symmetry of the color of quarks.

sublattices Sections of the primitive cell of a crystal. For instance, the Si lattice can be viewed as consisting of two interpenetrating face-centered cubic sublattices displaced along the body diagonal by one-quarter of the diagonal.

sublayer, inertial In a turbulent boundary layer, the region where inertial forces dominate.

sublayer, viscous In a turbulent boundary layer, the region immediately adjacent to the wall where viscous effects dominate. The sublayer thickness is approximately

$$\delta \approx \frac{5\nu}{u_*}.$$

sub-Poissonian statistics A typical photon counting experiment will measure a certain number of photons in time T. This is repeated over and over again until the statistical distribution of the number of photons detected in time T is built up, $P(n, T)$. For coherent light, this distribution can be calculated to be a Poissonian distribution, where the standard deviation Δn is equal to the square root of the mean photon number $\langle n \rangle$. For some light fields that cannot be modeled as classical stochastic processes, this distribution can be sub-Poissonian ($\Delta n \geq \sqrt{\langle n \rangle}$), which is indicative of a more regularly spaced sequence of photons. *See also* photon antibunching.

subrange, inertial The low end of the turbulent wave number spectrum where energy transfer takes place by inertial forces. Vortex stretching is the primary method of transfer.

subsonic flow　Flow in which the local Mach number is less than unity. The governing differential equations in subsonic flow are elliptic.

substitutional defects　Defects arising out of substitution of some atoms in a crystal by atoms of a different element although the basic structure remains the same.

Sudbury neutrino observatory (SNO)　The first detector capable of distinguishing electron neutrinos from muon or tauon neutrinos. The detector contains 1000 T of heavy water (D_2O) surrounded by 9500 photo multiplier tubes. Using heavy water gives an advantage over using ordinary water (Kamioka detector) because deuteron in heavy water is sensitive to the neutral current reaction:

$$\nu_e + d \rightarrow p + n + \nu_e.$$

A neutron realized in this reaction can be captured by another nucleus through a (n, γ) reaction. A scintillation counter can detect γ quanta. The minimum neutrino energy to activate this reaction is 2.22 MeV.

sum-frequency generation　When two laser beams of frequencies ω_1 and ω_2 are incident on a non-linear material, a new beam with frequency $\omega_{\text{sum}} = \omega_1 + \omega_2$ is generated. This occurs via simultaneous absorption of an incident photon from each field followed by emission of a photon at the sum-frequency.

summing over histories　Richard Feynman devised this method. This method of string theories has been fully developed by Stanley Mandelstam and Alexandar Polyakov.

sum rule　A formula which establishes the equality between some quantity or expression to the sum over all states of another quantity. The most prominent example is the Thomas–Reiche–Kuhn sum rule.

sunspots　Magnetic regions roughly the same diameter as the earth which appear as dark spots on the surface of the sun and can last anywhere from a few days to several weeks in the case of the larger ones. The temperature at the center of a *sunspot* is about 4500 K, whereas the photosphere is normally 6000 K. The number of *sunspots* varies cyclically with an 11 year period related to the solar magnetic cycle. During the *sunspot* cycle, the activity ranges from no *sunspots* near the time of minimum activity to hundreds near the time of maximum activity.

superallowed β^- decay　A special class of beta decay when the initial nuclear state is $J_i^\pi = 0^+$ to a final state $J_f^\pi = 0^+$ with the same isospin I. One example is $^{14}O \rightarrow {}^{14}N + e^+ + \nu_e$.

superconducting super collider　A huge, 52 miles in diameter, colliding-beam proton accelerator with superconducting magnets. Energy of a collision has to be 40 TeV.

superconductivity　A state of matter where the conductance of the matter is infinite at DC voltages. *Superconductivity* was discovered in 1911 by H. Kammerling Onnes, who found that certain elements like mercury, lead, and tin appeared to lose all electrical resistance when they were cooled below a certain temperature called the transition temperature. *Superconductivity* is characterized by zero DC resistance and perfect diamagnetism. The latter means that not only does a superconductor exclude all magnetic flux, but as a material in the normal state is cooled to below the transition temperature, any trapped flux is expelled. This latter phenomenon is called the Meissner effect. The existence of this effect implies that at high enough magnetic fields, when the superconductor is no longer able to expel the flux, the flux will penetrate the material and quench the *superconductivity*. The value of the magnetic field at which this happens is called the critical magnetic field.

There are two types of superconductors. One, in which the quenching occurs discontinuously (first order phase transition), is called a type I superconductor (such as mercury). Then there are those where the quenching occurs continuously and the phase transition is of second order. These are called type II superconductors. Flux starts to penetrate a type II superconductor at a critical field H_{c1}. Flux tubes penetrate the sample, each carrying a quantum of flux $h/2e$, where h is Planck's constant and e is the electron charge. This is called the Shubnikov phase. Then finally, at another critical field H_{c2}, the flux density in the material B reaches the value μH (μ = magnetic permeability in the normal state and H = applied magnetic field). At this point, the *superconductivity* is completely quenched.

Superconductors are also classified into low T_c and high T_c superconductors. The latter were discovered in 1986 by Bednorz and Müller. They are of type II and have a much higher transition temperature (T_c) than the low T_c type. The best known example is yttrium-barium-copper-oxide ($Y_1Ba_2Cu_3O_{7-\delta}$) with a transition temperature of around 92 K.

The phenomenon of *superconductivity* is explained by the Bardeen–Cooper–Schrieffer theory, which postulates that two electrons (or holes) of like charge develop an attraction (overcoming the Coulomb repulsion) as a result of the intercession of a third entity such as a phonon. These Cooper pairs carry current without resistance (or dissipation). Low T_c superconductors are amply described by this theory and it is not clear if high T_c superconductors can also be described by the same theory.

superconductors Substances exhibiting the rather unusual property of very low or negligible resistance to the flow of electric current below a certain temperature, the latter being known as the critical temperature. These substances include various alloys or compounds or metals and are repelled by magnetic fields. The critical temperature depends on the type of the substance.

supercritical field In heavy ion collisions it is possible to compound a nucleus with Z higher than Z critical (137). As result of this a *supercritical field* is created.

superdeformation (nuclei) For stable nuclei, departure from the equilibrium spherical form is generally small in the ground state. Extremely large deformations from spherical shape are called *superdeformations* and they are observed in excited configurations of medium weight nuclei produced by the fusion of two heavy ions in one. In this process, the formation of superdeformed bands (states with high values of J) is observed. An example is the ^{100}Mo $(^{36}S, 4n)^{132}Ce$ reaction. In this reaction, a 155 MeV ^{36}S beam is used on a target of ^{100}Mo. Superdeformed bands in ^{132}Ce are formed. Deformed nuclei de-excite through the emission of gamma rays.

superelastic collision A collision between a nucleus (or an atom) in an excited state and a nucleon (electron) in which the target system returns to the ground state and almost the entire excitation energy is transferred to the projectile.

superexchange A mechanism involving exchange interaction between two ions of an antiferromagnetic substance where two other ions of a different material, most commonly oxygen, play an intermediate role by forming couples with their spins resulting in the final coupling between the original ions through these opposite spins.

Superfish A particular computer program for computing various field parameters of accelerators such as induced voltages in accelerator rf cavities, mode frequencies, and shunt impedances for accelerating fields in resonant rf cavities (accelerator cavity losses depend on shunt impedance). *See also* more sophisticated MAFIA computer program.

superfluorescence Also known as Dicke superradiance. It is a superradiant process where N atoms are placed in an excited state and are spatially within one wavelength of one another. They may then radiate collectively, with a radiation rate proportional to N^2 rather than N.

supergravity The gauge theory of gravitation is the *supergravity* theory. Einstein's theory of gravity does not itself lend to quantization (problem divergences). Divergences are common in quantum theory of fields, but a renormalization procedure fails to solve this problem. *Supergravity* theory has better divergence behavior.

superheavy elements The heaviest close shell nucleus known is $^{208}Pb (Z = 82, N = 126)$. $Z = 114$ and 126 are strongly stabilized by shell effects. So far, $Z = 112$, and $A = 277$ are identified. The quest is continuing for elements with $Z > 112$ and $N \sim 184$. The element $Z = 112$, $N = 165 (A = 277)$ was created in Gesellschaft Fur Schwerionjenforschung lab in

Darmstadt, Germany using a beam of $^{70}Zn^{30}$ on a target of $^{208}Pb^{82}$.

superkamiokande A massive 50,000 T high-purity water Cernikov detector in a Japanese mine in Kamioka. This detector uses Cernikov radiation to detect solar neutrinos. A neutrino scattered by a charged particle will produce recoil and Cernikov light. For low energy neutrinos (coming from sun hydrogen burning), only scattering with electrons can produce such radiation (neutrinos with energy comparable to the electron rest mass energy of 0.5 MeV in a process of electron scattering can produce Cernikov light).

superlattice (1) Artificially periodically structured materials proposed by Tsu in 1969. A periodic variation of the composition of a material or the doping profile leads to a tunable periodicity. The introduction of the superlattice perturbs the bandstructure of the host materials, yielding a series of narrow sub-bands and forbidden gaps.

(2) Alternating layers of two different materials **A** and **B** result in a compositional *superlattice* structure. The structure has an additional spatial periodicity along the direction of alternation, over and above the inherent periodicity of the atomic lattice. This periodicity can be achieved by either compositional modulation or doping modulation in the case of a semiconductor. In the latter (called doping *superlattices* or n–i–p–i structures), the doping is alternated between n- and p-types. The resulting changes in the conduction and valence band profiles results in a periodic modulation of the potential energy seen by an electron or hole.

Compositional *superlattices* can be of four types depending on the relative alignments of the conduction and valence band edges. Note that type 2A *superlattices* result in semiconductors that are indirect gap in real space.

supermultiplet Multiplet comprising greater than three lines.

supernova *Supernovas* have a special role in the formation of matter because heavy elements are created in their explosions. In *supernova* explosions, shock waves created by a collapsing star core rebound and create ideal conditions for endothermic creation of elements beyond $A \sim 56$. In very massive stars (20-30 solar mass) under huge gravitational attraction collapse of stars becomes to collapse making huge explosion and ejecting matter in space. The rest of the *supernova* is a neutron star or black hole.

The mass of a *supernova* before explosion (Fowler, W.A. and Hoyle, F. Nucleosynthesis in massive stars and supernovae, Astrophysics Journal Supplement Series, 91, 201, 1964) is 57% ^{16}O rich mantle and the outer shell of 33% of H and ^4He. Under the influence of shock waves, different heavy ion reactions can happen, For example, $^{16}O + ^{16}O \rightarrow ^{28}Si + ^4He ^{28}Si + ^{28}Si \rightarrow ^{56}Ni + \gamma$.

The shock waves convert hydrogen into helium and helium into oxygen. Coulomb barriers for elements beyond nickel and iron are high because of a large number of protons. Most observed capturing neutrons make heavy elements. This process makes nuclei richer in neutrons followed by beta decay that keeps the formation in limits of valleys of stability.

supernova neutrinos Radiation of energy can take place in the formation of supernovas in several ways. Kinetic energy of matter ejected in space, gamma rays, positrons, and electron neutrinos are produced. Neutrinos and antineutrinos are produced in the process of annihilation of positrons and electrons. Another channel for this annihilation is the production of two gammas. Gammas have to brake through thick stellar mass and they are absorbed inside.

super-Poissonian statistics A typical photon counting experiment will measure a certain number of photons in time T. This is repeated over and over again until the statistical distribution of the number of photons detected in time T is built up, $P(n, T)$. For coherent light, this distribution can be calculated to be a Poissonian distribution, where the standard deviation Δn is equal to the square root of the mean photon number $\langle n \rangle$. For some light fields, including thermal light, this distribution can be super-Poissonian ($\Delta n \geq \sqrt{\langle n \rangle}$), which is indicative of a less regularly spaced sequence of photons. *See also* photon bunching.

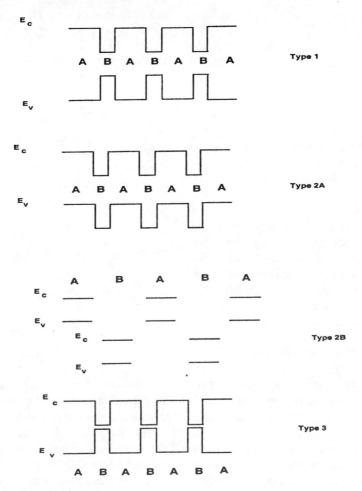

Energy band diagrams of 4 diferent types of compositional superlattices. The types are differentiated according to the relative alignments of the conduction and valence band edges of the two materials A and B constituting the superlattice

Four different types of superlattices.

superposition of states The most general solution to the Schrödinger equation (or any linear differential equation) is a linear sum of all possible solutions ($|n\rangle$), weighed by coefficients (C_n) that are determinable from initial conditions, $|\psi\rangle = \sum_{n=0}^{\infty} C_n |n\rangle$. Generally one uses eigenstates of any Hermitian operator. These eigenstates form a complete orthonormal basis set.

superposition principle States that the most general solution to a linear differential equation is a superposition of all possible solutions

super proton synchrotron Started operating at the peak energy of 400 GeV in 1976 at CERN. Fermilab has a more advanced version of this machine.

superradiance A high gain amplifier can emit with no incident laser field via the process of amplified spontaneous emission, or *superradiance*. In this process, a photon emitted by one atom molecule of the gain medium is then amplified via the process of stimulated emission. *See also* superfluorescence.

supersonic flow Flow in which the local Mach number is greater than unity. The governing differential equations in *supersonic flow* are hyperbolic. For the perturbed velocity field $\vec{u} = (u_\infty + u')\mathbf{i} + v'\mathbf{j} + w'\mathbf{k}$, a velocity potential

Φ is defined such that $\vec{u} = \nabla \Phi$. In the subsonic and supersonic regimes,

$$\left(1 - M_\infty^2\right) \frac{\partial^2 \phi}{\partial x^2} + \frac{\partial^2 \phi}{\partial y^2} + \frac{\partial^2 \phi}{\partial z^2} = 0 \,.$$

supersymmetric theories In *supersymmetric theories* is a symmetry that transforms bosons and fermions into one another (unifies particles with integer and half integer spins). There are an equal number of bosons and fermions for any given mass. Gravity with supersymmetry gives supergravity theories. A graviton is a particle (spin 3/2) which is responsible for supersymmetry in these theories. For every ordinary boson there is supersymetric spin 1/2 fermion. Every particle has supersymmetric particle identical except in spin (e.g., for a spin 1 photon, the supersymetric particle is 1/2 spin photino; every boson has a spin 1/2 supersymmetric fermion). Supersymmetry explains why at high energies, leptons, hadrons, and gauge bosons have smaller masses than normal.

superthermal electron, ion, or particle Many plasmas may be viewed as consisting of one or more bulk fluids in approximate thermal equilibrium plus various non-thermal components, such as resonantly accelerated particles or particles injected from an outside source. When particles in some non-thermal component have higher characteristic energies than those in the thermal bulk plasma, the particles are said to be superthermal. For example, in intense laser–plasma interactions, a laser impinging on a near-solid density target can produce *superthermal electrons* via the ponderomotive force, as well as a thermal blow-off plasma.

supplementary condition The condition that the state vector would behave as a state.

surface acoustic wave Acoustic wave that travels along the surface of a material. These usually decay rapidly into the bulk of the material, and the characteristic length of the decay is the wavelength. *Surface acoustic* wave devices are used in signal processing on a semiconductor chip. They are widely used in realizing tapped delay lines which are the mainstay of transversal filters.

surface electromagnetic wave Electromagnetic wave that travels along the surface of a material. These usually decay rapidly into the bulk of the material, and the characteristic length of the decay is the wavelength.

surface gravity waves Non-dispersive waves formed at the interface of a liquid and a gas. Solution of potential flow equations reveal that the wave frequency is

$$f = 2\pi \sqrt{gk \tanh kH}$$

where k is the wave number and H is the fluid depth; the phase speed $c = 2\pi f/k$ is

$$c = \sqrt{\frac{g}{k} \tanh kH} \,.$$

For deep water waves this reduces to

$$c = \sqrt{\frac{g}{k}}$$

and for shallow water waves this becomes

$$c = \sqrt{gH} \,.$$

Note that in the former case the phase speed does not depend upon the fluid depth, and in the latter case the phase speed is independent of wavelength, giving rise to the rarefaction phenomena of beaching waves tending to align themselves perpendicular to the shoreline. Particle motion in *surface gravity waves* are circular in nature.

surface phonon, plasmon, waves A flat vacuum–solid interface has solutions to the Laplace equation $\nabla^2 \phi = 0$ which propagate along the interface and decay exponentially from that interface when the dielectric function of the solid medium is equal to -1. Such waves are known as *surface waves*. For a dielectric, the condition $\varepsilon = -1$ is satisfied between the frequencies of the transverse and longitudinal optical phonon frequencies. This frequency in between is associated with the *surface phonon*. For a metallic medium, this *surface wave* is called *surface plasmon*.

surface states The states on the surface of a semiconductor to which electrons may be bound very closely.

surface tension Force acting at the interface of two or more immiscible fluids caused by intermolecular attractive forces. For an interface of curvature of radius R, the surface tension σ is proportional to the pressure jump across the interface

$$\sigma = \frac{R}{2}\Delta p .$$

The change in pressure arises from the curvature of the interface and the pressure on the convex side of the interface is lower.

surface waves Acoustic waves generated by earthquakes. These waves travel along a great circle, from the epicenter of the quake, close to the earth's surface. The plate on which the wave travels determines the wavelength of these waves, usually a fraction of the plate size.

susceptibility The *susceptibility* χ is defined by $\vec{P} = \epsilon_0 \chi \vec{E}$, where $\vec{P}$ is the polarization induced in a material under the influence of an external field $\vec{E}$. In general, the *susceptibility* is a tensor. It is scalar constant for a linear, isotropic, homogeneous material.

Sussex potential A special form of nuclei effective interaction that includes many-body correlations in Hartree–Fock nuclear structure computations. *Sussex potential* is not written in a functional form, but as a numerical description of the nucleon-nucleon interaction in the form of matrix elements in a basis of wave functions of shell model.

Sweet–Parker model An early theory for magnetic reconnection, proposed by Sweet (1958) and Parker (1963), in which plasma flows into a region where two sheets of oppositely-directed field lines are reconnecting (a resistive magnetohydrodynamics process); the magnetic energy released in the reconnection process is transferred to the plasma and expels it outwards perpendicular to the inflow direction. This type of reconnection process is a leading candidate for understanding solar flares, and is also important in some types of laboratory plasmas.

symmetric ordering An operator containing products of creation and annihilation operators is said to be *symmetrically ordered* if it is an equal admixture of terms with all creation operators acting to the left and annihilation operators to the right. For example, $A_{\text{symmetric}} = a^\dagger a + a a^\dagger$.

symmetries In a mathematical sense, when the solution of equations remains the same, even if some characteristic of the system they described is changed. If the change of some specific value of the system is equal in each point of space and solutions are unchanged we have global *symmetry* in respect to that characteristic. If some specific characteristic can be altered independently in each point of space, one can say that *symmetry* is local. For example, invariant to three space rotations, (O(3) group) is a continuous group and gives the conservation of angular momentum. Much *symmetry* is not related to ordinary space, but some internal space. It can be rotation in U(1) group gives conservation of charge in Maxwell's electromagnetic theory. Specific very important type of symmetries is gauge symmetries. In these types of *symmetries,* an independent transformation can be done in each point of time and space. *Symmetries* can be broken, i.e., for some direction in internal space a new phenomena can arise (ferromagnetism at some specific temperature). For example, a group of symmetry for electroweak interaction is SU(2)xU(1). At ordinary temperatures we observe two different forces (electromagnetic and weak), but at temperatures beyond 10^{15} degrees C there is no difference between these two forces. Similarly, at temperatures between 10^{30} and 10^{32} C, grand unified theory (SU(5); SO(10) or E6)are on scene (unification of electromagnetic, weak, strong interactions). At these temperatures (10^{30} and 10^{32} C), the effects of quantum gravity becomes important. These temperatures were present between 10^{-43} and 10^{-38} seconds after the Big Bang. Many grand unification theories incorporate supersymmetry (*symmetry* between bosons and fermions). Recent attempts include Einstein's theory of gravity.

symmetry group A group of particles that exhibits symmetry on a plot of the difference between the average charge of the group and the charge of an individual particle vs. hypercharge.

symmetry scars New observed phenomena in highly excited states of a nucleus. This phenomenon represents order in chaos.

SYNCH (also TRANSPORT, COMFORT, MAD) Special computer programs for periodic lattice accelerator design used to compute phase-space matching accelerator sections.

synchrocyclotron Cyclotron (cyclic accelerator) type of accelerator. To accelerate a particle to high energies, relativistic effects have to be taken into account. Resonant relativistic relations require that the frequency of the RF field has to be decreased or the magnetic field increased (or both) as the velocity of particles approaches the speed of light ($v \to c$).

Machines in which the magnetic field is constant, but with frequencies that are varied, are called *synchrocyclotrons*. Machines in which the magnetic field is changed (irrespectively of frequency) are called synchrotrons. In electron synchrotrons, frequency is kept constant; in proton synchrotrons both are varied.

Synchrotrons in the GeV range of energies have positioned magnets in the form of a ring. In some places of the ring, there are RF cavities that accelerate particles.

synchrotron radiation (1) Also known as cyclotron radiation, *synchrotron radiation* is emitted by charged particles whose trajectories are curved by magnetic fields, since the acceleration required to curve the particle's motion leads to the emission of electromagnetic radiation. A number of *synchrotron radiation* sources are presently in operation, using electron particle beams traveling through electron storage rings to provide X-ray light sources for various research applications.

(2) Moving in close synchrotron loops, charged particles emit intensive beams of ultraviolet and X-rays. This loss of energy must be compensated for by additional radiofrequency power in a synchrotron. This is a serious problem in the construction of large synchrotrons, when small beams of magnetic fields become large. These losses are known as beamstrahlung. These losses are the fourth power of beam energy for a given radius (10 GeV accelerator problem). This radiation is a valuable tool for biological and materials studies. These are the most intensive resource of X-rays and ultraviolet light.

synchrotrons See synchrocyclotron.

T

T_1 The lifetime, or inverse decay rate, of the population inversion of a two-level atom. Also known as $\gamma_\parallel$. In the radiatively broadened case, we have $T_1 = 2T_2$.

T_2 The inverse decay rate of the induced dipole moment of a two-level atom. Also known as $\gamma_\perp$. In general $1/T_1 = 1/2T_2 + 1/T_{\text{dephase}}$.

tachyon   A hypothetical particle that travels faster than light.

Tamm–Dancoff approximation   An approximate way of solving the Schrödinger equation for a system of many interacting particles (electrons or nucleons) by including states close in energy through nonperturbative methods and more remote excitations through perturbation theory.

Tamm–Dancoff method   A method of approximation to the wave function of an interacting particle system by considering superposition of several possible states, the latter number determining the degree of approximation being considered.

Tamm surface states   In 1932, Tamm demonstrated the existence of surface states of a special type near the surface of a crystal. James suggested that similar states could also exist near an interface between two different materials. An interface, like a surface, is a strong perturbation because of the discontinuity of the parameters of the material. The energy of such localized states can lie in both allowed and forbidden bands of the bulk dispersion relation. In the latter case, states localized at an interface will manifest as donor or acceptor impurities.

tandem accelerators   At Fermilab, two proton accelerators occupy a single tunnel (*see* Tevatron collider). The second one is proton synchrotron.

targeted radiotherapy   A method in radiotherapy of cancer that selectively exposes cancer cells using radionuclides conjugated to tumor seeking molecules. Radionuclides in use in this method are beta, alpha, or Auger electron emitters (example, 90Y, 131Y, 199 Au, 212 Bi, 125 I, etc.).

tau (τ^-)   Named after the Greek word $\tau\rho\iota\tau ov$ (third), it is the third charged lepton (after the electron and muon).

Heavy leptons, *tau* and antitau, have charges equal to -1, and masses of 1784 MeV. Their life-time is less than 510^{-12} s. The antiparticle is antitau (τ^+) and decays through weak interaction into electrons, muons, or other particles according to the Wainberg–Salam theory of weak interactions. For example, by weak interaction, tau lepton can decay to a tau neutrino and W^- boson. A W^- boson decays into a negative muon and a muon antineutrino.

tauon neutrino   Has a mass of less than 164 MeV and a charge of zero. They are not observed directly.

Taylor column   Column of fluid above a body in a rotating frame that appears to the surrounding flow as an extension of the body and effectively acts as a solid boundary. *See* Taylor–Proudman theorem.

Taylor–Couette instability   *See* Taylor–Couette vortices.

Taylor–Couette vortices   Counter-rotating toroidal vortices encountered in circular Couette flow above a critical Taylor number of 1708 (inner cylinder non-rotating). The vortices appear as discrete vortical bands and can be laminar or turbulent.

Taylor–Görtler vortices   Counter-rotating toroidal vortices encountered along in a boundary layer along a concave wall.

Taylor hypothesis   Assumption that fluctuations at a single point in a turbulent flow are caused by the advection of a frozen turbulent flow field past that point. Essentially, a temporal measurement of a quantity $q(t)$ is transformed to

Taylor number Dimensionless number used for stability criteria in a circular Couette flow;

$$\mathrm{Ta} = \frac{R_i (R_o - R_i)^3 (\Omega_i^2 - \Omega_o^2)}{\nu^2}$$

where R_i and R_o and Ω_i and Ω_o are the inner and outer cylinder radii and rotation rates respectively.

Taylor–Proudman theorem In an unstratified (homogeneous) rotating fluid, the velocity vector cannot vary in the direction of the rotation vector Ω such that

$$\frac{\partial \mathbf{u}}{\partial z} = 0.$$

Thus, the fluid will behave as if a structure extending only partly into the fluid layer extends completely through the layer. *See* Taylor column.

Taylor vortex Vortex satisfying the Navier-Stokes equation given by the tangential (circumferential) velocity field

$$u_\theta = \frac{H}{8\pi} \frac{r}{\nu t^2} e^{-r^2/4\nu t}$$

where H is the angular momentum of the vortex.

teleportation The transmission of a quantum object from a sender (A) to a remote receiver (B). If the object were classical, such as this page of the dictionary, then A could send the page by facsimile. However, if the letters in this page were quantum objects, Heisenberg's uncertainty principle would guarantee that the letters could not be transmitted with 100% fidelity. A quantum object cannot be copied (no-cloning theorem) and hence cannot be transmitted by usual methods.

Teleportation is a scheme to transmit a quantum object. A and B share two entangled quantum objects called an Einstein–Podolsky–Rosen pair. Their entangled wave functions form a set of four Bell basis states. To send a quantum object to B, A entangles its quantum object with its member of the pair and makes a Bell measurement to project it onto one of the four Bell basis states. The sender (A) then sends the result of the measurement (a classical object) to B using a classical channel of communication. Depending on the result, B makes one of four unitary transformations on its member of the EPR pair to regenerate the quantum object that A sent him. In this process, A destroyed its quantum object, but it has reappeared at B's end.

TEM modes TEM stands for transverse electromagnetic. *TEM modes* are propagating modes of the electromagnetic field where both the electric and magnetic fields are perpendicular to the direction of propagation.

temperature coefficient of reactivity The temperature gradient of reactivity of a thermal nuclear reactor ($\Delta \rho / \Delta T$, ρ- reactivity). The major contributor to the *temperature coefficient of reactivity* is thermal leakage because of decrease in moderator density and increase in fast neutrons. In liquid-metal-cooled fast breeder reactors, the Doppler effect in the resonance capture of neutrons, decrease in cooler density (sodium), and thermal expansion of the fuel are the main contributors to the *temperature coefficient of reactivity*.

temporal coherence The degree of *temporal coherence* for a light field is determined by the ability to predict the value of the electric field at a time t_1 if one knows the electric field at time t_2. A monochromatic wave is perfectly coherent, as knowledge of the frequency and initial value allows one to know the value of the field at any point in the future with no uncertainty.

tensor force A spin-dependent force between nucleons which has been introduced in analogy to the force between two magnets so that observed moments of the deuteron could be accounted for theoretically, in particular its magnetic dipole moment and the electric quadrupole moment.

term A group of $(2S + 1)(2L + 1)$ atomic states, usually close in energy, which share definite spin and orbital angular momentum quantum numbers S and L.

term splitting The difference in the energies of almost degenerate eigenstates of an atom, molecule or nucleus due to a small perturbation, such as the spin-orbit interaction.

ternary fission Fission consisting of three final nuclei.

TESSA-3 spectrometer (total energy suppression shield array) A special scintillation–bismuth germanate shielded germanium (BGO) detector at NSF Daresbury. This detector is suitable to investigate high-spin states of nucleons. Heavy ions are used to study these states, because by using them, sufficiently high orbital momentum can be achieved. Typical examples are ^{124}Sn $(^{48}Ca, 4n)$ ^{168}Hf and ^{100}Mo $(^{36}S, 4n)$ ^{132}Ce at energies ~ 4MeV/A. The emission of gamma rays follows these processes. Gammas or charged particles produce electron–hole pairs in semiconductor layers of detectors. These kinds of detectors have a single crystal of germanium doped as a diode. A reverse bias voltage is applied on that germanium and produced electrons and holes are attracted on opposite sides of the detector. The total amount of collected charge is proportional to the energy of incidence of the gamma ray.

test particle In calculations of plasma parameters such as the Debye length and electrical conductivity, it is often useful to analyze the Coulomb interactions of a sample plasma particle, or *test particle*, with the rest of the plasma. Such calculations are then said to use the *test particle* method.

tetrahedral coordination A crystal structure (e.g., diamond or zincblende structure) where each atom has its four nearest neighbors located at the vertices of a tetrahedron (with the reference atom being at the center of the tetrahedron).

tevatron collider A superconducting system that started to work in 1983. In 1990, this collider was operating at 900 GeV. Higgs particles with masses above 50 GeV can be detected in electron–positron collisions in this facility.

thermal electron, ion, or particle Many plasmas may be viewed as consisting of one or more bulk fluids in approximate thermal equilibrium plus various non-thermal components. It is then often convenient to treat particles from the thermal component (*thermal particles*) separately from those in the non-thermal component(s). For instance, magnetic fusion plasmas are often heated using superthermal neutral particle beams, in which case the ions in the plasma consist of a thermal (slowed-down) and superthermal (slowing-down) component.

thermalization Also known as slowing-down, this is the process (generally arising from collisions) by which fast (superthermal) particles give up energy to the plasma and slow down to thermal speeds.

thermal noise The electronic noise (temporal fluctuations) in the current through (or voltage over) a sample connected to a power supply. The noise arises from temporal fluctuations in the velocity of electrons interacting with scatterers in the sample. This type of noise has a temperature dependence. At low frequencies, the mean square fluctuation in the voltage $<V^2>$ over a sample is related to the absolute temperature T and the resistance R of the sample according to

$$<V^2> = 4kTR\Delta f$$

where Δf is the frequency window (bandwidth) within which the noise is measured.

Thermal noise is also referred to as Johnson noise or Nyquist noise.

thermal plasma A plasma in which all components (electrons, ions, and molecules) are in local thermodynamic equilibrium. The ion charge state distribution in this case is the Saha–Boltzmann distribution governed by the Saha equation.

thermal radiation Also known as heat radiation. When a material is at a nonzero temperature, it will radiate, typically at lower energies or higher wavelengths. In particular, objects in everyday life tend to emit infrared radiation. For a perfectly absorbing and reflecting surface (i.e., a blackbody), the emission spectrum is given by the Planck spectrum.

thermal reservoir When one couples a quantum system to its environment, and that environment is in thermal equilibrium at some temperature, one can assume that the large system (the reservoir, or environment) is unaffected by the actions of the small quantum system and use appropriate statistics to specify the state of the environment at all times.

thermionic emission The phenomenon of electron or hole emission over a potential barrier at a finite temperature. Such a barrier may exist at the interface of a metal and an insulator. The current density J associated with *thermionic emission* is given by the Richardson–Dushman law:

$$J = -\frac{qm}{2\pi^2\hbar^3}(kT)^2 e^{-W/kT}$$

where q is the charge of an electron (or hole), T is the absolute temperature, k is the Boltzmann constant, and W is the work function of the metal. Thus, if $\ln(J/T^2)$ is plotted against $1/kT$, the resulting curve will be a straight line with a slope of $-W$. Such a plot is used to experimentally measure the work function W.

thermodynamic equilibrium, plasma There is a very general result from statistical mechanics which states that, if a system is in thermodynamic equilibrium with another (or several other) system(s), all processes by which the systems can exchange energy must be exactly balanced by their reverse processes so that there is no net exchange of energy. For *plasmas in thermodynamic equilibrium,* one can view the plasma as an ion and electron system, and one sees that ionization must be balanced by recombination, Bremsstrahlung by absorption, line radiation by line absorption, etc. When thermodynamic equilibrium exists, the distribution function of particle energies and excited energy levels of the atoms can be obtained from the Maxwell–Boltzmann distribution, which is a function only of the temperature. The Saha equation is a special application of this. Because *thermodynamic equilibrium* is rarely achieved, especially in short-lived laboratory plasmas, one must generally also consider deviations from total equilibrium, leading to more complicated situations.

thermoelectric Materials that transport electricity efficiently while transporting heat not as efficiently. The figure of merit for a *thermoelectric* material is a dimensionless quantity defined as

$$ZT = \frac{S^2\sigma T}{\kappa}$$

where S is the Seebeck coefficient, σ is the electrical conductivity, κ is the thermal conductivity, and T is the absolute temperature.

thermoelectric effects The effect by which heat energy is converted directly into electrical energy and vice versa.

thermoluminescence The process of thermally releasing electrons (holes), trapped in localized states, which gives rise to photoluminescence upon subsequent recombination with holes (electrons). These electrons (holes) can often also be observed in electrical transport (thermally stimulated currents). The intriguing fact about the process is that a very small quantum energy (thermal, 25 meV at room temperature) is needed to produce emission of photons of several eV. *Thermoluminescence* applications has in dosimetry and as an infrared beam finder.

thermomagnetic effects Thermoelectric effects occurring in presence of magnetic field. *See* thermoelectric effect.

thermonuclear In nuclear physics, relating to processes which initialize the fusion of light nuclei because of their rapid motion at extremely high temperatures, leading to the release of fusion energy.

thermonuclear fusion (1) Describes fusion reactions achieved by heating the fuel into the plasma state to the point where ions have sufficient energy to fuse when they collide, typically requiring temperatures of at least 1 million K. *Thermonuclear fusion* converts a small amount of the mass of the reactants into energy via $E = mc^2$, and is the process by which most types of stars (including the sun) produce the energy to shine. In these stars, gravity compresses and heats the core stellar plasma until the power released from fusion balances the power radiated from the star; the star then reaches an equilibrium where *thermonuclear fusion* reactions sustain the internal pressure of the star in balance against the force of gravity. This prevents the star from collapsing, at least until it runs out of fusion fuel. On earth, controlled *thermonuclear fusion* reactions represent a possible long-term source of energy for humanity, though research remains decades away from economical fusion power. Uncontrolled fusion provides the immense power of thermonuclear weapons (hydrogen bombs). In controlled fusion research, the term thermonuclear is also used to characterize fusion reactions between thermal ions, as opposed to fusion reactions involving injected beam ions or other ions lying outside the thermal Maxwellian distribution.

(2) A process in which two nuclei interact and form a heavier nucleus. An example of this kind of reaction is a process that is investigated in fusion reactors. *See* tokamak.

thermonuclear reaction An exoenergetic nuclear reaction in which the nuclei of light elements in a gas at a very high temperature become energetic enough to combine with each other upon collision.

theta particle (meson) Discovered in the Crystal Ball collaboration among products of decay of psi particles. Ii has a mass of 1640 MeV and an angular moment of two. This particle could have double meson states (composed of two quarks and two antiquarks) or gluonium states.

theta pinch or thetatron A fast-pulsed pinch device in which an externally imposed current goes in the azimuthal/circumferential direction (generally in a solenoid) around a cylindrical plasma. Use of a fast-rising solenoidal current causes a rapidly increasing axial magnetic field which compresses and heats the plasma.

thin airfoil theory Linearized supersonic flow utilizing perturbations. For the perturbed velocity field $\vec{u} = (u_\infty + u')\mathbf{i} + v'\mathbf{j} + w'\mathbf{k}$, a velocity potential Φ is defined such that $\vec{u} = \nabla \Phi$.

In the transonic regime,

$$\left(1 - M_\infty^2\right) \frac{\partial^2 \phi}{\partial x^2} + \frac{\partial^2 \phi}{\partial y^2} + \frac{\partial^2 \phi}{\partial z^2}$$
$$= M_\infty^2 \left(\frac{\gamma + 1}{U_\infty} \frac{\partial \phi}{\partial x}\right) \frac{\partial^2 \phi}{\partial x^2}$$

while in the subsonic and supersonic regimes,

$$\left(1 - M_\infty^2\right) \frac{\partial^2 \phi}{\partial x^2} + \frac{\partial^2 \phi}{\partial y^2} + \frac{\partial^2 \phi}{\partial z^2} = 0.$$

For the linearized pressure coefficient, $C_p = -\frac{2u'}{u_\infty}$ and $v' = u_\infty \theta$, compressible corrections such as the subsonic Prandtl–Glauert rule,

$$C_p = \frac{C_{p_o}}{\sqrt{1 - M_\infty^2}} \qquad C_L = \frac{C_{L_o}}{\sqrt{1 - M_\infty^2}}$$

where C_{p_o} and C_{L_o} are the pressure and lift coefficients determined from incompressible flow, and the supersonic Prandtl–Glauert rule

$$C_p = \frac{2\theta}{\sqrt{M_\infty^2 - 1}} \qquad C_L = \frac{4\alpha}{\sqrt{M_\infty^2 - 1}}$$

$$C_D = \frac{4\alpha^2}{\sqrt{M_\infty^2 - 1}}$$

where α is the angle of attack of the thin airfoil.

third order susceptibility The susceptibility defined by $\vec{P} = \epsilon_0 \chi \vec{E}$ often has a dependence on the applied field. It is often useful to use a Taylor series expansion of the susceptibility in powers of the applied field. For an isotropic homogeneous material, this yields $\chi = \chi^{(1)} + \chi^{(2)} E + \chi^{(3)} E^2$. The factor $\chi^{(3)}$ is referred to as the *third order susceptibility*, as it results in a term in the polarization third order in the applied field. This factor is only nonzero for materials with no inversion symmetry. For a material that is not isotropic, the *third order susceptibility* is a tensor.

thixotropic fluid Non-Newtonian fluid in which the apparent viscosity decreases in time under a constant applied shear stress.

Thomas–Fermi equation A differential equation to calculate the electrostatic potential in the context of the Thomas-Fermi atom model:

$d^2\phi/dr^2 = \phi^{3/2}/r^{1/2}$, with boundary conditions $\phi(0) = 1$ and $\phi(\infty) = 0$.

Thomas–Fermi theory A generalization of Fermi-gas model in collective models of nuclear matter. In the Thomas–Fermi model, single-particle wave function is replaced by plane wave locally.

Thomas Jefferson National Accelerator Facility Has CEBAF (Continuous Electron Beam Accelerator Facility). This facility can examinate nuclei at scales smaller than the size of nucleons as research of quark-gluon degrees of freedom in nuclei, and electromagnetic response of nuclei [the first continuous beam electron accelerator at multi GeV energies (1-6 Gev)].

Thomas–Reiche–Kuhn sum rule This is an identity involving the transition matrix elements of an atom, $\sum_i \omega_{ij} |\langle i|d|j\rangle|^2 = 3\hbar e^2/2m$. Here, e and m are the charge and mass of an electron and ω_{ij} is the frequency difference between states $|i\rangle$ and $|j\rangle$. The dipole moment operator is $\vec{d} = e\vec{r}$.

Thomson effect The electricity generated in a single conductor, in the form of an emf, by maintaining a thermal gradient in it. Heating and/or cooling effect can then be produced by adjusting the flow of current along the thermal gradient.

Thomson scattering Scattering of electromagnetic radiation by free (or loosely bound) particles.

t'Hooft, Gerard Physicist from the University of Utrecht who notably contributed to the theory of electroweak forces, QED, gauge theories, etc. and won the Nobel Prize in physics.

Thouless number The conductance of a solid divided by the fundamental conductance $2e^2/h$ (e is the electronic charge and h is Planck's constant) is a dimensionless number called the *Thouless number*. It occurs in the theory of localization.

three-body problem In quantum mechanics, the problem of solving the equation of motion of three interacting quantum particles. The problem has no exact solution except for certain unphysical interactions.

three-body recombination In this atomic process occurring in relatively high density plasmas, two electrons (or an ion and an electron) interacting near an ion result in a recombination of one electron onto the ion, with the third particle carrying away the resulting energy. This process is the inverse of impact ionization.

three-j coefficients Expansion coefficients that occur when eigenfunctions of two individual angular momenta $\mathbf{j}_1$ and $\mathbf{j}_2$ are coupled to form eigenfunctions of the total angular momentum $\mathbf{J} = \mathbf{j}_1 + \mathbf{j}_2$. They are also called Wigner three-j symbols and are closely related, but not identical, to the Clebsch–Gordon coefficients.

three-level atom An atom that interacts with an electromagnetic field such that only three levels have significant population.

three-wave mixing A process in which two laser beams interact in a non-linear optical material, generating a third beam.

threshold dose A hypothetical dose below which ionizing radiation has no stochastic risk of cancer induction. Namely, below 0.1 Sv of whole body dose epidemiological studies have not observed statistical significant increase in the number of cancers (including leukemia). Extrapolation linear doses effects relationship from medium dose region (0.1–0.4 Sv) to low dose region (below 0.1 Sv, or according some authors below 0.2 Sv) is scientifically unjustified. Moreover, some authors claim hormesis (beneficial) effect of ionizing radiation in low dose range.

threshold gain The gain at which a laser turns on, where the gain per pass is equal to the loss. This is a well-defined concept for large lasers.

thyristor A device made of semiconductor for changing the direction of current in an electrical circuit.

tight binding model The model for computing the electron's energy in a crystal, based on the assumption of the center of the electronic orbit being located on some atom of the lattice structure and the electron having some perturbation.

tilt boundary A grain boundary is the junction of two single crystals with different crystallographic orientations. If the difference in the orientation is small, the boundary is referred to as a low-angle grain boundary. A *tilt boundary* is an example of a low angle grain boundary which is formed by a sequence of edge dislocations.

timeline A line formed of adjacent fluid parcels in a flow field at a given instant and observed over time.

time ordering To apply this, one places all creation operators to the left and all annihilation operators to the right and then arranges the annihilation operators so they act in order of increasing time. This is also applied to the creation operators acting to the left. For example, for $t_1 > t_2$, the expression $a^\dagger(t_2)a(t_2)a^\dagger(t_1)a(t_1)$ would be written as $a^\dagger(t_2)a^\dagger(t_1)a(t_1)a(t_2)$ after *time ordering* has been applied.

tipping angle Superfluorescence is triggered by vacuum fluctuations. Thus, it is never triggered in a semiclassical theory. This can be dealt with by assuming a stochastic tipping angle of the collective Bloch vector away from due north, whose statistics can be determined by the full quantum theory.

tip vortex Vortex created at the tip of a wing due to the difference in pressure between the upper and lower surfaces of the wing.

tokamak Acronym created from the Russian words, TOroidalnaya KAmera ee MAgnitnaya Katushka, or toroidal chamber and magnetic coil, which denotes a class of systems for the toroidal magnetic confinement of thermonuclear fusion plasmas. Originally designed by Andrei Sakharov (1921–1989) and developed in the USSR, the *tokamak* began setting performance records for magnetic confinement fusion systems in the late 1960s and remains the leading concept for magnetic confinement fusion today. The *tokamak* configuration is perhaps most easily visualized by considering a cylindrical vacuum tube (typically of D or O-shaped cross-section) which has been bent around a symmetry axis into a torus. A solenoid coil wound around the original tube provides a strong toroidal magnetic field (which can vary from about 0.1 to over 10 tesla). Magnetohydrodynamic equilibrium and stability are achieved through a combination of externally-driven toroidal plasma currents (up to tens of millions of amperes, forming the necessary poloidal magnetic field) and externally applied vertical magnetic fields. Perhaps the easiest way to produce the toroidal plasma current is to orient a second solenoid along the symmetry axis of the torus and use the toroidal plasma as a one-turn transformer secondary coil. The resulting ohmic heating is sufficient to produce temperatures on the order of one million K or more, depending on the plasma density and the capability of the toroidal field to confine the plasma. So far, ohmic heating has been insufficient to produce fusion energy, however, and this mode of operation is inherently pulsed, which has motivated research into alternative methods of plasma current drive.

Tollmien–Schlicting wave Unstable waves appearing in viscous boundary layers above a Reynolds number of 520. Also referred to as TS and TSS waves.

Tomonaga–Schwinger equation Equation of motion of a quantized field vector for determining its position on a surface that is very close to another such surface.

top quark (sixth quark, flavor-truth) Unobserved until 1990 (Liss, P.L. Tipton, The discovery of top quark, Sci. Am. 277, 54, 1997).

toroidal magnetic confinement This term describes a large number of magnetic confinement systems based on toroidal geometries, which are perhaps the simplest closed magnetic field configurations. These confinement systems use a combination of external field coils and internal plasma currents to produce toroidal and poloidal magnetic fields capable of confining plasmas as long as the currents and fields are sustained. The simplest such configuration, a solenoid coil bent into a torus, creates vertical particle drift motions and cannot confine a plasma, but the addition of various possible vertical and poloidal fields leads to a number of configurations with magnetohydrodynamically stable plasma equilibria. When such a system is symmetric about the major axis of the torus, it is said to be axisymmetric; this simplifies the analysis of such systems and also gives these systems unique physical properties.

toroidal pinch Perhaps the earliest proposed magnetic confinement fusion scheme (Thomson and Blackman, 1946, in the UK), this is a toroidal variant of the Z pinch, in which a transformer primary drives a rapidly increasing toroidal current in a plasma ring (the transformer secondary), and the pinch effect constricts the ring. The toroidal pinch suffers from magnetohydrodynamic instabilities which limit the confinement. Many of these can be ameliorated by adding a toroidal magnetic field, leading to the stabilized pinch class of devices (which need not actually be pinches in the strict sense), of which the tokamak and reversed-field pinch are two major examples.

Torricelli's theorem The velocity of a liquid jet discharged from an orifice in a tank is a function of the height h of the free surface of the fluid above the orifice

$$U = \sqrt{2gh}$$

where both the jet and free surface are open to the atmosphere.

total angular momentum The vector sum of the two kinds of angular momentum of an atom, viz. that associated with the orbital motion of the electron and the other with the spin motion of the electron.

Trace Sometimes known as "spur". The result of adding the matrix elements along the diagonal.

trace The *trace* of a matrix is defined as the sum of its diagonal elements. It is invariant under a similarity transformation. It is also cyclic, i.e., $\text{Tr}(ABC) = \text{Tr}(CAB) = \text{Tr}(BCA)$.

trailing vortex wake Wake of vortices behind an aircraft or other lifting body generated from the lifting surfaces. The wake is created by the roll-up of the vortex sheet into discreet vortices and consists of at least one counter-rotating vortex pair. Also referred at as a wake vortex and wake turbulence. *See* downwash and vortex pair.

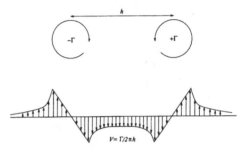

Trailing vortex wake with downwash behind a wing.

transferred electron effect The effect whereby electrons in a semiconductor with multiple conduction band valleys are transferred from one conduction valley to another under the influence of an external electric field that imparts additional energy to the electrons. The Ridley–Watkins–Hilsum–Gunn effect is an example of this effect.

transformation theory The systematic study of transformations which, when applied to the Hamiltonian of a quantum system do not change the values of certain observables.

transistor A semiconductor device sandwiched between p-type and n-type, very widely used in electrical/electronic circuits as amplifier, oscillator, detector, etc.

transit broadening When a beam of atoms crosses an optical cavity, some are leaving and some are entering. This can be modeled as a group of atoms stationary in the cavity mode with additional dephasing decay of the dipole moment. This is due to one atom leaving with a nonzero dipole moment and another entering in the ground state with no dipole moment. This effectively dephases the dipole moment of the atoms in the cavity mode.

transition Regime of flow which is between laminar and turbulent characterized by periods of intermittency where the flow field rapidly changes from one regime to the other and back again.

transition (the liquid phase of nuclear matter) At densities lower than inside normal atomic nuclei, nuclear matter theoretically has to go from a liquid to a gas phase. This phase should occur at a temperature of 10^11 K or 15 MeV. This *transition* is quantum in nature.

transition frequency The point of intersection on the frequency response plot, of the constant amplitude asymptote and the constant velocity line.

transition matrix elements For a given interaction Hamiltonian H_I, the transition matrix elements are defined as $H_I^{ij} = \langle i|H_I|f\rangle$. In the Schrödinger picture, this yields $H_I^{ij} = \int_{-\infty}^{\infty} \psi_i^* H_I \psi_f$.

transition probability In quantum mechanics, the probability that a quantum system will make a transition from one state to another.

transition rate (R) The rate at which the population of one energy level is transferred to another via some external influence. For periodic excitation using time-dependent perturbation theory, one has Fermi's Golden rule, which yields $R = (2\pi/\hbar)|\langle f|V_{\text{int}}|i\rangle|^2 \times \rho(E_f - E_i = \hbar\omega_0)$. Here, i and f denote initial and final states, V_{int} and ω_0 denote the amplitude and frequency of the excitation, and $E_{i,f}$ is the energy of the initial and final states.

translation operator The *translation operator*, when acting on a scalar function, is defined via $\mathcal{T}(a)\psi(x) = \psi(x + a)$.

transmission coefficient The ratio of transmitted to incident energy flux that occurs when a quantum wave hits a semitransparent obstacle.

transmission electron diffraction An electron beam will be diffracted by the periodic arrangement of atoms in a solid it traverses. If the optics of a TEM are slightly changed, then the diffraction pattern, rather than the image of the surface, can be projected on to a screen. If the crystal is large with respect to the beam size, spots will be produced on the screen which bear information about the crystal structure. For nearly perfect crystals, lines (called *Kikuchi lines*) will also be seen and can be used to determine crystal orientation. Samples with crystallites smaller than the beam size and with random orientation will show rings.

transmission matrix A matrix relating the transmitted wave amplitudes, transmitting through a structure to the incident wave amplitudes.

$$\begin{pmatrix} B_1 \\ B_2 \\ \cdots \\ B_n \end{pmatrix} = \begin{pmatrix} t_{11} & t_{12} & \cdots & t_{1n} \\ t_{21} & t_{22} & \cdots & t_{2n} \\ \cdots & \cdots & \cdots & \cdots \\ t_{n1} & t_{n2} & \cdots & t_{nn} \end{pmatrix} \begin{pmatrix} A_1 \\ A_2 \\ \cdots \\ A_n \end{pmatrix}$$

where As are the amplitudes of the incident modes, Bs are the amplitudes of the transmitted modes, and ts are the elements of the *transmission matrix*.

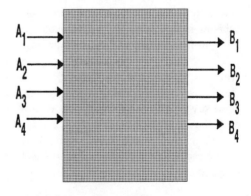

Depiction of a transmission matrix.

transonic flow The flow regime $0.8 < M < 1.2$ where the flow may contain both subsonic and supersonic flow. For the perturbed velocity field $\vec{u} = (u_\infty + u')\mathbf{i} + v'\mathbf{j} + w'\mathbf{k}$, a velocity potential Φ is defined such that $\vec{u} = \nabla\Phi$. In the transonic regime,

$$\left(1 - M_\infty^2\right)\frac{\partial^2 \phi}{\partial x^2} + \frac{\partial^2 \phi}{\partial y^2} + \frac{\partial^2 \phi}{\partial z^2}$$
$$= M_\infty^2 \left(\frac{\gamma + 1}{U_\infty}\frac{\partial \phi}{\partial x}\right)\frac{\partial^2 \phi}{\partial x^2}.$$

transport, in plasmas The problem of understanding the motions of particles in a plasma (and the related flows of energy, momentum, and other physical quantities) is extremely important in many, if not all, areas of plasma research. The theory of *transport in plasmas* is highly complex, but an understanding of transport is vital to controlled fusion research (where insufficient energy confinement is a major obstacle to producing fusion energy), plasma astrophysics (where radiation transport through plasmas often plays a dominant role), and many other areas including high energy-density plasmas, plasma processing of materials, space plasmas, and more. Since plasmas are many-body systems, it is not possible to follow all six degrees of freedom of each particle in the plasma, and consequently, statistical methods and fluid theories must be employed, though even these often prove barely tractable for realistic situations. The wide variety of possible plasma conditions (spanning over 30 orders of magnitude in density and over six orders of magnitude in temperature) leads to a wide range of phenomena, including flows, turbulence, waves and non-linear wave-particle interactions, and shocks. Specific approximations are generally needed to treat specific classes of plasma conditions over specific time and distance scales. Some key topics in plasma transport research include the determination of transport coefficients such as viscosity and diffusivity, and related parameters such as electrical conductivity and particle and energy confinement times.

transversality condition In electrodynamics, the condition that electromagnetic fields have only transversal components $\nabla \cdot A = 0$.

transverse charge The effective charge associated with the absorption induced by transverse optical phonons. It is also referred to as Born effective charge.

transverse delta function This has an integral representation $\delta_{ij}^T = (1/2\pi)^3 \int d^3k \left[\delta_{ij} - (k_i k_j/k^2)\right] \exp(i\vec{k} \cdot \vec{r})$. Here, $\vec{k}$ is the wave vector, and i and j represent Cartesian coordinate indices.

transverse form factor With total angular momentum $J > 0$, nuclei have usually nonzero magnet moments. It can interact with an intrinsic magnetic dipole of an electron. This gives an additional term in the expression for the cross-section for elastic scattering of electrons on nuclei, called the *transverse form factor*.

transverse Laplacian This is defined as $\nabla_T^2 = \partial^2/\partial x^2 + \partial^2/\partial y^2$. Here we have assumed z as the longitudinal axis.

transverse modes Generally, these are Gaussian modes of a cavity, transverse electromagnetic modes. Their exact nature depends on the geometry of the cavity. For rectangular cavities, they are given in terms of Hermite polynomials, and for cylindrical cavities they are given in terms of Bessel functions.

transverse vibration The vibration in a system where the displacement happens in a direction normal to the direction of motion of the wave.

trap A device for spatially localizing a collection of atoms or molecules. Typically constructed using a combination of laser beams and magnets or electrostatic forces.

trapped particles *See* mirror effect, banana orbit.

trapping An electron in a solid, which is otherwise free to move around in the solid, may be attracted and bound to an impurity. This capturing of the electron by the impurity is called *trapping*. Traps can emit the trapped electron if they are thermally or optically excited.

traveling wave A wave in which energy is transmitted from one part of a medium to another.

triad A chord consisting of three tones, one being for the given tone while its major or minor is augmented or diminished.

triclinic Bravais lattice There are seven crystal symmetries corresponding to the seven point group symmetries of the Bravais lattice. Triclinic is one of them.

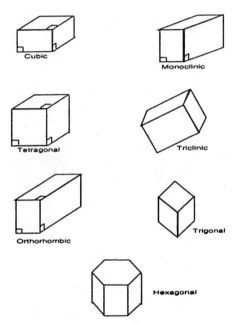

Seven types of Bravais lattices.

trigonal Bravais lattice One of the seven crystal symmetries of Bravais lattices.

triple α process A stellar helium burning process:

$$^4He + {}^4He + {}^4He \rightarrow {}^{12}C + \gamma .$$

This process provides the opportunity to produce heavier elements than ^8Be in helium burning stars.

triplet states The three states of a spectral line split into three components when the degeneracy is removed by applying an appropriate field.

TRISTAN An electron–positron colliding machine located in Japan (60GeV in the center of mass). *See also* storage rings.

tritium (**1**) ^3He is made from two protons and one neutron. It is an example of a three-body hypothetical force. Two-body forces act between nucleons 1-2, 2-3, and 3-1. After taking away the sum of interactions those three pairs, if there is still some residual force present are called a three-body force. This additional part is much weaker than two body forces, and it is neglected in calculations.

(**2**) The heavy hydrogen isotope consisting of a proton and two neutrons. Unlike the lighter isotopes (protium and deuterium), *tritium* is radioactive (a weak beta emitter) with a half-life of 12.3 years. *Tritium* is of interest in fusion energy research since the deuterium–*tritium* fusion reaction has the highest reaction rate at the plasma densities and temperatures which are presently achievable. The *tritium* nucleus is also known as a triton.

Troyon limit This denotes the maximum achievable ratio of plasma pressure to magnetic pressure (beta limit) for the tokamak plasma configuration to maintain magnetohydrodynamic equilibrium. Exceeding this limit generally results in plasma instabilities and disruptions.

$T_{dephase}$ The inverse decay rate of the induced dipole moment of a two-level atom that is due solely to transit broadening, collisional broadening, or other elastic processes that cause the dipole moment to dephase.

tunnel diode A diode device in which a quantum effect causes carriers to pass through a sharp barrier.

tunnel effect The ability of a particle to pass through a region of finite extent in which the particle's potential energy is greater than its total energy; this is a quantum-mechanical phenomenon which would be impossible according to classical mechanics.

tunneling The ability of a quantum particle to penetrate a barrier even if its energy is less than the energy height of the barrier. *Tunneling* comes about because the wave function of a particle and its first spatial derivative must be continuous at the interface of the barrier. Thus, if the wave function is nonzero at the interface, it cannot immediately vanish inside the barrier and must extend some distance into the barrier before it decays to zero. Since the squared magnitude of the wave function is the probability of finding the quantum particle anywhere, this means that there is a nonzero probability of finding the particle inside the barrier. If the barrier is thin enough, the wave function may not decay to zero before the particle exits the barrier. In this case, the particle can go through the barrier and find itself on the other side. This phenomenon is called *tunneling*. It refers to the fact that a barrier which is opaque to a classical particle may be transparent or translucent to a quantum particle.

turbine A device that extracts energy from a moving fluid.

turbomachine Any of a number of devices that adds (pump) or extracts (turbine) energy from a moving fluid via a rotating shaft.

turbulence A concise description of *turbulence* is nearly impossible. Simply put, it is a state of fluid motion characterized by seemingly random three-dimensional behavior. Most real flows are *turbulent* and all flows become *turbulent* once a given critical value (usually the Reynolds number) is exceeded, often after transition from stable to unstable regimes. *Turbulent* flow has vorticity, diffusivity, and dissipation, is highly non-linear and possibly chaotic, and is characterized by irregular fluctuations of velocity and pressure in all three dimensions. Some common characteristics of turbulence include unsteadiness, where the field contains various temporal scales across a wide spectrum of frequencies, randomness, where the unsteady fluctuations are impossible to accurately predict, three-dimensionality, where motion occurs in all three dimensions on both the small and large scales, vorticity, where stretching of vortical filaments in the flow dissipate energy from large to small scales, intermittency, where flow behavior may change suddenly over time and then return to its previous state, mixing, where convective mixing leads to rapid diffusion of the fluid across the flow field, and non-linearity, where the flow characteristics may change radically for a small change in input parameters such as Reynolds number and initial or boundary conditions. These characteristics are not necessarily all-inclusive. From scaling arguments, the number of degrees of freedom in an arbitrary flow can be shown to depend on Re as

$$N \sim \text{Re}^{9/4}$$

showing that for $\text{Re} = 10^3 \rightarrow N \sim 10^6$ and $\text{Re} = 10^6 \rightarrow N \sim 10^{12}$. Thus, the number of degrees of freedom quickly outpaces any reasonable ability to calculate the behavior exactly from a deterministic standpoint.

In general, *turbulence* can be grossly categorized as one of three types of turbulent flows: grid like, free-shear layer like, or wall layer like. In the former case, the flow is a turbulent flow field, often isotropic and homogeneous, that decays in space and time. This type of *turbulent* flow occurs in the wake of a grid from the interaction of multiple *turbulent* wakes. In the case of free-shear layer flows, interaction between flows of varying velocities result in several regions that may have different *turbulent* scales or qualities. This occurs in turbulent jets or wakes. In the final case, the flow can best be stated as a *turbulent* boundary layer, though this is a gross oversimplification. Basic analysis of *turbulent* flows requires decomposing the flow variables (velocity, pressure, etc.) into mean and fluctuating portions

$$q(t) \equiv \bar{q} + q'(t)$$

where $\bar{q}$ is averaged over some time (and thus, free of small scale temporal fluctuations) and $q'(t)$ is the time varying quantity. The various methods of analyzing *turbulent* data are too numerous and complicated to mention here. *Turbulence* is commonly considered the penultimate problem in modern fluid dynamics.

twinning Plastic deformation of a crystal that results in a partial displacement of neighboring planes. The deformed part of the crystal becomes the mirror image of the undeformed part.

twist boundary A twist boundary is an example of a low angle grain boundary formed by a sequence of screw dislocations.

two-body force A force between two particles which is not affected by the existence of other particles in the vicinity, such as a gravitational force or a Coulomb force between charged particles.

two-body problem The problem of predicting the motions of two objects obeying Newton's laws of motion and exerting forces on each other according to some specified law, such as Newton's law of gravitation, given their masses and their positions and velocities at some initial time.

two-component neutrino theory A theory according to which the neutrino and antineutrino have exactly zero rest mass, and the neutrino spin is always antiparallel to its motion. while the antineutrino spin is parallel to its motion.

two-level atom An atom that interacts with an electromagnetic field such that only two levels have significant population.

two-photon absorption A system with two energy levels separated by energy E can make a transition between those two states by absorbing or emitting two photons (nearly coincident) whose individual energies add to E, i.e., $E_1 + E_2 = E$. The cross-section, or probablility of this occurring, is proportional to the square of the incident light.

two-photon coherent state A particular squeezed state in which the squeezing operator $S(z) = \exp\left[(1/2)[z^*a^2 - za^{\dagger 2}\right]$ acts on a coherent state $|\alpha\rangle$. The name refers to the fact that this state has a nonzero photon occupation number only for even numbers of photons.

two-time correlation function A two time correlation function is a measure of the predictability of the system. One typically encounters functions like $\langle O^\dagger(t)O(t+\tau)\rangle$. This function is a measure of our knowledge of that variable (or quantum operator) at time $t + \tau$ given that we know its value at t.

Tyndall effect The phenomenon of light scattering by a sol that comprises very small particles. The sol appears fluorescent and cloudy, and the light becomes polarized.

U

U(1) symmetry   Group of symmetry associated with circle rotation. In gauge theory an invariant of equations to this group in each point of space-time (locally) gives a description of electromagnetic interaction. This invariant gives gauge particle photons (spin 1).

U (mass unit) $u = $ mass of $^{12}C/12 = 1$ kg/$N_A = 1.660540210^{-27}$ kg.

ultrahigh energy densities   (relativistic heavy ion collider, RHIC) Major new facility in nuclear physics, the study of matter at the highest energy densities and most energetic collisions of heavy nuclei. This allows the investigation of matter properties similar to those in cores of neutron stars and big bang, as well as expected transitions to a new phase of nuclear matter (phase in which quarks and gluons are no longer confined within nucleons and mesons).

ultralarge-scale integrated circuits   Electronic circuits where more than 1,000,000 functional devices (e.g., transistors) are integrated on a single chip.

ultrashort pulses   Pulses in which the pulse duration is comparable to the period of oscillation of the electric field.

ultraviolet   Refers to electromagnetic radiation with a wavelength below that of visible light but above that of X-rays, typically in the wavelength range of 0.6–380 nm.

Umklapp processes   Scattering of a particle from one Brillouin zone into another. The net change in the wave vector of the particle is then required to be large. Thus, *Umklapp processes* are caused by spatially localized scattering potentials that have large wave vector Fourier components.

uncertainty principle   A concept expressing the limitations of the possibility of simultaneous accurate measurements of two conjugate physical observables imposed by the wave–particle duality of quantum systems. The concept leads to Heisenberg's uncertainty relations, e.g., $\Delta E \cdot \Delta t \geq \hbar/2, \Delta x \cdot \Delta p_x \geq \hbar/2$. Here, Δ symbolizes the inaccuracy of the determination of the attached variable.

undepleted pump approximation   It is common in non-linear optics for several beams to interact in a crystal, resulting in an exchange of energy from one beam to another. In many situations, one of the beams is a very strong pump beam and it gives energy to another weaker beam, perhaps through some parametric amplification process. If the pump beam is very strong and gives only a small percentage of its energy to another beam, it can be treated as a reservoir with a constant electric field amplitude.

unified theory   Grand *unified theory* without gravity (SU (5), SO (10) or E_6). These large symmetries can brake on SU(3) for QCD and SU(2)xU(1) for electro weak theory.

uniform flow   Flow in which the velocity is constant across streamlines.

unitary group   The group of unitary transformations on a complex vector space.

unitary matrix   Matrix representing a unitary transformation. Its inverse is identical to its conjugate transpose.

unitary symmetry   In the theory of strong interactions this is an approximate symmetry which is the basis of the quark model following which all hadrons are built from three quarks.

unitary transformation   A linear transformation on a vector space which preserves inner products and norms. As states of quantum systems are represented by vectors in a complex vector space (unitary space), changes from one representation to another are effected by *unitary transformations*. Likewise, the changes between the different pictures of quantum mechanics (i.e., Heisenberg, Schrödinger, interaction) are also accomplished by *unitary transformations*. *Unitary transformations* are expressed by linear operators whose adjoint is equal to its inverse.

unit cell Symmetric properties of crystal can be shown by a *unit cell*. For example, a body-centered cubic *unit cell* has body-centered symmetry. One *unit cell* can be divided into several primitive cells. After a translation operation, the cell can also fill in all the crystal space.

universal conductance fluctuations The conductance of a sample placed in a magnetic field at low temperatures exhibits reproducible fluctuations as the magnetic field is scanned. These are called magnetofingerprints and are related to the configuration of elastic scatterers in the sample which scatter electrons and holes but do not randomize their phases. The rms value of the fluctuations is of universal quantity e^2/h (= 40 μSiemens).

unmagnetized plasma A plasma with no background magnetic field, or one in which the background magnetic field is negligible. This is the same as saying that if the plasma beta is sufficiently larger than unity, the role of the magnetic field is unimportant.

unpolarized light Light for which the electric field components along two orthogonal axes are uncorrelated. Also light which is 50% transmitted by a polarizer regardless of the orientation of the polarizer.

Unstable Beam Facility Institute for Nuclear Study, University of Tokyo This facility can produce an environment similar to the environment responsible for the formation of elements in stars. Neutron reach elements are important in the synthesis of elements beyond $A \sim 56$ in supernovas. They can produce superheavy elements (beyond ^{208}Pb are unstable because of Coulomb repulsion among protons).

unstable resonator A cavity in which a ray will not eventually repeat its path, but will leave the cavity. Used mainly for high power lasers where the gain per pass is large.

unstable state A state which will eventually decay to a lower-lying energy state.

unsteady flow Flow in which the flow variables (velocity, pressure, etc.) are a function of time such that $\mathbf{u} = u(t)$.

upsilon meson Υ Was discovered in Fermilab (1977). This is an unstable massive meson (bottomonium state bb, beauty quarks). The mass is about 10 proton masses. This particle has pointed to the new fifth heavy quark. Three bound states of bottomonium exist. In 1980, a fourth bottonium state was discovered at 10.58 GeV.

URMEL *See* Superfish.

V

vacancy A missing atom in a crystal. It is called a point defect or a Schottky defect.

vacuum A vacuum has structure as a consequence of the uncertainty principle. The product of uncertainty about energy and time is not smaller than some numerical constant. For some event confinement in some short time interval, there is high uncertainty about its energy. This means that in some short period of time a vacuum can have some nonzero energy in a form of creation and annihilation some particle and its antiparticle, or in the appearance and disappearance of some physical field (electrical or chromo-electrical). This represents a variation of the quantum field (for example, a sea of quark-antiquark pairs). These particles are present only as fluctuation of fields produced by other particles. These fluctuations are usually too small to be observed. A *vacuum* is investigated by heating (up to 1500 billion degrees) colliding pairs of heavy ions at high energies.

vacuum arc Also known as a cathodic arc, the *vacuum arc* is a device for creating a *plasma* from solid metal. An arc is struck on the metal, and the arc's high power density vaporizes and ionizes the metal, creating a plasma which sustains the arc. The *vacuum arc* is different from a high-pressure arc because the metal vapor itself is ionized, rather than an ambient gas. The *vacuum arc* is used in industry for creating metal and metal compound coatings.

vacuum fluctuations The ground, or vacuum, state of an electromagnetic field (or harmonic oscillator) has an average electric field (or displacement) of zero, but a nonzero value for the square of the field (square of the displacement). This results in a nonzero variance of the field (or displacement), known as *vacuum fluctuations*.

vacuum polarization Fluctuations in the vacuum state of all the field modes with which an atom interacts can induce a fluctuating polarization.

vacuum pressure *See* pressure, vacuum.

vacuum–Rabi splitting When an atom and cavity mode are coupled together with the Jaynes–Cummings coupling constant g, the one-quantum energy states (with $E = 3/2\hbar\omega$) are split. The new states are mixtures of the bare states and are displaced by $\pm g$. The result is that spontaneous emission of an atom in a small cavity may result in a doublet structure in the spectrum.

vacuum state A common name for the ground state of an electromagnetic field or harmonic oscillator.

valence band Energy states corresponding to the energies of the valency electrons. This band is located below the conduction band.

valence bond Covalent bond.

valence electrons The electrons in a crystal belong to one of three types. The first is core electrons, which are closest to the positively charged nuclei and remain tightly bound to the nuclei. They can never carry current. The second is *valence electrons*, which are in the outermost shells of the atom and are loosely bound. They participate in chemical bonding. Thermal excitations at nonzero temperatures break bonds and free corresponding valence electrons. The third type is free electrons (or conduction electrons), which are not bound to any nucleus and hence can carry current.

valence nucleon Nucleons in a shell model are divided into core and *valence* (active) *nucleons*. Core nucleons are assumed inactive, except they provide the binding energy to the *valence nucleons*. The core is one of the closed shell nuclei and can be treated as a vacuum state of the problem. The Hamiltonian of the nuclei system can be written as the sum of single-particle Hamiltonians for all active nucleons.

valley of stability Space of stable nuclei with proton number $Z = 1$ to $Z = 82$ (lead). For the first order of approximation, stable nuclei have $N = Z$.

Van Allen radiation belts Plasma regions in the Earth's magnetosphere (or in other magnetospheres) in which charged particles are trapped by the magnetic mirror effect. These zones are named after James A. Van Allen, who discovered them in 1958.

van Cittert–Zernike theorem This theorem expresses the field correlation at two points, generated by a spatially incoherent, quasi-monochromatic planar source.

Van der Meer, Simon Author of a stochastic cooling scheme that provided the opportunity to build the UA1 detector (with Carlo Rubbia) and discover intermediate W and Z bosons. Van der Meer and Rubbia received the Nobel Prize in 1984.

Van der Pauw's method A method to measure the resistivity and Hall coefficient of a thin film material. The film is cut into a cloverleaf pattern, and a point contact is made to each leaf. The resistivity and Hall coefficient are determined by applying a current between two of the leads and measuring the voltage between the other two leads in the presence of a magnetic field applied normal to the plane of the leaf. Measurements are taken with all possible combinations of the leads and the resistivity, and Hall coefficients are extracted from formulas relating the measured currents and voltages.

Van der Waals equation An equation of state for a real gas, and is given by

$$(P + a/v^2)(v - b) = RT$$

P being the pressure of the gas, v its volume/mole, T is the temperature of the gas in absolute scale, R is called the universal gas constant per mole, a and b are constants. a and b are actually correction terms, a for the attractive forces between molecules and b for the finite size of molecules.

van der Waals force (1) An attractive force between nucleons. Nuclear forces can arise from quark–quark interaction by analogy with molecules.

(2) Forces of electrostatic origin that exist between molecules and atoms. When two atoms are brought close together, they polarize each other because of the electrostatic interaction between the nuclei and electron clouds of the two atoms. At very close distances, the net force between the atoms is repulsive. At slightly larger distances, it becomes attractive and then decays to zero at even larger distances. It is the *van der Waals forces* that hold the atoms and molecules together in solids.

(3) Forces that arise between two electrically neutral objects that each have no net electric dipole moment. The fluctuating dipole of one object induces a dipole in the other, and a dipole–dipole force occurs.

van Hove singularities Critical points in the energy–wave vector dispersion relations of electrons (i.e., critical values of the wave vector) at which the density of states diverges to infinity.

The spin-resolved density of states in energy $D(E)$ is given by

$$D(E) = (1/2\pi)^n \frac{\partial E}{\partial k^n}$$

where n is the dimension of the sample ($n = 1, 2$, or 3). For example, in a one-dimensional solid, the *van Hove singularities* will occur whenever the derivative $\partial k/\partial E$ diverges. This happens at the center of the Brillouin zone and at the edges.

Van Vleck paramagnetism Paramagnetism that is independent of temperature but with a small positive susceptibility.

variance The *variance* of a fluctuating variable O is give by $\Delta O = \sqrt{\langle O^2 \rangle - \langle O \rangle^2}$.

variational method Theoretical approach to finding upper bounds on the energy of low-lying levels of a given symmetry for quantum systems. The method also yields an approximation for the state function which is usually obtained by introducing a trial function with one or more parameters which are varied to minimize the energy integral. According to the type of parameters,

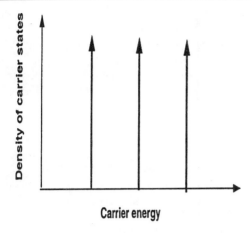

Density of states vs. energy in an quasi-zero-dimensional structure called a quantum dot. The density of states diverges at sub-band edges and is zero everywhere else. The subband energies correspond to van Hove singularities.

one distinguishes linear variation methods (Ritz variational principle) from non-linear variations which require iterative techniques.

variational principle *See* variational method.

vector coupling coefficients Transformation coefficients that occur when the products of the eigenfunctions of two angular momenta are coupled to the eigenfunctions of the sum of the two angular momenta. *See also* Clebsch–Gordon coefficients, Wigner coefficients, and three-j coefficients.

vector model of atomic or nuclear structure An intuitive model to represent the structure of the angular momentum features in atoms or nuclei, in which spin and orbital angular momenta of the electrons (or nucleons) are symbolized by vectors upon which special addition rules are superimposed to account for the way angular momenta add in quantum mechanics.

vector particles Boson particles with spins equal to one (they obey Bose–Einstein statistics).

vector potential As the divergence of the magnetic field $\vec{B}$ is zero, it can be written as the curl of another vector field, $\vec{B} = \vec{\nabla} \times \vec{A}$, where $\vec{A}$ is referred to as the *vector potential*. It is not uniquely specified by this definition, as any other *vector potential* $\vec{A}'$ obtained by a gauge transformation of $\vec{A}$ yields the same magnetic field.

Vegard's law This law stipulates empirically that the lattice constant of a ternary compound is a linear function of the alloy composition and can be found by linearly extrapolating between the lattice constants of its binary constituents. Hence, the lattice constant of a ternary compound $A_xB_{1-x}C$ is found from the lattice constants of the binary constituents as

$$l_{ABC} = l_{BC} + (l_{AC} - l_{BC})x$$

where l stands for the lattice constant.

velocity modulation transistor A field effect transistor operates on the following principle: The current flowing between two terminals (called source and drain) can be modulated by an electrostatic field (or potential) applied at a third terminal (called the gate). The current is proportional to the conductance of the conducting channel between the source and drain (at a fixed source-to-drain bias) and the gate potential changes this conductance.

The conductance is given by

$$G = \rho\mu$$

where ρ is the charge density in the conducting channel and μ is the mobility of the carriers contributing to the charge. Ordinary field effect transistors change the conductance by changing ρ with the gate potential. A *velocity modulation transistor* changes μ. The gate potential attracts the charges towards the surface of the channel where the mobility is lower because of surface scattering. This reduces the conductance and drops the source-to-drain current (switching the transistor off). The advantage of this approach is that the switching time is not limited by the transit time of charges in the channel. Instead, it depends on the velocity relaxation time which is typically sub-picoseconds in technologically important semiconductors at room temperature.

velocity of light In a vacuum, the speed of light is defined to be 2.998×10^8 m/s. It is also given by $c = 1/\sqrt{\epsilon_0 \mu_0}$, where ϵ_0 is the permitivity of free space and μ_0 is the permeability

velocity overshoot

of free space. Inside a medium, the *velocity of light* is reduced by the index of refraction of the medium $v_{\text{light}} = c/n$.

velocity overshoot   When a high electric field is applied to a solid, the drift velocity of electrons or holes rapidly rises, reaches a peak, and then drops to the steady-state value. This is known as *velocity overshoot*, whereby the velocity can temporarily exceed the steady-state value. This happens because the scattering rate increases when the electrons or holes become hot (their energy increases). The time taken for the energy to increase is roughly the so-called energy-relaxation time, whereas the time taken for the velocity to respond to the electric field is the momentum relaxation time. The former can be much larger than the latter. Hence the velocity responds much faster than the energy, causing the overshoot.

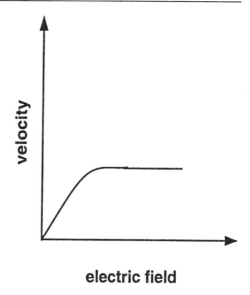

The drift velocity of charge carriers in a solid vs. applied electric field. The velocity at first rises linearly with the field and then saturates to a fixed value.

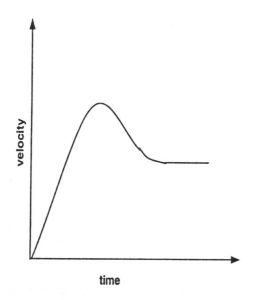

Temporal response of the drift velocity of electrons to a suddenly applied strong electric field. The velocity overshoots the steady-state velocity momentarily and then settles down to the steady-state value gradually.

velocity potential   Scalar function ϕ which satisfies both

$$u \equiv \frac{\partial \phi}{\partial x}$$

and

$$v \equiv \frac{\partial \phi}{\partial y}$$

which exists for all irrotational flows. The *velocity potential* also satisfies the Laplace equation

$$\nabla^2 \phi = \frac{\partial^2 \phi}{\partial x^2} + \frac{\partial^2 \phi}{\partial y^2} = 0$$

exactly.

velocity saturation   When an electric field is applied to a solid, an ordered drift motion of electrons and holes is superimposed on the random motion of these entities. Whereas the random motion results in no resultant drift velocity, the ordered motion gives rise to a net drift velocity and a current.

When an electric field is applied to a solid, the electrons and holes in the solid are accelerated. However, the scattering of the electrons and holes due to static scatterers such as impurities and dynamic scatterers such as phonons (lattice vibrations) retards the electrons. Finally, a steady-state velocity is reached where the accelerating force due to the electric field just balances the decelerating force due to scattering.

In the Drude model, scattering is viewed as a frictional force which is proportional to the velocity. Hence, Newton's law predicts

$$m\frac{dv}{dt} + \frac{v}{\tau} = q\mathcal{E}$$

where v is the velocity, t is the time, τ is a characteristic scattering time, q is the charge of the electron or hole, and $\mathcal{E}$ is the applied electric field. The second term on the left side is the frictional force due to scattering.

In a steady-state (time-derivative = 0), the velocity is found to be given by

$$v = \frac{q\tau}{m}\mathcal{E}$$

which predicts that the velocity is linearly proportional to the electric field. Indeed, the drift velocity is found to be proportional to the electric field (the proportionality constant is called the mobility, which can be written down from the above equation) if the electric field is small. At high electric fields, the dependence is non-linear because the characteristic scattering time τ becomes a function of the electric field $\mathcal{E}$. In fact, in many materials like silicon, the velocity saturates to a constant value at high electric fields. This phenomenon is known as *velocity saturation*.

It must be mentioned that in some materials like GaAs, the velocity never saturates but instead exhibits non-monotonic behavior as a function of the electric field. The velocity first rises with the applied electric field, reaches a peak, and then drops. This non-monotonic behavior can arise from various sources. In GaAs, it is caused by the Ridley–Hilsum–Gunn effect associated with the transfer of electrons from one conduction band valley to another. The negative differential mobility associated with such non-monotonic behavior has found applications in high frequency oscillators.

vena contracta   The region just downstream of the discharge of a liquid jet emanating from an orifice. The jet slightly contracts in the area after leaving the orifice due to momentum effects.

venturi   A nozzle consisting of a converging–diverging duct. Often used in gases to accelerate a flow from subsonic to superonic. *See* converging–diverging nozzle.

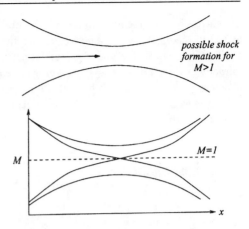

Possible flow states in a venturi.

venturi meter   A flow-rate meter utilizing a venturi. Measurement of the pressure difference upstream of the venturi and at the venturi throat can be used to determine the flow rate using empirical relations.

vertex detector   Detector designed to measure particle traces as precisely as possible near the vertex or site of collision.

vertical cavity surface emitting lasers (VCSEL)   A laser is a device that emits coherent light based on amplification via stimulated emission of photons. There are two conditions that must be satisfied for a laser to operate: the medium comprising the laser must exhibit optical gain or amplification (meaning it emits more photons than it absorbs; alternately, one can view the absorption coefficient as being negative), and there has to be a cavity which acts like a feedback loop so that the closed-loop optical gain can be infinite (an infinite gain amplifier is an oscillator that produces an output without an input).

The above two conditions are referred to as the Bernard–Durrefourg conditions.

The cavity is the structure within which the laser light is repeatedly reflected and amplified. The walls of the cavity are partial mirrors that allow some of the light to escape (most of it is reflected).

The *vertical cavity surface light emitting laser (VCSEL)* is a laser to which the cavity is vertically placed and light is emitted from the top surface which is one of the walls. It is often realized by a quantum well laser which consists of a narrow bandgap semiconductor (with a high refractive index) sandwiched between two semiconductor layers with a wider gap and smaller refractive index. The narrow gap layer is called a quantum well which traps both electrons and holes as well as photons. The quantum well thus acts as a cavity.

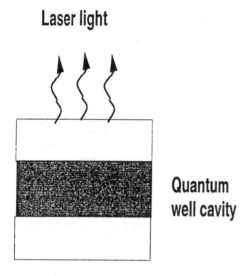

Cross-sectional view of a quantum well based vertical cavity surface emitting laser.

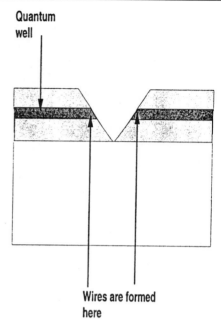

V-groove quantum wires.

very large-scale integrated circuits Electronic circuits where more than 10,000 functional devices (e.g., transistors) are integrated on a single chip.

V-groove wire A V-shaped groove is etched into a quantum well. Electrons accumulate near the edge of the groove and constitute two parallel one-dimensional conductors (quantum wires).

vibrational energy The energy content of the vibrational degrees of freedom of a molecular state. Because of the interaction with rotational and electronic degrees of freedom, it is not a directly measurable quantity except in certain simple circumstances.

vibrational level An energy level of a molecule which is a member of a vibrational progression and is characterized by a vibrational quantum number.

vibrational model of a nucleus This model describes a nucleus as a drop of fluid. Properties of a nucleus can be described as phenomena of the surface tension of the drop and the volume energy of the drop. The spherical shape of the nucleus is the state of equilibrium (potential energy is minimum). The spherical model is a simple one; spherical nuclei have no rotational degrees of freedom. Many nuclei are deformed and rotational degrees of freedom have to be included. The vibrational quantum of energy is called a phonon. *See also* shape vibrations of nuclei.

vibrational quantum number A quantum number ν indicating the vibrational motion of nuclei in a molecule neglecting rotational and electronic excitation so that the vibrational energy can be approximately given as $\hbar\omega(\nu+1/2)$, where $\hbar$ is Planck's constant and ω is the vibrational frequency (multiplied by 2π).

vibrational spectrum Also called vibrational progression. The part of a sequence of molecular spectral lines which results from transitions between vibrational levels of a molecule and which resembles the spectrum of a harmonic quantum oscillator.

vibration of strings In string theory, particles (quanta) have extensions and they can vibrate (analogous to ordinary strings). That is different from standard theories where particles (quanta) are point like. The harmonics (normal vibrations) are determined by the tension of the strings. Each vibrational mode of strings corresponds to some particle. The vibrational frequency of the mode of the string determines the energy of that particle and, hence, its mass. The familiar particles are understood as different modes of a single string. Superstring theory combines string theory with supersymmetric mathematical structures. In such a way, the problem of combining gravity and quantum mechanic is overcome. This allows the consideration of all four forces as a manifestation of one underlying principle. The vibrational frequencies of a string are determined by its tension. This energy is extremely high 10^{19} GeV.

violet cell A solar cell with a shallow p–n junction which has a high spectral response in the violet region of the solar spectrum.

virtual mass *See* added mass.

virtual process A process which has the potential to interfere with a real physical process although it is not observable by itself. The interference may be constructive or destructive and is usually expressed in the framework of perturbation theory.

virtual quantum Also called virtual particle. A particle or photon which, in an intermediate state, acts as the agent of an interaction (e.g., the Coulomb interaction) and does not satisfy the energy–momentum relation of a free particle. It cannot be directly observed.

virtual state An unstable state of an excited atom, molecule, or nucleus with a lifetime that far exceeds typical single particle time scales, e.g., the time it takes an electron to traverse the linear dimension of a molecule.

viscoelastic fluid Non-Newtonian fluid in which the fluid partially or completely returns to its original state once the deforming stress is removed.

viscosity A measure of a fluid's resistance to motion due primarily to friction of the fluid molecules. *See* absolute viscosity and kinematic viscosity.

visibility of fringes A measure of the depth of a fringe. It is defined as $V = (I_{max} - I_{min})/(I_{max} + I_{min})$. It is equal to 1 for perfect fringes and 0 for no fringes.

vitreous state The state of a supercooled liquid appearing in the form of glass, the viscosity being very high.

Voigt profile The line shape of a transition that is simultaneously homogeneously and Doppler broadened, $S(\omega) = (4ln2/\pi)^{1/2} (e^{b^2}/\delta \omega_D \, erfc(b))$. The parameter $b = (4ln2)^{1/2} \delta\omega_0/\delta\omega_D$ where $\delta\omega_0$ is the homogeneous width and $\delta\omega_D$ is the Doppler width.

Volterra dislocation A dislocation affected by cutting a material in the form of a ring and putting it back together after the cut surfaces are dislocated.

Voronoy polyhedron A generalized Wigner–Seitz cell chosen about a lattice point where the set of lattice points do not necessarily form a Bravais lattice.

vortex A structure that has a circulatory or rotational motion. A *vortex* can be either rotational or irrotational depending upon the local value of vorticity. An irrotational *vortex* of strength Γ can be most easily represented by the tangential (circumferential) velocity field

$$u_\theta = \frac{\Gamma}{2\pi r}$$

where r is the distance from the center of the *vortex* and

$$\Gamma = \oint_C \mathbf{u} \cdot d\mathbf{l}$$

or using Stoke's theorem

$$\Gamma = \int_S \omega \cdot d\mathbf{S}$$

where S is the area of integration inside of C. Thus, the fluid is irrotational everywhere except at the center of the *vortex*. Common vortical representations include the Rankine and Lamb–Oseen vortices.

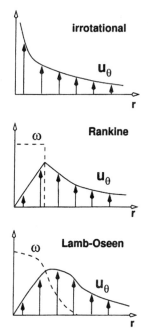

Common types of vortices.

vortex line (1) A curve such that its tangent at any point gives the direction of the local vorticity vector. *Vortex lines* obey the Helmholtz vortex theorems such that a *vortex line* can only end at a solid boundary or form a closed loop (vortex ring).

(2) When a type II superconductor is subjected to a magnetic field whose strength is intermediate between the lower and upper critical fields, the superconductor exists in a mixed state which is neither completely superconducting nor completely normal. Rather, the sample consists of a complicated structure of normal and superconducting regions. The magnetic field partially penetrates the sample in the form of thin filaments of flux. Within each filament, the field is high and the material is normal (not superconducting). Outside the filament, the material remains superconducting and the field decays exponentially with distance, with a decay constant equal to the London penetration depth. Circulating around each film is a vortex of screening current which is called a *vortex line*.

vortex pair A pair of vortices, either two-dimensional or three-dimensional, separated by a distance b, which move under mutually induced motion. For a case of same-signed (co-rotating) vortices, the motion is circular about a common center of vorticity (similar to planetary motion about a center of gravity). For the case of opposite-signed (counter-rotating) vortices, the direction of motion is perpendicular to the line connecting the centers of the *vortex pair*. If the vortices are of equal strength (circulation) Γ, then the motion is a straight line. In either case, the induced velocity of the vortices is $U = \Gamma/2\pi b$. Three-dimensional *vortex pairs* may experience long-wavelength (Crowe) or short-wavelength instabilities.

vortex ring A line vortex whose ends link to form a ring. Due to the velocity induction from one part of the vortex on every other part, the *vortex ring* translates much like an opposite-signed vortex pair. Due to the three-dimensional nature, it experiences a short-wavelength instability.

vortex sheet An infinite number of vortex filaments generated by a discontinuity in velocity. The junction between the velocity jump forms the sheet. Though the sheet may be idealized as infinitesimal, in reality the sheet or velocity change has a finite thickness. *Vortex sheets* result from Kelvin–Helmholtz formations and flow over wings.

vortex street See Kármán vortex street.

vortex wake The wake behind a body consisting of vortices created at the three-dimensional boundaries. For a rectangular wing, vortices are created at the wing-tips. For a delta wing, vortices are created at the leading edge. Corners also generate vortical wake structures. *See* trailing vortex wake.

vorticity Kinematic definition relating the amount of rotation in a flow field given by the curl of the velocity vector

$$\omega = \nabla \times \mathbf{u}.$$

In an irrotational or potential flow, $\omega = 0$.

W

Wafer scale integration The concept of using every area — no matter how small — on a chip to perform some useful circuit function (e.g., computation or signal processing). The entire surface of the chip is therefore utilized for a giant circuit.

waist For a Gaussian beam inside an optical cavity, that is, one whose transverse intensity has a Gaussian distribution of $I \propto e^{-2(x^2+y^2)/w^2(z)}$, one refers to the minimal value of the spot size $w(z)$ as the beam *waist*, where the radius of curvature is infinite.

waiting time distribution ($W(\tau)$) Gives the probability of a photon emission at time τ given that aproton emission happened at $t = 0$ and no other emission occurred in the intervening time.

wake Region behind a body in a viscous flow where the flow field has a velocity deficit due to momentum loss in the boundary layer. In an irrotational upstream flow, vorticity generation in the boundary layer creates a *wake* which is rotational (nonzero vorticity), resulting in a flow field downstream of the body with irrotational and rotational portions. *Wakes* are generally classified as laminar or turbulent, but can also be related to a wave phenomenon as well (*see* Kelvin wedge). In surface flow (such as a ship), both turbulent and wave *wakes* are present, each with a distinct shape. Boundary layer formation and separation have a large impact on the characteristics of the subsequent *wake*.

Wake fields Produced in accelerators by electromagnetic interaction of charged beam particles and metallic surfaces of the beam chamber. These fields can change trajectory of beam particles. *Wake fields* depend on geometry and material of the chamber.

wake vortex Any vortex in the wake of a flow whose generation is linked to the existence of the wake itself. Prevalent in lift-generating and juncture flows.

wall energy Energy of the boundary between domains in any ferromagnetic substance that are oppositely directed, measured per unit area.

wall layer The region in a boundary layer immediately adjacent to the wall containing both the viscous sublayer and the overlap region.

Wannier functions The wave function of an electron possessing a momentum $\hbar k$ in a crystal can be written as

$$\psi_{\vec{k}}(\vec{r}) = e^{i\vec{k}\cdot\vec{r}} u_{\vec{k}(\vec{r})}$$

where the function $u_{\vec{k}(\vec{r})}$ is the Bloch function that is periodic in space and has the same period as that of the crystal lattice. The above equation is the statement of Bloch theorem.

Since the statement of the Bloch theorem implies that

$$u_{\vec{k}(\vec{r}+n\vec{R})} = u_{\vec{k}(\vec{r})}$$

where n is an integer and $\vec{R}$ is the lattice vector whose magnitude is the lattice constant, it is easy to see that the wave function of an electron in a crystal obeys the relation

$$\psi_{\vec{k}}\left(\vec{r} + n\vec{R}\right) = e^{i\vec{k}\cdot n\vec{R}} \psi_{\vec{k}}(\vec{r}) .$$

The Bloch function can be written as

$$u_{\vec{k}(\vec{r})} = \sum_n e^{i\vec{k}\cdot n\vec{R}} \phi\left(\vec{r} - n\vec{R}\right)$$

where the functions $\phi(\vec{r} - n\vec{R})$ are called *Wannier functions*. They are orthonormal in that

$$\int \phi\left(\vec{r} - n\vec{R}\right) \phi\left(\vec{r} - m\vec{R}\right) d\vec{r} = \delta_{mn}$$

where the δ is a Krönicker delta.

wave Any of a number of information and energy transmitting motions which do not transmit mass. Different types of fluid *waves* include sound *waves* and shock *waves* which are longitudinal compressive *waves* and surface *waves*. In fluid dynamics, *waves* are either dispersive or non-dispersive.

wave equation

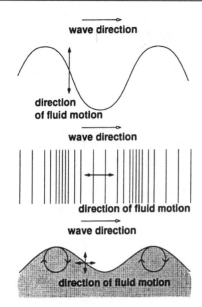

Transverse, longitudinal, and surface waves.

wave equation The classical *wave equation*, or Helmholtz equation, is one that relates the second time derivative of a variable to its second spatial derivative via $\partial^2 E \partial x^2 - (1/v^2)\partial^2 E \partial t^2 = 0$, where v is the wave velocity. The solution to this equation is any function $E(kx - vt)$, where $k = 2\pi/\lambda$ is the wave number. This is also known as D'Alembert's equation. This term is also used for other equations that have wavelike solutions, for example the Schrödinger equation.

wave function The function $\Psi(\vec{r}, t)$ that satisfies the Schrödinger equation in the position representation. It can also be defined as the projection of the state vector onto a position eigenstate, $\Psi(\vec{r}, t) \equiv \langle x|\Psi(\vec{r}, t)\rangle$.

wavelength The distance from peak to peak of a wave disturbance.

wave mechanics There are two popular representations in non-relativistic quantum mechanics: the matrix representation attributed to Heisenberg and the wave representation attributed to Schrödinger. The backbone of the latter is the Schrödinger equation which has the mathematical form of a wave equation. The wave function can be viewed as the amplitude of a scalar wave in time and space as described by the Schrödinger equation.

wave mixing If n beams are incident on a non-linear medium producing a new beam, the process is referred to as $n + 1$ *wave mixing*.

wave number The *wave number* is designated by k, and is equal to 2π divided by the wavelength λ.

wave packet A wave that is spatially localized. This *wave packet* can be formed by a superposition of monochromatic waves using Fourier's theorem.

wave–particle duality The observation that, depending on the experimental setup, quantum particles can behave sometimes as waves and sometimes as particles. Likewise, electromagnetic radiation can exhibit particle properties as well as the expected wave nature. The dual aspect of matter waves is expressed by the de Broglie relations and quantified in Heisenberg's uncertainty relations.

wave vector A vector whose magnitude is the wave number, pointing in the direction of propagation of a plane electromagnetic wave.

wave vector space The momentum space for the wave vector, the latter acting normal to the wave front.

W-boson (gauge bosons of weak interaction) The charged intermediate bosons (weak interaction) discovered in January 1983, and several months later Z neutral.

The discovery was made in CERN using an antiproton-proton collider. *W-bosons* have a mass of 82 Gev. The mass of Z is 92 GeV. These particles were predicted by the Glashow–Salam–Weinberg (GSW) electroweak theory.

weak interactions This kind of interaction is mediated by the W-mesons. These bosons change the flavor of quarks, but not color. The range of *weak interaction* is extremely short — only 10^{-3} fm, which is three orders of magnitude less than the long-range part of nuclear force. In nuclear physics, this interaction can be considered a zero-range or contact interaction. W-bosons carry charges and they change the charge state of a particle. Z-bosons are a source of neutral weak current and are responsible for the neutrino-electron scattering type of reaction ($\nu + e^- \rightarrow \nu + e^-$).

weak link A tunneling barrier between two conductors. This is a highly resistive connection between the two conductors and a charge carrier can tunnel through this region from one conductor to another. A Josephson junction consists of two superconductors with a *weak link* interposed between them.

weak localization This is a quantum mechanical correction to the conductivity of two-dimensional electron gases. The conductivity of a two-dimensional solid can be viewed in the transmission framework that was established by Rolf Landauer (Landauer's formalism applies to one- and three-dimensional solids as well). The more a solid transmits electrons, the more current it passes at a constant voltage and the more conductive it is. Similarly, more reflection (due to scattering of electrons within the solid) leads to higher resistance. There is a special set of reflected trajectories that can be grouped pairwise into time-reversed pairs which correspond to two paths that trace out exactly the same region of space inside the solid but in opposite directions. These paths always interfere constructively (since they are exactly in phase) and, hence reinforce the resistance. Thus, the resistance is always a little more than it would have been otherwise. Since electrons can maintain phase coherence only if they suffer no inelastic collisions, low temperatures are a pre-requisite for observing this additional quantum mechanical contribution to the resistance. A manifestation of weak localization is seen at low temperatures when a sample is subjected to a transverse magnetic field. The resistance of the sample decreases as the quantum mechanical correction gradually goes to zero with increasing magnetic field (negative magnetoresistance). The magnetic field introduces a phase shift between the time-reversed trajectories (called the Aharonov–Bohm phase shift), which depends on the magnetic field and the area enclosed by the time-reversed pair. Since different pairs enclose different trajectories, different pairs interfere differently and the net interference gradually averages to zero. Thus, the resistance gradually drops to the classical value as the magnetic field is increased.

weakly ionized plasma A plasma in which only a small fraction of the atoms are ionized, as opposed to a highly ionized plasma, in which nearly all atoms are ionized, or a fully ionized plasma, in which all atoms are stripped of all electrons nearly all the time.

Weber number Ratio of inertial forces to surface tension important in free-surface flow

$$We \equiv \frac{\rho U^2 L}{\sigma}$$

where σ is the surface tension.

Webster effect When a bipolar junction transistor is operated at high current levels (high collector and emitter currents), the carriers (electrons in the case of npn transistors and holes in the case of pnp transistors) that enter the base from the emitter raise the majority carrier concentration in the base to maintain charge neutrality. This effectively decreases the emitter injection efficiency, which is the ratio of current injected from the emitter to the base to the current injected from the base to the emitter. As a result, the current gain of the transistor decreases.

Weiner–Khintchine theorem This theorem defines the spectral density of a stationary random process $\Gamma(t)$ via $S(\omega) = (1/2\pi) \int_{-\infty}^{\infty} \Gamma(\tau) e^{i\omega \tau} d\tau$.

Weiner process A stochastic process that is Gaussian distributed. In numerical simulations of stochastic differential equations, the Weiner increment is given by $dW = B\sqrt{dt}$ where B is the standard deviation of the Gaussian distribution and is physically related to a damping rate involved in the problem being modeled.

Wein's displacement law This law states that $\lambda_{max} T = 0.2898 \times 10^{-2}$, where T is the temperature of a blackbody radiator, and λ_{max} is the wavelength at which the blackbody spectrum is maximized.

weir A dam used in an open channel over which water flows which is used for flow measurement by measuring the height of the fluid flowing over the dam. For low upstream velocities, the flow rate for a sharp-crested *weir* is given by

$$Q = \frac{2}{3} C_d \cdot \text{width} \sqrt{2g} \cdot (\text{height})^{1.5}$$

where C_d is an empirical discharge coefficient. Various types of *weirs* are sharp-crested, broad-crested, triangular, trapezoidal, proportional (Suttro wier), and ogee spillways.

Weissenberg method A photographic method of studying the crystal structure by X-rays. The single crystal is rotated and the X-ray beam is allowed to fall on it at right angles to the axis of rotation and the photographic film moves parallel to the axis. The crystal is screened in such a way that only one layer line is exposed at one time.

Weisskopf–Wigner approximation In treating spontaneous emission using perturbation theory, an approximation that leads to exponential decay of probability of being in the excited state.

Weiss law The inverse dependence of susceptibility on absolute temperature

$$\chi \propto \frac{1}{T}$$

while the susceptibility of ferromagnets empirically follows the dependence

$$\chi \propto \frac{1}{T - \theta_c}$$

where θ_c is the Curie temperature.

Weiss oscillations The electrical conductivity of a periodic two-dimensional array of potential barriers (called an antidot lateral surface superlattice) oscillates in a magnetic field. The peaks or troughs occur whenever the cyclotron radius associated with the motion of an electron in a magnetic field is commensurate with the period of the lattice.

Weizsäcker–Williams method The method allows a collision between two particles by allowing one particle at rest while the other passes by, and thereby generates bremstrahlung radiation. This is measured.

Wentzel–Kramers–Brillouin method (WKB method) Semiclassical approximation of quantum wave functions and energy levels based on an expansion of the wave function in powers of Planck's constant.

Werner–Wheeler method An approximate method to compute parameters of cylindrically symmetric small deformation of nuclei using irrotational-flow model.

Weyl ordering Also known as symmetric ordering.

whistler A plasma wave which propagates parallel to the magnetic field produced by currents outside the plasma at a frequency less than that of the electron cyclotron frequency, and which is circularly polarized, rotating about the magnetic field in the same sense as the electron gyromotion. The *whistler* is also known as the electron cyclotron wave. The *whistler* was discovered accidentally during World War I by large ground-loop antennas intended for spying on enemy telephone signals. Ionospheric *whistlers* are produced by distant lightning and get their name because of a characteristic descending audio-frequency tone, which is a result of the plasma dispersion relation for the wave, lower frequencies travel somewhat slower and therefore arrive at the detector later.

white noise This is a stochastic process that has a constant spectral density, that is all frequencies are equally represented in terms of intensity.

Wiedemann–Franz law An empirical law of 1853 that postulates that the ratio of thermal to electrical conductivity of a metal is proportional to the absolute temperature T with a proportionality constant that is about the same for all metals.

Wiggler magnets Specific combinations of short bending magnets with alternating field used in electron accelerators to produce coherent and incoherent photon beams or to manipulate electron beam properties. They are used to produce very intense beams of synchrotron radiation, or to pump a free electron laser. There are two designs of *Wiggler magnets:* flat design with planar magnetic field components, and helical design in which transverse component rotates along the magnetic axle.

Wigner distribution function (1) A quasi-probability function used in quantum optics. It is defined as the Fourier transformation of a symmetrically ordered characteristic function by $W(\alpha) = \frac{1}{\pi^2} \int \exp(\eta^*\alpha - \eta\alpha^*) \text{Tr}[\rho e^{\eta a^\dagger - \eta^* a}] d^2\eta$. Here, ρ is the density operator of some open quantum system, and alpha is a complex variable. This function always exists, but is not always positive.

(2) A quantum mechanical function which is a quantum mechanical equivalent of the Boltzmann distribution function. The latter describes the classical probability of finding a particle at a given region of space with a given momentum at a given instant of time. It is difficult, however, to interpret the *Wigner distribution function* as a probability since it can be complex and even negative.

There is a Wigner equation that describes the evolution of the Wigner function in time and (real and momentum) space. The *Wigner distribution function* can be used to calculate transport variables such as current density, carrier density, energy density, etc. within a quantum mechanical formalism. Therein lies its utility.

Wigner–Eckart theorem (1) Describes coupling of the angular momentum. The matrix element of an operator rank of k between states with angular momentum J and J' is

$$\langle JM|T_{kq}|J'M'\rangle = (-1)^{J-M} \begin{pmatrix} J & k & J' \\ -M & q & M' \end{pmatrix} \langle J\|T_k\|J'\rangle$$

$<J\|T_k\|J'>$ is the reduced matrix element, its invariant under rotation of the coordinate system.

(2) A theorem in the quantum theory of angular momentum which states that the matrix elements of a spherical tensor operator can be factored into two parts, one which expresses the geometry and another which contains the relevant information about the physical properties of the states involved. The first factor is a vector coupling coefficient and the second is a reduced matrix element independent of the magnetic quantum numbers.

Wigner–Seitz cell (1) The smallest volume of space in a crystal, which when repeated in all directions without overlapping, reproduces the complete crystal without leaving any void is called the primitive unit cell. Integral multiples of the primitive cell are also unit cells, since by repeating them in space one can reproduce the crystal. However, they are not primitive because they are not the smallest such unit. The *Wigner–Seitz cell* is a primitive cell chosen about a lattice point in a crystal such that any region within the cell is closer to the chosen lattice point than to any other lattice point in the crystal.

(2) When all lines, each of which connects a lattice point to its nearest lattice points, are bisected, the cell enclosed by all bisects is defined as the *Wigner-Seitz cell*. After a translation operation, the cell can also fill in all the crystal space.

Wigner–Seitz method The method estimates the band structure by evaluating the energy levels of electrons, based on the assumption of spherical symmetry for electrons around the ion.

Wigner theorem It predicts the conservation of electron-spin angular momentum.

Wigner three-j symbol *See* three-j coefficients.

Woods–Saxon potential form Represents the radial distribution of nuclear density with a diffused edge in the form:

$$\rho(r) = \frac{\rho_0}{1 + \exp\{(r-c)/z\}},$$

where ρ_0 is the nuclear matter density (roughly 310^{14} mass of water), z is a parameter that measures the diffuseness of the nuclear surface with a typical value of 0.5 fm (related to the thickness of the surface region, 1fm= 10^{-15}m), and c is the distance from the center to the point in which density drops on half value.

work function The energy difference between the Fermi energy and vacuum energy of electrons in a metal. It is the minimum energy that must be supplied to the metal to release an electron from the metal into free space. The *work function* W is directly observed in photoemission experiments. The photon energy $h\nu$ required to photo-emit an electron is related to the kinetic energy (K.E.) of the released electron by the relation

$$K.E. = h\nu - W .$$

Wronskian A mathematical functional of functions used in quantum mechanics. The *Wronskian* of two functions ϕ_1 and ϕ_2 (where the ϕs themselves are functions of a quantity x) is defined as

$$W(\phi_1, \phi_2) = \frac{\partial \phi_1}{\partial x}\phi_2 - \frac{\partial \phi_2}{\partial x}\phi_1 .$$

X

xenon poisoning   Neutron capture in ^{135}Xe produces a large negative effect on reactivity of thermal fission reactors.

X-particles   In unified gauge theories, charged quark-leptons can be carried by X or Y bosons. High-energy quarks and leptons are inter-convertible. The observation of proton (lifetime over 10^{30} years) decay would be support the unified theory.

X pinch   A variant of the Z pinch *plasma* that is made using two (or more) fine wires (typically 5-50 mm diameter), which cross and touch at a single point (forming an X shape). Using a pulsed power device, large currents are sent through the wires in a very short time. The currents in each individual wire add at the cross point of the wires, where the total current exceeds the threshold for formation of a Z pinch. When this Z pinch collapses, one or more short (1 ns or less), intense bursts of x rays are emitted from a region that can be submicron in diameter. The *X pinch* is especially valuable for direct or monochromatic x-ray backlighting (radiography).

X-rays   (1) Electromagnetic waves with wavelengths in the range between 10^{-4} nm and 10 nm.

(2) Refers to electromagnetic radiation with a wavelength below that of visible and ultraviolet light. Typically in the wavelength range of 10^{-4} to 1 nm.

(3) Invisible electromagnetic radiation with frequencies much larger than the frequency of visible light. Since the wavelength of *X-rays* is comparable to the lattice constant in several crystals, X-ray diffraction is used as a means to study crystal properties.

W.L. Bragg found in 1913 that X-rays diffracted off the surface of crystalline solid produced characteristic interference fringes that were absent in the case of a liquid. He explained this phenomenon by considering a crystal as being made out of parallel planes of ions spaced a distance d apart. Bright interference fringes (corresponding to constructive interference of X-rays reflected off two different planes of ions) occur if the following condition is met

$$n\lambda = 2d \sin\theta$$

where n is an integer, λ is the wavelength of the incident X-ray, d is the distance between two successive lattice planes, and θ is the angle of incidence.

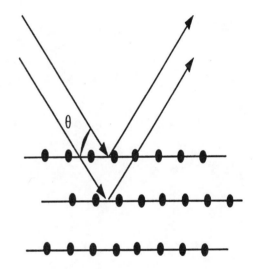

Ray diagram to explain the Bragg condition.

The von Laue explanation of the interference fringes is slightly different from the Bragg interpretation. Here, sectioning of the crystal into parallel planes of ions (so-called lattice planes) is not required, nor does one need to assume specular reflections from the lattice planes. Instead, one regards the crystal as being constructed out of atoms placed at the sites of a Bravais lattice, each of which can absorb and re-radiate the incident X-ray. Sharp interference fringes will be observed only in directions and at wavelengths for which the rays radiated from all lattice points interfere constructively.

Y

Yang–Mills particles The particles that intermediate in gauge interaction are named after C.N. Yang of the State University of New York at Stony-Brook and Robert L. Mills of Ohio State University. Photons in the theory of electromagnetism are an example of *Yang–Mills particles*.

yin-yang Coil *See* baseball coils.

Young's interference experiment In this experiment, a coherent field is incident on a screen with two slits. Behind the slits, the intensity of the light forms an interference pattern, evidence of the wave nature of light.

Young's modulus When an elastic material is subjected to a stress, a strain results in the material. If the stress is not too large, the stress and strain are linearly related according to the relation:

Stress = Young's Modulus × Strain .

Thus, *Young's modulus* is the ratio of the stress to the strain.

yrast band A rotational band consisting of the lowest energy member of each spin that is formed in composite systems created by collision of heavy ions.

Yukawa, Hideki Postulated that the existence of meson exchange creates a strong force between the proton and neutron. The first discovered particle of this kind was the pion (1947). In 1935, Yukawa suggested the existence of a vector boson as an intermediate particle in weak interaction.

Yukawa meson A particle postulated by Yukawa as the agent of the strong, short-range forces between nucleons. This particle, now identified as a pion, has to have a finite rest mass to account for the short range of the nuclear force. Pions account for only part of the nuclear force.

Yukawa potential A simple potential function of the form $V(r) = V_0 \exp(-r/\mu)/r$, where r is the distance between the nucleons and V_0 and μ are constants, is in use to parameterize relevant features of the nuclear force, such as strength and range of the force respectively.

Z

Z (neutral current) Neutral intermediate boson discovered in proton-antiproton collisions.

Zeeman effect The splitting of spectral lines of atomic or molecular radiation due to the presence of a static magnetic field. One distinguishes between the normal *Zeeman effect* for systems with zero spin and the anomalous *Zeeman effect*, which involves both an orbital and a spin magnetic moment. The latter effect changes into the Paschen–Back effect in very strong magnetic fields.

Zener breakdown When a p–n junction diode, consisting of a junction of p- and n-type doped semiconductors, is strongly reverse-biased (meaning a large positive voltage is applied to the n-type material and a large negative voltage to the p-type material), the edge of the valence band in the p-type material can rise above the edge of the conduction band in the n-type material. Electrons from the mostly filled valence band of the p-type material can then tunnel into the mostly empty conduction band of the n-type material leading to a large reverse-biased current. The voltage over the diode remains surprisingly constant once breakdown occurs. Hence, Zener diodes are widely used in voltage regulators.

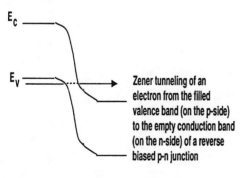

Energy band diagram of a reverse biased p–n junction diode undergoing Zener breakdown.

zero-coupled pair and approximation Pairs of identical nucleons in nuclei in the ground state of a nucleus prefer to couple to angular momentum zero. For those nucleons the dipole contribution of magnetic moments vanishes.

zero-point energy The ground state of the harmonic oscillator has a nonzero ground state energy of $\hbar\omega/2$ according to quantum mechanics. This is due to the inability to simultaneously specify the position and momentum of the oscillator due to the uncertainty principle. The oscillator cannot simultaneously have a zero displacement from equilibrium (for zero potential energy) and zero momentum (for zero kinetic energy).

zero-point vibration The vibration corresponding to the energy left over in matter at the temperature of absolute zero.

Z function *See* plasma dispersion function.

Zhukhovski airfoil Any airfoil generated using a Zhukhovski transformation, resulting in a cusped trailing edge where the lines forming the upper and lower surfaces are tangent to each other at the trailing point of the airfoil. This results in a finite velocity at the trailing edge.

Zhukhovski transformation Conformal transformation in which the boundary of an airfoil in real space is mapped into a circle about which the potential flow field can easily be determined. The basic transformation is given by

$$z = \zeta + \frac{b^2}{\zeta}$$

where $z(x, y)$ represents the plane of the airfoil and $\zeta(\xi, \eta)$ represents the plane of the circle; b is a constant. To transform a circle into a cambered airfoil, the center of the circle is offset from the origin of the ζ-plane by some finite amount. From this transformation, it can be shown that the circulation about an airfoil is given by

$$\Gamma = \pi U_\infty c \sin(\alpha)$$

and the lift coefficient is given by

$$C_L = 2\pi\alpha$$

where the chord length of the airfoil c in the z-plane is approximately four times the radius of the circle in the ζ-plane.

zinc–blende structure A lattice consisting of two interpenetrating face-centered cubic Bravais lattices, displaced along the body diagonal of the cubic cell by one-quarter the length of the diagonal, is called the diamond lattice. If the lattice points are occupied by two different kinds of atoms, then the diamond lattice is called zinc-blende. An example of a compound that exhibits the zinc-blende crystal type is GaAs.

zone folding In a superlattice, the reduced Brillouin zone breaks up into smaller zones. The number of smaller zones is the ratio of the period of the superlattice to the lattice constant of the constituent materials. Each of these smaller zones is called a minizone, and the energy bands within the minizones are called minibands. It is as if the original minizone has been folded into itself several times to create the first minizone.

Z pinch A type of pinch device in which the externally-driven pinching current goes in the z-direction, parallel to the axis of the cylindrical plasma. Parallel current filaments attract one another, imploding the pinch plasma. *Z pinch* devices have been studied for centuries, but became especially interesting in the 1950s as candidates for magnetic confinement fusion. The pinch plasmas themselves are too magnetohydrodynamically unstable to produce fusion energy. However, present-day *Z pinch* devices are excellent sources of intense X-rays, producing peak X-ray powers greater than 100 TW from cylindrical pinch plasmas 10 to 20 mm long and 2 mm in diameter. Among several possible applications, these X-rays might be useful for producing inertial confinement fusion.

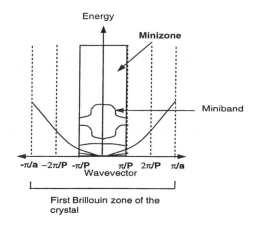

Folding of the Brillouin zone of a crystal into a minizone when an artificial periodicity is imposed on the crystal by incorporating it in a superlattice.